NEW FOR 2014

Cambridge IGCSE® Biology
Third Edition

D G Mackean
Dave Hayward

HODDER EDUCATION
AN HACHETTE UK COMPANY

Unless otherwise acknowledged, the questions and answers that appear in this book and CD were written by the author.

Although every effort has been made to ensure that website addresses are correct at time of going to press, Hodder Education cannot be held responsible for the content of any website mentioned in this book. It is sometimes possible to find a relocated web page by typing in the address of the home page for a website in the URL window of your browser.

Hachette UK's policy is to use papers that are natural, renewable and recyclable products and made from wood grown in sustainable forests. The logging and manufacturing processes are expected to conform to the environmental regulations of the country of origin.

Orders: please contact Bookpoint Ltd, 130 Milton Park, Abingdon, Oxon OX14 4SB. Telephone: (44) 01235 827720. Fax: (44) 01235 400454. Lines are open 9.00–5.00, Monday to Saturday, with a 24-hour message answering service. Visit our website at www.hoddereducation.com

® IGCSE is the registered trademark of Cambridge International Examinations

© DG Mackean 2002 and Dave Hayward 2014

First published in 2002 by

Hodder Education

An Hachette UK Company

London NW1 3BH

Second edition published 2009

This third edition published 2014

Impression number   5 4 3 2 1

Year   2018 2017 2016 2015

All rights reserved. Apart from any use permitted under UK copyright law, no part of this publication may be reproduced or transmitted in any form or by any means, electronic or mechanical, including photocopying and recording, or held within any information storage and retrieval system, without permission in writing from the publisher or under licence from the Copyright Licensing Agency Limited. Further details of such licences (for reprographic reproduction) may be obtained from the Copyright Licensing Agency Limited, Saffron House, 6–10 Kirby Street, London EC1N 8TS.

The drawings are by DG Mackean, whose copyright they are unless otherwise stated, and whose permission should be sought before they are reproduced or adapted in other publications.

Cover photo © mathisa – Fotolia

First edition layouts by Jenny Fleet

Original illustrations by DG Mackean, prepared and adapted by Wearset Ltd

Additional illustrations by Ethan Danielson, Richard Draper and Mike Humphries

Natural history artwork by Chris Etheridge

Full colour illustrations on pages 7–10 by Pamela Haddon

Third edition typeset in 11/13pt ITC Galliard Std by Integra Software Services Pvt. Ltd., Pondicherry, India

Printed and bound in Italy

A catalogue record for this title is available from the British Library

ISBN 978 1444 176 469

Contents

	Acknowledgements	vi
	To the student	viii
1	**Characteristics and classification of living organisms**	**1**
	Characteristics of living organisms	1
	Concept and use of a classification system	2
	Features of organisms	6
	Dichotomous keys	21
2	**Organisation and maintenance of the organism**	**24**
	Cell structure and organisation	24
	Levels of organisation	29
	Size of specimens	33
3	**Movement in and out of cells**	**36**
	Diffusion	36
	Osmosis	40
	Active transport	48
4	**Biological molecules**	**51**
	Biological molecules	51
	Proteins	53
	Structure of DNA	54
	Water	55
5	**Enzymes**	**59**
	Enzyme action	59
6	**Plant nutrition**	**66**
	Photosynthesis	66
	Leaf structure	77
	Mineral requirements	81
7	**Human nutrition**	**86**
	Diet	86
	Alimentary canal	95
	Mechanical digestion	98
	Chemical digestion	100
	Absorption	103
8	**Transport in plants**	**110**
	Transport in plants	110
	Water uptake	114
	Transpiration	116
	Translocation	121

9 Transport in animals — 124
Transport in animals — 124
Heart — 125
Blood and lymphatic vessels — 132
Blood — 136

10 Diseases and immunity — 142
Pathogens and transmission — 142
Defences against diseases — 148

11 Gas exchange in humans — 156
Gas exchange in humans — 156

12 Respiration — 165
Respiration — 165
Aerobic respiration — 165
Anaerobic respiration — 169

13 Excretion in humans — 174
Excretion — 174

14 Co-ordination and response — 180
Nervous control in humans — 181
Sense organs — 186
Hormones in humans — 190
Homeostasis — 192
Tropic responses — 197

15 Drugs — 205
Drugs — 205
Medicinal drugs — 205
Misused drugs — 207

16 Reproduction — 213
Asexual reproduction — 213
Sexual reproduction — 219
Sexual reproduction in plants — 221
Sexual reproduction in humans — 232
Sex hormones in humans — 241
Methods of birth control in humans — 243
Sexually transmitted infections (STIs) — 245

17 Inheritance — 250
Inheritance — 250
Chromosomes, genes and proteins — 250
Mitosis — 254
Meiosis — 255
Monohybrid inheritance — 259

18	**Variation and selection**	**270**
	Variation	270
	Adaptive features	274
	Selection	279
19	**Organisms and their environment**	**284**
	Energy flow	284
	Food chains and food webs	285
	Nutrient cycles	292
	Population size	296
20	**Biotechnology and genetic engineering**	**305**
	Biotechnology and genetic engineering	305
	Biotechnology	305
	Genetic engineering	310
21	**Human influences on ecosystems**	**316**
	Food supply	316
	Habitat destruction	320
	Pollution	324
	Conservation	334

Examination questions	**347**
Answers to numerical questions	**384**
Index	**385**

Acknowledgements

I am grateful to Eleanor Miles and Nina Konrad at Hodder Education for their guidance and encouragement. I would also like to thank Andreas Schindler for his skill and persistance in tracking down suitable photographs, and Sophie Clark, Charlotte Piccolo and Anne Trevillion were invaluable in editing the text and CD.
With special thanks to Margaret Mackean for giving her blessing to the production of this new edition.
The publishers would like to thank the following for permission to reproduce copyright material:

Examination questions

All the examination questions used in this book are reproduced by permission of Cambridge International Examinations.

Artwork and text acknowledgements

Figure 3.27 from J.K. Brierley, Plant Physiology (The Association for Science Education, 1954); **Figure 4.4** from J.Bonner and A.W. Galston, Principles of Plant Physiology (W.H. Freeman and Co., 1952); **Figure 6.27** from S.B.Verma and N.J. Rosenberg, Agriculture and the atmospheric carbon dioxide build-up , (Span, 22 February 1979); **Figure 7.4** from World Resources Report 1998-9; **Table 7.2** from National Nutrient Database, Agricultural Research Service, United States Department of Agriculture; **Figure 7.18** from John Besford, Good Mouthkeeping; or how to save your children's teeth and your own while you're about it (Oxford University Press, 1984); **Figure 9.12** and **Figure 15.6** from Royal College of Physicians (1977), Smoking or Health. The third report from the Royal College of Physicians of London (London: Pitman Medical); **Figure 10.2** from World Resources Report 1998-9; **Figure 10.8** (after) Brian Jones, Introduction to Human and Social Biology, 2/e (John Murray, 1985); **Table p.173** from Donald Emslie-Smith et.al., Textbook of Physiology, 11th Revised Edition (Churchill Livingstone, 1988); **Figure 16.58** from G.W. Corner, The Hormones in Human Reproduction (Princeton University Press, 1942); **Figure 19.12** from Robert H. Whittaker, Communities and Ecosystems, 2nd editon (Macmillan College Textbooks, 1975); **Figures 19.27**, **19.28** and **19.30** from Trevor Lewis and L.R. Taylor, Introduction to Experimental Ecology (Academic Press, 1967); **Figure 19.22** from James Bonner, The World's People and the World's Food Supply (Carolina Biology Readers Series, 1980), copyright © Carolina Biological Supply Company, Burlington, North Carolina; **Figure 19.24** from F.M. Burnett, Natural History of Infectious Disease, 3rd edition (Cambridge University Press, 1962); **Figure 21.15** from W.E. Shewell-Cooper, The ABC of Soils (English Universities Press, 1959); **Figure 21.8** from Clive A. Edwards, Soil Pollutants and Soil Animals (Scientific American, 1969), copyright © 1969 by Scientific American Inc.; **Figure 21.30** from J.E. Hansen and S. Ledeboff, New Scientist (22 October 1985).
Every effort has been made to trace or contact all rights holders. The publishers will be pleased to rectify any omissions or errors brought to their notice at the earliest opportunity.

Photo acknowledgements

p.3 *tl* © Reddogs – Fotolia, *tr* © Riverwalker – Fotolia; **p.4** *tl* © Science Photo Library/Alamy, *tr* © Premium Stock Photography GmbH/Alamy, *bl* © Simon Colmer/Alamy, *br* © Premium Stock Photography GmbH/Alamy; **p.5** *l* © Eric Gevaert – Fotolia, *cl* © Eric Isselée – Fotolia, *c* © Tom Brakefield/Stockbyte/Thinkstock, *cr* © uzuri71/iStockphoto/Thinkstock, *r* © Philip Date – Fotolia; **p.14** © Nature Picture Library/Britain On View/Getty Images; **p.15** © allocricetulus – Fotolia; **p.16** © YPetukhova – Fotolia; **p.17** © Ed Reschke/Photolibrary/Getty Images; **p.18** *tl* © Natural Visions/Alamy, *bl* © Nigel Cattlin / Alamy, *br* © Biophoto Associates/Science Photo Library; **p.24** © Biophoto Associates/Science Photo Library; **p.25** © Biophoto Associates/Science Photo Library; **p.27** © Medical-on-Line/Alamy; **p.28** *tl* © Dr. Martha Powell/Visuals Unlimited/ Getty Images, *br* © Robert Harding Picture Library Ltd/Alamy; **p.29** *l* © Biophoto Associates/Science Photo Library, *r* © Biophoto Associates/ Science Photo Library; **p.41** © Nigel Cattlin/Alamy; **p.44** *bl* © inga spence/Alamy, *r* © London News Pictures/Rex Features; **p.45** *tl* © Mark Extance/REX, *tr* © Science Photo Library/Alamy, *br* © Gonzalo Arroyo Moreno/Getty Images; **p.46** *tr* © D.G. Mackean, *br* © J.C. Revy, Ism/ Science Photo Library; **p.47** © J.C. Revy, Ism/Science Photo Library; **p.52** © Biophoto Associates/Science Photo Library; **p.54** © Dr A. Lesk, Laboratory Of Molecular Biology/Science Photo Library; **p.56** © Science Source/Science Photo Library; **p.57** © A. Barrington Brown/Science Photo Library; **p.65** © D.G. Mackean; **p.72** © Natural Visions/Alamy; **p.76** © Dr Tim Wheeler, University of Reading; **p.78** *tl* © Sidney Moulds/ Science Photo Library, *bl* © Dr Geoff Holroyd/Lancaster University; **p.81** © Gene Cox; **p.83** © Dilston Physic Garden/Colin Cuthbert/Science Photo Library; **p.89** © Romeo Gacad/AFP/Getty Images; **p.94** © Medical-on-Line/Alamy; **p.95** © Jeff Rotman / Alamy; **p.105** © David Scharf/ Science Photo Library; **p.108** © Okea – Fotolia; **p.112** *tr* © Biophoto Associates/Science Photo Library, *br* © Biophoto Associates/Science Photo Library; **p.113** © Biophoto Associates/Science Photo Library; **p.114** © D.G. Mackean; **p.120** © Rolf Langohr – Fotolia; **p.122** © imageBROKER/ Alamy; **p.127** © ACE STOCK LIMITED/Alamy; **p.128** © Biophoto Associates/Science Photo Library; **p.133** © Biophoto Associates/Science Photo Library; **p.137** © Biophoto Associates/Science Photo Library; **p.138** © Andrew Syred/Science Photo Library; **p.146** *tr* © tomalu – Fotolia, *bl* © David R. Frazier Photolibrary, Inc./Alamy; **p.148** © RioPatuca/Alamy; **p.150** © PhotoEuphoria/iStock/Thinkstock; **p.151** © Juan Mabromata/ AFP/Getty Images; **p.158** © Biophoto Associates/Science Photo Library; **p.160** © Philip Harris Education/www.findel-education.co.uk; **p.163** © Steve Gschmeissner/Science Photo Library/SuperStock; **p.176** © Biophoto Associates/Science Photo Library; **p.178** © Ken Welsh/Design Pics/ Corbis; **p.180** © Jason Oxenham/Getty Images; **p.183** © Biophoto Associates/Science Photo Library; **p.191** © Biophoto Associates/Science Photo Library; **p.193** © Biophoto Associates/Science Photo Library; **p.194** © milphoto – Fotolia; **p.197** © D.G. Mackean; **p.198** © D.G. Mackean; **p.199** © D.G. Mackean; **p.202** © D.G. Mackean; **p.210** *all* © Biophoto Associates/Science Photo Library; **p.212** © Michel Lipchitz/ AP/Press Association Images; **p.214** *l* © Biophoto Associates/Science Photo Library, *tr* © P. Morris/ Ardea, *cr* © Kurt Holter – Fotolia, *br* © SyB – Fotolia; **p.215** *tl* © Chris Howes/Wild Places Photography/Alamy, *tr* © photonewman/iStock/Getty Images; **p.217** *all* © D.G. Mackean; **p.218** *tr* © Rosenfield Image Ltd/Science Photo Library, *br* © Science Pictures Limited/Science Photo Library; **p.222** © D.G. Mackean; **p.223** *tl* © Ami Images/Science Photo Library, *tr* © Power And Syred/Science Photo Library; **p.224** © lu-photo – Fotolia; **p.225** © blickwinkel/Alamy; **p.231** *all* © D.G. Mackean; **p.232** © D.G. Mackean; **p.235** *l* © John Walsh/Science Photo Library, *r* © Biophoto Associates/Science Photo Library; **p.237** © London Fertility Centre; **p.238** *l* © Edelmann/Science Photo Library, *r* © Hannes Hemann/DPA/Press Association Images; **p.239** *l* © GOUNOT3B SCIENTIFIC/ BSIP/SuperStock, *r* © Keith/Custom Medical Stock Photo/Science Photo Library; **p.251** © SMC Images/Oxford Scientific/Getty Images; **p.255** © Ed Reschke/Photolibrary/Getty Images; **p.257** © Manfred Kage/Science Photo Library; **p.259** © Biophoto Associates/Science Photo Library; **p.263** © Philip Harris Education/www.findel-education.co.uk; **p.270** With permission from East Malling Research; **p.273** © Biophoto Associates/ Science Photo Library; **p.275** *l* © Valery Shanin – Fotolia, *r* © outdoorsman – Fotolia; **p.276** *bl* © Marco Uliana – Fotolia, *r* © Kim Taylor/Warren Photographic, *br* © Wolfgang Kruck – Fotolia; **p.277** *l* © paolofusacchia – Fotolia, *r* NO CREDIT; **p.278** *tl* © shaiith – Fotolia, *bl* © Robert Harding Picture Library Ltd/Alamy, *tr* © Imagestate Media (John Foxx), *br* © Jon Bertsch/Visuals Unlimited/Science Photo Library; **p.279** © Biophoto Associates/Science Photo Library; **p.280** *l* © Bill Coster IN/Alamy, *cl* © Bill Coster IN/Alamy, *cr* © Michael W. Tweedie/Science Photo Library, *r* © Michael W. Tweedie/Science Photo Library; **p.281** *l* © Karandaev – Fotolia, *r* © Joachim Opelka – Fotolia; **p.282** © Sir Ralph Riley; **p.286** *tl* © D.P. Wilson/Flpa/Minden Pictures/Getty Images, *tr* © Wim van Egmond/Visuals Unlimited, Inc./Science Photo Library, *bl* © lightpoet – Fotolia; **p.288** *tl* © Colin Green, *tr* © Colin Green, *bl* © Mohammed Huwais/AFP/GettyImages, *insert* © Environmental Investigations Agency; **p.291** © Marcelo Brodsky/Science Photo Library; **p.292** © Marvin Dembinsky Photo Associates / Alamy; **p.293** © buFka – Fotolia; **p.295** © Dr Jeremy

Burgess/Science Photo Library; **p.298** © Ecosphere Associates Inc, Tuscon, Arizona; **p.300** © Mark Edwards/Still Pictures/Robert Harding; **p.302** © AndreAnita/iStock/Thinkstock; **p.306** © Martyn F. Chillmaid/Science Photo Library; **p.309** © Dr. Ariel Louwrier, StressMarq Biosciences Inc.; **p.310** © Julia Kamlish/Science Photo Library; **p.311** l © Visuals Unlimited/Corbis, r © Martyn F. Chillmaid/Science Photo Library; **p.312** l © Dung Vo Trung/Sygma/Corbis, r © adrian arbib/Alamy; **p.316** © Photoshot Holdings Ltd/Alamy; **p.317** l © D.G. Mackean, tr © by paul – Fotolia, br © sergbob – Fotolia; **p.318** l © Nigel Cattlin/Alamy, r © Biophoto Associates/Science Photo Library; **p.321** tl © Nigel Cattlin/Alamy, cl © Pietro D'Antonio – Fotolia, bl © epa european pressphoto agency b.v./Alamy, tr © paul abbitt rml/Alamy; **p.322** © Biophoto Associates/Science Photo Library; **p.323** © Simon Fraser/Science Photo Library; **p.326** l © GAMMA/Gamma-Rapho via Getty Images, r © J Svedberg/Ardea.com; **p.327** l © Photoshot Holdings Ltd/Alamy, r © Roy Pedersen – Fotolia; **p.328** tl © Mike Goldwater/Alamy, tr © Thomas Nilsen/Science Photo Library, br © P.Baeza,Publiphoto Diffusion/Science Photo Library; **p.329** © Simon Fraser/Science Photo Library; **p.334** © Alex Bartel/Science Photo Library; **p.335** © David R. Frazier/Science Photo Library; **p.336** l © James Holmes/Zedcor/Science Photo Library, r © Sicut Enterprises Limited/ww.sicut.co.uk; **p.337** l © Andrey Kekyalyaynen/Alamy, r © Dr David J.Patterson/Science Photo Library; **p.338** l © Imagestate Media (John Foxx), r © NHPA/Photoshot; **p.339** tr © KeystoneUSA-ZUMA/Rex Features, br © photobypixie777 – Fotolia; **p.340** © OAPhotography – Fotolia; **p.342** © Johannes Graupner/IGB; **p.343** l © Jack Hobhouse / Alamy, tr © Derek Croucher/Alamy, br © wildpik/Alamy; **p.350** © PHOTOTAKE Inc./Alamy; **p.351** © Science Photo Library/Alamy; **p.353** tr © eyewave – Fotolia, br © Svetlana Kuznetsova – Fotolia; **p.360** © Dr Jeremy Burgess/Science Photo Library; **p.365** © PHOTOTAKE Inc./Alamy

t = top, *b* = bottom, *l* = left, *c* = centre

Every effort has been made to contact copyright holders, and the publishers apologise for any omissions which they will be pleased to rectify at the earliest opportunity.

To the student

Cambridge IGCSE® Biology Third Edition aims to provide an up-to-date and comprehensive coverage of the Core and Extended curriculum in Biology, specified in the current Cambridge International Examinations IGCSE® syllabus.

This third edition has been completely restructured to align the chapters in the book with the syllabus. Each chapter starts with the syllabus statements to be covered in that chapter, and ends with a checklist, summarising the important points covered. The questions included at the end of each chapter are intended to test your understanding of the text you have just read. If you cannot answer the question straightaway, read that section of text again with the question in mind. There are past paper examination questions at the end of the book.

To help draw attention to the more important words, scientific terms are printed in bold the first time they are used. As you read through the book, you will notice three sorts of shaded area in the text.

Material highlighted in green is for the Cambridge IGCSE Extended curriculum.

Areas highlighted in yellow contain material that is not part of the Cambridge IGCSE syllabus. It is extension work and will not be examined.

> Questions are highlighted by a box like this.

The accompanying Revision CD-ROM provides invaluable exam preparation and practice. We want to test your knowledge with interactive multiple choice questions that cover both the Core and Extended curriculum. These are organised by chapter.

Together, the textbook and CD-ROM will provide you with the information you need for the Cambridge IGCSE syllabus. I hope you enjoy using them.

I am indebted to Don Mackean for a substantial amount of the content of this textbook. Since 1962, he has been responsible for writing excellent Biology books to support the education of countless students, as well as providing an extremely useful source of information and inspiration for your teachers and their teachers. Don's diagrams, many of which are reproduced in this book, are legendary.

Dave Hayward

1 Characteristics and classification of living organisms

Characteristics of living organisms
Listing and describing the characteristics of living organisms
Concept and use of a classification system
How organisms are classified, using common features
Defining species
Using the binomial system of naming species
Features of organisms
Identifying the main features of cells

The five-kingdom classification scheme

The basic features of plants and animals
The main features of groups in the animal kingdom
The main features of groups in the plant kingdom
The main features of viruses

Dichotomous keys
Use of keys based on easily identifiable features
Construction of dichotomous keys

● Characteristics of living organisms

Key definitions
Movement is an action by an organism causing a change of position or place (see Chapter 14).
Respiration describes the chemical reactions in cells that break down nutrient molecules and release energy (see Chapter 12).
Sensitivity is the ability to detect and respond to changes in the environment (see Chapter 14).
Growth is a permanent increase in size (see Chapter 16).
Reproduction is the processes that make more of the same kind of organism (see Chapter 16). Single-celled organisms and bacteria may simply keep dividing into two. Multicellular plants and animals may reproduce sexually or asexually.
Excretion is the removal from organisms of toxic materials and substances in excess of requirements (see Chapter 13).
Nutrition is the taking in of materials for energy, growth and development (see Chapters 6 and 7).

All living organisms, whether they are single-celled or multicellular, plants or animals, show the characteristics included in the definitions above: movement, respiration, sensitivity, growth, reproduction, excretion and nutrition.

One way of remembering this list of the characteristics of living things is by using the mnemonic **MRS GREN**. The letters stand for the first letters of the characteristics.

Mnemonics work by helping to make the material you are learning more meaningful. They give a structure which is easier to recall later. This structure may be a word, or a name (such as MRS GREN) or a phrase. For example, 'Richard of York gave battle in vain' is a popular way of remembering the colours of the rainbow in the correct sequence.

Key definitions
If you are studying the extended syllabus you need to learn more detailed definitions of some of the characteristics of living things.

Movement is an action by an organism or part of an organism causing a change of position or place.
Most single-celled creatures and animals move about as a whole. Fungi and plants may make movements with parts of their bodies (see Chapter 14).
Respiration describes the chemical reactions in cells that break down nutrient molecules and release energy for metabolism. Most organisms need oxygen for this (see Chapter 12).
Sensitivity is the ability to detect or sense stimuli in the internal or external environment and to make appropriate responses (see Chapter 14).
Growth is a permanent increase in size and dry mass by an increase in cell number or cell size or both (see Chapter 16). Even bacteria and single-celled creatures show an increase in size. Multicellular organisms increase the numbers of cells in their bodies, become more complicated and change their shape as well as increasing in size (see 'Sexual reproduction in humans' in Chapter 16).
Excretion is the removal from organisms of the waste products of metabolism (chemical reactions in cells including respiration), toxic materials and substances in excess of requirements (see Chapter 13).
Respiration and other chemical changes in the cells produce waste products such as carbon dioxide. Living organisms expel these substances from their bodies in various ways (see Chapter 13).
Nutrition is the taking in of materials for energy, growth and development. Plants require light, carbon dioxide, water and ions. Animals need organic compounds and ions and usually need water (see Chapters 6 and 7).
Organisms can take in the materials they need as solid food, as animals do, or they can digest them first and then absorb them, like fungi do, or they can build them up for themselves, like plants do. Animals, using ready-made organic molecules as their food source, are called heterotrophs and form the consumer levels of food chains. Photosynthetic plants are called autotrophs and are usually the first organisms in food chains (see Chapters 6 and 19).

Concept and use of a classification system

> **Key definitions**
> A **species** is a group of organisms that can reproduce to produce fertile offspring.
> The **binomial system** is an internationally agreed system in which the scientific name of an organism is made up of two parts showing the genus and the species.

You do not need to be a biologist to realise that there are millions of different organisms living on the Earth, but it takes a biologist to sort them into a meaningful order, i.e. to **classify** them.

There are many possible ways of classifying organisms. You could group all aquatic organisms together or put all black and white creatures into the same group. However, these do not make very meaningful groups; a seaweed and a porpoise are both aquatic organisms, a magpie and a zebra are both black and white; but neither of these pairs has much in common apart from being living organisms and the latter two being animals. These would be **artificial systems** of classification.

A biologist looks for a **natural system** of classification using important features which are shared by as large a group as possible. In some cases it is easy. Birds all have wings, beaks and feathers; there is rarely any doubt about whether a creature is a bird or not. In other cases it is not so easy. As a result, biologists change their ideas from time to time about how living things should be grouped. New groupings are suggested and old ones abandoned.

Species

The smallest natural group of organisms is the **species**. A species can be defined as a group of organisms that can reproduce to produce fertile offspring.

Members of a species also often resemble each other very closely in appearance, unless humans have taken a hand in the breeding programmes. All cats belong to the same species but there are wide variations in the appearance of different breeds (see 'Variation' in Chapter 18). An American Longhair and a Siamese may look very different but they have few problems in breeding together. Robins, blackbirds and sparrows are three different species of bird. Apart from small variations, members of a species are almost identical in their anatomy, physiology and behaviour.

Closely related species are grouped into a **genus** (plural: **genera**). For example, stoats, weasels and polecats are grouped into the genus *Mustela*.

Binomial nomenclature

Species must be named in such a way that the name is recognised all over the world.

'Cuckoo flower' and 'Lady's smock' are two common names for the same wild plant. If you are not aware that these are alternative names this could lead to confusion. If the botanical name, *Cardamine pratensis*, is used, however, there is no chance of error. The Latin form of the name allows it to be used in all the countries of the world irrespective of language barriers.

People living in Britain are familiar with the appearance of a blackbird – a very common garden visitor. The male has jet black plumage, while the female is brown. Its scientific name is *Turdus merula* and the adult is about 24 cm long (see Figure 1.1). However, someone living in North America would describe a blackbird very differently. For example, the male of one species, *Agelaius phoeniceus*, has black plumage with red shoulder patches and yellow flashes, while the female is speckled brown. It is about the size of a sparrow – only about 20 cm long (see Figure 1.2). A British scientist could get very confused talking to an American scientist about a blackbird! Again, the use of the scientific name avoids any confusion.

The **binomial system** of naming species is an internationally agreed system in which the scientific name of an organism is made up of two parts showing the genus and the species. Binomial means 'two names'; the first name gives the genus and the second gives the species. For example, the stoat and weasel are both in the genus *Mustela* but they are different species; the stoat is *Mustela erminea* and the weasel is *Mustela nivalis*.

The name of the genus (the generic name) is always given a capital letter and the name of the species (the specific name) always starts with a small letter.

Frequently, the specific name is descriptive, for example *edulis* means 'edible', *aquatilis* means 'living in water', *bulbosus* means 'having a bulb', *serratus* means 'having a jagged (serrated) edge'.

Concept and use of a classification system

Figure 1.1 *Turdus merula* ♂

Figure 1.2 *Agelaius phoeniceus* ♂

If you are studying the extended syllabus you need to be able to explain why it is important to classify organisms. By classifying organisms it is possible to identify those most at risk of extinction. Strategies can then be put in place to conserve the threatened species. Apart from the fact that we have no right to wipe out species forever, the chances are that we will deprive ourselves not only of the beauty and diversity of species, but also of potential sources of valuable products such as drugs. Many of our present-day drugs are derived from plants (e.g. quinine and aspirin) and there may be many more sources as yet undiscovered. We are also likely to deprive the world of genetic resources (see 'Conservation' in Chapter 21).

By classifying organisms it is also possible to understand evolutionary relationships. Vertebrates all have the presence of a vertebral column, along with a skull protecting a brain, and a pair of jaws (usually with teeth). By studying the anatomy of different groups of vertebrates it is possible to gain an insight into their evolution.

The skeletons of the front limb of five types of vertebrate are shown in Figure 1.3. Although the limbs have different functions, such as grasping, flying, running and swimming, the arrangement and number of the bones is almost the same in all five. There is a single top bone (the humerus), with a ball and socket joint at one end and a hinge joint at the other. It makes a joint with two other bones (the radius and ulna) which join to a group of small wrist bones. The limb skeleton ends with five groups of bones (the hand and fingers), although some of these groups are missing in the bird.

The argument for evolution says that, if these animals are not related, it seems very odd that such a similar limb skeleton should be used to do such different things as flying, running and swimming. If, on the other hand, all the animals came from the same ancestor, the ancestral skeleton could have changed by small stages in different ways in each group. So we would expect to find that the basic pattern of bones was the same in all these animals. There are many other examples of this kind of evidence among the vertebrate animals.

Classification is traditionally based on studies of **morphology** (the study of the form, or outward appearance, of organisms) and **anatomy** (the study of their internal structure, as revealed by dissection). Aristotle was the first known person to attempt to devise a system of classification based on morphology and anatomy. He placed organisms in a hierarchy according to the complexity of their structure and function. Indeed, some of his ideas still existed just 200 years ago. He separated animals into two groups: those with blood and those without, placing invertebrates into the second group and vertebrates into the first. However, he was not aware that some invertebrates do have a form of haemoglobin. Using blood as a common feature would put earthworms and humans in the same group! Earthworm blood is red: it contains haemoglobin, although it is not contained in red blood cells.

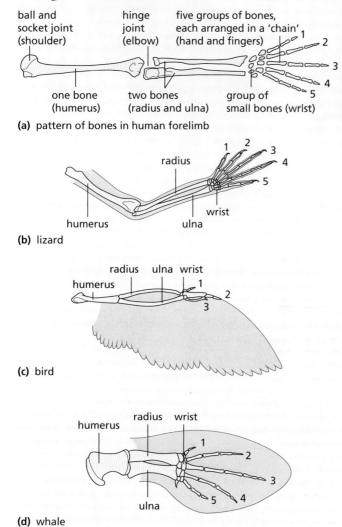

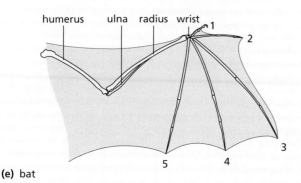

Figure 1.3 Skeletons of five vertebrate limbs

Plants have been classified according to their morphology, but appearances can be deceptive. The London Plane tree and the British Sycamore were considered to be closely related because of the similarity in their leaf shape, as shown in Figure 1.4.

Figure 1.4 Leaves of the British Sycamore (left) and London Plane (right)

However, a closer study of the two species exposes major differences: leaf insertion (how they are arranged on a branch) in London Plane is alternate, while it is opposite in the Sycamore. Also, their fruits are very different, as shown in Figure 1.5.

Figure 1.5 Fruits of the British Sycamore (left) and London Plane (right)

The scientific name of the London Plane is *Platanus acerifolia* (meaning 'leaves like an Acer'); that of the British Sycamore is *Acer pseudoplatanus* ('pseudo' means 'false'). They do not even belong in the same genus.

The use of **DNA** has revolutionised the process of classification. Eukaryotic organisms contain chromosomes made up of strings of genes. The chemical which forms these genes is called DNA (which is short for deoxyribonucleic acid). The DNA is made up of a sequence of bases, coding for amino acids and, therefore, proteins (see Chapters 4 and 17). Each species has a distinct number of chromosomes and a unique sequence of bases in its DNA, making it identifiable and distinguishable from other species. This helps particularly when different species are very similar morphologically (in appearance) and anatomically (in internal structure).

The process of biological classification called **cladistics** involves organisms being grouped together according to whether or not they have one or more shared unique characteristics derived from the group's last common ancestor, which are not present in more distant ancestors. Organisms which share a more recent ancestor (and are, therefore, more closely related) have DNA base sequences that are more similar than those that share only a distant ancestor.

Human and primate evolution is a good example of how DNA has been used to clarify a process of evolution. Traditional classification of primates (into monkeys, apes and humans) was based on their anatomy, particularly their bones and teeth. This put humans on a separate branch, while grouping the other apes together into one family called Pongidae.

However, genetic evidence using DNA provides a different insight – humans are more closely related to chimpanzees (1.2% difference in the genome – the complete set of genetic material of the organism) and gorillas (1.6% different) than to orang-utans (3.1% different). Also, chimpanzees are closer to humans than to gorillas (see Figure 1.6).

Bonobos and chimps are found in Zaire and were only identified as different species in 1929. The two species share the same percentage difference in the genome from humans.

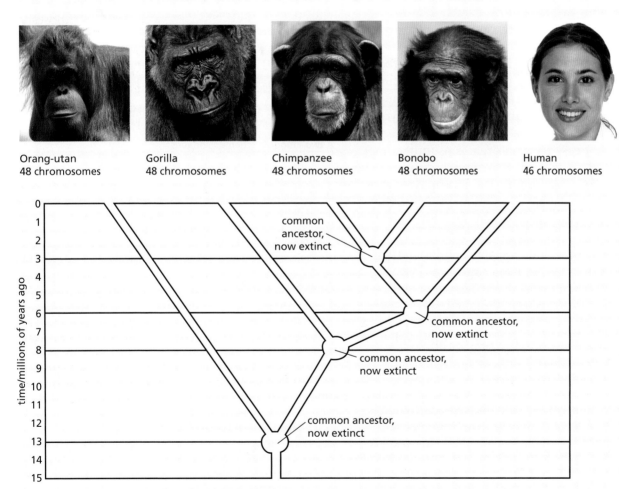

Figure 1.6 Classification of primates, based on DNA evidence

Features of organisms

All living organisms have certain features in common, including the presence of cytoplasm and cell membranes, and DNA as genetic material.

All living organisms also contain **ribosomes** in the cytoplasm, floating freely or attached to membranes called **rough endoplasmic reticulum (ER)**. Ribosomes are responsible for protein synthesis.

The Whittaker five-kingdom scheme

The largest group of organisms recognised by biologists is the kingdom. But how many kingdoms should there be? Most biologists used to favour the adoption of two kingdoms, namely **Plants** and **Animals**. This, however, caused problems in trying to classify fungi, bacteria and single-celled organisms which do not fit obviously into either kingdom. A scheme now favoured by many biologists is the Whittaker five-kingdom scheme consisting of **Animal**, **Plant**, **Fungus**, **Prokaryote** and **Protoctist**.

It is still not easy to fit all organisms into the five-kingdom scheme. For example, many protoctista with chlorophyll (the protophyta) show important resemblances to some members of the algae, but the algae are classified into the plant kingdom.

Viruses are not included in any kingdom – they are not considered to be living organisms because they lack cell membranes (made of protein and lipid), cytoplasm and ribosomes and do not demonstrate the characteristics of living things: they do not feed, respire, excrete or grow. Although viruses do reproduce, this only happens inside the cells of living organisms, using materials provided by the host cell.

This kind of problem will always occur when we try to devise rigid classification schemes with distinct boundaries between groups. The process of evolution would hardly be expected to result in a tidy scheme of classification for biologists to use.

Extension work

As scientists learn more about organisms, classification schemes change. Genetic sequencing has provided scientists with a different way of studying relationships between organisms. The **three-domain scheme** was introduced by Carl Woese in 1978 and involves grouping organisms using differences in ribosomal RNA structure. Under this scheme, organisms are classified into three domains and six kingdoms, rather than five. The sixth kingdom is created by splitting the Prokaryote kingdom into two. The domains are:

1 **Archaea**: containing ancient prokaryotic organisms which do not have a nucleus surrounded by a membrane. They have an independent evolutionary history to other bacteria and their biochemistry is very different to other forms of life.
2 **Eubacteria**: prokaryotic organisms which do not have a nucleus surrounded by a membrane.
3 **Eukarya**: organisms that have a membrane-bound nucleus. This domain is further subdivided into the kingdoms Protoctist, Fungus, Plant and Animal.

A summary of the classification schemes proposed by scientists is shown in Figure 1.7.

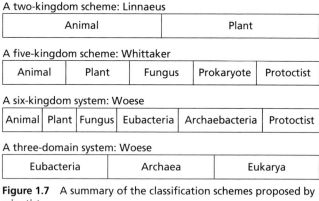

Figure 1.7 A summary of the classification schemes proposed by scientists

An outline classification of plants and animals follows and is illustrated in Figures 1.8–1.11.

The plant kingdom

These are made up of many cells – they are multicellular. Plant cells have an outside wall made of cellulose. Many of the cells in plant leaves and stems contain chloroplasts with photosynthetic pigments, e.g. chlorophyll. Plants make their food by photosynthesis.

Features of organisms

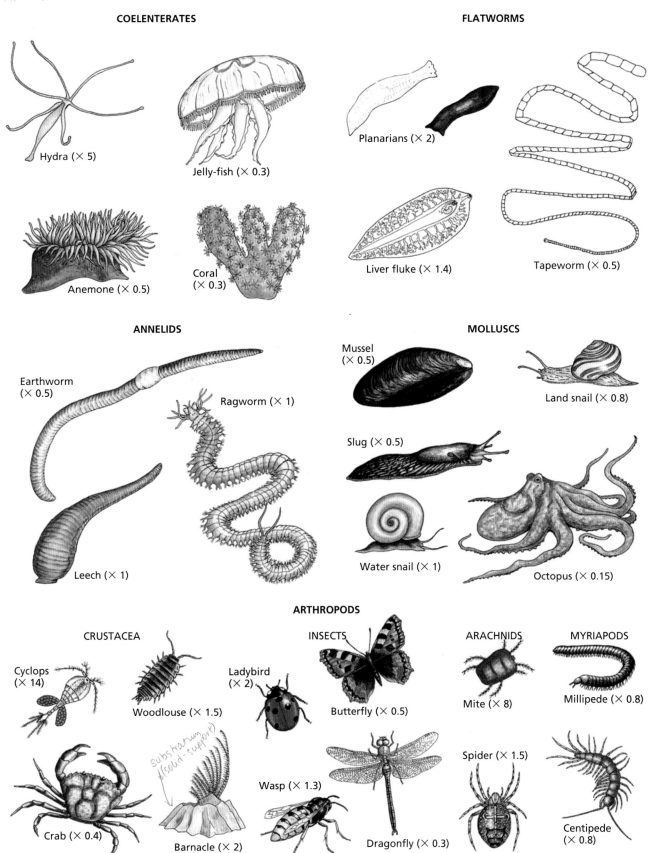

Figure 1.8 The animal kingdom; examples of five invertebrate groups (phyla)

1 CHARACTERISTICS AND CLASSIFICATION OF LIVING ORGANISMS

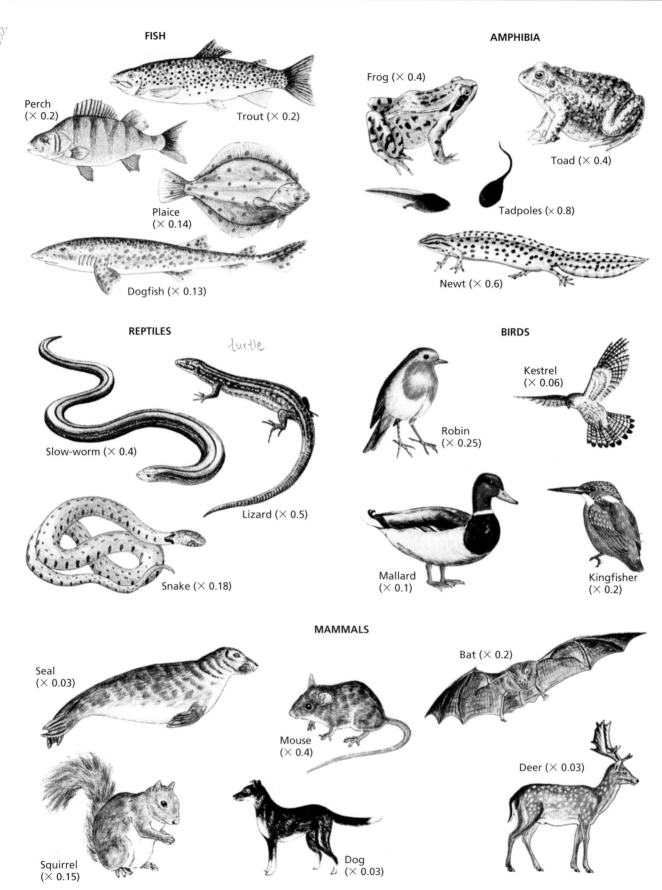

Figure 1.9 The animal kingdom; the vertebrate classes

Features of organisms

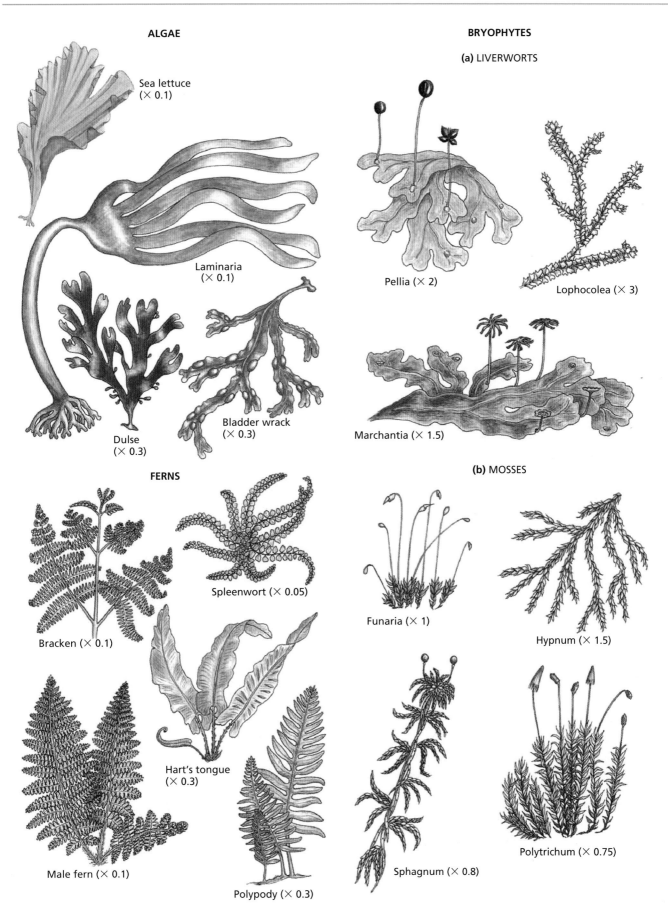

Figure 1.10 The plant kingdom; plants that do not bear seeds

1 CHARACTERISTICS AND CLASSIFICATION OF LIVING ORGANISMS

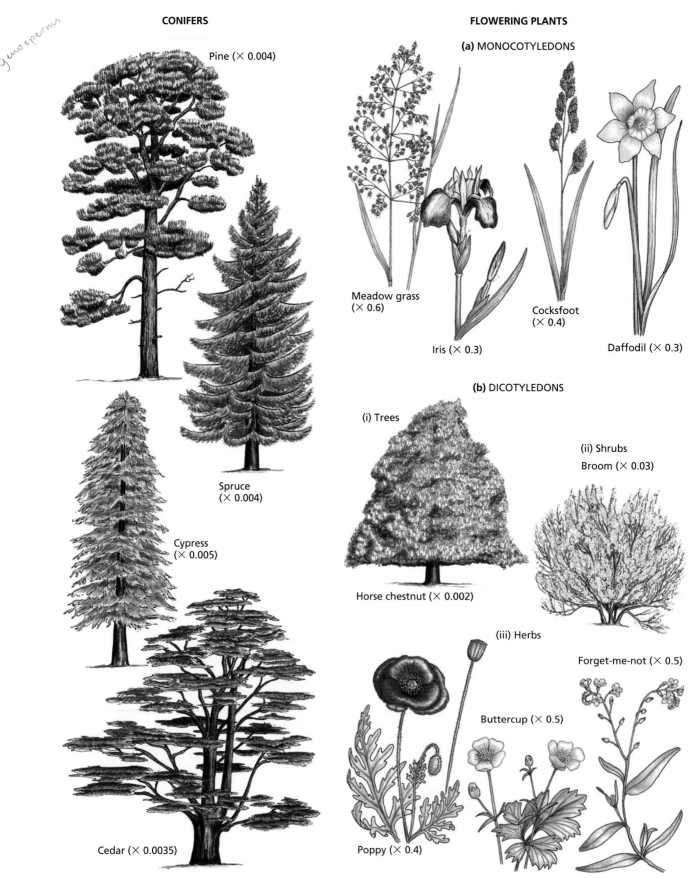

Figure 1.11 The plant kingdom; seed-bearing plants

The animal kingdom

Animals are multicellular organisms whose cells have no cell walls or chloroplasts. Most animals ingest solid food and digest it internally.

Animal kingdom
(Only eight groups out of 23 are listed here.) Each group is called a phylum (plural = phyla).

*{
    Coelenterates (sea anemones, jellyfish)
    Flatworms
    Nematode worms
    Annelids (segmented worms)
    Arthropods
        CLASS
        Crustacea (crabs, shrimps, water fleas)
        Insects
        Arachnids (spiders and mites)
        Myriapods (centipedes and millipedes)
    Molluscs (snails, slugs, mussels, octopuses)
    Echinoderms (starfish, sea urchins)

    Vertebrates
        CLASS
        Fish
        Amphibia (frogs, toads, newts)
        Reptiles (lizards, snakes, turtles)
        Birds
        Mammals
        (Only four subgroups out of about 26 are listed.)
            Insectivores
            Carnivores
            Rodents
            Primates

*All the organisms which do not have a vertebral column are often referred to as invertebrates. Invertebrates are not a natural group, but the term is convenient to use.

Arthropods

The arthropods include the crustacea, insects, centipedes and spiders (see Figure 1.8 on page 7). The name arthropod means 'jointed limbs', and this is a feature common to them all. They also have a hard, firm external skeleton, called a **cuticle**, which encloses their bodies. Their bodies are segmented and, between the segments, there are flexible joints which permit movement. In most arthropods, the segments are grouped together to form distinct regions, the head, thorax and abdomen. Table 1.1 outlines the key features of the four classes of arthropod.

Crustacea

Marine crustacea are crabs, prawns, lobsters, shrimps and barnacles. Freshwater crustacea are water fleas, *Cyclops*, the freshwater shrimp (*Gammarus*) and the water louse (*Asellus*). Woodlice are land-dwelling crustacea. Some of these crustacea are illustrated in Figure 1.8 on page 7.

Like all arthropods, crustacea have an exoskeleton and jointed legs. They also have two pairs of antennae which are sensitive to touch and to chemicals, and they have **compound eyes**. Compound eyes are made up of tens or hundreds of separate lenses with light-sensitive cells beneath. They are able to form a crude image and are very sensitive to movement.

Typically, crustacea have a pair of jointed limbs on each segment of the body, but those on the head segments are modified to form antennae or specialised mouth parts for feeding (see Figure 1.12).

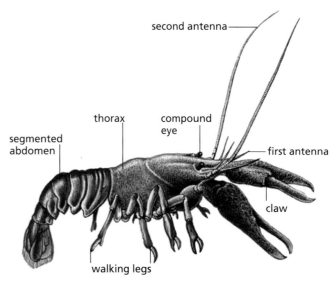

Figure 1.12 External features of a crustacean (lobster ×0.2)

Insects

The insects form a very large class of arthropods. Bees, butterflies, mosquitoes, houseflies, earwigs, greenfly and beetles are just a few of the subgroups in this class.

Insects have segmented bodies with a firm exoskeleton, three pairs of jointed legs, compound eyes and, typically, two pairs of wings. The segments are grouped into distinct head, thorax and abdomen regions (see Figure 1.13).

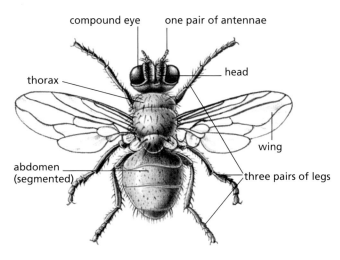

Figure 1.13 External features of an insect (greenbottle, ×5). Flies, midges and mosquitoes have only one pair of wings.

Insects differ from crustacea in having wings, only one pair of antennae and only three pairs of legs. There are no limbs on the abdominal segments.

The insects have very successfully colonised the land. One reason for their success is the relative impermeability of their cuticles, which prevents desiccation even in very hot, dry climates.

Arachnids

These are the spiders, scorpions, mites and ticks. Their bodies are divided into two regions, the cephalothorax and the abdomen (see Figure 1.14). They have four pairs of limbs on the cephalothorax, two pedipalps and two chelicerae. The pedipalps are used in reproduction; the chelicerae are used to pierce their prey and paralyse it with a poison secreted by a gland at the base. There are usually several pairs of simple eyes.

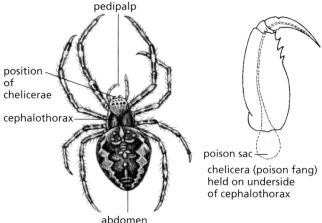

Figure 1.14 External features of an arachnid (×2.5)

Myriapods

These are millipedes and centipedes. They have a head and a segmented body which is not obviously divided into thorax and abdomen. There is a pair of legs on each body segment but in the millipede the abdominal segments are fused in pairs and it looks as if it has two pairs of legs per segment (see Figure 1.15).

As the myriapod grows, additional segments are formed. The myriapods have one pair of antennae and simple eyes. Centipedes are carnivorous; millipedes feed on vegetable matter.

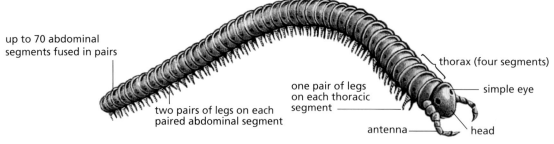

Figure 1.15 External features of a myriapod (×2.5)

Table 1.1 Key features of the four classes of arthropods

Insects	Arachnids	Crustacea	Myriapods
e.g. dragonfly, wasp	e.g. spider, mite	e.g. crab, woodlouse	e.g. centipede, millipede
• three pairs of legs	• four pairs of legs	• five or more pairs of legs	• ten or more pairs of legs (usually one pair per segment)
• body divided into head, thorax and abdomen	• body divided into cephalothorax and abdomen	• body divided into cephalothorax and abdomen	• body not obviously divided into thorax and abdomen
• one pair of antennae		• two pairs of antennae	• one pair of antennae
• one pair of compound eyes	• several pairs of simple eyes	• one pair of compound eyes	• simple eyes
• usually have two pairs of wings	• chelicerae for biting and poisoning prey	• exoskeleton often calcified to form a carapace (hard)	

Vertebrates

Vertebrates are animals which have a vertebral column. The vertebral column is sometimes called the spinal column or just the spine and consists of a chain of cylindrical bones (vertebrae) joined end to end.

Each vertebra carries an arch of bone on its dorsal (upper) surface. This arch protects the spinal cord (see Chapter 14), which runs most of the length of the vertebral column. The front end of the spinal cord is expanded to form a brain which is enclosed and protected by the skull.

The skull carries a pair of jaws which, in most vertebrates, have rows of teeth.

The five classes of vertebrates are fish, amphibia, reptiles, birds and mammals. Table 1.2 summarises the key features of these classes.

Body temperature

Fish, amphibia and reptiles are often referred to as 'cold-blooded'. This is a misleading term. A fish in a tropical lagoon or a lizard basking in the sun will have warm blood. The point is that these animals have a variable body temperature which, to some extent, depends on the temperature of their surroundings. Reptiles, for example, may control their temperature by moving into sunlight or retreating into shade but there is no internal regulatory mechanism.

So-called 'warm-blooded' animals, for the most part, have a body temperature higher than that of their surroundings. The main difference, however, is that these temperatures are kept more or less constant despite any variation in external temperature. There are internal regulatory mechanisms (see Chapter 14) which keep the body temperature within narrow limits.

It is better to use the terms **poikilothermic** (variable temperature) and **homoiothermic** (constant temperature). However, to simplify the terms, 'cold blooded' and 'warm blooded' will be referred to in this section.

The advantage of homoiothermy is that an animal's activity is not dependent on the surrounding temperature. A lizard may become sluggish if the surrounding temperature falls. This could be a disadvantage if the lizard is being pursued by a homoiothermic predator whose speed and reactions are not affected by low temperatures.

Fish

Fish are poikilothermic (cold blooded) vertebrates. Many of them have a smooth, streamlined shape which offers minimal resistance to the water through which they move (see Figure 1.16). Their bodies are covered with overlapping scales and they have fins which play a part in movement.

Fish breathe by means of filamentous gills which are protected by a bony plate, the operculum.

Fish reproduce sexually but fertilisation usually takes place externally; the female lays eggs and the male sheds sperms on them after they have been laid.

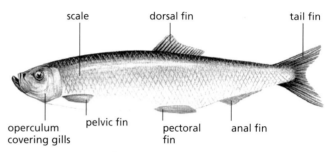

Figure 1.16 Herring (*Clupea*, ×0.3)

Amphibia

Amphibia are poikilothermic (cold blooded) vertebrates with four limbs and no scales. The class includes frogs, toads and newts. The name, amphibian, means 'double life' and refers to the fact that the organism spends part of its life in water and part on the land. In fact, most frogs, toads and newts spend much of their time on the land, in moist situations, and return to ponds or other water only to lay eggs.

The external features of the common frog are shown in Figure 1.17. Figure 1.9 on page 8 shows the toad and the newt.

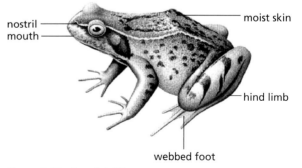

Figure 1.17 *Rana* (×0.75)

The toad's skin is drier than that of the frog and it has glands which can exude an unpleasant-tasting chemical which discourages predators. Newts differ from frogs and toads in having a tail. All three groups are carnivorous.

Amphibia have four limbs. In frogs and toads, the hind feet have a web of skin between the toes. This offers a large surface area to thrust against the water when the animal is swimming. Newts swim by a wriggling, fish-like movement of their bodies and make less use of their limbs for swimming.

Amphibia have moist skins with a good supply of capillaries which can exchange oxygen and carbon dioxide with the air or water. They also have lungs which can be inflated by a kind of swallowing action. They do not have a diaphragm or ribs.

Frogs and toads migrate to ponds where the males and females pair up. The male climbs on the female's back and grips firmly with his front legs (see Figure 1.18). When the female lays eggs, the male simultaneously releases sperms over them. Fertilisation, therefore, is external even though the frogs are in close contact for the event.

Figure 1.18 Frogs pairing. The male clings to the female's back and releases his sperm as she lays the eggs.

Reptiles

Reptiles are land-living vertebrates. Their skins are dry and the outer layer of epidermis forms a pattern of scales. This dry, scaly skin resists water loss. Also the eggs of most species have a tough, parchment-like shell. Reptiles, therefore, are not restricted to damp habitats, nor do they need water in which to breed.

Reptiles are poikilothermic (cold blooded) but they can regulate their temperature to some extent. They do this by basking in the sun until their bodies warm up. When reptiles warm up, they can move about rapidly in pursuit of insects and other prey.

The reptiles include lizards, snakes, turtles, tortoises and crocodiles (see Figure 1.19 and Figure 1.9 on page 8).

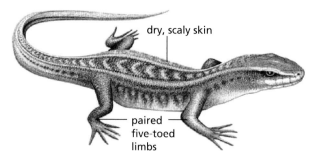

Figure 1.19 *Lacerta* (×1.5)

Apart from the snakes, reptiles have four limbs, each with five toes. Some species of snake still retain the vestiges of limbs and girdles.

Male and female reptiles mate, and sperms are passed into the female's body. The eggs are, therefore, fertilised internally before being laid. In some species, the female retains the eggs in the body until they are ready to hatch.

Birds

Birds are homoiothermic (warm blooded) vertebrates.

The vertebral column in the neck is flexible but the rest of the vertebrae are fused to form a rigid structure. This is probably an adaptation to flight, as the powerful wing muscles need a rigid frame to work against.

The epidermis over most of the body produces a covering of feathers but, on the legs and toes, the epidermis forms scales. The feathers are of several kinds. The fluffy down feathers form an insulating layer close to the skin; the contour feathers cover the body and give the bird its shape and colouration; the large quill feathers on the wing are essential for flight.

Birds have four limbs, but the forelimbs are modified to form wings. The feet have four toes with claws which help the bird to perch, scratch for seeds or capture prey, according to the species.

The upper and lower jaws are extended to form a beak which is used for feeding in various ways.

Figure 1.20 shows the main features of a bird.

In birds, fertilisation is internal and the female lays hard-shelled eggs in a nest where she incubates them.

Figure 1.20 The main features of a pigeon (×0.14)

Mammals

Mammals are homoiothermic (warm blooded) vertebrates with four limbs. They differ from birds in having hair rather than feathers. Unlike the other vertebrates they have a diaphragm which plays a part in breathing (see Chapter 11). They also have mammary glands and suckle their young on milk.

A sample of mammals is shown in Figure 1.9 on page 8 and Figure 1.21 illustrates some of the mammalian features.

Humans are mammals. All mammals give birth to fully formed young instead of laying eggs. The eggs are fertilised internally and undergo a period of development in the uterus (see 'Sexual reproduction in humans' in Chapter 16).

Figure 1.21 Mammalian features. The furry coat, the external ear pinnae and the facial whiskers (vibrissae) are visible mammalian features in this gerbil.

The young may be blind and helpless at first, e.g. cats, or they may be able to stand up and move about soon after birth, e.g. sheep and cows. In either case, the youngster's first food is the milk which it sucks from the mother's teats. The milk is made in the mammary glands and contains all the nutrients that the offspring need for the first few weeks or months, depending on the species.

As the youngsters get older, they start to feed on the same food as the parents. In the case of carnivores, the parents bring the food to the young until they are able to fend for themselves.

Table 1.2 Key features of the five classes of vertebrates

Vertebrate class	Fish	Amphibia	Reptiles	Birds	Mammals
Examples	herring, perch, also sharks	frog, toad, newt	lizard, snake	robin, pigeon	mouse
Body covering	scales	moist skin	dry skin, with scales	feathers, with scales on legs	fur
Movement	fins (also used for balance)	four limbs, back feet are often webbed to make swimming more efficient	four legs (apart from snakes)	two wings and two legs	four limbs
Reproduction	produce jelly-covered eggs in water	produce jelly-covered eggs in water	produce eggs with a rubbery, waterproof shell; laid on land	produce eggs with a hard shell; laid on land	produce live young
Sense organs	eyes; no ears; lateral line along body for detecting vibrations in water	eyes; ears	eyes; ears	eyes; ears	eyes; ears with a pinna (external flap)
Other details	cold blooded; gills for breathing	cold blooded; lungs and skin for breathing	cold blooded; lungs for breathing	warm blooded; lungs for breathing; beak	warm blooded; lungs for breathing; females have mammary glands to produce milk to feed young; four types of teeth

The plant kingdom

It is useful to have an overview of the classification of the plant kingdom, although only two groups (ferns and flowering plants) will be tested in the examination.

Plant kingdom

DIVISION

Red algae
Brown algae } seaweeds and filamentous forms; mostly aquatic
Green algae

Bryophytes (no specialised conducting tissue)

    CLASS
    Liverworts
    Mosses

Vascular plants (well-developed xylem and phloem)

    CLASS
    Ferns

    { Conifers (seeds not enclosed in fruits) } Sometimes called, collectively, 'seed-bearing plants'
    { Flowering plants (seeds enclosed in fruits) }

        SUBCLASS
        Monocotyledons (grasses, lilies)
        Dicotyledons (trees, shrubs, herbaceous plants)

            FAMILY
            e.g. Ranunculaceae (one of about 70 families)

                GENUS
                e.g. *Ranunculus*

                    SPECIES
                    e.g. *Ranunculus bulbosus*
                    (bulbous buttercup)

Ferns

Ferns are land plants with quite highly developed structures. Their stems, leaves and roots are very similar to those of the flowering plants.

The stem is usually entirely below ground and takes the form of a structure called a **rhizome**. In bracken, the rhizome grows horizontally below ground, sending up leaves at intervals. The roots which grow from the rhizome are called adventitious roots (see 'Transport in plants' in Chapter 8). This is the name given to any roots which grow directly from the stem rather than from other roots.

The stem and leaves have sieve tubes and water-conducting cells similar to those in the xylem and phloem of a flowering plant (see Chapter 8). For this reason, the ferns and seed-bearing plants are sometimes referred to as vascular plants, because they all have vascular bundles or vascular tissue. Ferns also have multicellular roots with vascular tissue.

The leaves of ferns vary from one species to another (see Figure 1.22, and Figure 1.10 on page 9), but they are all several cells thick. Most of them have an upper and lower epidermis, a layer of palisade cells and a spongy mesophyll similar to the leaves of a flowering plant.

Figure 1.22 Young fern leaves. Ferns do not form buds like those of the flowering plants. The midrib and leaflets of the young leaf are tightly coiled and unwind as it grows.

Ferns produce gametes but no seeds. The zygote gives rise to the fern plant, which then produces single-celled spores from numerous **sporangia** (spore capsules) on its leaves. The sporangia are formed on the lower side of the leaf but their position depends on the species of fern. The sporangia are usually arranged in compact groups (see Figure 1.23).

Features of organisms

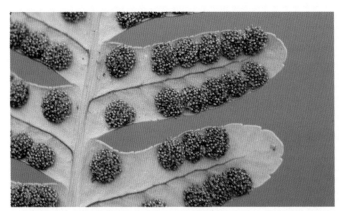

Figure 1.23 Polypody fern. Each brown patch on the underside of the leaf is made up of many sporangia.

Flowering plants

Flowering plants reproduce by seeds which are formed in flowers. The seeds are enclosed in an ovary. The general structure of flowering plants is described in Chapter 8. Examples are shown in Figure 1.11 on page 10. Flowering plants are divided into two subclasses: monocotyledons and dicotyledons. Monocotyledons (monocots for short), are flowering plants which have only one cotyledon in their seeds. Most, but not all, monocots also have long, narrow leaves (e.g. grasses, daffodils, bluebells) with parallel leaf veins (see Figure 1.24(a)).

The dicotyledons (dicots for short), have two cotyledons in their seeds. Their leaves are usually broad and the leaf veins form a branching network (see Figure 1.24(b)).

The key features of monocots and dicots are summarised in Table 1.3.

Table 1.3 Summary of the key features of monocots and dicots

Feature	Monocotyledon	Dicotyledon
leaf shape	long and narrow	broad
leaf veins	parallel	branching
cotyledons	one	two
grouping of flower parts (petals, sepals and carpels)	threes	fives

In addition to knowing the features used to place animals and plants into the appropriate kingdoms, you also need to know the main features of the following kingdoms: Fungus, Prokaryote and Protoctist.

The fungi kingdom

Most fungi are made up of thread-like hyphae (see Figure 1.25), rather than cells, and there are many nuclei distributed throughout the cytoplasm in their hyphae (see Figure 1.26).

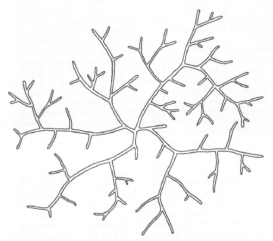

Figure 1.25 The branching hyphae form a mycelium.

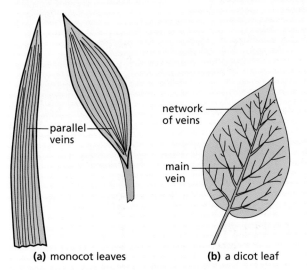

(a) monocot leaves (b) a dicot leaf

Figure 1.24 Leaf types in flowering plants

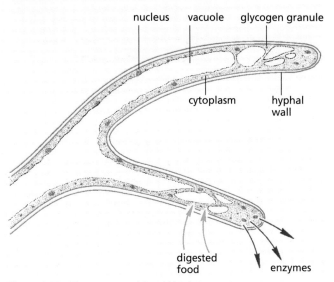

Figure 1.26 The structure of fungal hyphae

17

The fungi include fairly familiar organisms such as mushrooms, toadstools, puffballs and the bracket fungi that grow on tree trunks (Figure 1.27). There are also the less obvious, but very important, mould fungi which grow on stale bread, cheese, fruit or other food. Many of the mould fungi live in the soil or in dead wood. The yeasts are single-celled fungi similar to the moulds in some respects.

Some fungal species are parasites, as is the bracket fungus shown in Figure 1.27. They live in other organisms, particularly plants, where they cause diseases which can affect crop plants, such as the mildew shown in Figure 1.28. (See also Chapter 10.)

Figure 1.27 A parasitic fungus. The 'brackets' are the reproductive structures. The mycelium in the trunk will eventually kill the tree.

Figure 1.28 Mildew on wheat. Most of the hyphae are inside the leaves, digesting the cells, but some grow out and produce the powdery spores seen here.

The Prokaryote kingdom

These are the bacteria and the blue-green algae. They consist of single cells but differ from other single-celled organisms because their chromosomes are not organised into a nucleus.

Bacterial structure

Bacteria (singular: bacterium) are very small organisms consisting of single cells rarely more than 0.01 mm in length. They can be seen only with the higher powers of the microscope.

Their cell walls are made, not of cellulose, but of a complex mixture of proteins, sugars and lipids. Some bacteria have a **slime capsule** outside their cell wall. Inside the cell wall is the cytoplasm, which may contain granules of glycogen, lipid and other food reserves (see Figure 1.29).

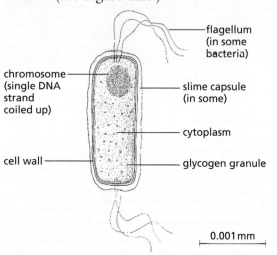

Figure 1.29 Generalised diagram of a bacterium

Each bacterial cell contains a single chromosome, consisting of a circular strand of DNA (see Chapter 4 and 'Chromosomes, genes and proteins' in Chapter 17). The chromosome is not enclosed in a nuclear membrane but is coiled up to occupy part of the cell, as shown in Figure 1.30.

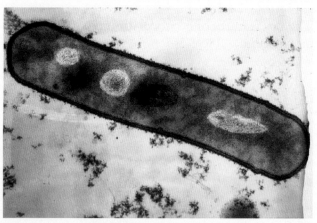

Figure 1.30 Longitudinal section through a bacterium (×27 000). The light areas are coiled DNA strands. There are three of them because the bacterium is about to divide twice (see Figure 1.31).

Individual bacteria may be spherical, rod-shaped or spiral and some have filaments, called **flagella**, projecting from them. The flagella can flick and so move the bacterial cell about.

Features of organisms

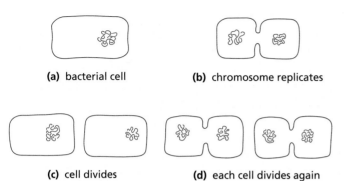

Figure 1.31 Bacterium reproducing. This is asexual reproduction by cell division (see 'Asexual reproduction' in Chapter 16 and 'Mitosis' in Chapter 17).

The Protoctist kingdom

These are single-celled (unicellular) organisms which have their chromosomes enclosed in a nuclear membrane to form a nucleus. Some examples are shown in Figure 1.32.

Some of the protoctista, e.g. *Euglena*, possess chloroplasts and make their food by photosynthesis. These protoctista are often referred to as unicellular 'plants' or **protophyta**. Organisms such as *Amoeba* and *Paramecium* take in and digest solid food and thus resemble animals in their feeding. They may be called unicellular 'animals' or **protozoa**.

Amoeba is a protozoan which moves by a flowing movement of its cytoplasm. It feeds by picking up bacteria and other microscopic organisms as it goes. *Vorticella* has a contractile stalk and feeds by creating a current of water with its cilia. The current brings particles of food to the cell. *Euglena* and *Chlamydomonas* have chloroplasts in their cells and feed, like plants, by photosynthesis.

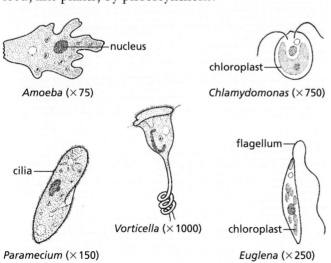

Figure 1.32 Protoctista. *Chlamydomonas* and *Euglena* have chloroplasts and can photosynthesise. The others are protozoa and ingest solid food.

Viruses

There are many different types of virus and they vary in their shape and structure. All viruses, however, have a central core of RNA or DNA (see Chapter 4) surrounded by a protein coat. Viruses have no nucleus, cytoplasm, cell organelles or cell membrane, though some forms have a membrane outside their protein coats.

Virus particles, therefore, are not cells. They do not feed, respire, excrete or grow and it is debatable whether they can be classed as living organisms. Viruses do reproduce, but only inside the cells of living organisms, using materials provided by the host cell.

A generalised virus particle is shown in Figure 1.33. The nucleic acid core is a coiled single strand of RNA. The coat is made up of regularly packed protein units called **capsomeres** each containing many protein molecules. The protein coat is called a **capsid**.

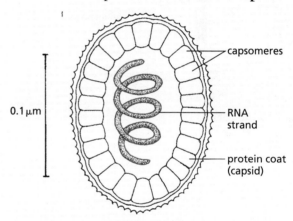

Figure 1.33 Generalised structure of a virus

● Extension work

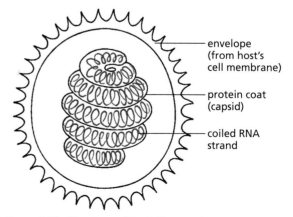

Figure 1.34 Structure of the influenza virus

Outside the capsid, in the influenza virus and some other viruses, is an envelope which is probably derived from the cell membrane of the host cell (Figure 1.34).

Ideas about classification

From the earliest days, humans must have given names to the plants and animals they observed, particularly those that were useful as food or medicine. Over the years, there have been many attempts to sort plants and animals into related groups. Aristotle's 'Ladder of Nature' (Figure 1.35) organised about 500 animal species into broad categories.

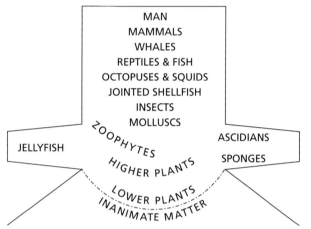

Figure 1.35 Aristotle's 'Ladder of Nature'

The 16th-century herbalists, such as John Gerard, divided the plant world into 'kindes' such as grasses, rushes, grains, irises and bulbs. Categories such as 'medicinal plants' and 'sweet-smelling plants', however, did not constitute a 'natural' classification based on structural features. The herbalists also gave the plants descriptive Latin names, e.g. *Anemone tenuifoliaflorecoccinea* ('the small-leaved scarlet anemone'). The first name shows a recognition of relationship to *Anemone nemorumfloreplenoalbo* ('the double white wood anemone'), for example. This method of naming was refined and popularised by Carl Linnaeus (see below).

John Ray (1625–1705)

Ray was the son of a blacksmith who eventually became a Fellow of the Royal Society. He travelled widely in Britain and Europe making collections of plants, animals and rocks.

In 1667 and 1682 he published a catalogue of British plants based on the structure of their flowers, seeds, fruits and roots. He was the first person to make a distinction between monocots and dicots. Ray also published a classification of animals, based on hooves, toes and teeth. Ultimately he devised classificatory systems for plants, birds, mammals, fish and insects. In doing this, he brought order out of a chaos of names and systems.

At the same time he studied functions, adaptations and behaviour of organisms.

In 1691 he claimed that fossils were the mineralised remains of extinct creatures, possibly from a time when the Earth was supposedly covered by water. This was quite contrary to established (but varied) views on the significance of fossils. Some thought that the fossils grew and developed in the rocks, others supposed that God had put them there 'for his pleasure' and still others claimed that the Devil put them in the rocks to 'tempt, frighten or confuse'. A more plausible theory was that a huge flood had washed marine creatures on to the land.

Despite Ray's declaration, the modern idea of the significance of fossils was not generally accepted until Darwin's day (see 'Selection' in Chapter 18).

Carl Linnaeus (1707–1778)

Linnaeus was a Swedish naturalist who initially graduated in medicine but became interested in plants. He travelled in Scandinavia, England and Eastern Europe, discovering and naming new plant species.

In 1735 he published his *Systema Naturae*, which accurately described about 7700 plant species and classified them, largely on the basis of their reproductive structures (stamens, ovaries, etc., see 'Sexual reproduction in plants' in Chapter 16). He further grouped species into genera, genera into classes, and classes into orders. ('Phyla' came later.) He also classified over 4000 animals, but rather less successfully, into mammals, birds, insects and worms.

Linnaeus refined and popularised the binomial system of naming organisms, in which the first name represents the genus and the second name the species. (See 'Concept and use of a classification system' earlier in this chapter.) This system is still the official starting point for naming or revising the names of organisms.

Although the classificatory system must have suggested some idea of evolution, Linnaeus steadfastly rejected the theory and insisted that no species created by God had ever become extinct.

Dichotomous keys

Dichotomous keys are used to identify unfamiliar organisms. They simplify the process of identification. Each key is made up of pairs of contrasting features (dichotomous means two branches), starting with quite general characteristics and progressing to more specific ones. By following the key and making appropriate choices it is possible to identify the organism correctly.

Figure 1.36 shows an example of a dichotomous key that could be used to place an unknown vertebrate in the correct class. Item 1 gives you a choice between two alternatives. If the animal is poikilothermic (cold blooded), you move to item 2 and make a further choice. If it is homoiothermic (warm blooded), you move to item 4 for your next choice.

The same technique may be used for assigning an organism to its class, genus or species. However, the important features may not always be easy to see and you have to make use of less fundamental characteristics.

VERTEBRATE CLASSES

1	Poikilothermic	2
	Homoiothermic	4
2	Has fins but no limbs	**Fish**
	Has four limbs	3
3	Has no scales on body	**Amphibian**
	Has scales	**Reptile**
4	Has feathers	**Bird**
	Has fur	**Mammal**

Figure 1.36 A dichotomous key for vertebrate classes

Figure 1.37 is a key for identifying some of the possible invertebrates to be found in a compost heap. Of course, you do not need a key to identify these familiar animals but it does show you how a key can be constructed.

You need to be able to develop the skills to construct simple dichotomous keys, based on easily identifiable features. If you know the main characteristics of a group, it is possible to draw up a systematic plan for identifying an unfamiliar organism. One such plan is shown in Figure 1.38 (on the next page).

INHABITANTS OF A COMPOST HEAP

1	Has legs	2
	No legs	5
2	More than six legs	3
	Six legs	4
3	Short, flattened grey body	**Woodlouse**
	Long brown/yellow body	**Centipede**
4	Pincers on last segment	**Earwig**
	Hard wing covers	**Beetle**
5	Body segmented	**Earthworm**
	Body not segmented	6
6	Has a shell	**Snail**
	No shell	**Slug**

Figure 1.37 A dichotomous key for some invertebrates in a compost heap

Figure 1.39 (overleaf) shows five different items of laboratory glassware. If you were unfamiliar with the resources in a science lab you may not be able to name them. We are going to create a dichotomous key to help with identification. All the items have one thing in common – they are made of glass. However, each has features which make it unique and we can devise questions based on these features. The first task is to study the items, to work out what some of them have in common and what makes them different from others. For example, some have a pouring spout, others have graduations marked on the glass for measuring, some have a neck (where the glass narrows to form a thinner structure), some can stand without support because they have a flat base, and so on.

The first question should be based on a feature which will split the group into two. The question is going to generate a 'yes' or 'no' answer. For each of the two sub-groups formed, a further question based on the features of some of that sub-group should then be formulated. Figure 1.40 (overleaf) shows one possible solution.

This is not the only way that a dichotomous key could be devised for the laboratory glassware shown. Construct your own key and test it for each object.

1 CHARACTERISTICS AND CLASSIFICATION OF LIVING ORGANISMS

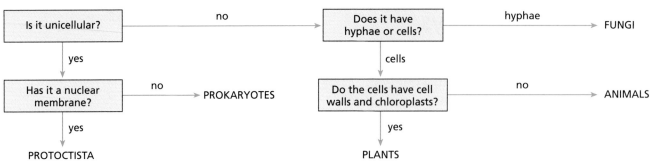

Figure 1.38 Identification plan

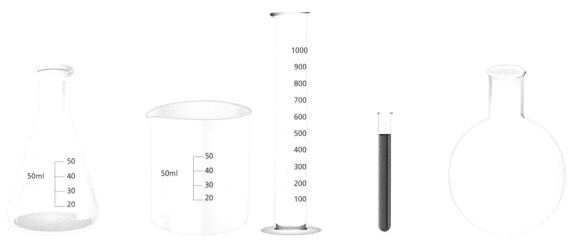

Figure 1.39 Items of laboratory glassware

1 Has it got a pouring spout?
   Yes 2
   No 3

2 Has it got a broad base?
   Yes .... **Beaker**
   No .... **Measuring cylinder**

3 Has it got straight sides for the whole of its length?
   Yes .... **Boiling tube**
   No 4

4 Has it got sloping sides?
   Yes .... **Conical flask**
   No .... **Round-bottomed flask**

Figure 1.40 Dichotomous key for identifying laboratory glassware

Questions

Core
1. Why do you think poikilothermic (cold blooded) animals are slowed down by low temperatures? (See Chapter 5.)
2. Which vertebrate classes:
    a are warm-blooded
    b have four legs
    c lay eggs
    d have internal fertilisation
    e have some degree of parental care?
3. Figure 1.32 on page 19 shows some protoctista. Using only the features shown in the drawings, construct a dichotomous key that could be used to identify these organisms.
4. Construct a dichotomous key that would lead an observer to distinguish between the following plants: daffodil, poppy, buttercup, meadow grass, iris (see Figure 1.11, page 10). (There is more than one way.)
    Why this is an 'artificial' key rather than a 'natural' key?

Extended
5. Classify the following organisms: beetle, sparrow, weasel, gorilla, bracken, buttercup.
    For example, butterfly: Kingdom, animal; Group, arthropod; Class, insect.
6. The white deadnettle is *Lamium album*; the red deadnettle is *Lamium purpureum*. Would you expect these two plants to cross-pollinate successfully?
7. If a fire destroys all the above-ground vegetation, the bracken (a type of fern) will still grow well in the next season. Suggest why this should be so.
8. Which kingdoms contain organisms with:
    a many cells
    b nuclei in their cells
    c cell walls
    d hyphae
    e chloroplasts?

Checklist

After studying Chapter 1 you should know and understand the following:

- The seven characteristics of living things are movement, respiration, sensitivity, growth, reproduction, excretion and nutrition.
- A species is a group of organisms that can reproduce to produce fertile offspring.
- The binomial system is an internationally agreed system in which the scientific name of an organism is made up of two parts showing the genus and the species.
- Classification is a way of sorting organisms into a meaningful order, traditionally using morphology and anatomy, but recently also using DNA.
- All living organisms have certain features in common, including the presence of cytoplasm and cell membranes, and DNA as genetic material.
- Animals get their food by eating plants or other animals.
- Arthropods have a hard exoskeleton and jointed legs.
- Crustacea mostly live in water and have more than three pairs of legs.
- Insects mostly live on land and have wings and three pairs of legs.
- Arachnids have four pairs of legs and poisonous mouth parts.
- Myriapods have many pairs of legs.
- Vertebrates have a spinal column and skull.
- Fish have gills, fins and scales.
- Amphibia can breathe in air or in water.
- Reptiles are land animals; they lay eggs with leathery shells.
- Birds have feathers, beaks and wings; they are homoiothermic (warm-blooded).
- Mammals have fur, they suckle their young and the young develop inside the mother.
- Keys are used to identify unfamiliar organisms.
- Dichotomous means two branches, so the user is given a choice of two possibilities at each stage.

- Prokaryotes are microscopic organisms; they have no proper nucleus.
- Protoctists are single-celled organisms containing a nucleus.
- Fungi are made up of thread-like hyphae. They reproduce by spores.
- Plants make their food by photosynthesis.
- Ferns have well-developed stems, leaves and roots. They reproduce by spores.
- Seed-bearing plants reproduce by seeds.
- Flowering plants have flowers; their seeds are in an ovary which forms a fruit.
- Monocots have one cotyledon in the seed; dicots have two cotyledons in the seed.
- Viruses do not possess the features of a living organism.

2 Organisation and maintenance of the organism

Cell structure and organisation
Plant and animal cell structures
Functions of structures

Ribosomes, rough ER and mitochondria
Mitochondria and respiration

Levels of organisation
Specialised cells and their functions
Definitions and examples of tissues, organs and systems

Size of specimens
Calculations of magnification and size, using millimetres

Calculations of magnification using micrometres

● Cell structure and organisation

Cell structure

If a very thin slice of a plant stem is cut and studied under a microscope, it can be seen that the stem consists of thousands of tiny, box-like structures. These structures are called **cells**. Figure 2.1 is a thin slice taken from the tip of a plant shoot and photographed through a microscope. Photographs like this are called **photomicrographs**. The one in Figure 2.1 is 60 times larger than life, so a cell which appears to be 2 mm long in the picture is only 0.03 mm long in life.

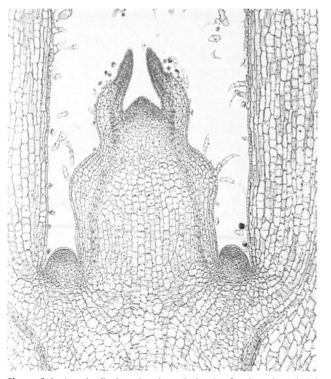

Figure 2.1 Longitudinal section through the tip of a plant shoot (×60). The slice is only one cell thick, so light can pass through it and allow the cells to be seen clearly.

Thin slices of this kind are called **sections**. If you cut *along the length* of the structure, you are taking a **longitudinal section** (Figure 2.2(b)). Figure 2.1 shows a longitudinal section, which passes through two small developing leaves near the tip of the shoot, and two larger leaves below them. The leaves, buds and stem are all made up of cells. If you cut *across* the structure, you make a **transverse section** (Figure 2.2(a)).

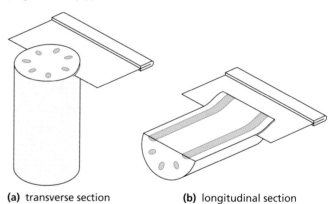

(a) transverse section   (b) longitudinal section

Figure 2.2 Cutting sections of a plant stem

It is fairly easy to cut sections through plant structures just by using a razor blade. To cut sections of animal structures is more difficult because they are mostly soft and flexible. Pieces of skin, muscle or liver, for example, first have to be soaked in melted wax. When the wax goes solid it is then possible to cut thin sections. The wax is dissolved away after making the section.

When sections of animal structures are examined under the microscope, they, too, are seen to be made up of cells but they are much smaller than plant cells and need to be magnified more. The photomicrograph of kidney tissue in Figure 2.3 has been magnified 700 times to show the cells clearly. The sections are often treated with dyes, called **stains**, in order to make the structures inside the cells show up more clearly.

Cell structure and organisation

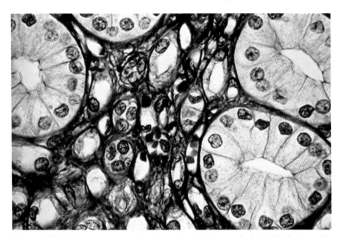

Figure 2.3 Transverse section through a kidney tubule (×700). A section through a tube will look like a ring (see Figure 2.14(b)). In this case, each 'ring' consists of about 12 cells.

Making sections is not the only way to study cells. Thin strips of plant tissue, only one cell thick, can be pulled off stems or leaves (Experiment 1, page 28). Plant or animal tissue can be squashed or smeared on a microscope slide (Experiment 2, page 29) or treated with chemicals to separate the cells before studying them.

There is no such thing as a typical plant or animal cell because cells vary a great deal in their size and shape depending on their function. Nevertheless, it is possible to make a drawing like Figure 2.4 to show features which are present in most cells. *All cells* have a **cell membrane**, which is a thin boundary enclosing the **cytoplasm**. Most cells have a **nucleus**.

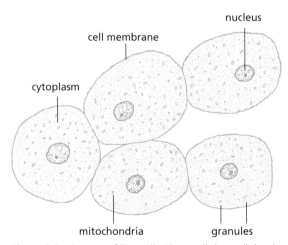

Figure 2.4 A group of liver cells. These cells have all the characteristics of animal cells.

Cytoplasm

Under the ordinary microscope (light microscope), cytoplasm looks like a thick liquid with particles in it. In plant cells it may be seen to be flowing about. The particles may be food reserves such as oil droplets or granules of starch. Other particles are structures known as **organelles**, which have particular functions in the cytoplasm. In the cytoplasm, a great many chemical reactions are taking place which keep the cell alive by providing energy and making substances that the cell needs.

The liquid part of cytoplasm is about 90% water with molecules of salts and sugars dissolved in it. Suspended in this solution there are larger molecules of fats (lipids) and proteins (see Chapter 4). Lipids and proteins may be used to build up the cell structures, such as the membranes. Some of the proteins are **enzymes** (see Chapter 5). Enzymes control the rate and type of chemical reactions which take place in the cells. Some enzymes are attached to the membrane systems of the cell, whereas others float freely in the liquid part of the cytoplasm.

Cell membrane

This is a thin layer of cytoplasm around the outside of the cell. It stops the cell contents from escaping and also controls the substances which are allowed to enter and leave the cell. In general, oxygen, food and water are allowed to enter; waste products are allowed to leave and harmful substances are kept out. In this way the cell membrane maintains the structure and chemical reactions of the cytoplasm.

Nucleus (plural: nuclei)

Most cells contain one nucleus, which is usually seen as a rounded structure enclosed in a membrane and embedded in the cytoplasm. In drawings of cells, the nucleus may be shown darker than the cytoplasm because, in prepared sections, it takes up certain stains more strongly than the cytoplasm. The function of the nucleus is to control the type and quantity of enzymes produced by the cytoplasm. In this way it regulates the chemical changes which take place in the cell. As a result, the nucleus determines what the cell will be, for example, a blood cell, a liver cell, a muscle cell or a nerve cell.

The nucleus also controls cell division, as shown in Figure 2.5. A cell without a nucleus cannot reproduce. Inside the nucleus are thread-like structures called **chromosomes**, which can be seen most easily at the time when the cell is dividing (see Chapter 17 for a fuller account of chromosomes).

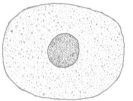
(a) Animal cell about to divide.

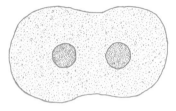
(b) The nucleus divides first.

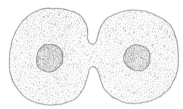
(c) The daughter nuclei separate and the cytoplasm pinches off between the nuclei.

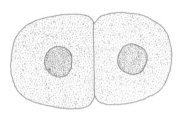
(d) Two cells are formed – one may keep the ability to divide, and the other may become specialised.

Figure 2.5 Cell division in an animal cell

Plant cells

A few generalised animal cells are represented by Figure 2.4, while Figure 2.6 is a drawing of two palisade cells from a plant leaf. (See 'Leaf structure' in Chapter 6.)

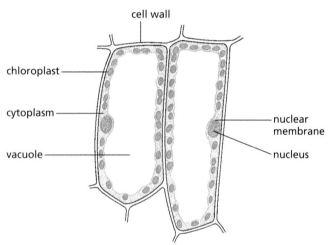

Figure 2.6 Palisade cells from a leaf

Plant cells differ from animal cells in several ways.

1 Outside the cell membrane they all have a **cell wall** which contains cellulose and other compounds. It is non-living and allows water and dissolved substances to pass through. The cell wall is not selective like the cell membrane. (Note that plant cells *do* have a cell membrane but it is not easy to see or draw because it is pressed against the inside of the cell wall (see Figure 2.7).)

Under the microscope, plant cells are quite distinct and easy to see because of their cell walls. In Figure 2.1 it is only the cell walls (and in some cases the nuclei) which can be seen. Each plant cell has its own cell wall but the boundary between two cells side by side does not usually show up clearly. Cells next to each other therefore appear to be sharing the same cell wall.

2 Most mature plant cells have a large, fluid-filled space called a **vacuole**. The vacuole contains **cell sap**, a watery solution of sugars, salts and sometimes pigments. This large, central vacuole pushes the cytoplasm aside so that it forms just a thin lining inside the cell wall. It is the outward pressure of the vacuole on the cytoplasm and cell wall which makes plant cells and their tissues firm (see 'Osmosis' in Chapter 3). Animal cells may sometimes have small vacuoles in their cytoplasm but they are usually produced to do a particular job and are not permanent.

3 In the cytoplasm of plant cells are many organelles called **plastids**. These are not present in animal cells. If they contain the green substance **chlorophyll**, the organelles are called **chloroplasts** (see Chapter 6). Colourless plastids usually contain starch, which is used as a food store. (Note: the term *plastid* is **not** a syllabus requirement.)

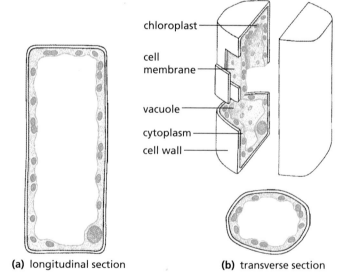

(a) longitudinal section (b) transverse section

Figure 2.7 Structure of a palisade mesophyll cell. It is important to remember that, although cells look flat in sections or in thin strips of tissue, they are in fact three-dimensional and may seem to have different shapes according to the direction in which the section is cut. If the cell is cut across it will look like (b); if cut longitudinally it will look like (a).

The shape of a cell when seen in a transverse section may be quite different from when the same cell is seen in a longitudinal section and Figure 2.7 shows why this is so. Figures 8.4(b) and 8.4(c) on page 112 show the appearance of cells in a stem vein as seen in transverse and longitudinal section.

Table 2.1 Summary: the parts of a cell

	Name of part	Description	Where found	Function (supplement only)
Animal and plant cells	cytoplasm	jelly-like, with particles and organelles in	enclosed by the cell membrane	contains the cell organelles, e.g. mitochondria, nucleus site of chemical reactions
	cell membrane	a partially permeable layer that forms a boundary around the cytoplasm	around the cytoplasm	prevents cell contents from escaping controls what substances enter and leave the cell
	nucleus	a circular or oval structure containing DNA in the form of chromosomes	inside the cytoplasm	controls cell division controls cell development controls cell activities
Plant cells only	cell wall	a tough, non-living layer made of cellulose surrounding the cell membrane	around the outside of plant cells	prevents plant cells from bursting allows water and salts to pass through (freely permeable)
	vacuole	a fluid-filled space surrounded by a membrane	inside the cytoplasm of plant cells	contains salts and sugars helps to keep plant cells firm
	chloroplast	an organelle containing chlorophyll	inside the cytoplasm of some plant cells	traps light energy for photosynthesis

When studied at much higher magnifications with the **electron microscope**, the cytoplasm of animal and plant cells no longer looks like a structureless jelly but appears to be organised into a complex system of membranes and vacuoles. Organelles present include the **rough endoplasmic reticulum**, a network of flattened cavities surrounded by a membrane, which links with the nuclear membrane. The membrane holds **ribosomes**, giving its surface a rough appearance. Rough endoplasmic reticulum has the function of producing, transporting and storing proteins. Ribosomes can also be found free in the cytoplasm. They build up the cell's proteins (see Chapter 4).

Mitochondria are tiny organelles, which may appear slipper-shaped, circular or oval when viewed in section. In three dimensions, they may be spherical, rod-like or elongated. They have an outer membrane and an inner membrane with many inward-pointing folds. Mitochondria are most numerous in regions of rapid chemical activity and are responsible for producing energy from food substances through the process of aerobic respiration (see Chapter 12).

Note that prokaryotes do not possess mitochondria or rough endoplasmic reticulum in their cytoplasm.

Figure 2.8(a) is a diagram of an animal cell magnified 10 000 times. Figure 2.8(b) is an electron micrograph of a liver cell. Organelles in the cytoplasm can be seen clearly. They have recognisable shapes and features.

Figure 2.8(c) is an electron micrograph of a plant cell. In addition to the organelles already named and described, other organelles are also present such as chloroplasts and a cell wall.

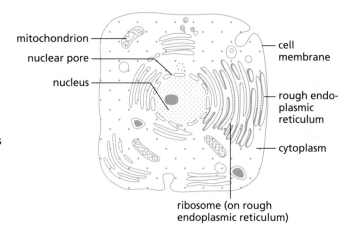

(a) diagram of a liver cell (×10 000)

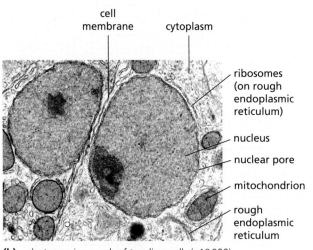

(b) electron micrograph of two liver cells (×10 000)

Figure 2.8 Cells at high magnification

2 ORGANISATION AND MAINTENANCE OF THE ORGANISM

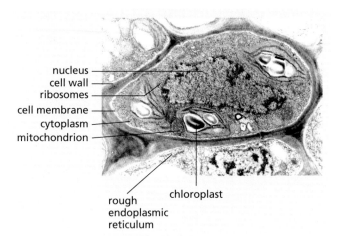

(c) electron micrograph of a plant cell (×6000)

Figure 2.8 Cells at high magnification (continued)

Practical work

Looking at cells

1 Plant cells – preparing a slide of onion epidermis cells

The onion provides a very useful source of epidermal plant tissue which is one cell thick, making it relatively easy to set up as a temporary slide. The onion is made up of fleshy leaves. On the incurve of each leaf there is an epidermal layer which can be peeled off (Figure 2.9(a)).

- Using forceps, peel a piece of epidermal tissue from the incurve of an onion bulb leaf.
- Place the epidermal tissue on a glass microscope slide.
- Using a scalpel, cut out a 1 cm square of tissue (discarding the rest) and arrange it in the centre of the slide.
- Add two to three drops of iodine solution. (This will stain any starch in the cells and provides a contrast between different components of the cells.)
- Using forceps, a mounted needle or a wooden splint, support a coverslip with one edge resting near to the onion tissue, at an angle of about 45° (Figure 2.9(b)).
- Gently lower the coverslip over the onion tissue, trying to avoid trapping any air bubbles. (Air bubbles will reflect light when viewing under the light microscope, obscuring the features you are trying to observe.)
- Leave the slide for about 5 minutes to allow the iodine stain to react with the specimen. The iodine will stain the cell nuclei pale yellow and the starch grains blue.
- Place the slide on to the microscope stage, select the lowest power objective lens and focus on the specimen. Increase the magnification using the other objective lenses. Under high power, the cells should look similar to those shown in Figure 2.10.
- Make a large drawing of **one** cell and label the following parts: cell wall, cell membrane, cytoplasm, nucleus.

An alternative tissue is rhubarb epidermis (Figure 2.9(c)). This can be stripped off from the surface of a stalk and treated in the same way as the onion tissue. If red epidermis from rhubarb stalk is used, you will see the red cell sap in the vacuoles.

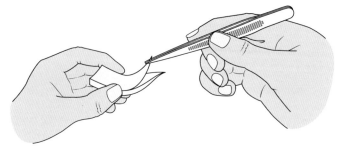

(a) peel the epidermis from the inside of an onion bulb leaf

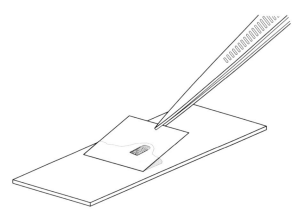

(b) place the epidermis on to the slide, adding 2–3 drops of iodine solution and carefully lowering a coverslip on to it

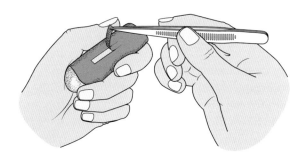

(c) alternatively, peel a strip of red epidermis from a piece of rhubarb skin

Figure 2.9 Looking at plant cells

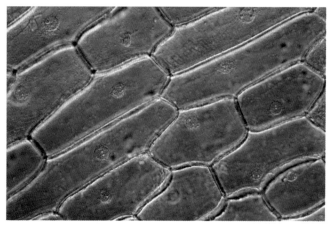

Figure 2.10 Onion epidermis cells

2 Plant cells – preparing cells with chloroplasts

- Using forceps, remove a leaf from a moss plant.
- Place the leaf in the centre of a microscope slide and add one or two drops of water.
- Place a coverslip over the leaf.
- Examine the leaf cells with the high power objective of a microscope. The cells should look similar to those shown in Figure 2.11.

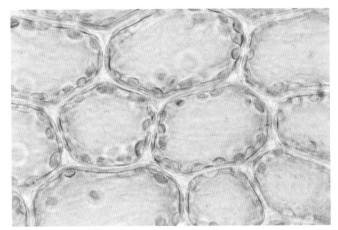

Figure 2.11 Cells in a moss leaf (×500). The vacuole occupies most of the space in each cell. The chloroplasts are confined to the layer of cytoplasm lining the cell wall.

3 Animal cells – preparing human cheek cells

Human cheek cells are constantly being rubbed off inside the mouth as they come in contact with the tongue and food. They can therefore be collected easily for use in a temporary slide.

Note: The Department of Education and Science and, subsequently, Local Authorities, used to recommend that schools should not use the technique which involves studying the epithelial cells which appear in a smear taken from the inside of the cheek. This was because of the very small risk of transmitting the AIDS virus. However, this guidance has now changed. A document, *Safety in Science Education* (1996) by the DfEE in Britain states that official government guidance on cheek cells has been effectively reversed, indicating that the use of cotton buds is now 'permitted' together with appropriate precautions to treat contaminated items with disinfectant or by autoclaving.

- Rinse your mouth with water to remove any fragments of food.
- Take a cotton bud from a freshly opened pack. Rub the cotton bud lightly on the inside of your cheek and gums to collect some cheek cells in saliva.
- Rub the cotton bud on to the centre of a clean microscope slide, to leave a sample of saliva. Repeat if the sample is too small. Then drop the cotton bud into a container of absolute alcohol or disinfectant.
- Add two to three drops of methylene blue dye. (This will stain parts of the cheek cells to make nuclei more visible.)
- Using forceps, a mounted needle or wooden splint, support a coverslip with one edge resting near to the cheek cell sample, at an angle of about 45°. Gently lower the coverslip over the tissue, trying to avoid trapping any air bubbles. (Air bubbles will reflect light when viewing under the light microscope, obscuring the features you are trying to observe.)
- Leave the slide for a few minutes to allow the methylene blue stain to react with the specimen.
- Place the slide on to the microscope stage, select the lowest power objective lens and focus on the specimen. Increase the magnification using the other objective lenses. Under high power, the cells should look similar to those shown in Figure 2.12, but less magnified.
- Make a large drawing of **one** cell and label the following parts: cell membrane, cytoplasm, nucleus.
- Place your used slide in laboratory disinfectant before washing.

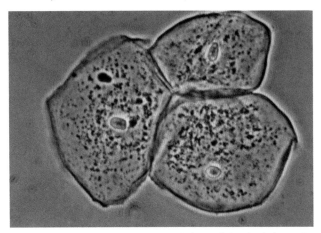

Figure 2.12 Cells from the lining epithelium of the cheek (×1500)

An alternative method of obtaining cells is to press some transparent sticky tape on to a well-washed wrist. When the tape is removed and studied under the microscope, cells with nuclei can be seen. A few drops of methylene blue solution will stain the cells and make the nuclei more distinct.

● Levels of organisation

Specialisation of cells

Most cells, when they have finished dividing and growing, become specialised. When cells are specialised:

- they do one particular job
- they develop a distinct shape
- special kinds of chemical change take place in their cytoplasm.

The changes in shape and the chemical reactions enable the cell to carry out its special function. Red blood cells and root hair cells are just two examples of specialised cells. Figure 2.13 shows a variety of specialised cells.

The specialisation of cells to carry out particular functions in an organism is sometimes referred to as **'division of labour'** within the organism. Similarly, the special functions of mitochondria, ribosomes and other cell organelles may be termed division of labour within the cell.

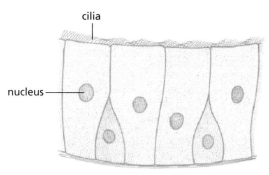

(a) ciliated cells
These cells form the lining of the nose and windpipe, and the tiny cytoplasmic 'hairs', called cilia, are in a continual flicking movement which creates a stream of fluid (mucus) that carries dust and bacteria through the bronchi and trachea, away from the lungs.

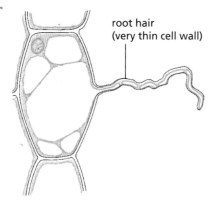

(b) root hair cell
These cells absorb water and mineral salts from the soil. The hair-like projection on each cell penetrates between the soil particles and offers a large absorbing surface. The cell membrane is able to control which dissolved substances enter the cell.

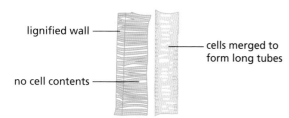

(c) xylem vessels
These cells transport mineral ions from the roots to the leaves. A substance called lignin impregnates and thickens the cell walls making the cells very strong and impermeable. This gives the stem strength. The lignin forms distinctive patterns in the vessels – spirals, ladder shapes, reticulate (net-like) and pitted. Xylem vessels are made up of a series of long xylem cells joined end-to-end (Figure 8.4(a)). Once a region of the plant has stopped growing, the end walls of the cells are digested away to form a continuous, fine tube (Figure 8.4(c)). The lignin thickening prevents the free passage of water and nutrients, so the cytoplasm in the cells dies. Effectively, the cells form long, thin, strong straws.

Figure 2.13 Specialised cells (not to scale)

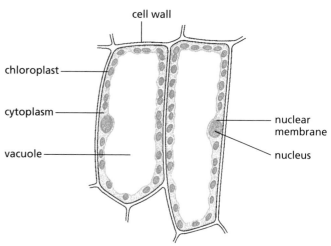

(d) palisade mesophyll cells
These are found underneath the upper epidermis of plant leaves. They are columnar (quite long) and packed with chloroplasts to trap light energy. Their function is to make food for the plant by photosynthesis using carbon dioxide, water and light energy.

(e) nerve cells
These cells are specialised for conducting electrical impulses along the fibre, to and from the brain and spinal cord. The fibres are often very long and connect distant parts of the body to the CNS, e.g. the foot and the spinal column. Chemical reactions cause the impulses to travel along the fibre.

Levels of organisation

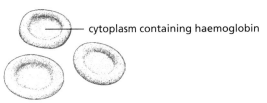

(f) red blood cells

These cells are distinctive because they have no nucleus when mature. They are tiny disc-like cells which contain a red pigment called haemoglobin. This readily combines with oxygen and their function is the transport of oxygen around the body.

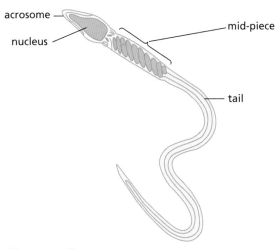

(g) sperm cell

Sperm cells are male sex cells. The front of the cell is oval shaped and contains a nucleus which carries genetic information. There is a tip, called an acrosome, which secretes enzymes to digest the cells around an egg and the egg membrane. Behind this is a mid-piece which is packed with mitochondria to provide energy for movement. The tail moves with a whip-like action enabling the sperm to swim. Their function is reproduction, achieved by fertilising an egg cell.

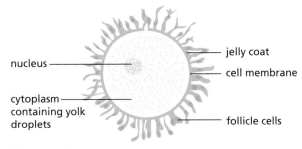

(h) egg cell

Egg cells (ova, singular: ovum) are larger than sperm cells and are spherical. They have a large amount of cytoplasm, containing yolk droplets made up of protein and fat. The nucleus carries genetic information. The function of the egg cell is reproduction.

Figure 2.13 Specialised cells (not to scale) (continued)

Tissues and organs

There are some microscopic organisms that consist of one cell only (see 'Features of organisms' in Chapter 1). These can carry out all the processes necessary for their survival. The cells of the larger plants and animals cannot survive on their own. A muscle cell could not obtain its own food and oxygen. Other specialised cells have to provide the food and oxygen needed for the muscle cell to live. Unless these cells are grouped together in large numbers and made to work together, they cannot exist for long.

Tissues

A **tissue**, such as bone, nerve or muscle in animals, and epidermis, xylem or pith in plants, is made up of many hundreds of cells often of a single type. The cells of each type have a similar structure and function so that the tissue itself can be said to have a particular function; for example, muscles contract to cause movement, xylem carries water in plants. Figure 2.14 shows how some cells are arranged to form simple tissues.

> **Key definition**
> A **tissue** is a group of cells with similar structures, working together to perform a shared function.

Organs

Organs consist of several tissues grouped together to make a structure with a special function. For example, the stomach is an organ which contains tissues made from epithelial cells, gland cells and muscle cells. These cells are supplied with food and oxygen brought by blood vessels. The stomach also has a nerve supply. The heart, lungs, intestines, brain and eyes are further examples of organs in animals. In flowering plants, the root, stem and leaves are the organs. The tissues of the leaf include epidermis, palisade tissue, spongy tissue, xylem and phloem (see Chapter 8).

> **Key definition**
> An **organ** is a structure made up of a group of tissues, working together to perform a specific function.

Organ systems

An **organ system** usually refers to a group of organs whose functions are closely related. For example, the heart and blood vessels make up the **circulatory system**; the brain, spinal cord and nerves make up the **nervous system** (Figure 2.15). In a flowering plant, the stem, leaves and buds make up a system called the **shoot** (Figure 8.1 on page 110).

> **Key definition**
> A **system** is a group of organs with related functions, working together to perform a body function.

2 ORGANISATION AND MAINTENANCE OF THE ORGANISM

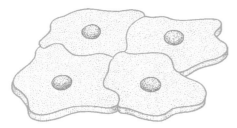

(a) cells forming an epithelium
A thin layer of tissue, e.g. the lining of the mouth cavity. Different types of epithelium form the internal lining of the windpipe, air passages food canal, etc., and protect these organs from physical or chemical damage.

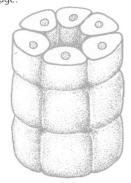

(b) cells forming a small tube
e.g. a kidney tubule (see p. 177). Tubules such as this carry liquids from one part of an organ to another.

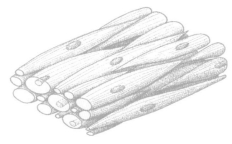

(c) one kind of muscle cell
Forms a sheet of muscle tissue. Blood vessels, nerve fibres and connective tissues will also be present. Contractions of this kind of muscle help to move food along the food canal or close down small blood vessels.

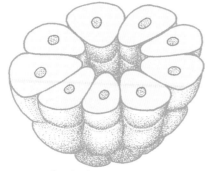

(d) cells forming part of a gland
The cells make chemicals which are released into the central space an carried away by a tubule such as shown in **(b)**. Hundreds of cell groups like this would form a gland like the salivary gland.

Figure 2.14 How cells form tissues

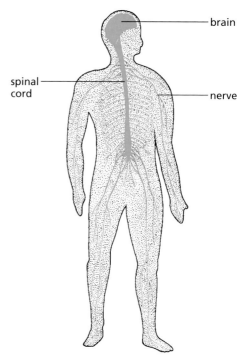

(a) nervous system

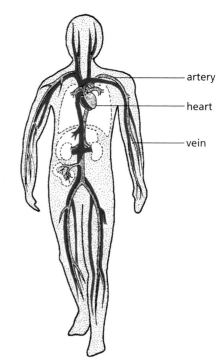

(b) circulatory system

Figure 2.15 Two examples of systems in the human body

Organisms

An **organism** is formed by the organs and systems working together to produce an independent plant or animal.

An example in the human body of how cells, tissues and organs are related is shown in Figure 2.16.

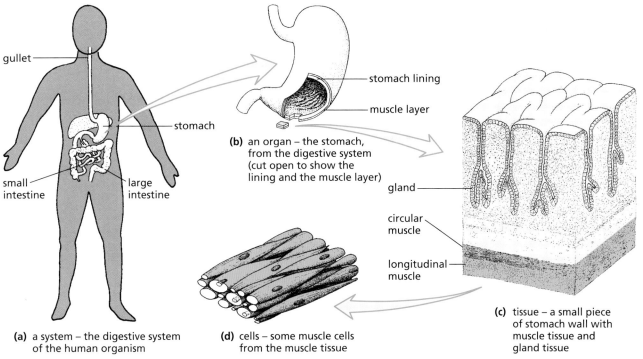

Figure 2.16 An example of how cells, tissues and organs are related

● Size of specimens

The light microscope

Most cells cannot be seen with the naked eye. A **hand lens** has a magnification of up to ×20, but this is not sufficient to observe the detail in cells. The **light microscope** (Figure 2.17) has two convex lenses, providing magnifications of up to ×1500, although most found in school laboratories will only magnify to ×400. The eyepiece lens is usually ×10 and there is a choice of objective lenses (typically ×4, ×10 and ×40), set in a nosepiece which can be rotated. Light, provided by a mirror or a bulb, is projected through the specimen mounted on a microscope slide. It passes through the objective and eyepieces lenses and the image is magnified so that detail of the specimen can be seen. Coarse and fine focus knobs are used to sharpen the image. Specimens are mounted on microscope slides, which may be temporary or permanent preparations. Temporary slides are quick to prepare, but the specimens dry out quite rapidly, so they cannot be stored successfully. A coverslip (a thin piece of glass) is carefully laid over the specimen. This helps to keep it in place, slows down dehydration and protects the objective lens from moisture or stains. A permanent preparation usually involves dehydrating the specimen and fixing it in a special resin such as Canada Balsam. These types of slides can be kept for many years.

Calculating magnification

A lens is usually marked with its magnifying power. This indicates how much larger the image will be, compared to the specimen's actual size. So, if the lens is marked ×10, the image will be ten times greater than the specimen's real size. Since a light microscope has two lenses, the magnification of both of these lenses needs to be taken into account. For example, if the specimen is viewed using a ×10 eyepiece lens and ×40 objective lens, the total magnification will be 10 × 40 = 400.

2 ORGANISATION AND MAINTENANCE OF THE ORGANISM

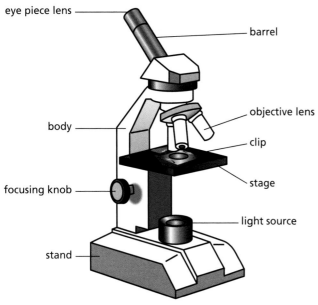

Figure 2.17 A light microscope

$$\text{Magnification} = \frac{\text{observed size of the image (or drawing)}}{\text{actual size of the specimen}}$$

When performing this type of calculation, make sure that the units of both sizes are the same. If they are different, convert one to make them the same. For example, if the actual size is in millimetres and the observed size is in centimetres, convert the centimetres to millimetres. (There are 10 millimetres in a centimetre.)

You may be required to calculate the actual size of a specimen, given a drawing or photomicrograph and a magnification.

$$\text{Actual size of the specimen} = \frac{\text{observed size of the image (or drawing)}}{\text{magnification}}$$

When you state the answer, make sure you quote the units (which will be the same as those used for measuring the observed size).

When the image is drawn, the drawing is usually much larger than the image, so the overall magnification of the specimen is greater still.

Organelles in cells are too small to be measured in millimetres. A smaller unit, called the **micrometre** (micron or μm) is used. Figure 2.18 shows a comparison of the sizes of a range of objects. The scale is in **nanometres** because of the tiny size of some of the objects. There are 1000 nanometres in 1 micrometre. (Note: the term nanometre is **not** a syllabus requirement.)

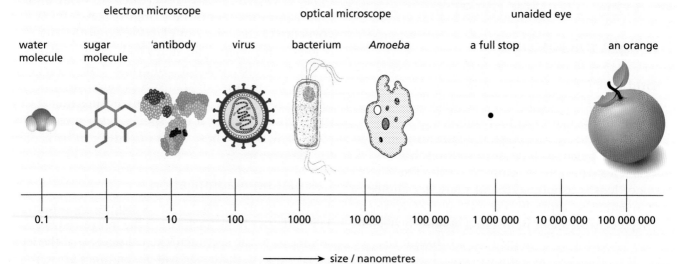

Figure 2.18 Comparing the sizes of a range of objects

There are

1 000 000 micrometres in a metre

10 000 micrometres in a centimetre

1000 micrometres in a millimetre.

Remember to make sure that the units of both sizes used in a calculation involving magnification are the same. So, if the actual size is in micrometres and the observed size is in millimetres, convert the millimetres to micrometres.

Questions

Core
1. a What structures are usually present in both animal and plant cells?
   b What structures are present in plant cells but not in animal cells?
2. What cell structure is largely responsible for controlling the entry and exit of substances into or out of the cell?
3. In what way does the red blood cell shown in Figure 2.13(f) differ from most other animal cells?
4. How does a cell membrane differ from a cell wall?
5. Why does the cell shown in Figure 2.7(b) appear to have no nucleus?
6. a In order to see cells clearly in a section of plant tissue, which magnification would you have to use?

   A ×5
   B ×10
   C ×100
   D ×1000

   b What is the approximate width (in millimetres) of one of the largest cells in Figure 2.3?
7. In Figure 2.3, the cell membranes are not always clear. Why is it still possible to decide roughly how many cells there are in each tubule section?
8. a Study Figure 8.7 on page 113 and identify examples of tissues and an organ.
   b Study Figure 7.13 on page 97 and identify examples of tissues and an organ.

Checklist

After studying Chapter 2 you should know and understand the following:

- Nearly all plants and animals are made up of thousands or millions of microscopic cells.
- All cells contain cytoplasm enclosed in a cell membrane.
- Most cells have a nucleus.
- Many chemical reactions take place in the cytoplasm to keep the cell alive.
- The nucleus directs the chemical reactions in the cell and also controls cell division.
- Plant cells have a cellulose cell wall and a large central vacuole.
- Cells are often specialised in their shape and activity to carry out particular jobs.
- Large numbers of similar cells packed together form a tissue.
- Different tissues arranged together form organs.
- A group of related organs makes up a system.
- The magnification of a specimen can be calculated if the actual size and the size of the image are known.

- Cytoplasm contains organelles such as mitochondria, chloroplasts and ribosomes.
- The magnification and size of biological specimens can be calculated using millimetres or micrometres.

3 Movement in and out of cells

Diffusion
Definition
Importance of diffusion of gases and solutes
Movement of substances in and out of cells

  Kinetic energy of molecules and ions
  Factors that influence diffusion

Osmosis
Movement of water through the cell membrane
Plant support

  Definition of osmosis and other terms associated with the process
  The effect of different solutions on tissues

Water potential
The uptake of water by plants
The importance of turgor pressure to plant support

Active transport
Definition of active transport
Movement of molecules and ions against a concentration gradient, using energy from respiration

The importance of active transport to the uptake of glucose

Cells need food materials which they can oxidise for energy or use to build up their cell structures. They also need salts and water, which play a part in chemical reactions in the cell. Finally, they need to get rid of substances such as carbon dioxide, which, if they accumulated in the cell, would upset some of the chemical reactions or even poison the cell.

Substances may pass through the cell membrane either passively by diffusion or actively by some form of active transport.

● Diffusion

Key definition
Diffusion is the net movement of molecules and ions from a region of their higher concentration to a region of their lower concentration down a concentration gradient, as a result of their random movement.

The molecules of a gas such as oxygen are moving about all the time. So are the molecules of a liquid or a substance such as sugar dissolved in water. As a result of this movement, the molecules spread themselves out evenly to fill all the available space (Figure 3.1).

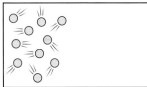

molecules moving about   become evenly distributed

Figure 3.1 Diffusion

This process is called **diffusion**. One effect of diffusion is that the molecules of a gas, a liquid or a dissolved substance will move from a region where there are a lot of them (i.e. concentrated) to regions where there are few of them (i.e. less concentrated) until the concentration everywhere is the same. Figure 3.2(a) is a diagram of a cell with a high concentration of molecules (e.g. oxygen) outside and a low concentration inside. The effect of this difference in concentration is to make the molecules diffuse into the cell until the concentration inside and outside is the same, as shown in Figure 3.2(b).

 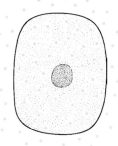

(a) greater concentration outside cell

(b) concentrations equal on both sides of the cell membrane

Figure 3.2 Molecules entering a cell by diffusion

Whether this will happen or not depends on whether the cell membrane will let the molecules through. Small molecules such as water (H_2O), carbon dioxide (CO_2) and oxygen (O_2) can pass through the cell membrane fairly easily. So diffusion tends to equalise the concentration of these molecules inside and outside the cell all the time.

When a cell uses oxygen for its aerobic respiration, the concentration of oxygen inside the cell falls and so oxygen molecules diffuse into the cell until the concentration is raised again. During tissue respiration, carbon dioxide is produced and so its concentration inside the cell increases. Once again diffusion takes place, but this time the molecules move out of the cell. In this way, diffusion can explain how a cell takes in its oxygen and gets rid of its carbon dioxide.

The importance of diffusion of gases and solutes

Gases

Most living things require a reliable source of oxygen for respiration. This moves into the organism by diffusion down a concentration gradient. Small animals with a large surface area to volume ratio may obtain oxygen through their body surface. Larger animals rely on gas exchange organs such as lungs or gills, which provide a large surface area for gas exchange, and a circulatory system to transport the oxygen to all their cells. Carbon dioxide, produced during aerobic respiration, is potentially toxic if it builds up. It is removed using the same mechanisms, again by diffusion.

Photosynthetic plants need carbon dioxide for making their food. This diffuses through the stomata in the leaves (see Chapter 8) into the air spaces in the mesophyll, eventually reaching the palisade cells. Oxygen, produced during photosynthesis, along with water vapour from the transpiration stream, diffuses out of the leaf through the stomata. The rate of diffusion of water vapour depends on the temperature, humidity and wind speed (see 'Water uptake' in Chapter 8). Any oxygen needed for respiration (some is generated by photosynthesis) and carbon dioxide produced (some is used up by photosynthesis) also diffuses through the stomata of the leaves.

Nitrogen is the commonest gas in the atmosphere. (78% of the air is nitrogen.) Nitrogen gas also enters the bloodstream by diffusion, but it is not used by the body. It is an inert (unreactive) gas so, under normal circumstances, it causes no problems. However, divers are at risk. As a diver swims deeper, the surrounding water pressure increases and this in turns raises the pressure in the diver's air tank. An increase in nitrogen pressure in the air tank results in more nitrogen diffusing into the diver's tissues, the amount going up the longer the diver stays at depth. Nitrogen is not used by the body tissues, so it builds up. When the diver begins to return to the surface of the water, the pressure decreases and the nitrogen can come out of solution, forming bubbles in the blood if the diver ascends too quickly. These bubbles can block blood flow and become lodged in joints resulting in a condition called **decompression sickness**, or '**the bends**'. Unless the diver rises slowly in planned stages, the effect of the nitrogen bubbles is potentially lethal and can only be overcome by rapid recompression.

Solutes

Mineral ions in solution, such as nitrates and magnesium, are thought to diffuse across the tissues of plant roots, but most are absorbed into the roots by active transport.

In the ileum, water-soluble vitamins such as vitamin B and vitamin C are absorbed into the bloodstream by diffusion.

In the kidneys, some solutes in the renal capsule, such as urea and salts, pass back into the bloodstream by diffusion. Initially, glucose is reabsorbed by diffusion, but active transport is also involved. Dialysis machines (see Chapter 13) use diffusion to remove small solutes (urea, uric acid and excess salts) from the blood.

Rates of diffusion

Molecules and ions in liquids and gases move around randomly using **kinetic energy** (energy from movement). The speed with which a substance diffuses through a cell wall or cell membrane will depend on temperature and many other conditions including the distance it has to diffuse, the difference between its concentration inside and outside the cell, the size of its molecules or ions and the surface area across which the diffusion is occurring.

Surface area

If 100 molecules diffuse through 1 mm^2 of a membrane in 1 minute, it is reasonable to suppose that an area of 2 mm^2 will allow twice as many through in the same time. Thus the rate of diffusion into a cell will depend on the cell's surface area. The greater the surface area, the faster is the total diffusion. Cells which are involved in rapid absorption, such as those in the kidney or the intestine, often have their 'free' surface membrane formed into hundreds of tiny projections called **microvilli** (see Figure 3.3) which increase the absorbing surface.

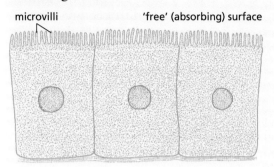

Figure 3.3 Microvilli

The shape of a cell will also affect the surface area. For example, the cell in Figure 3.4(a) has a greater surface area than that in Figure 3.4(b), even though they each have the same volume.

Figure 3.4 Surface area. The cells both have the same volume but the cell in (a) has a much greater surface area.

Temperature

An increase in temperature causes an increase in the kinetic energy which molecules and ions possess. This enables them to move faster, so the process of diffusion speeds up.

Concentration gradient

The bigger the difference in the concentration of a substance on either side of a membrane, the faster it will tend to diffuse. The difference is called a **concentration gradient** or **diffusion gradient** (Figure 3.5). If a substance on one side of a membrane is steadily removed, the diffusion gradient is maintained. When oxygen molecules enter a red blood cell they combine with a chemical (haemoglobin) which takes them out of solution. Thus the concentration of free oxygen molecules inside the cell is kept very low and the diffusion gradient for oxygen is maintained.

Figure 3.5 Concentration gradient

Distance

Cell membranes are all about the same thickness (approximately $0.007\,\mu m$) but plant cell walls vary in their thickness and permeability. Generally speaking, the thicker the wall, the slower the rate of diffusion. When oxygen diffuses from the alveoli of the lungs into red blood cells, it has to travel through the cell membranes of the alveoli, the blood capillaries and the red blood cells in addition to the cytoplasm of each cell. This increased distance slows down the diffusion rate.

Size of molecules or ions

In general, the larger the molecules or ions, the slower they diffuse. However, many ions and molecules in solution attract water molecules around them (see p. 43) and so their effective size is greatly increased. It may not be possible to predict the rate of diffusion from the molecular size alone.

Controlled diffusion

Although for any one substance, the rate of diffusion through a cell membrane depends partly on the concentration gradient, the rate is often faster or slower than expected. Water diffuses more slowly and amino acids diffuse more rapidly through a membrane than might be expected. In some cases this is thought to happen because the ions or molecules can pass through the membrane only by means of special pores. These pores may be few in number or they may be open or closed in different conditions.

In other cases, the movement of a substance may be speeded up by an enzyme working in the cell membrane. So it seems that 'simple passive' diffusion, even of water molecules, may not be so simple or so passive after all where cell membranes are concerned.

When a molecule gets inside a cell there are a great many structures and processes which may move it from where it enters to where it is needed. Simple diffusion is unlikely to play a very significant part in this movement.

Practical work

Experiments on diffusion

1 Diffusion and surface area

- Use a block of starch agar or gelatine at least 3 cm thick. Using a ruler and a sharp knife, measure and cut four cubes

from the jelly with sides of 3.0 cm, 2.0 cm, 1.0 cm and 0.5 cm.
- Place the cubes into a beaker of methylene blue dye or potassium permanganate solution.
- After 15 minutes, remove the cubes with forceps and place them on to a white tile.
- Cut each of the cubes in half and measure the depth to which the dye has diffused.
- Calculate the surface area and volume of each cube and construct a table of your data. Remember to state the units in the heading for each column.

Question
Imagine that these cubes were animals, with the jelly representing living cells and the dye representing oxygen. Which of the 'animals' would be able to survive by relying on diffusion through their surface to provide them with oxygen?

Taking it further
Try cutting different shapes, for example cutting a block 3.0 cm long, 1.0 cm wide and 0.5 cm deep. What type of animal would this represent? (Refer to Figure 1.7 on page 6.) Research how this type of animal obtains its oxygen.

2 Diffusion and temperature
- Set up two beakers with equal volumes of hot water and iced water.
- Add a few grains of potassium permanganate to each beaker and observe how rapidly the dissolved dye spreads through each column of water. An alternative is to use tea bags.

Question
Give an explanation for the results you observed.

3 Diffusion and concentration gradients and distance
- Push squares of wetted red litmus paper with a glass rod or wire into a wide glass tube which is at least 30 cm long and corked at one end, so that they stick to the side and are evenly spaced out, as shown in Figure 3.6. (It is a good strategy to mark 2 cm intervals along the outside of the tube, starting at 10 cm from one end, with a permanent marker or white correction fluid before inserting the litmus paper.)
- Close the open end of the tube with a cork carrying a plug of cotton wool saturated with a strong solution of ammonia. Start a stop watch.
- Observe and record the time when each square of litmus starts to turn blue in order to determine the rate at which the alkaline ammonia vapour diffuses along the tube.
- Repeat the experiment using a dilute solution of ammonia.
- Plot both sets of results on a graph, labelling each plot line.

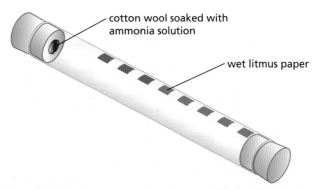

Figure 3.6 Experiment to measure the rate of diffusion of ammonia in air

Questions
1. Which ammonia solution diffused faster? Can you explain why?
2. Study your graph. What happened to the rate of diffusion as the ammonia travelled further along the tube? Can you explain why?

4 Diffusion and particle size
- Take a 15 cm length of dialysis tubing which has been soaked in water and tie a knot tightly at one end.
- Use a dropping pipette to partly fill the tubing with a mixture of 1% starch solution and 1% glucose solution.
- Rinse the tubing and test-tube under the tap to remove all traces of starch and glucose solution from the outside of the dialysis tubing.
- Put the tubing in a boiling tube and hold it in place with an elastic band as shown in Figure 3.7.
- Fill the boiling tube with water and leave for 30 minutes.
- Use separate teat pipettes to remove samples of liquid from the dialysis tubing and the boiling tube. Test both samples with iodine solution and Benedict's reagent.

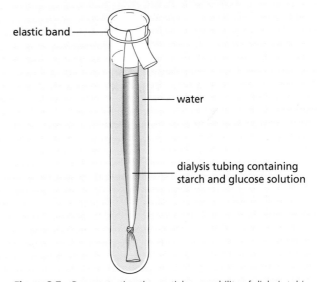

Figure 3.7 Demonstrating the partial permeability of dialysis tubing

3 MOVEMENT IN AND OUT OF CELLS

Result
The liquid inside the dialysis tubing goes blue with iodine solution and may give a positive Benedict's test, but the sample from the boiling tube only gives a positive Benedict's test.

Interpretation
The blue colour is characteristic of the reaction which takes place between starch and iodine, and is used as a test for starch. A positive Benedict's test gives a colour change from blue to cloudy green, yellow or brick red (see Chapter 4). The results show that glucose molecules have passed through the dialysis tubing into the water but the starch molecules have not moved out of the dialysis tubing. This is what we would expect if the dialysis tubing was partially permeable on the basis of its pore size. Starch molecules are very large (see Chapter 4) and probably cannot get through the pores. Glucose molecules are much smaller and can, therefore, get through.

● Osmosis

If a dilute solution is separated from a concentrated solution by a **partially permeable** membrane, water diffuses across the membrane from the dilute to the concentrated solution. This is known as **osmosis** and is shown in Figure 3.8.

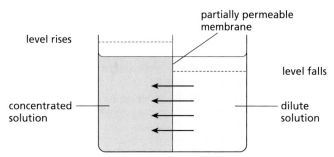

Figure 3.8 Osmosis. Water will diffuse from the dilute solution to the concentrated solution through the partially permeable membrane. As a result, the liquid level will rise on the left and fall on the right.

A partially permeable membrane is porous but allows water to pass through more rapidly than dissolved substances.

Since a dilute solution contains, in effect, more water molecules than a concentrated solution, there is a diffusion gradient which favours the passage of water from the dilute solution to the concentrated solution.

In living cells, the cell membrane is partially permeable and the cytoplasm and vacuole (in plant cells) contain dissolved substances. As a consequence, water tends to diffuse into cells by osmosis if they are surrounded by a weak solution, e.g. fresh water. If the cells are surrounded by a stronger solution, e.g. sea water, the cells may lose water by osmosis. These effects are described more fully later.

Animal cells

In Figure 3.9 an animal cell is shown very simply. The coloured circles represent molecules in the cytoplasm. They may be sugar, salt or protein molecules. The blue circles represent water molecules.

The cell is shown surrounded by pure water. Nothing is dissolved in the water; it has 100% concentration of water molecules. So the concentration of free water molecules outside the cell is greater than that inside and, therefore, water will diffuse into the cell by osmosis.

The membrane allows water to go through either way. So in our example, water can move into or out of the cell.

The cell membrane is partially permeable to most of the substances dissolved in the cytoplasm. So although the concentration of these substances inside may be high, they cannot diffuse freely out of the cell.

The water molecules move into and out of the cell, but because there are more of them on the outside, they will move in faster than they move out. The liquid outside the cell does not have to be 100% pure water. As long as the concentration of water outside is higher than that inside, water will diffuse in by osmosis.

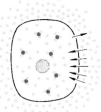 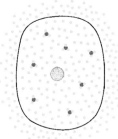

(a) There is a higher concentration of free water molecules outside the cell than inside, so water diffuses into the cell.

(b) The extra water makes the cell swell up.

Figure 3.9 Osmosis in an animal cell

Water entering the cell will make it swell up and, unless the extra water is expelled in some way, the cell will burst.

Conversely, if the cells are surrounded by a solution which is more concentrated than the cytoplasm, water will pass out of the cell by osmosis and the cell will shrink. Excessive uptake or loss of water by osmosis may damage cells.

For this reason, it is very important that the cells in an animal's body are surrounded by a liquid which has the same concentration as the liquid inside the cells. The liquid outside the cells is called **tissue fluid** (see 'Blood and lymphatic vessels' in Chapter 9) and its concentration depends on the concentration of the blood. In vertebrates, the concentration of the blood is monitored by the brain and adjusted by the kidneys, as described in Chapter 13.

By keeping the blood concentration within narrow limits, the concentration of the tissue fluid remains more or less constant (see 'Homeostasis' in Chapter 14) and the cells are not bloated by taking in too much water or dehydrated by losing too much.

Plant cells

The cytoplasm of a plant cell and the cell sap in its vacuole contain salts, sugars and proteins which effectively reduce the concentration of free water molecules inside the cell. The cell wall is freely permeable to water and dissolved substances but the cell membrane of the cytoplasm is partially permeable. If a plant cell is surrounded by water or a solution more dilute than its contents, water will pass into the vacuole by osmosis. The vacuole will expand and press outwards on the cytoplasm and cell wall. The cell wall of a mature plant cell cannot be stretched, so there comes a time when the inflow of water is resisted by the inelastic cell wall, as shown in Figure 3.10.

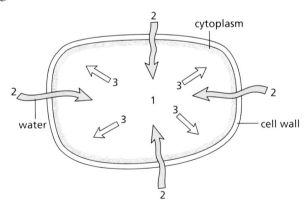

1 since there is effectively a lower concentration of water in the cell sap
2 water diffuses into the vacuole
3 and makes it push out against the cell wall

Figure 3.10 Osmosis in a plant cell

This has a similar effect to inflating a soft bicycle tyre. The tyre represents the firm cell wall, the floppy inner tube is like the cytoplasm and the air inside corresponds to the vacuole. If enough air is pumped in, it pushes the inner tube against the tyre and makes the tyre hard.

When plant cells have absorbed a maximum amount of water by osmosis, they become very rigid, due to the pressure of water pressing outwards on the cell wall. The end result is that the stems and leaves are supported. If the cells lose water there is no longer any water pressure pressing outwards against the cell walls and the stems and leaves are no longer supported. At this point, the plant becomes limp and wilts (see Figure 3.11).

(a) plant wilting **(b)** plant recovered after watering

Figure 3.11 Wilting

Practical work

Experiments on osmosis

Some of the experiments use 'Visking' dialysis tubing. It is made from cellulose and is partially permeable, allowing water molecules to diffuse through freely, but restricting the passage of dissolved substances to varying extents. It is used in kidney dialysis machines because it lets the small molecules of harmful waste products, such as urea, out of the blood but retains the blood cells and large protein molecules (Chapter 13).

1 Osmosis and water flow

- Take a 20 cm length of dialysis tubing which has been soaked in water and tie a knot tightly at one end.
- Place 3 cm³ of a strong sugar solution in the tubing using a plastic syringe and add a little coloured dye.
- Fit the tubing over the end of a length of capillary tubing and hold it in place with an elastic band. Push the capillary tubing into the dialysis tubing until the sugar solution enters the capillary.
- Now clamp the capillary tubing so that the dialysis tubing is totally immersed in a beaker of water, as shown in Figure 3.12.
- Watch the level of liquid in the capillary tubing over the next 10–15 minutes.

3 MOVEMENT IN AND OUT OF CELLS

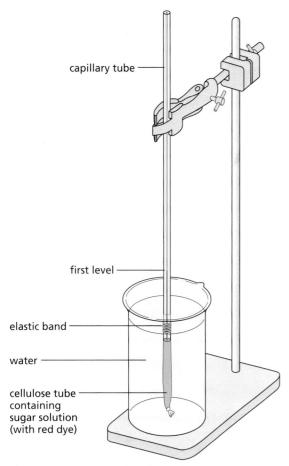

Figure 3.12 Demonstration of osmosis

Result
The level of liquid in the capillary tube rises.

Interpretation
Water must be passing into the sugar solution from the beaker. This is what you would expect when a concentrated solution is separated from water by a partially permeable membrane.

A process similar to this might be partially responsible for moving water from the roots to the stem of a plant.

2 The effects of water and sugar solution on potato tissue

- Push a No.4 or No.5 cork borer into a large potato.
  Caution: Do not hold the potato in your hand but use a board as in Figure 3.13(a).
- Push the potato tissue out of the cork borer using a pencil as in Figure 3.13(b). Prepare a number of potato cylinders in this way and choose the two longest. (They should be at least 50 mm long.) Cut these two accurately to the same length, e.g. 50, 60 or 70 mm. Measure carefully.
- Label two test-tubes A and B and place a potato cylinder in each. Cover the potato tissue in tube A with water; cover the tissue in B with a 20% sugar solution.

- Leave the tubes for 24 hours.
- After this time, remove the cylinder from tube A and measure its length. Notice also whether it is firm or flabby. Repeat this for the potato in tube B, but rinse it in water before measuring it.

(a) place the potato on a board

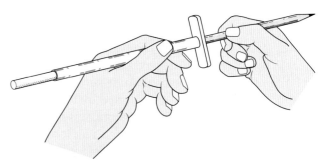

(b) push the potato cylinder out with a pencil

Figure 3.13 Obtaining cylinders of potato tissue

Result
The cylinder from tube A should have gained a millimetre or two and feel firm. The cylinder from tube B should be a millimetre or two shorter and feel flabby.

Interpretation
The cells of the potato in tube A have absorbed water by osmosis, causing an increase in the length of the potato cylinder.

In tube B, the sugar solution is stronger than the cell sap of the potato cells, so these cells have lost water by osmosis, resulting in the potato cylinder becoming flabby and shorter.

An alternative to measuring the potato cores is to weigh them before and after the 24 hours' immersion in water or sugar solution. The core in tube A should gain weight and that in tube B should lose weight. It is important to blot the cores dry with a paper towel before weighing them.

Whichever method is used, it is a good idea to pool the results of the whole class since the changes may be quite small. A gain in length of 1 or 2 mm might be due to an error in measurement, but if most of the class record an increase in length, then experimental error is unlikely to be the cause.

Osmosis

Key definition
Osmosis is the net movement of water molecules from a region of higher water potential (a dilute solution) to a region of lower water potential (a concentrated solution) through a partially permeable membrane.

How osmosis works

When a substance such as sugar dissolves in water, the sugar molecules attract some of the water molecules and stop them moving freely. This, in effect, reduces the concentration of water molecules. In Figure 3.14 the sugar molecules on the right have 'captured' half the water molecules. There are more free water molecules on the left of the membrane than on the right, so water will diffuse more rapidly from left to right across the membrane than from right to left.

The partially permeable membrane does not act like a sieve in this case. The sugar molecules can diffuse from right to left but, because they are bigger and surrounded by a cloud of water molecules, they diffuse more slowly than the water, as shown in Figure 3.15.

Artificial partially permeable membranes are made from cellulose acetate in sheets or tubes and used for **dialysis**. The pore size can be adjusted during manufacture so that large molecules cannot get through at all.

The **cell membrane** behaves like a partially permeable membrane. The partial permeability may depend on pores in the cell membrane but the processes involved are far more complicated than in an artificial membrane and depend on the structure of the membrane and on living processes in the cytoplasm. The cell membrane contains lipids and proteins. Anything which denatures proteins, for example, heat, also destroys the structure and the partially permeable properties of a cell membrane. If this happens, the cell will die as essential substances diffuse out of the cell and harmful chemicals diffuse in.

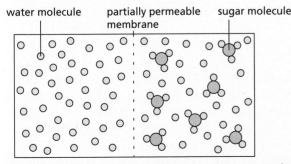

Figure 3.14 The diffusion gradient for water. There are more free water molecules on the left, so more will diffuse from left to right than in the other direction. Sugar molecules will diffuse more slowly from right to left.

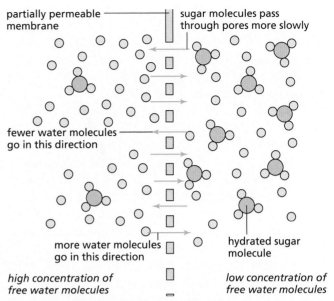

Figure 3.15 The diffusion theory of osmosis

Water potential

The **water potential** of a solution is a measure of whether it is likely to lose or gain water molecules from another solution. A dilute solution, with its high proportion of free water molecules, is said to have a higher water potential than a concentrated solution, because water will flow from the dilute to the concentrated solution (from a high potential to a low potential). Pure water has the highest possible water potential because water molecules will flow from it to any other aqueous solution, no matter how dilute. When adjacent cells contain sap with different water potentials, a water potential gradient is created. Water will move from a cell with a higher water potential (a more dilute solution) to a cell with a lower water potential (a more concentrated solution). This is thought to be one way in which water moves from root hair cells through to the xylem of a plant root (see Figure 8.11 on page 115).

The importance of water potential and osmosis in the uptake of water by plants

A plant cell with the vacuole pushing out on the cell wall is said to be **turgid** and the vacuole is exerting **turgor pressure** on the inelastic cell wall.

If all the cells in a leaf and stem are turgid, the stem will be firm and upright and the leaves held out straight. If the vacuoles lose water for any reason, the cells will lose their turgor and become **flaccid**. (See Experiment 4 'Plasmolysis' on page 46.) If a plant has flaccid cells, the leaves will be limp and the stem will droop. A plant which loses water to this extent is said to be 'wilting' (see Figure 3.11).

Root hair cells are in contact with water trapped between soil particles. When the water potential of the cell sap is lower than that of the soil water, the water will enter the cells by osmosis providing the plant with the water it needs. (This process is described in more detail in 'Water uptake' in Chapter 8.)

When a farmer applies chemical fertilisers to the soil, the fertilisers dissolve in the soil water. Too much fertiliser can lower the osmotic potential of the soil water. This can draw water out of the plant root hair cells by osmosis, leading to wilting and death of crop plants.

Irrigation of crops can have a similar effect. Irrigation which provides just enough water for the plant can lead to a build-up of salts in the soil. The salts will eventually cause the soil water to have a lower water potential than the plant root cells. Crops can then no longer be grown on the land, because they wilt and die because of water loss by osmosis. Much agricultural land in hot countries has become unusable due to the side-effects of irrigation (Figure 3.16).

Figure 3.16 An irrigation furrow

Some countries apply salt to roads in the winter to prevent the formation of ice (Figure 3.17). However, vehicle wheels splash the salt on to plants at the side of the road. The build-up of salts in the roadside soil can kill plants living there, due to water loss from the roots by osmosis.

Figure 3.17 Salt gritter at work to prevent ice formation on a road

The importance of water potential and osmosis in animal cells and tissues

It is vital that the fluid which bathes cells in animals, such as tissue fluid or blood plasma, has the same water potential as the cell contents. This prevents any net flow of water into or out of the cells. If the bathing fluid has a higher water potential (a weaker concentration) than the cells, water will move into the cells by osmosis causing them to swell up. As animal cells have no cell wall and the membrane has little strength, water would continue to enter and the cells will eventually burst (a process called **haemolysis** in red blood cells). Single-celled animals such as *Amoeba* (see Figure 1.32 on page 19) living in fresh water obviously have a problem. They avoid bursting by possessing a **contractile vacuole**. This collects the water as it enters the cell and periodically releases it through the cell membrane, effectively baling the cell out. When surgeons carry out operations on a patient's internal organs, they sometimes need to rinse a wound. Pure water cannot be used as this would enter any cells it came into contact with and cause them to burst. A saline solution, with the same water potential as tissue fluid, has to be used.

In England in 1995, a teenager called Leah Betts (Figure 3.18) collapsed after taking an Ecstasy tablet. One of the side-effects of taking Ecstasy is that the brain thinks the body is dehydrating so the person becomes very thirsty. Leah drank far too much water: over 7 litres (12 pints) in 90 minutes. Her kidneys could not cope and the extra water in her system diluted her blood. Her brain cells took in water by osmosis, causing them to swell up and burst. She died hours later.

Figure 3.18 Poster campaign featuring Leah Betts to raise awareness of the dangers of taking the drug ecstasy.

Diarrhoea is the loss of watery faeces. It is caused when water cannot be absorbed from the contents of the large intestine, or when extra water is secreted into the large intestine due to a viral or bacterial infection. For example, the cholera bacterium produces a toxin (poison) that causes the secretion of chloride ions into the small intestine. This lowers the water potential of the gut contents, so water is drawn into the intestine by osmosis. The result is the production of watery faeces. Unless the condition is treated, dehydration and loss of salts occur, which can be fatal. Patients need rehydration therapy. This involves the provision of frequent sips of water and the use of rehydration drinks. These usually come in sachets available from pharmacists and supermarkets. The contents are dissolved in water and drunk to replace the salts and glucose that are lost through dehydration.

During physical activity, the body may sweat in order to maintain a steady temperature. If liquids are not drunk to compensate for water loss through sweating, the body can become dehydrated. Loss of water from the blood results in the plasma becoming more concentrated (its water potential decreases). Water is then drawn out of the red blood cells by osmosis. The cells become **plasmolysed**. Their surface area is reduced, causing them to be less effective in carrying oxygen. The shape of the cells is known as being **crenated** (see Figure 3.19).

People doing sport sometimes use sports drinks (Figure 3.20) which are **isotonic** (they have the same water potential as body fluids). The drinks contain water, salts and glucose and are designed to replace lost water and salts, as well as providing energy, without creating osmotic problems to body cells. However, use of such drinks when not exercising vigorously can lead to weight gain in the same way as the prolonged use of any sugar-rich drink.

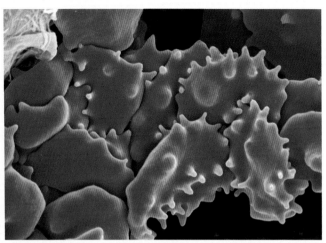

Figure 3.19 Plasmolysed red blood cells

Figure 3.20 People may use isotonic sports drinks.

Practical work

Further experiments on osmosis

3 **Osmosis and turgor**

- Take a 20 cm length of dialysis tubing which has been soaked in water and tie a knot tightly at one end.
- Place 3 cm³ of a strong sugar solution in the tubing using a plastic syringe (Figure 3.21(a)) and then knot the open end of the tube (Figure 3.21(b)). The partly-filled tube should be quite floppy (Figure 3.21(c)).

3 MOVEMENT IN AND OUT OF CELLS

- Place the tubing in a test-tube of water for 30–45 minutes.
- After this time, remove the dialysis tubing from the water and note any changes in how it looks or feels.

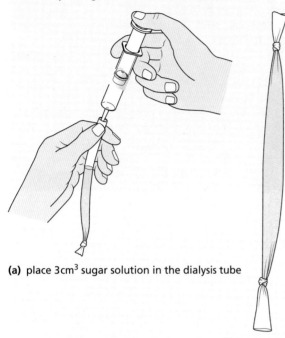

(a) place 3cm^3 sugar solution in the dialysis tube

(b) knot tightly, after expelling the air bubbles

(c) the partly filled tube should be flexible enough to bend

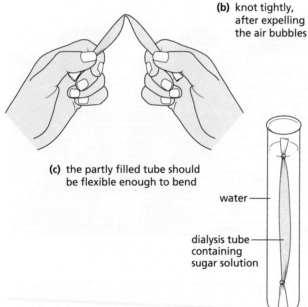

Figure 3.21 Experiment to illustrate turgor in a plant cell

Result
The tubing will become firm, distended by the solution inside.

Interpretation
The dialysis tubing is partially permeable and the solution inside has fewer free water molecules than outside. Water has, therefore, diffused in and increased the volume and the pressure of the solution inside.

This is a crude model of what is thought to happen to a plant cell when it becomes turgid. The sugar solution represents the cell sap and the dialysis tubing represents the cell membrane and cell wall combined.

4 Plasmolysis

- Peel a small piece of epidermis (the outer layer of cells) from a red area of a rhubarb stalk (see Figure 2.9(c) on page 28).
- Place the epidermis on a slide with a drop of water and cover with a coverslip (see Figure 2.9(b)).
- Put the slide on a microscope stage and find a small group of cells.
- Place a 30% solution of sugar at one edge of the coverslip with a pipette and then draw the solution under the coverslip by placing a piece of blotting paper on the opposite side, as shown in Figure 3.22.
- Study the cells you identified under the microscope and watch for any changes in their appearance.

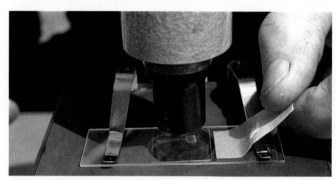

Figure 3.22 Changing the water for sugar solution

Result
The red cell sap will appear to shrink and get darker and pull the cytoplasm away from the cell wall leaving clear spaces. (It is not possible to see the cytoplasm but its presence can be inferred from the fact that the red cell sap seems to have a distinct outer boundary in those places where it has separated from the cell wall.) Figure 3.23 shows the turgid and plasmolysed cells.

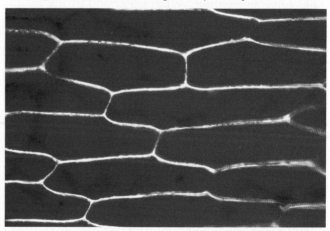

(a) Turgid cells (×100). The cells are in a strip of epidermis from a rhubarb stalk. The cytoplasm is pressed against the inside of the cell wall by the vacuole.

Figure 3.23 Demonstration of plasmolysis in rhubarb cells

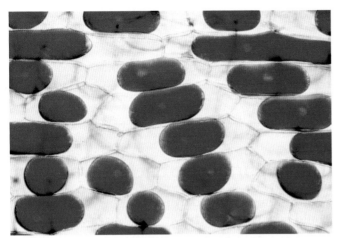

(b) Plasmolysed cells (×100). The same cells as they appear after treatment with sugar solution. The vacuole has lost water by osmosis, shrunk and pulled the cytoplasm away from the cell wall.

Figure 3.23 Demonstration of plasmolysis in rhubarb cells (continued)

Interpretation
The interpretation in terms of osmosis is outlined in Figure 3.24. The cells are said to be **plasmolysed**.

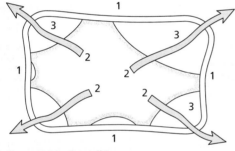

1 the solution outside the cell is more concentrated than the cell sap
2 water diffuses out of the vacuole
3 the vacuole shrinks, pulling the cytoplasm away from the cell wall, leaving the cell flaccid

Figure 3.24 Plasmolysis

The plasmolysis can be reversed by drawing water under the coverslip in the same way that you drew the sugar solution under. It may need two or three lots of water to flush out all the sugar. If you watch a group of cells, you should see their vacuoles expanding to fill the cells once again.

Rhubarb is used for this experiment because the coloured cell sap shows up. If rhubarb is not available, the epidermis from a red onion scale can be used.

5 The effects of varying the concentration of sucrose solution on potato tissue

- Push a No.4 or No.5 cork borer into a large potato. **Caution:** Do not hold the potato in your hand, but use a board as in Figure 3.13(a) on page 42.
- Push the potato tissue out of the cork borer using a pencil as in Figure 3.13(b). Prepare six potato cylinders in this way and cut them all to the same length. (They should be at least 50 mm long.) Measure them carefully.
- Label six test-tubes with the concentration of sucrose solution in them (e.g. $0.0\,mol\,dm^{-3}$, $0.2\,mol\,dm^{-3}$, $0.4\,mol\,dm^{-3}$, $0.6\,mol\,dm^{-3}$, $0.8\,mol\,dm^{-3}$ and $1.0\,mol\,dm^{-3}$) and place them in a test-tube rack.
- Add the same volume of the correct sucrose solution to each test-tube.
- Weigh a cylinder of potato, record its mass and place it in the first test-tube. Repeat until all the test-tubes have been set up.
- Leave the tubes for at least 30 minutes.
- After this time, remove the potato cylinder from the first tube, surface dry the potato and re-weigh it. Notice also whether it is firm or flabby. Repeat this for the other potato cylinders.
- Calculate the change in mass and the percentage change in mass for each cylinder.

$$\text{Percentage change in mass} = \frac{\text{change in mass}}{\text{mass at start}} \times 100$$

- Plot the results on a graph with sucrose concentration on the horizontal axis and percentage change in mass on the vertical axis.
  Note: there will be negative as well as positive percentage changes in mass, so your graph axes will have to allow for this.

Result
The cylinders in the weaker sucrose solutions will have gained mass and feel firm. One of the cylinders may have shown no change in mass. The cylinders in the more concentrated sucrose solutions will have lost mass and feel limp.

Interpretation
If the cells of the potato have absorbed water by osmosis, there will be an increase in the mass of the potato cylinder. This happens when the external solution has a higher water potential than that inside the potato cells. (The sucrose solution is less concentrated than the contents of the potato cells.) Water molecules move into each cell through the cell membrane. The water molecules move from a higher water potential to a lower water potential. The cells become turgid, so the cylinder feels firm.

If the cells of the potato have lost water by osmosis, there will be a decrease in mass of the potato cylinder. This happens when the external solution has a lower water potential than that inside the potato cells. (The sucrose solution is more concentrated than the contents of the potato cells.) Water molecules move out of each cell through the cell membrane. The water molecules move from a higher water potential to a lower water potential. The cells become plasmolysed or flaccid, so the cylinder feels flabby.

3 MOVEMENT IN AND OUT OF CELLS

Question
Study your graph. Can you predict the sucrose concentration which would be equivalent to the concentration of the cell sap in the potato cells?

6 Partial permeability

- Take a 15 cm length of dialysis tubing which has been soaked in water and tie a knot tightly at one end.
- Use a dropping pipette to partly fill the tubing with 1% starch solution.
- Put the tubing in a test-tube and hold it in place with an elastic band as shown in Figure 3.25.
- Rinse the tubing and test-tube under the tap to remove all traces of starch solution from the outside of the dialysis tube.
- Fill the test-tube with water and add a few drops of iodine solution to colour the water yellow.
- Leave for 10–15 minutes.
- After this time, observe any changes in the solution in the test-tube.

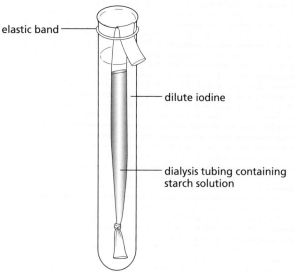

Figure 3.25 Experiment to demonstrate the effect of a partially permeable membrane

Result
The starch inside the dialysis tubing goes blue but the iodine outside stays yellow or brown.

Interpretation
The blue colour is characteristic of the reaction which takes place between starch and iodine, and is used as a test for starch (see Chapter 4). The results show that iodine molecules have passed through the dialysis tubing into the starch but the starch molecules have not moved out into the iodine. This is what we would expect if the dialysis tubing were partially permeable on the basis of its pore size. Starch molecules are very large and probably cannot get through the pores. Iodine molecules are much smaller and can, therefore, get through.

Note: This experiment illustrates that movement of water is not necessarily involved and the pore size of the membrane makes it genuinely partially permeable with respect to iodine and starch.

● Active transport

> **Key definition**
> **Active transport** is the movement of particles through a cell membrane from a region of lower concentration to a region of higher concentration using the energy from respiration.

The importance of active transport

If diffusion were the only method by which a cell could take in substances, it would have no control over what went in or out. Anything that was more concentrated outside would diffuse into the cell whether it was harmful or not. Substances which the cell needed would diffuse out as soon as their concentration inside the cell rose above that outside it. The cell membrane, however, has a great deal of control over the substances which enter and leave the cell.

In some cases, substances are taken into or expelled from the cell against the concentration gradient. For example, sodium ions may continue to pass out of a cell even though the concentration outside is greater than inside. The cells lining the small intestine take up glucose against a concentration gradient. The processes by which substances are moved against a concentration gradient are not fully understood and may be quite different for different substances but they are all generally described as **active transport**.

Anything which interferes with respiration, such as a lack of oxygen or glucose, prevents active transport taking place. This indicates that active transport needs a supply of energy from respiration. Figure 3.26 shows a possible model to explain active transport.

The carrier molecules shown in Figure 3.26 are protein molecules. As shown in (b), they are responsible for transporting substances across the membrane during active transport.

In some cases, a combination of active transport and controlled diffusion seems to occur. For example, sodium ions are thought to get into a cell by diffusion through special pores in the membrane and are expelled by a form of active transport. The reversed diffusion gradient for sodium ions created in this way is very important in the conduction of nerve impulses in nerve cells.

Active transport

Epithelial cells in the villi of the small intestine have the role of absorbing glucose against a concentration gradient. The cells contain numerous mitochondria in which respiration takes place. The chemical energy produced is converted into kinetic energy for the movement of the glucose molecules. The same type of process occurs in the cells of the kidney tubules for the reabsorption of glucose molecules into the bloodstream against their concentration gradient.

Plants need to absorb mineral salts from the soil, but these salts are in very dilute solution. Active transport enables the cells of plant roots to take up salts from this dilute solution against the concentration gradient. Again, chemical energy from respiration is converted into kinetic energy for movement of the salts.

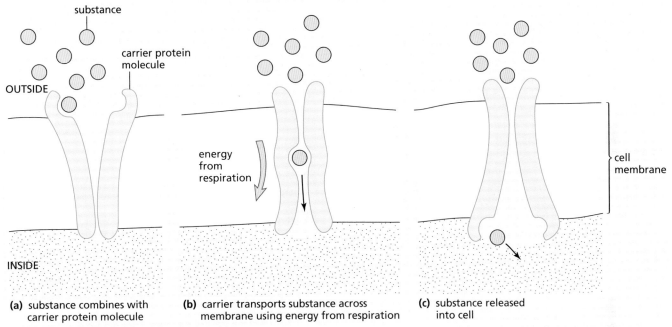

Figure 3.26 A theoretical model to explain active transport

Questions

Core

1 A 10% solution of copper sulfate is separated by a partially permeable membrane from a 5% solution of copper sulfate.
   Will water diffuse from the 10% solution to the 5% solution or from the 5% solution to the 10% solution? Explain your answer.
2 If a fresh beetroot is cut up, the pieces washed in water and then left for an hour in a beaker of water, little or no red pigment escapes from the cells into the water. If the beetroot is boiled first, the pigment does escape into the water. Bearing in mind the properties of a living cell membrane, offer an explanation for this difference.
3 In Experiment 1 (Figure 3.12), what do you think would happen in these cases?
   a A much stronger sugar solution was placed in the cellulose tube.
   b The beaker contained a weak sugar solution instead of water.
   c The sugar solution was in the beaker and the water was in the cellulose tube?
4 In Experiment 1, the column of liquid accumulating in the capillary tube exerts an ever-increasing pressure on the solution in the dialysis tubing. Bearing this in mind and assuming a very long capillary, at what stage would you expect the net flow of water from the beaker into the dialysis tubing to cease?

Extended

5 When doing experiments with animal tissues they are usually bathed in Ringer's solution, which has a concentration similar to that of blood or tissue fluid. Why do you think this is necessary?
6 Why does a dissolved substance reduce the number of 'free' water molecules in a solution?
7 When a plant leaf is in daylight, its cells make sugar from carbon dioxide and water. The sugar is at once turned into starch and deposited in plastids.
   What is the osmotic advantage of doing this? (Sugar is soluble in water; starch is not.)

3 MOVEMENT IN AND OUT OF CELLS

8 In Experiment 3 (Figure 3.21), what might happen if the cellulose tube filled with sugar solution was left in the water for several hours?

9 In Experiment 4, Figure 3.24 explains why the vacuole shrinks. Give a brief explanation of why it swells up again when the cell is surrounded by water.

10 An alternative interpretation of the results of Experiment 6 might be that the dialysis tubing allowed molecules (of any size) to pass in but not out. Describe an experiment to test this possibility and say what results you would expect:
   a if it were correct
   b if it were false.

11 Look at Figure 9.25 on page 136. The symbol O_2 represents an oxygen molecule.
Explain why oxygen is entering the cells drawn on the left but leaving the cells on the right.

12 Look at Figure 11.5 on page 158. It represents one of the small air pockets (an alveolus) which form the lung.
   a Suggest a reason why the oxygen and carbon dioxide are diffusing in opposite directions.
   b What might happen to the rate of diffusion if the blood flow were to speed up?

13 List the ways in which a cell membrane might regulate the flow of substances into the cell.

14 What is your interpretation of the results shown by the graph in Figure 3.27?

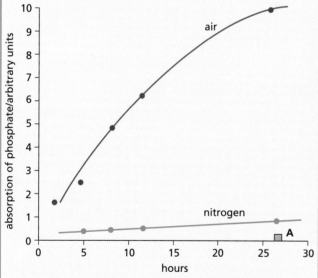

Figure 3.27 The absorption of phosphate ions in air and in nitrogen by roots of beech. **A** represents the concentration of phosphate in external solution

Checklist

After studying Chapter 3 you should know and understand the following:

- Diffusion is the result of molecules of liquid, gas or dissolved solid moving about.
- The molecules of a substance diffuse from a region where they are very concentrated to a region where they are less concentrated.
- Substances may enter cells by simple diffusion, controlled diffusion or active transport.
- Osmosis is the diffusion of water through a partially permeable membrane, from a dilute solution of salt or sugar to a concentrated solution because the concentrated solution contains fewer free water molecules.
- Cell membranes are partially permeable and cytoplasm and cell sap contain many substances in solution.
- Cells take up water from dilute solutions but lose water to concentrated solutions because of osmosis.
- Osmosis maintains turgor in plant cells.
- Active transport involves the movement of substances against their concentration gradient.
- Active transport requires energy.

- Kinetic energy of molecules and ions results in their diffusion.
- Osmosis involves the diffusion of water from a region of higher water potential to a region of lower water potential through a partially permeable membrane.
- The meanings of the terms *turgid*, *turgor pressure*, *plasmolysis* and *flaccid*.
- The importance of water potential and osmosis to animal and plant cells.
- Turgor pressure in cells provides support in plants.
- Active transport is important as it allows movement of substances across membranes against a concentration gradient.

4 Biological molecules

Biological molecules
The chemical elements that make up carbohydrates, fats and proteins
The sub-units that make up biological molecules
Food tests for starch, reducing sugars, proteins, fats and oils, vitamin C
The role of water as a solvent

The shape of proteins and their functions
The structure of DNA
Roles of water as a solvent in organisms

● Biological molecules

Carbon is an element present in all biological molecules. Carbon atoms can join together to form chains or ring structures, so biological molecules can be very large (macromolecules), often constructed of repeating sub-units (monomers). Other elements always present are oxygen and hydrogen. Nitrogen is sometimes present. When macromolecules are made of long chains of monomers held together by chemical bonds, they are known as **polymers** (poly means 'many'). Examples are polysaccharides (chains of single sugar units such as glucose), proteins (chains of amino acids) and nucleic acids (chains of nucleotides). Molecules constructed of lots of small units often have different properties from their sub-units, making them suitable for specific functions in living things. For example, glucose is very soluble and has no strength, but cellulose (a macromolecule made of glucose units) is insoluble and very tough – ideal for the formation of cell walls around plant cells.

Cells need chemical substances to make new cytoplasm and to produce energy. Therefore the organism must take in food to supply the cells with these substances. Of course, it is not quite as simple as this; most cells have specialised functions (Chapter 2) and so have differing needs. However, all cells need water, oxygen, salts and food substances and all cells consist of water, proteins, lipids, carbohydrates, salts and vitamins or their derivatives.

Carbohydrates

These may be simple, soluble sugars or complex materials like starch and cellulose, but all carbohydrates contain carbon, hydrogen and oxygen only. A commonly occurring simple sugar is **glucose**, which has the chemical formula $C_6H_{12}O_6$.

The glucose molecule is often in the form of a ring, represented as

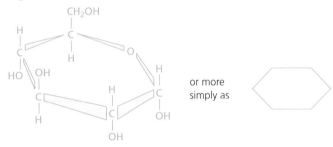

Figure 4.1 Glucose molecule showing ring structure

Two molecules of glucose can be combined to form a molecule of maltose $C_{12}H_{22}O_{11}$ (Figure 4.2).

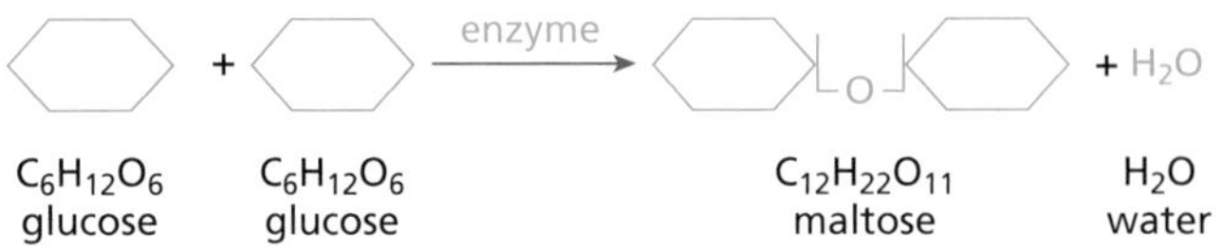

Figure 4.2 Formation of maltose

Sugars with a single carbon ring are called **monosaccharides**, e.g. glucose and fructose. Those sugars with two carbon rings in their molecules are called **disaccharides**, e.g. maltose and sucrose. Mono- and disaccharides are readily soluble in water.

When many glucose molecules are joined together, the carbohydrate is called a **polysaccharide**. **Glycogen** (Figure 4.3) is a polysaccharide that forms a food storage substance in many animal cells. The **starch** molecule is made up of hundreds of glucose molecules joined together to form long chains. Starch is an important storage substance in the plastids of plant cells. Plastids are important organelles in plant cells. They are the sites where molecules like starch are made and stored. One familiar example of a plastid is the chloroplast. **Cellulose** consists of even longer chains of glucose molecules. The chain molecules are grouped together to form microscopic fibres, which are laid down in layers to form the cell wall in plant cells (Figures 4.4 and 4.5).

Polysaccharides are not readily soluble in water.

4 BIOLOGICAL MOLECULES

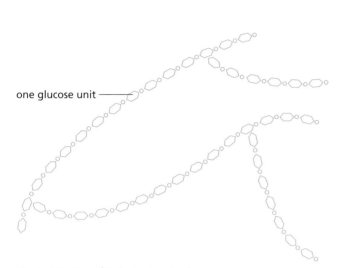

Figure 4.3 Part of a glycogen molecule

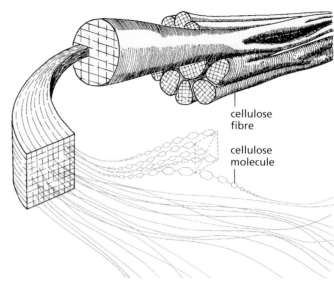

Figure 4.4 Cellulose. Plant cell walls are composed of long, interwoven and interconnected cellulose fibres, which are large enough to be seen with the electron microscope. Each fibre is made up of many long-chain cellulose molecules.

Figure 4.5 Scanning electron micrograph of a plant cell wall (×20 000) showing the cellulose fibres

Fats

Fats are a solid form of a group of molecules called **lipids**. When lipids are liquid they are known as oils. Fats and oils are formed from carbon, hydrogen and oxygen only. A molecule of fat (or oil) is made up of three molecules of an organic acid, called a **fatty acid**, combined with one molecule of **glycerol**.

$$\text{glycerol} \begin{cases} H_2-C-O- \\ | \\ H-C-O- \\ | \\ H_2-C-O- \end{cases} \begin{matrix} \text{fatty acid} \\ \\ \text{fatty acid} \\ \\ \text{fatty acid} \end{matrix}$$

Drawn simply, fat molecules can be represented as in Figure 4.6.

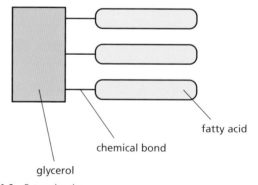

Figure 4.6 Fat molecule

Lipids form part of the cell membrane and the internal membranes of the cell such as the nuclear membrane. Droplets of fat or oil form a source of energy when stored in the cytoplasm.

Proteins

Some proteins contribute to the structures of the cell, e.g. to the cell membranes, the mitochondria, ribosomes and chromosomes. These proteins are called **structural proteins**.

There is another group of proteins called **enzymes**. Enzymes are present in the membrane systems, in the mitochondria, in special vacuoles and in the fluid part of the cytoplasm. Enzymes control the chemical reactions that keep the cell alive (see Chapter 5).

Although there are many different types of protein, all contain carbon, hydrogen, oxygen and nitrogen, and many contain sulfur. Their molecules are made up of long chains of simpler chemicals called **amino acids** (Figure 4.7).

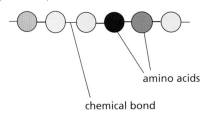

Figure 4.7 Protein molecule (part of)

Vitamins

This is a category of substances which, in their chemical structure at least, have little in common. Plants can make their own vitamins. Animals have to obtain many of their vitamins ready-made. Vitamins, or substances derived from them, play a part in chemical reactions in cells – for example those which involve a transfer of energy from one compound to another. If cells are not supplied with vitamins or the substances needed to make them, the cell physiology is thrown out of order and the whole organism suffers. One example of a vitamin is ascorbic acid (vitamin C) (see 'Diet' in Chapter 7).

Water

Most cells contain about 75% water and will die if their water content falls much below this. Water is a good solvent and many substances move about the cells in a watery solution.

Synthesis and conversion in cells

Cells are able to build up (synthesise) or break down their proteins, lipids and carbohydrates, or change one to another. For example, animal cells synthesise glycogen from glucose by joining glucose molecules together (Figure 4.3); plant cells synthesise starch and cellulose from glucose. All cells can make proteins from amino acids and they can build up fats from glycerol and fatty acids. Animal cells can change carbohydrates to lipids, and lipids to carbohydrates; they can also change proteins to carbohydrates but they cannot make proteins unless they are supplied with amino acids. Plant cells, on the other hand, can make their own amino acids starting from sugars and salts. The cells in the green parts of plants can even make glucose starting from only carbon dioxide and water (see 'Photosynthesis' in Chapter 6).

● Proteins

There are about 20 different amino acids in animal proteins, including alanine, leucine, valine, glutamine, cysteine, glycine and lysine. A small protein molecule might be made up from a chain consisting of a hundred or so amino acids, e.g. glycine–valine–valine–cysteine–leucine–glutamine–, etc. Each type of protein has its amino acids arranged in a particular sequence.

The chain of amino acids in a protein takes up a particular shape as a result of cross-linkages. Cross-linkages form between amino acids that are not neighbours, as shown in Figure 4.8. The shape of a protein molecule has a very important effect on its reactions with substances, as explained in 'Enzymes' in Chapter 5.

For example, the shape of an enzyme molecule creates an **active site**, which has a complementary shape to the substrate molecule on which it acts. This makes enzymes very specific in their action (they usually only work on one substrate).

Antibodies are proteins produced by white blood cells called lymphocytes. Each antibody has a binding site, which can lock onto pathogens such as bacteria. This destroys the pathogen directly, or marks it so that it can be detected by other white blood cells called phagocytes. Each pathogen has **antigens** on its surface that are a particular shape, so specific antibodies with complementary shapes to the antigen are needed (see Chapter 10, page 149).

4 BIOLOGICAL MOLECULES

When a protein is heated to temperatures over 50 °C, the cross-linkages in its molecules break down; the protein molecules lose their shape and will not usually regain it even when cooled. The protein is said to have been **denatured**. Because the shape of the molecules has been altered, the protein will have lost its original properties.

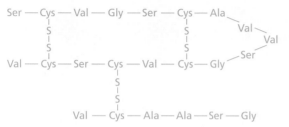

Figure 4.8 A small imaginary protein made from only five different kinds of amino acid. Note that cross-linkage occurs between cysteine molecules with the aid of sulfur atoms.

Egg-white is a protein. When it is heated, its molecules change shape and the egg-white goes from a clear, runny liquid to a white solid and cannot be changed back again. The egg-white protein, albumen, has been denatured by heat.

Proteins form enzymes and many of the structures in the cell, so if they are denatured the enzymes and the cell structures will stop working and the cell will die. Whole organisms may survive for a time above 50 °C depending on the temperature, the period of exposure and the proportion of the cells that are damaged.

● Structure of DNA

A DNA molecule is made up of long chains of nucleotides, formed into two strands. A **nucleotide** is a 5-carbon sugar molecule joined to a phosphate group ($-PO_3$) and an organic base (Figure 4.9). In DNA the sugar is deoxyribose and the organic base is either **adenine** (A), **thymine** (T), **cytosine** (C) or **guanine** (G).

Note: for exam purposes, it is only necessary to be able state the letters, *not* the names of these bases.

The nucleotides are joined by their phosphate groups to form a long chain, often thousands of nucleotides long. The phosphate and sugar molecules are the same all the way down the chain but the bases may be any one of the four listed above (Figure 4.10).

The DNA in a chromosome consists of two strands (chains of nucleotides) held together by chemical bonds between the bases. The size of the molecules ensures that A (adenine) always pairs with T (thymine) and C (cytosine) pairs with G (guanine). The double strand is twisted to form a helix (like a twisted rope ladder with the base pairs representing the rungs) (Figures 4.11 and 4.12).

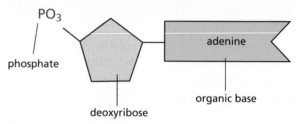

Figure 4.9 A nucleotide (adenosine monophosphate)

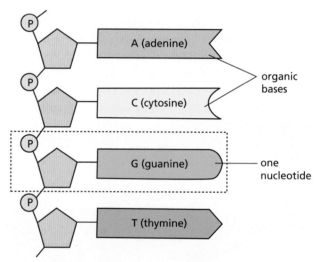

Figure 4.10 Part of a DNA molecule with four nucleotides

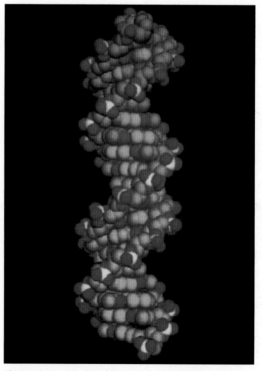

Figure 4.11 Model of the structure of DNA

Water

Water molecules take part in a great many vital chemical reactions. For example, in green plants, water combines with carbon dioxide to form sugar (see Chapter 6). In animals, water helps to break down and dissolve food molecules (see 'Chemical digestion' in Chapter 7). Blood is made up of cells and a liquid called **plasma**. This plasma is 92% water and acts as a transport medium for many dissolved substances, such as carbon dioxide, urea, digested food and hormones. Blood cells are carried around the body in the plasma.

Water also acts as a transport medium in plants. Water passes up the plant from the roots to the leaves in xylem vessels and carries with it dissolved mineral ions. Phloem vessels transport sugars and amino acids in solution from the leaves to their places of use or storage (see Chapter 8).

Water plays an important role in excretion in animals. It acts as a powerful solvent for excretory materials, such as nitrogenous molecules like urea, as well as salts, spent hormones and drugs. The water has a diluting effect, reducing the toxicity of the excretory materials.

The physical and chemical properties of water differ from those of most other liquids but make it uniquely effective in supporting living activities. For example, water has a high capacity for heat (high thermal capacity). This means that it can absorb a lot of heat without its temperature rising to levels that damage the proteins in the cytoplasm. However, because water freezes at 0 °C most cells are damaged if their temperature falls below this and ice crystals form in the cytoplasm. (Oddly enough, rapid freezing of cells in liquid nitrogen at below −196 °C does not harm them).

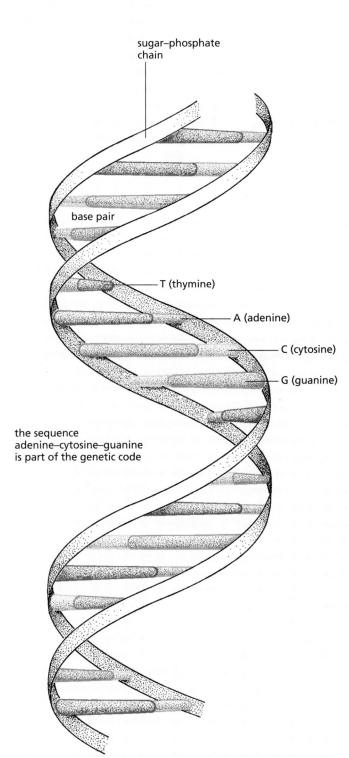

Figure 4.12 The drawing shows part of a DNA molecule schematically

Table 4.1 Summary of the main nutrients

Nutrient	Elements present	Examples	Sub-units
carbohydrate	carbon, hydrogen, oxygen	starch, glycogen, cellulose, sucrose	glucose
fat/oil (oils are liquid at room temperature, but fats are solid)	carbon, hydrogen, oxygen (but lower oxygen content than carbohydrates)	vegetable oils, e.g. olive oil; animal fats, e.g. cod liver oil, waxes	fatty acids and glycerol
protein	carbon, hydrogen, oxygen, nitrogen, sometimes sulfur or phosphorus	enzymes, muscle, haemoglobin, cell membranes	amino acids (about 20 different forms)

4 BIOLOGICAL MOLECULES

● Extension work

DNA

In 1869, a chemist working on cell chemistry discovered a compound that contained nitrogen and phosphorus (as well as carbon). This was an unusual combination. The substance seemed to originate from nuclei and was at first called 'nuclein' and then 'nucleic acid'. Subsequent analysis revealed the bases adenine, thymine, cytosine and guanine in nucleic acid, together with a carbohydrate later identified as deoxyribose. In the early 1900s, the structure of nucleotides (base–sugar–phosphate, Figure 4.9) was determined and also how they linked up to form deoxyribonucleic acid (DNA).

In the 1940s, a chemist, Chargaff, showed that, in a sample of DNA, the number of adenines (A) was always the same as the number of thymines (T). Similarly, the amounts of cytosine (C) and guanine (G) were always equal. This information was to prove crucial to the work of Crick and Watson in determining the structure of DNA.

Francis Crick was a physicist and **James Watson** (from the USA) a biologist. They worked together in the Cavendish Laboratory at Cambridge in the 1950s. They did not do chemical analyses or experiments, but used the data that was available from X-ray crystallography and the chemistry of nucleotides to try out different models for the structure of DNA.

The regular pattern of atoms in a crystal causes a beam of X-rays to be scattered in such a way that the structure of the molecules in the crystal can be determined (Figure 4.13(a)). The scattered X-rays are directed on to a photographic plate which, when developed, reveals images similar to the one in Figure 4.13(b).

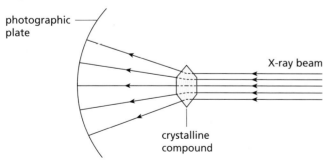

(a) simplified representation of the scattering of X-rays by crystalline structures

Figure 4.13 X-ray crystallography

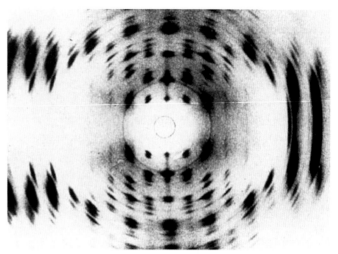

(b) one of the X-ray images produced by X-rays scattered by DNA. The number and positions of the dark areas allows the molecular structure to be calculated.

By precise measurements of the spots on the photograph and some very complex mathematics, the molecular structure of many compounds could be discovered.

It proved possible to obtain DNA in a crystalline form and subject it to X-ray analysis. Most of the necessary X-ray crystallography was carried out by **Maurice Wilkins** and **Rosalind Franklin** at King's College, London.

Crick and Watson assembled models on a trial-and-error basis. The suitability of the model was judged by how well it conformed to the X-ray measurements and the chemical properties of the components.

The evidence all pointed to a **helical** structure (like a spiral staircase). At first they tried models with a core of three or four nucleotide chains twisted around each other and with the bases attached to the outside.

These models did not really fit the X-ray data or the chemical structures of the nucleotides. Watson tried a two-chain helical model with the bases pointing inwards. Initially he paired adenine (A) with adenine (A), cytosine (C) with cytosine (C), etc. But thymine (T) and cytosine (C) were smaller molecules than adenine (A) and guanine (G) and this pairing would distort the double helix.

This is where Chargaff's work came to the rescue. If there were equal numbers of adenine (A) and thymine (T), and equal numbers of cytosine (C) and guanine (G), it was likely that this pairing of bases, large plus small, would fit inside the sugar–phosphate double helix without distortion.

The X-ray data confirmed that the diameter of the helix would allow this pairing and the chemistry of the bases would allow them to hold together. The outcome is the model of DNA shown in Figures 4.10, 4.11 and 4.12.

Crick, Watson and Wilkins were awarded the Nobel Prize for medicine and physiology in 1962. Rosalind Franklin died in 1958, so her vital contribution was not formally rewarded.

Figure 4.14 Crick (right) and Watson with their model of the DNA molecule

Practical work

Food tests

1 Test for starch

- Shake a little starch powder in a test-tube with some warm water to make a suspension.
- Add 3 or 4 drops of **iodine solution**. A dark blue colour should be produced.

Note: it is also possible to use iodine solution to test for starch in leaves, but a different procedure is used (see Chapter 6).

2 Test for reducing sugar

- Heat a little glucose solution with an equal volume of **Benedict's solution** in a test-tube. The heating is done by placing the test-tube in a beaker of boiling water (see Figure 4.15), or warming it gently over a blue Bunsen flame. However, if this second technique is used, the test-tube should be moved constantly in and out of the Bunsen flame to prevent the liquid boiling and shooting out of the tube. The solution will change from clear blue to cloudy green, then yellow and finally to a red precipitate (deposit) of copper(I) oxide.

3 Test for protein (Biuret test)

- To a 1% solution of albumen (the protein of egg-white) add 5 cm³ dilute sodium hydroxide (**CARE**: this solution is caustic), followed by 5 cm³ 1% copper sulfate solution. A purple colour indicates protein. If the copper sulfate is run into the food solution without mixing, a violet halo appears where the two liquids come into contact.

4 Test for fat

- Shake two drops of cooking oil with about 5 cm³ ethanol in a dry test-tube until the fat dissolves.
- Pour this solution into a test-tube containing a few cm³ water. A milky white emulsion will form. This shows that the solution contained some fat or oil.

5 Test for vitamin C

- Draw up 2 cm³ fresh lemon juice into a plastic syringe.
- Add this juice drop by drop to 2 cm³ of a 0.1% solution of DCPIP (a blue dye) in a test-tube. The DCPIP will become colourless quite suddenly as the juice is added. The amount of juice added from the syringe should be noted down.
- Repeat the experiment but with orange juice in the syringe. If it takes more orange juice than lemon juice to decolourise the DCPIP, the orange juice must contain less vitamin C.

Application of the food tests

The tests can be used on samples of food such as milk, potato, raisins, onion, beans, egg-yolk or peanuts to find out what food materials are present. The solid samples are crushed in a mortar and shaken with warm water to extract the soluble products. Separate samples of the watery mixture of crushed food are tested for starch, glucose or protein as described above. To test for fats, the food must first be crushed in ethanol, not water, and then filtered. The clear filtrate is poured into water to see if it goes cloudy, indicating the presence of fats.

4 BIOLOGICAL MOLECULES

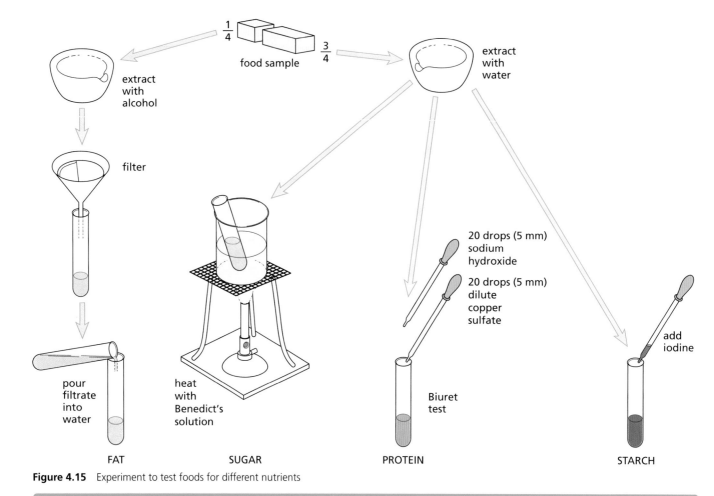

Figure 4.15 Experiment to test foods for different nutrients

Question

Core

1. a What do the chemical structures of carbohydrates and fats have in common?
   b How do their chemical structures differ?
   c Suggest why there are many more different proteins than there are carbohydrates.

Checklist

After studying Chapter 4 you should know and understand the following:

- Living matter is made up of a number of important types of molecules, including proteins, lipids and carbohydrates.
- All three types of molecule contain carbon, hydrogen and oxygen atoms; proteins also contain nitrogen and sometimes phosphorus or sulfur.
- Carbohydrates are made from monosaccharide units, often glucose.
- Carbohydrates are used as an energy source; glycogen and starch make good storage molecules. Cellulose gives plant cell walls their strength.
- Proteins are built up from amino acids joined together by chemical bonds.
- Lipids include fats, fatty acids and oils.
- Fats are made from fatty acids and glycerol.
- Proteins and lipids form the membranes outside and inside the cell.
- Food tests are used to identify the main biological molecules.
- Water is important in living things as a solvent.

- In different proteins the 20 or so amino acids are in different proportions and arranged in different sequences.
- The structure of a protein molecule enables it to carry out specific roles as enzymes and antibodies.
- DNA is another important biological molecule. It has a very distinctive shape, made up of nucleotides containing bases.
- Water has an important role as a solvent in organisms.

5 Enzymes

Enzyme action	Description of enzyme action
Definitions of catalyst and enzyme	Active site
The importance of enzymes in living organisms	Explanation of the effect of temperature and pH on enzyme molecules
The specific nature of enzymes	Specificity
The effects of pH and temperature on enzyme activity	
Complementary shape of enzyme and substrate	

> **Key definitions**
> A **catalyst** is a substance that increases the rate of a chemical reaction and is not changed by the reaction.
> An **enzyme** is a protein that functions as a biological catalyst.

Enzymes are proteins that act as **catalysts**. They are made in all living cells. Enzymes, like catalysts, can be used over and over again because they are not used up during the reaction and only a small amount is needed to speed the reaction up (Figure 5.1).

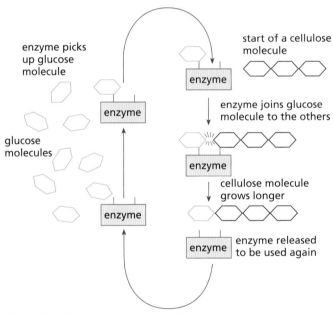

Figure 5.1 Building up a cellulose molecule

● Enzyme action

How an enzyme molecule might work to join two other molecules together and so form a more complicated substance (the product) is shown in Figure 5.2.

An example of an enzyme-controlled reaction such as this is the joining up of two glucose molecules to form a molecule of maltose. You can see that the enzyme and substrate molecules have **complementary** shapes (like adjacent pieces of a jigsaw) so they fit together. Other substrate molecules would not fit into this enzyme as they would have the 'wrong' shape. For example, the substrate molecule in Figure 5.2(b) would not fit the enzyme molecule in Figure 5.2(a). The product (substance AB in Figure 5.2(a)) is released by the enzyme molecule and the enzyme is then free to repeat the reaction with more substrate molecules. Molecules of the two substances might have combined without the enzyme being present, but they would have done so very slowly (it could take hours or days to happen without the enzyme). By bringing the substances close together, the enzyme molecule makes the reaction take place much more rapidly. The process can be extremely fast: it has been found that catalase, a very common enzyme found in most cells, can break down 40 000 molecules of hydrogen peroxide every second! A complete chemical reaction takes only a few seconds when the right enzyme is present.

As well as enzymes being responsible for joining two substrate molecules together, such as two glucose molecules to form maltose, they can also create long chains. For example, hundreds of glucose molecules can be joined together, end to end, to form a long molecule of starch to be stored in the plastid of a plant cell. The glucose molecules can also be built up into a molecule of cellulose to be added to the cell wall. Protein molecules are built up by enzymes, which join together tens or hundreds of amino acid molecules. These proteins are added to the cell membrane, to the cytoplasm or to the nucleus of the cell. They may also become the proteins that act as enzymes.

5 ENZYMES

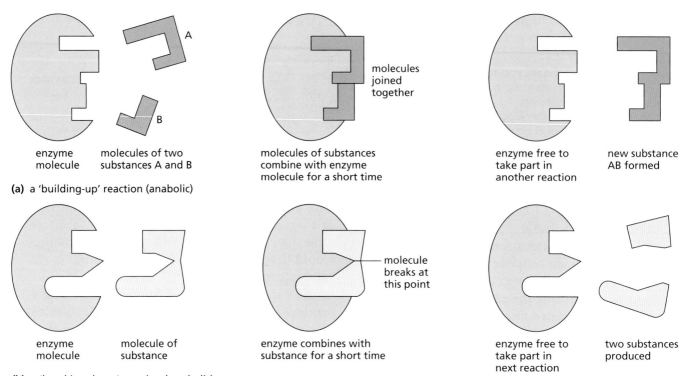

(a) a 'building-up' reaction (anabolic)

(b) a 'breaking-down' reaction (catabolic)

Figure 5.2 Possible explanation of enzyme action

Enzymes and temperature

A rise in temperature increases the rate of most chemical reactions; a fall in temperature slows them down. However, above 50 °C the enzymes, being proteins, are denatured and stop working.

Figure 5.2 shows how the shape of an enzyme molecule could be very important if it has to fit the substances on which it acts. Above 50 °C the shapes of enzymes are permanently changed and the enzymes can no longer combine with the substances.

This is one of the reasons why organisms may be killed by prolonged exposure to high temperatures. The enzymes in their cells are denatured and the chemical reactions proceed too slowly to maintain life.

One way to test whether a substance is an enzyme is to heat it to boiling point. If it can still carry out its reactions after this, it cannot be an enzyme. This technique is used as a 'control' (see 'Aerobic respiration' in Chapter 12) in enzyme experiments.

Enzymes and pH

Acid or alkaline conditions alter the chemical properties of proteins, including enzymes. Most enzymes work best at a particular level of acidity or alkalinity (pH), as shown in Figure 5.3.

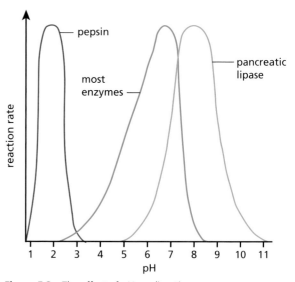

Figure 5.3 The effect of pH on digestive enzymes

The protein-digesting enzyme in your stomach, for example, works well at an acidity of pH 2. At this pH, the enzyme amylase, from your saliva, cannot work at all. Inside the cells, most enzymes will work best in neutral conditions (pH 7). The pH or temperature at which an enzyme works best is often called its **optimum** pH or temperature. Conditions in the duodenum are slightly alkaline: the optimum pH for pancreatic lipase is pH 8.

Although changes in pH affect the activity of enzymes, these effects are usually reversible, i.e. an enzyme that is inactivated by a low pH will resume its normal activity when its optimum pH is restored.

Rates of enzyme reactions

As explained above, the rate of an enzyme-controlled reaction depends on the temperature and pH. It also depends on the concentrations of the enzyme and its substrate. The more enzyme molecules produced by a cell, the faster the reaction will proceed, provided there are enough substrate molecules available. Similarly, an increase in the substrate concentration will speed up the reaction if there are enough enzyme molecules to cope with the additional substrate.

An enzyme-controlled reaction involves three groups of molecules, although the product may be two or more different molecules:

$$\text{substrate} \xrightarrow{\text{enzyme}} \text{product}$$

The substance on which an enzyme acts is called its **substrate** and the molecules produced are called the **products**. Thus, the enzyme sucrase acts on the substrate sucrose to produce the monosaccharide products glucose and fructose.

Reactions in which large molecules are built up from smaller molecules are called **anabolic** reactions (Figure 5.2(a)). When the enzyme combines with the substrate, an **enzyme-substrate complex** is formed temporarily.

Figure 5.2(b) shows an enzyme speeding up a chemical change, but this time it is a reaction in which the molecule of a substance is split into smaller molecules. Again, when the enzyme combines with the substrate, an **enzyme-substrate complex** is formed temporarily. Try chewing a piece of bread, but keep it in your mouth without swallowing it. Eventually you should detect the food tasting sweeter, as maltose sugar is formed. If starch is mixed with water it will break down very slowly to sugar, taking several years. In your saliva there is an enzyme called **amylase** that can break down starch to sugar in minutes or seconds. In cells, many of the 'breaking-down' enzymes are helping to break down glucose to carbon dioxide and water in order to produce energy (Chapter 12).

Reactions that split large molecules into smaller ones are called **catabolic** reactions.

Intra- and extracellular enzymes

All enzymes are made inside cells. Most of them remain inside the cell to speed up reactions in the cytoplasm and nucleus. These are called **intracellular enzymes** ('intra' means 'inside'). In a few cases, the enzymes made in the cells are let out of the cell to do their work outside. These are **extracellular enzymes** ('extra' means 'outside'). Fungi and bacteria (see 'Features of organisms' in Chapter 1) release extracellular enzymes in order to digest their food. A mould growing on a piece of bread releases starch-digesting enzymes into the bread and absorbs the soluble sugars that the enzyme produces from the bread. In the digestive systems of animals ('Alimentary canal' in Chapter 7), extracellular enzymes are released into the stomach and intestines in order to digest the food.

Enzymes are specific

This means simply that an enzyme which normally acts on one substance will not act on a different one. Figure 5.2(a) shows how the shape of an enzyme can control what substances it combines with. The enzyme in Figure 5.2(a) has a shape called the **active site**, which exactly fits the substances on which it acts, but will not fit the substance in Figure 5.2(b). So, the shape of the active site of the enzyme molecule and the substrate molecule are **complementary**. Thus, an enzyme which breaks down starch to maltose will not also break down proteins to amino acids. Also, if a reaction takes place in stages, e.g.

starch $\longrightarrow$ maltose (stage 1)

maltose $\longrightarrow$ glucose (stage 2)

a different enzyme is needed for each stage.

The names of enzymes usually end with **-ase** and they are named according to the substance on which they act, or the reaction which they speed up. For example, an enzyme that acts on proteins may be called a **protease**; one that removes hydrogen from a substance is a **dehydrogenase**.

Enzymes and temperature

Figure 5.4 shows the effect of temperature on an enzyme-controlled reaction.

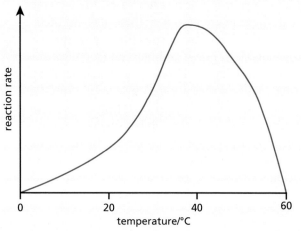

Figure 5.4 Graph showing the effect of temperature on the rate of an enzyme-controlled reaction

Generally, a rise of 10 °C will double the rate of an enzyme-controlled reaction in a cell, up to an optimum temperature of around 37 °C (body temperature). This is because the enzyme and substrate molecules are constantly moving, using kinetic energy. The reaction only occurs when the enzyme and substrate molecules come into contact with each other. As the temperature is increased, the molecules gain more kinetic energy, so they move faster and there is a greater chance of collisions happening. Therefore the rate of reaction increases. Above the optimum temperature the reaction will slow down. This is because enzyme molecules are proteins. Protein molecules start to lose their shape at higher temperatures, so the active site becomes deformed. Substrate molecules cannot fit together with the enzyme, stopping the reaction. Not all the enzyme molecules are affected straight away, so the reaction does not suddenly stop – it is a gradual process as the temperature increases above 37 °C. Denaturation is a permanent change in the shape of the enzyme molecule. Once it has happened the enzyme will not work any more, even if the temperature is reduced below 37 °C. An example of a protein denaturing is the cooking of egg-white (made of the protein albumin). Raw egg-white is liquid, transparent and colourless. As it is heated, it turns solid and becomes opaque and white. It cannot be changed back to its original state or appearance.

Enzymes and pH

Extremes of pH may denature some enzymes irreversibly. This is because the active site of the enzyme molecule can become deformed (as it does when exposed to high temperatures). As a result, the enzyme and substrate molecules no longer have complementary shapes and so will not fit together.

Practical work

Tests for proteins, fats and carbohydrates are described in Chapter 4. Experiments on the digestive enzymes amylase and pepsin are described in Chapter 7.

1 Extracting and testing an enzyme from living cells
In this experiment, the enzyme to be extracted and tested is **catalase** and the substrate is hydrogen peroxide (H_2O_2). Certain reactions in the cell produce hydrogen peroxide, which is poisonous. Catalase makes the hydrogen peroxide harmless by breaking it down to water and oxygen.

$$2H_2O_2 \xrightarrow{\text{catalase}} 2H_2O + O_2$$

- Grind a small piece of liver with about 20 cm³ water and a little sand in a mortar. This will break open the liver cells and release their contents.
- Filter the mixture and share it between two test-tubes, A and B. The filtrate will contain a great variety of substances dissolved out from the cytoplasm of the liver cells, including many enzymes. Because enzymes are specific, however, only one of these, catalase, will act on hydrogen peroxide.
- Add some drops of the filtrate from test-tube A to a few cm³ of hydrogen peroxide in a test-tube. You will see a vigorous reaction as the hydrogen peroxide breaks down to produce oxygen. (The oxygen can be tested with a glowing splint.)
- Now boil the filtrate in tube B for about 30 seconds. Add a few drops of the boiled filtrate to a fresh sample of hydrogen peroxide. There will be no reaction because boiling has denatured the catalase.
- Next, shake a little manganese(IV) oxide powder in a test-tube with some water and pour this into some hydrogen peroxide. There will be a vigorous reaction similar to the one with the liver extract. If you now boil some manganese(IV) oxide with water and add this to hydrogen peroxide, the reaction will still occur. Manganese(IV) oxide is a catalyst but it is not an enzyme because heating has not altered its catalytic properties.
- The experiment can be repeated with a piece of potato to compare its catalase content with that of the liver. The piece of potato should be about the same size as the liver sample.

Enzyme action

● Extension work

Investigate a range of plant tissues to find out which is the best source of catalase. Decide how to make quantitative comparisons (observations which involve measurements). Possible plant tissues include potato, celery, apple and carrot.

2 The effect of temperature on an enzyme reaction

Amylase is an enzyme that breaks down starch to a sugar (maltose).

- Draw up 5 cm³ of 5% amylase solution in a plastic syringe (or graduated pipette) and place 1 cm³ in each of three test-tubes labelled A, B and C.
- Rinse the syringe thoroughly and use it to place 5 cm³ of a 1% starch solution in each of three test-tubes labelled 1, 2 and 3.
- To each of tubes 1 to 3, add six drops only of dilute iodine solution using a dropping pipette.

- Prepare three water baths by half filling beakers or jars with:
  a ice and water, adding ice during the experiment to keep the temperature at about 10 °C
  b water from the cold tap at about 20 °C
  c warm water at about 35 °C by mixing hot and cold water.
- Place tubes 1 and A in the cold water bath, tubes 2 and B in the water at room temperature, and tubes 3 and C in the warm water.
- Leave them for 5 minutes to reach the temperature of the water (Figure 5.5).
- After 5 minutes, take the temperature of each water bath, then pour the amylase from tube A into the starch solution in tube 1 and return tube 1 to the water bath.
- Repeat this with tubes 2 and B, and 3 and C.
- As the amylase breaks down the starch, it will cause the blue colour to disappear. Make a note of how long this takes in each case.

Questions

1 At what temperature did the amylase break down starch most rapidly?
2 What do you think would have been the result if a fourth water bath at 90 °C had been used?

3 The effect of pH on an enzyme reaction

- Label five test-tubes 1 to 5 and use a plastic syringe (or graduated pipette) to place 5 cm³ of a 1% starch solution in each tube.
- Add acid or alkali to each tube as indicated in the table below. Rinse the syringe when changing from sodium carbonate to acid.

Tube	Chemical	Approximate pH	
1	1 cm³ sodium carbonate solution (0.05 mol dm⁻³)	9	(alkaline)
2	0.5 cm³ sodium carbonate solution (0.05 mol dm⁻³)	7–8	(slightly alkaline)
3	nothing	6–7	(neutral)
4	2 cm³ ethanoic (acetic) acid (0.1 mol dm⁻³)	6	(slightly acid)
5	4 cm³ ethanoic (acetic) acid (0.1 mol dm⁻³)	3	(acid)

- Place several rows of iodine solution drops in a cavity tile.
- Draw up 5 cm³ of 5% amylase solution in a clean syringe and place 1 cm³ in each tube. Shake the tubes and note the time (Figure 5.6).
- Use a clean dropping pipette to remove a small sample from each tube in turn and let one drop fall on to one of the iodine drops in the cavity tile. Rinse the pipette in a beaker of water between each sample. Keep on sampling in this way.
- When any of the samples fails to give a blue colour, this means that the starch in that tube has been completely broken down to sugar by the amylase. Note the time when this happens for each tube and stop taking samples from

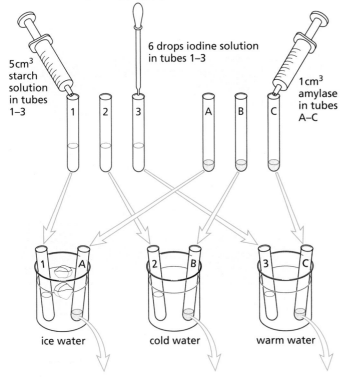

note the time and add the amylase to the starch solution

Figure 5.5 Experiment to investigate the effect of temperature on an enzyme reaction

5 ENZYMES

that tube. Do not continue sampling for more than about 15 minutes, but put a drop from each tube on to a piece of pH paper and compare the colour produced with a colour chart of pH values.

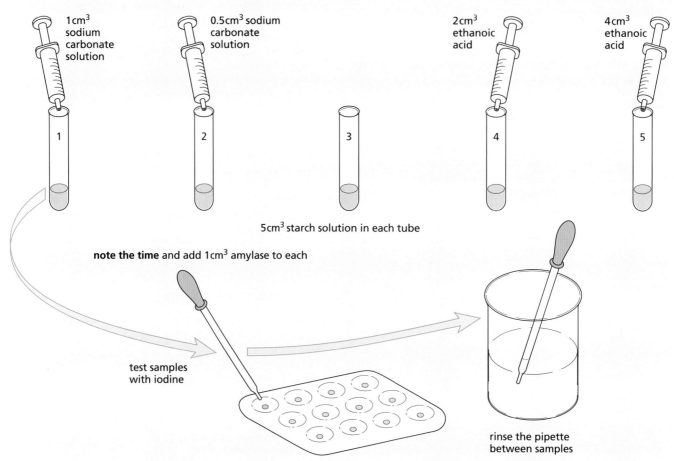

Figure 5.6 Experiment to investigate the effect of pH on an enzyme reaction

Questions

1. At what pH did the enzyme, amylase, work most rapidly?
2. Is this its optimum pH?
3. Explain why you might have expected the result that you got.
4. Your stomach pH is about 2. Would you expect starch digestion to take place in the stomach?

Questions

Extended

1. Which of the following statements apply both to enzymes and to any other catalysts?
   a Their activity is stopped by high temperature.
   b They speed up chemical reactions.
   c They are proteins.
   d They are not used up during the reaction.
2. How would you expect the rate of an enzyme-controlled reaction to change if the temperature was raised:
   a from 20 °C to 30 °C
   b from 35 °C to 55 °C?
   Explain your answers.
3. There are cells in your salivary glands that can make an extracellular enzyme, amylase. Would you expect these cells to make intracellular enzymes as well? Explain your answer.

4 Apple cells contain an enzyme that turns the tissues brown when an apple is peeled and left for a time. Boiled apple does not go brown (Figure 5.7). Explain why the boiled apple behaves differently.

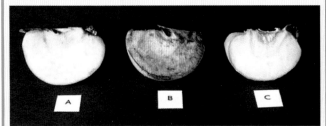

Figure 5.7 Experiment to investigate enzyme activity in an apple. Slice A has been freshly cut. B and C were cut 2 days earlier but C was dipped immediately in boiling water for 1 minute.

Checklist

After studying Chapter 5 you should know and understand the following:

- Catalysts are substances that increase the rate of chemical reactions and are not changed in the process.
- Enzymes are proteins that function as biological catalysts.
- Enzymes are important in all organisms because they maintain a reaction speed needed to sustain life.
- The substance on which an enzyme acts is called the substrate. After the reaction, a product is formed.
- An enzyme and its substrate have complementary shapes.
- Enzymes are affected by pH and temperature and are denatured above 50 °C.

- Different enzymes may accelerate reactions which build up or break down molecules.
- Each enzyme acts on only one substance (breaking down) or a pair of substances (building up).
- Enzymes tend to be very specific in the reactions they catalyse, due to the complementary shape of the enzyme and its substrate.
- Changes in temperature affect the kinetic energy of enzyme molecules and their shape.
- Enzymes can be denatured by changes in temperature and pH.

6 Plant nutrition

Photosynthesis
Definition of photosynthesis
Word equation
Investigations into the necessity for chlorophyll, light and carbon dioxide for photosynthesis, using appropriate controls
Investigations into the effects of varying light intensity, carbon dioxide concentration and temperature on the rate of photosynthesis

Balanced chemical equation
Use and storage of the products of photosynthesis
Definition of limiting factors
Role of glasshouses in creating optimal conditions for photosynthesis

Leaf structure
Identify the main tissues in a leaf

Adaptations of leaves for photosynthesis

Mineral requirements
The importance of nitrate ions and magnesium ions

Explaining the effects of mineral deficiencies on plant growth

● Photosynthesis

Key definition
Photosynthesis is the process by which plants manufacture carbohydrates from raw materials using energy from light.

All living organisms need food. They need it as a source of raw materials to build new cells and tissues as they grow. They also need food as a source of energy. Food is a kind of 'fuel' that drives essential living processes and brings about chemical changes (see 'Diet' in Chapter 7 and 'Aerobic respiration' in Chapter 12). Animals take in food, digest it and use the digested products to build their tissues or to produce energy.

Plants also need energy and raw materials but, apart from a few insect-eating species, plants do not appear to take in food. The most likely source of their raw materials would appear to be the soil. However, experiments show that the weight gained by a growing plant is far greater than the weight lost by the soil it is growing in. So there must be additional sources of raw materials.

Jean-Baptiste van Helmont was a Dutch scientist working in the 17th century. At that time very little was known about the process of photosynthesis. He carried out an experiment using a willow shoot. He planted the shoot in a container with 90.8 kg of dry soil and placed a metal grill over the soil to prevent any accidental gain or loss of mass. He left the shoot for 5 years in an open yard, providing it with only rainwater and distilled water for growth. After 5 years he reweighed the tree and the soil (see Figure 6.1) and came to the conclusion that the increase in mass of the tree (74.7 kg) was due entirely to the water it had received. However, he was unaware that plants also take in mineral salts and carbon dioxide, or that they use light as a source of energy.

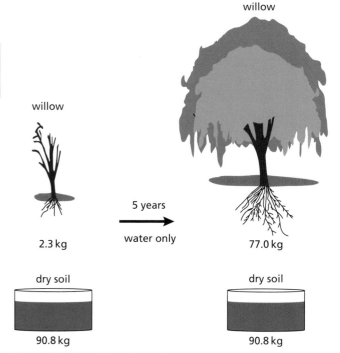

Figure 6.1 Van Helmont's experiment

A **hypothesis** to explain the source of food in a plant is that it *makes it* from air, water and soil salts. Carbohydrates (Chapter 4) contain the elements carbon, hydrogen and oxygen, as in glucose ($C_6H_{12}O_6$). The carbon and oxygen could be supplied by carbon dioxide (CO_2) from the air, and the hydrogen could come from the water (H_2O) in the soil. The nitrogen and sulfur needed for making proteins (Chapter 4) could come from nitrates and sulfates in the soil.

This building-up of complex food molecules from simpler substances is called **synthesis** and it needs enzymes and energy to make it happen. The enzymes are present in the plant's cells and the energy for the first stages in the synthesis comes from sunlight. The process is, therefore, called **photosynthesis** ('photo' means 'light'). There is evidence to suggest that the green substance, **chlorophyll**, in the chloroplasts of plant cells, plays a part in photosynthesis. Chlorophyll absorbs sunlight and makes the energy from sunlight available for chemical reactions. Thus, in effect, the function of chlorophyll is to convert light energy to chemical energy.

A chemical equation for photosynthesis would be

$$\text{carbon dioxide} + \text{water} \xrightarrow[\text{chlorophyll}]{\text{light energy}} \text{glucose} + \text{oxygen}$$

In order to keep the equation simple, glucose is shown as the food compound produced. In reality, the glucose is rapidly converted to sucrose for transport around the plant, then stored as starch or converted into other molecules.

Practical work

Experiments to investigate photosynthesis

The design of biological experiments is discussed in Chapter 12 'Aerobic respiration', and this should be revised before studying the next section.

A hypothesis is an attempt to explain certain observations. In this case the hypothesis is that plants make their food by photosynthesis. The equation shown above is one way of stating the hypothesis and is used here to show how it might be tested.

$$6CO_2 + 6H_2O \xrightarrow[\text{chlorophyll}]{\text{sunlight}} C_6H_{12}O_6 + 6O_2$$

uptake of carbon dioxide | uptake of water | production of sugar (or starch) | release of oxygen

If photosynthesis is occurring in a plant, then the leaves should be producing sugars. In many leaves, as fast as sugar is produced it is turned into starch. Since it is easier to test for starch than for sugar, we regard the production of starch in a leaf as evidence that photosynthesis has taken place.

The first three experiments described below are designed to see if the leaf can make starch without chlorophyll, sunlight or carbon dioxide, in turn. If the photosynthesis hypothesis is sound, then the lack of any one of these three conditions should stop photosynthesis, and so stop the production of starch. But, if starch production continues, then the hypothesis is no good and must be altered or rejected.

In designing the experiments, it is very important to make sure that only *one* variable is altered. If, for example, the method of keeping light from a leaf also cuts off its carbon dioxide supply, it would be impossible to decide whether it was the lack of light or lack of carbon dioxide that stopped the production of starch. To make sure that the experimental design has not altered more than one variable, a **control** is set up in each case. This is an identical situation, except that the condition missing from the experiment, e.g. light, carbon dioxide or chlorophyll, is present in the control (see 'Aerobic respiration' in Chapter 12).

Destarching a plant

If the production of starch is your evidence that photosynthesis is taking place, then you must make sure that the leaf does not contain any starch at the beginning of the experiment. This is done by **destarching** the leaves. It is not possible to remove the starch chemically, without damaging the leaves, so a plant is destarched simply by leaving it in darkness for 2 or 3 days. Potted plants are destarched by leaving them in a dark cupboard for a few days. In the darkness, any starch in the leaves will be changed to sugar and carried away from the leaves to other parts of the plant. For plants in the open, the experiment is set up on the day before the test. During the night, most of the starch will be removed from the leaves. Better still, wrap the leaves in aluminium foil for 2 days while they are still on the plant. Then test one of the leaves to see that no starch is present.

Testing a leaf for starch

Iodine solution (yellow/brown) and starch (white) form a deep blue colour when they mix. The test for starch, therefore, is to add iodine solution to a leaf to see if it goes blue. However, a living leaf is impermeable to iodine and the chlorophyll in the leaf masks any colour change. So, the leaf has to be treated as follows:

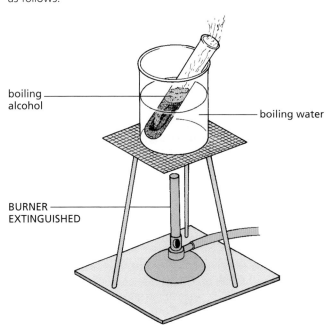

Figure 6.2 Experiment to remove chlorophyll from a leaf

- Heat some water to boiling point in a beaker and then **turn off the Bunsen flame**.
- Use forceps to dip a leaf in the hot water for about 30 seconds. This kills the cytoplasm, denatures the enzymes and makes the leaf more permeable to iodine solution.
- **Note**: make sure the Bunsen flame is extinguished before starting the next part of the procedure, as ethanol is flammable.

- Push the leaf to the bottom of a test-tube and cover it with ethanol (alcohol). Place the tube in the hot water (Figure 6.2). The alcohol will boil and dissolve out most of the chlorophyll. This makes colour changes with iodine easier to see.
- Pour the green alcohol into a spare beaker, remove the leaf and dip it once more into the hot water to soften it.
- Spread the decolourised leaf flat on a white tile and drop iodine solution on to it. The parts containing starch will turn blue; parts without starch will stain brown or yellow with iodine.

1 Is chlorophyll necessary for photosynthesis?

It is not possible to remove chlorophyll from a leaf without killing it, and so a variegated leaf, which has chlorophyll only in patches, is used. A leaf of this kind is shown in Figure 6.3(a). The white part of the leaf serves as the experiment, because it lacks chlorophyll, while the green part with chlorophyll is the control. After being destarched, the leaf – still on the plant – is exposed to daylight for a few hours. Remove a leaf from the plant; draw it carefully to show where the chlorophyll is (i.e. the green parts) and test it for starch as described above.

Result

Only the parts that were previously green turn blue with iodine. The parts that were white stain brown (Figure 6.3(b)).

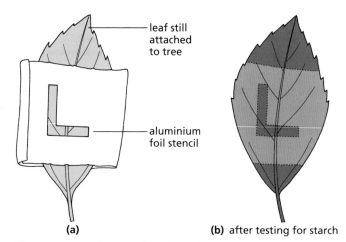

(a)   **(b)** after testing for starch

Figure 6.4 Experiment to show that light is necessary

Interpretation

As starch has not formed in the areas that received no light, it seems that light is needed for starch formation and thus for photosynthesis.

You could argue that the aluminium foil had stopped carbon dioxide from entering the leaf and that it was shortage of carbon dioxide rather than absence of light which prevented photosynthesis taking place. A further control could be designed, using transparent material instead of aluminium foil for the stencil.

3 Is carbon dioxide needed for photosynthesis?

- Water two destarched potted plants and enclose their shoots in polythene bags.
- In one pot place a dish of soda-lime to absorb the carbon dioxide from the air (the experiment). In the other place a dish of sodium hydrogencarbonate solution to produce carbon dioxide (the control), as shown in Figure 6.5.
- Place both plants in the light for several hours and then test a leaf from each for starch.

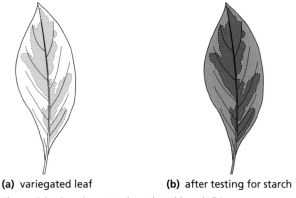

(a) variegated leaf   **(b)** after testing for starch

Figure 6.3 Experiment to show that chlorophyll is necessary

Interpretation

Since starch is present only in the parts that originally contained chlorophyll, it seems reasonable to suppose that chlorophyll is needed for photosynthesis.

It must be remembered, however, that there are other possible interpretations that this experiment has not ruled out; for example, starch could be made in the green parts and sugar in the white parts. Such alternative explanations could be tested by further experiments.

2 Is light necessary for photosynthesis?

- Cut a simple shape from a piece of aluminium foil to make a stencil and attach it to a destarched leaf (Figure 6.4(a)).
- After 4 to 6 hours of daylight, remove the leaf and test it for starch.

Result

Only the areas which had received light go blue with iodine (Figure 6.4(b)).

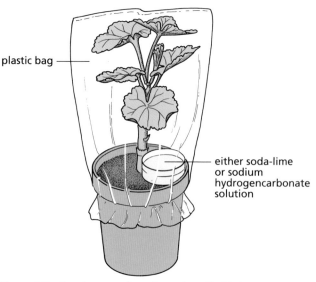

Figure 6.5 Experiment to show that carbon dioxide is necessary

Result
The leaf that had no carbon dioxide does not turn blue. The one from the polythene bag containing carbon dioxide does turn blue.

Interpretation
The fact that starch was made in the leaves that had carbon dioxide, but not in the leaves that had no carbon dioxide, suggests that this gas must be necessary for photosynthesis. The control rules out the possibility that high humidity or high temperature in the plastic bag prevents normal photosynthesis.

4 Is oxygen produced during photosynthesis?

- Place a short-stemmed funnel over some Canadian pondweed in a beaker of water.
- Fill a test-tube with water and place it upside-down over the funnel stem (Figure 6.6). (The funnel is raised above the bottom of the beaker to allow the water to circulate.)
- Place the apparatus in sunlight. Bubbles of gas should appear from the cut stems and collect in the test-tube.
- Set up a control in a similar way but place it in a dark cupboard.
- When sufficient gas has collected from the plant in the light, remove the test-tube and insert a glowing splint.

Result
The glowing splint bursts into flames.

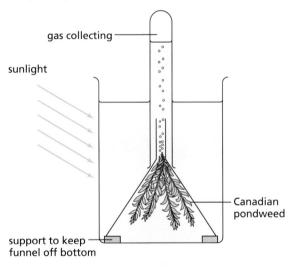

Figure 6.6 Experiment to show that oxygen is produced

Interpretation
The relighting of a glowing splint does not prove that the gas collected in the test-tube is *pure* oxygen, but it does show that it contains extra oxygen and this must have come from the plant. The oxygen is given off only in the light.

Note that water contains dissolved oxygen, carbon dioxide and nitrogen. These gases may diffuse in or out of the bubbles as they pass through the water and collect in the test-tube. The composition of the gas in the test-tube may not be the same as that in the bubbles leaving the plant.

Controls

When setting up an experiment and a control, which of the two procedures constitutes the 'control' depends on the way the prediction is worded. For example, if the prediction is that 'in the absence of light, the pondweed will not produce oxygen', then the 'control' is the plant in the light. If the prediction is that 'the pondweed in the light will produce oxygen', then the 'control' is the plant in darkness. As far as the results and interpretation are concerned, it does not matter which is the 'control' and which is the 'experiment'.

The results of the four experiments support the hypothesis of photosynthesis as stated at the beginning of this chapter and as represented by the equation. Starch formation (our evidence for photosynthesis) does not take place in the absence of light, chlorophyll or carbon dioxide, and oxygen production occurs only in the light.

If starch or oxygen production had occurred in the absence of any one of these conditions, we should have to change our hypothesis about the way plants obtain their food. Bear in mind, however, that although our results support the photosynthesis theory, they do not prove it. For example, it is now known that many stages in the production of sugar and starch from carbon dioxide do not need light (the 'light-independent' reaction).

5 What is the effect of changing light intensity on the rate of photosynthesis? (Method 1)

In this investigation, the rate of production of bubbles by a pond plant is used to calculate the rate of photosynthesis.

- Prepare a beaker of water or a boiling tube, into which a spatula end of sodium hydrogencarbonate has been stirred (this dissolves rapidly and saturates the water with carbon dioxide, so CO_2 is not a limiting factor).
- Collect a fresh piece of Canadian pondweed and cut one end of the stem, using a scalpel blade.
- Attach a piece of modelling clay or paperclip to the stem and put it into the beaker (or boiling tube).
- Set up a light source 10 cm away from the beaker and switch on the lamp (Figure 6.7). Bubbles should start appearing from the cut end of the plant stem. Count the number of bubbles over a fixed time e.g. 1 minute and record the result. Repeat the count.

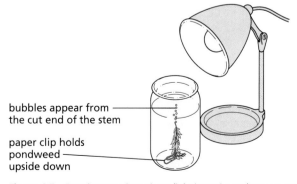

Figure 6.7 Experiment to investigate light intensity and oxygen production

6 PLANT NUTRITION

- Now move the light source so that it is 20 cm from the beaker. Switch on the lamp and leave it for a few minutes, to allow the plant to adjust to the new light intensity. Count the bubbles as before and record the results.
- Repeat the procedure so that the numbers of bubbles for at least five different distances have been recorded. Also, try switching off the bench lamp and observe any change in the production of bubbles.
- There is a relationship between the distance of the lamp from the plant and the light intensity received by the plant. Light intensity = $\frac{1}{D^2}$ where D = distance.
- Convert the distances to light intensity, then plot a graph of light intensity/arbitrary units (*x*-axis) against rate of photosynthesis/bubbles per minute (*y*-axis).

Note: in this investigation another variable, which could affect the rate of photosynthesis, is the heat given off from the bulb. To improve the method, another beaker of water could be placed between the bulb and the plant to act as a heat filter while allowing the plant to receive the light.

- If the bubbles appear too rapidly to count, try tapping a pen or pencil on a sheet of paper at the same rate as the bubbles appear and get your partner to slide the paper slowly along for 15 seconds. Then count the dots (Figure 6.8).

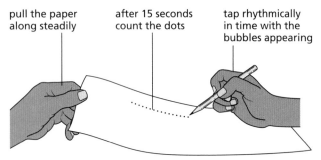

Figure 6.8 Estimating the rate of bubble production

Result
The rate of bubbling should decrease as the lamp is moved further away from the plant. When the light is switched off, the bubbling should stop.

Interpretation
Assuming that the bubbles contain oxygen produced by photosynthesis, as the light intensity is increased the rate of photosynthesis (as indicated by the rate of oxygen bubble production) increases. This is because the plant uses the light energy to photosynthesise and oxygen is produced as a waste product. The oxygen escapes from the plant through the cut stem. We are assuming also that the bubbles do not change in size during the experiment. A fast stream of small bubbles might represent the same volume of gas as a slow stream of large bubbles.

6 What is the effect of changing light intensity on the rate of photosynthesis? (Method 2)
This alternative investigation uses leaf discs from land plants (Figure 6.9).

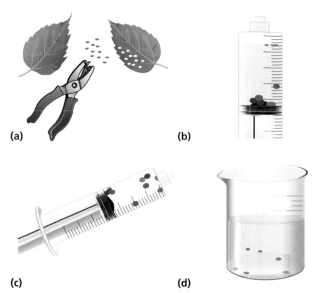

Figure 6.9 Using leaf discs to investigate the effect of light intensity on photosynthesis

- Use a cork borer or paper hole punch to cut out discs from a fresh, healthy leaf such as spinach, avoiding any veins (Figure 6.9(a)). The leaves contain air spaces. These cause the leaf discs to float when they are placed in water.
- At the start of the experiment, the air needs to be removed from the discs. To do this place about 10 discs into a large (10 cm³) syringe and tap it so the discs fall to the bottom (opposite the plunger end).
- Place one finger over the hole at the end of the syringe barrel. Fill the barrel with water, then replace the plunger.
- Turn the syringe so the needle end is facing up and release your finger.
- Gently push the plunger into the barrel of the syringe to force out any air from above the water (Figure 6.9(b)).
- Now replace your finger over the syringe hole and withdraw the plunger to create a vacuum.
- Keep the plunger withdrawn for about 10 seconds. This sucks out all the air from the leaf discs. They should then sink to the bottom (Figure 6.9(c)). Release the plunger.
- Repeat the procedure if the discs do not all sink.
- Remove the discs from the syringe and place them in a beaker, containing water, with a spatula of sodium hydrogencarbonate dissolved in it (Figure 6.9(d)).
- Start a stopwatch and record the time taken for each of the discs to float to the surface. Ignore those that did not sink. Calculate an average time for the discs to float.
- Repeat the method, varying the light intensity the discs are exposed to in the beaker (see Experiment 5 for varying the light intensity produced by a bench lamp).

Result
The greater the light intensity, the quicker the leaf discs float to the surface.

Interpretation
As the leaf discs photosynthesise they produce oxygen, which is released into the air spaces in the disc. The oxygen makes the

discs more buoyant, so as the oxygen accumulates, they float to the surface of the water. As light intensity increases, the rate of photosynthesis increases.

7 What is the effect of changing carbon dioxide concentration on the rate of photosynthesis?

Sodium hydrogencarbonate releases carbon dioxide when dissolved in water. Use the apparatus shown in Figure 6.10.

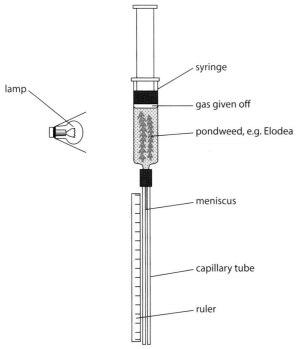

Figure 6.10 Apparatus for investigating the effect of changing carbon dioxide concentration on the rate of photosynthesis

- To set this up, remove the plunger from the 20 cm³ syringe and place two or three pieces of pondweed, with freshly cut stems facing upwards, into the syringe barrel. Hold a finger over the end of the capillary tube and fill the syringe with distilled water.
- Replace the plunger, turn the apparatus upside down and push the plunger to the 20 cm³ mark, making sure that no air is trapped.
- Arrange the apparatus as shown in Figure 6.10 and move the syringe barrel until the meniscus is near the top of the graduations on the ruler. The bulb should be a fixed distance from the syringe, e.g. 10 cm.
- Switch on the lamp and measure the distance the meniscus moves over 3 minutes. Repeat this several times, then calculate an average.
- Repeat the procedure using the following concentrations of sodium hydrogencarbonate solution: 0.010, 0.0125, 0.0250, 0.0500 and 0.1000 mol dm⁻³.
- Plot a graph of the concentration of sodium hydrogencarbonate solution (*x*-axis) against the mean distance travelled by the meniscus (*y*-axis).

Result
The higher the concentration of sodium hydrogencarbonate solution, the greater the distance moved by the meniscus.

Interpretation
As the concentration of available carbon dioxide is increased, the distance travelled by the meniscus also increases. The movement of the meniscus is caused by oxygen production by the pondweed due to photosynthesis. So an increase in carbon dioxide increases the rate of photosynthesis.

8 What is the effect of changing temperature on the rate of photosynthesis?

Use the methods described in Experiments 5 or 6, but vary the temperature of the water instead of the light intensity.

Questions

1. Which of the following are needed for starch production in a leaf?
carbon dioxide, oxygen, nitrates, water, chlorophyll, soil, light
2. In Experiment 1 (concerning the need for chlorophyll), why was it not necessary to set up a separate control experiment?
3. What is meant by 'destarching' a leaf? Why is it necessary to destarch leaves before setting up some of the photosynthesis experiments?
4. In Experiment 3 (concerning the need for carbon dioxide), what were the functions of:
    a the soda-lime
    b the sodium hydrogencarbonate
    c the polythene bag?
5. a Why do you think pondweed, rather than a land plant, is used for Experiment 4 (concerning production of oxygen)?
    b In what way might this choice make the results less useful?
6. A green plant makes sugar from carbon dioxide and water. Why is it not suitable to carry out an experiment to see if depriving a plant of water stops photosynthesis?
7. Does the method of destarching a plant take for granted the results of Experiment 2? Explain your answer.

You need to be able to state the balanced chemical equation for photosynthesis.

$$6CO_2 + 6H_2O \xrightarrow[\text{chlorophyll}]{\text{light energy}} C_6H_{12}O_6 + 6O_2$$

The process of photosynthesis

Although the details of photosynthesis vary in different plants, the hypothesis as stated in this chapter has stood up to many years of experimental testing and is universally accepted. The next section describes how photosynthesis takes place in a plant.

The process takes place mainly in the cells of the leaves (Figure 6.11) and is summarised in

6 PLANT NUTRITION

Figure 6.12. In land plants water is absorbed from the soil by the roots and carried in the water vessels of the veins, up the stem to the leaf. Carbon dioxide is absorbed from the air through the stomata (pores in the leaf, see 'Leaf structure' later in this chapter). In the leaf cells, the carbon dioxide and water are combined to make sugar. The energy for this reaction comes from sunlight that has been absorbed by the green pigment **chlorophyll**. The chlorophyll is present in the chloroplasts of the leaf cells and it is inside the **chloroplasts** that the reaction takes place. Chloroplasts (Figure 6.12(d)) are small, green structures present in the cytoplasm of the leaf cells. Chlorophyll is the substance that gives leaves and stems their green colour. It is able to absorb energy from light and use it to split water molecules into hydrogen and oxygen (the 'light' or 'light-dependent' reaction). The oxygen escapes from the leaf and the hydrogen molecules are added to carbon dioxide molecules to form sugar (the 'dark' or 'light-independent' reaction). In this way the light energy has been transferred into the chemical energy of carbohydrates as they are synthesised.

Figure 6.11 All the reactions involved in producing food take place in the leaves. Notice how little the leaves overlap

There are four types of chlorophyll that may be present in various proportions in different species. There are also a number of photosynthetic pigments, other than chlorophyll, which may mask the colour of chlorophyll even when it is present, e.g. the brown and red pigments that occur in certain seaweeds.

The plant's use of photosynthetic products

The glucose molecules produced by photosynthesis are quickly built up into starch molecules and added to the growing starch granules in the chloroplast. If the glucose concentration was allowed to increase in the mesophyll cells of the leaf, it could disturb the osmotic balance between the cells (see 'Osmosis' in Chapter 3). Starch is a relatively insoluble compound and so does not alter the osmotic potential of the cell contents.

The starch, however, is steadily broken down to sucrose (Chapter 4) and this soluble sugar is transported out of the cell into the food-carrying cells (see Chapter 8) of the leaf veins. These veins will distribute the sucrose to all parts of the plant that do not photosynthesise, e.g. the growing buds, the ripening fruits, the roots and the underground storage organs.

The cells in these regions will use the sucrose in a variety of ways (Figure 6.13).

Respiration

The sugar can be used to provide energy. It is oxidised by respiration (Chapter 12) to carbon dioxide and water, and the energy released is used to drive other chemical reactions such as the building-up of proteins described below.

Storage

Sugar that is not needed for respiration is turned into starch and stored. Some plants store it as starch grains in the cells of their stems or roots. Other plants, such as the potato or parsnip, have special storage organs (tubers) for holding the reserves of starch (see 'Asexual reproduction' in Chapter 16). Sugar may be stored in the fruits of some plants; grapes, for example, contain a large amount of glucose.

Synthesis of other substances

As well as sugars for energy and starch for storage, the plant needs cellulose for its cell walls, lipids for its cell membranes, proteins for its cytoplasm and pigments for its flower petals, etc. All these substances are built up (synthesised) from the sugar molecules and other molecules produced in photosynthesis.

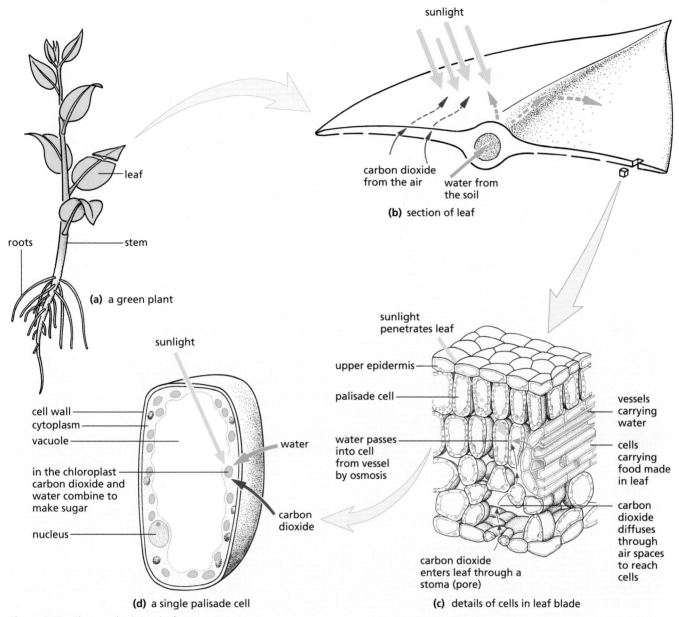

Figure 6.12 Photosynthesis in a leaf

By joining hundreds of glucose molecules together, the long-chain molecules of cellulose (Chapter 4, Figure 4.4) are built up and added to the cell walls.

Amino acids (see Chapter 4) are made by combining **nitrogen** with sugar molecules or smaller carbohydrate molecules. These amino acids are then joined together to make the proteins that form the enzymes and the cytoplasm of the cell. The nitrogen for this synthesis comes from **nitrates** which are absorbed from the soil by the roots.

Some proteins also need **sulfur** molecules and these are absorbed from the soil in the form of **sulfates** (SO_4). **Phosphorus** is needed for DNA (Chapter 4) and for reactions involving energy release. It is taken up as phosphates (PO_4).

The chlorophyll molecule needs magnesium (Mg). This metallic element is also obtained from salts in the soil.

Many other elements, e.g. iron, manganese, boron, are also needed in very small quantities for healthy growth. These are often referred to as **trace elements**.

6 PLANT NUTRITION

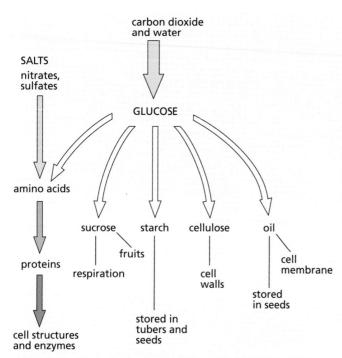

Figure 6.13 Green plants can make all the materials they need from carbon dioxide, water and salts.

The metallic and non-metallic elements are all taken up in the form of their ions by the plant roots.

All these chemical processes, such as the uptake of salts and the building-up of proteins, need energy from respiration to make them happen.

Gaseous exchange in plants

Air contains the gases nitrogen, oxygen, carbon dioxide and water vapour. Plants and animals take in or give out these last three gases and this process is called **gaseous exchange**.

You can see from the equation for photosynthesis that one of its products is oxygen. Therefore, in daylight, when photosynthesis is going on in green plants, they will be taking in carbon dioxide and giving out oxygen. This exchange of gases is the opposite of that resulting from respiration (Chapter 12) but it must not be thought that green plants do not respire. The energy they need for all their living processes – apart from photosynthesis – comes from respiration, and this is going on all the time, using up oxygen and producing carbon dioxide.

During the daylight hours, plants are photosynthesising as well as respiring, so that all the carbon dioxide produced by respiration is used up by photosynthesis. At the same time, all the oxygen needed by respiration is provided by photosynthesis. Only when the rate of photosynthesis is faster than the rate of respiration will carbon dioxide be taken in and the excess oxygen given out (Figure 6.14).

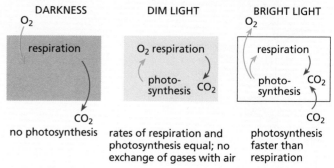

Figure 6.14 Respiration and photosynthesis

Compensation point

As the light intensity increases during the morning and fades during the evening, there will be a time when the rate of photosynthesis exactly matches the rate of respiration. At this point, there will be no net intake or output of carbon dioxide or oxygen. This is the **compensation point**. The sugar produced by photosynthesis exactly compensates for the sugar broken down by respiration.

Practical work

How will the gas exchange of a plant be affected by being kept in the dark and in the light?

This investigation makes use of hydrogencarbonate indicator, which is a test for the presence of carbon dioxide. A build-up of carbon dioxide turns it from pink/red to yellow. A decrease in carbon dioxide levels causes the indicator to turn purple.

- Wash three boiling tubes first with tap water, then with distilled water and finally with hydrogencarbonate indicator (the indicator will change colour if the boiling tube is not clean).
- Then fill the three boiling tubes to about two thirds full with hydrogencarbonate indicator solution.
- Add equal-sized pieces of Canadian pondweed to tubes 1 and 2 and seal all the tubes with stoppers.
- Expose tubes 1 and 3 to light using a bench lamp and place tube 2 in a black box, or a dark cupboard, or wrap it in aluminium foil (Figure 6.15). After 24 hours note the colour of the hydrogencarbonate indicator in each tube.

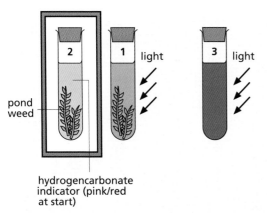

Figure 6.15 Experiment to compare gas exchange in plants kept in the dark and in the light

Result
The indicator in tube 3 (the control) which was originally pink/red should not change colour; that in tube 2 (plant in the dark) should turn yellow; and in tube 1 (plant in the light) the indicator should be purple.

Interpretation
Hydrogencarbonate indicator is a mixture of dilute sodium hydrogencarbonate solution with the dyes cresol red and thymol blue. It is a pH indicator in equilibrium with the carbon dioxide, i.e. its original colour represents the acidity produced by the carbon dioxide in the air. An increase in carbon dioxide makes it more acidic and it changes colour from orange/red to yellow. A decrease in carbon dioxide makes it less acid and causes a colour change to purple.

The results, therefore, provide evidence that in the light (tube 1) aquatic plants use up more carbon dioxide in photosynthesis than they produce in respiration. In darkness (tube 2) the plant produces carbon dioxide (from respiration). Tube 3 is the control, showing that it is the presence of the plant that causes a change in the solution in the boiling tube.

The experiment can be criticised on the grounds that the hydrogencarbonate indicator is not a specific test for carbon dioxide but will respond to any change in acidity or alkalinity. In tube 1 there would be the same change in colour if the leaf produced an alkaline gas such as ammonia, and in tube 2 any acid gas produced by the leaf would turn the indicator yellow. However, knowledge of the metabolism of the leaf suggests that these are less likely events than changes in the carbon dioxide concentration.

Effects of external factors on rate of photosynthesis

The rate of the light reaction will depend on the light intensity. The brighter the light, the faster will water molecules be split in the chloroplasts. The 'dark' reaction will be affected by temperature. A rise in temperature will increase the rate at which carbon dioxide is combined with hydrogen to make carbohydrate.

Limiting factors

> **Key definition**
> A **limiting factor** is something present in the environment in such short supply that it restricts life processes.

If you look at Figure 6.16(a), you will see that an increase in light intensity does indeed speed up photosynthesis, but only up to a point. Beyond that point, any further increase in light intensity has only a small effect. This limit on the rate of increase could be because all available chloroplasts are fully occupied in light absorption. So, no matter how much the light intensity increases, no more light can be absorbed and used. Alternatively, the limit could be imposed by the fact that there is not enough carbon dioxide in the air to cope with the increased supply of hydrogen atoms produced by the light reaction. Or, it may be that low temperature is restricting the rate of the 'dark' reaction.

Figure 6.16(b) shows that, if the temperature of a plant is raised, then the effect of increased illumination is not limited so much. Thus, in Figure 6.16(a), it seems likely that the increase in the rate of photosynthesis could have been limited by the temperature. Any one of the external factors – temperature, light intensity or carbon dioxide concentration – may limit the effects of the other two. A temperature rise may cause photosynthesis to speed up, but only to the point where the light intensity limits further increase. In such conditions, the external factor that restricts the effect of the others is called the **limiting factor**.

Since there is only 0.03% of carbon dioxide in the air, it might seem that a shortage of carbon dioxide could be an important limiting factor. Indeed, experiments do show that an increase in carbon dioxide concentration does allow a faster rate of photosynthesis. However, recent work in plant physiology has shown that the extra carbon dioxide affects reactions other than photosynthesis.

The main effect of extra carbon dioxide is to slow down the rate of oxidation of sugar by a process called **photorespiration** and this produces the same effect as an increase in photosynthesis.

Although carbon dioxide concentration limits photosynthesis only indirectly, artificially high levels of carbon dioxide in greenhouses do effectively increase yields of crops (Figure 6.17).

Greenhouses also maintain a higher temperature and so reduce the effect of low temperature as a limiting factor, and they clearly optimise the light reaching the plants.

Parts of the world such as tropical countries often benefit from optimum temperatures and rainfall for crop production. However, greenhouses are still often used because they allow the growers to control how much water and nutrients the plants receive and they can also reduce crop damage by insect pests and disease. Sometimes rainfall is too great to benefit the plants. In an experiment in the Seychelles in the wet season of 1997, tomato crops in an open field yielded 2.9 kg m^{-2}. In a greenhouse, they yielded 6.5 kg m^{-2}.

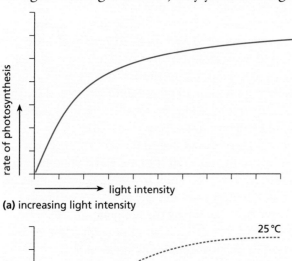
(a) increasing light intensity

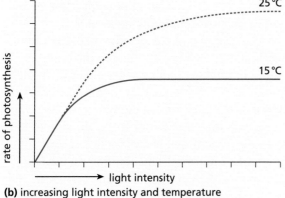
(b) increasing light intensity and temperature

Figure 6.16 Limiting factors in photosynthesis

The concept of limiting factors does not apply only to photosynthesis. Adding fertiliser to the soil, for example, may increase crop yields, but only up to the point where the roots can take up all the nutrients and the plant can build them into proteins, etc. The uptake of mineral ions is limited by the absorbing area of the roots, rates of respiration, aeration of the soil and availability of carbohydrates from photosynthesis.

Figure 6.17 Carrot plants grown in increasing concentrations of carbon dioxide from left to right

Currently there is debate about whether athletic performance is limited by the ability of the heart and lungs to supply oxygenated blood to muscles, or by the ability of the muscles to take up and use the oxygen.

The role of the stomata

The **stomata** (Figure 6.20) in a leaf may affect the rate of photosynthesis according to whether they are open or closed. When photosynthesis is taking place, carbon dioxide in the leaf is being used up and its concentration falls. At low concentrations of carbon dioxide, the stomata will open. Thus, when photosynthesis is most rapid, the stomata are likely to be open, allowing carbon dioxide to diffuse into the leaf. When the light intensity falls, photosynthesis will slow down and the build-up of carbon dioxide from respiration will make the stomata close. In this way, the stomata are normally regulated by the rate of photosynthesis rather than photosynthesis being limited by the stomata. However, if the stomata close during the daytime as a result of excessive water loss from the leaf, their closure will restrict photosynthesis by preventing the inward diffusion of atmospheric carbon dioxide.

Normally the stomata are open in the daytime and closed at night. Their closure at night, when intake of carbon dioxide is not necessary, reduces the loss of water vapour from the leaf (see 'Transpiration' in Chapter 8).

Leaf structure

The relationship between a leaf and the rest of the plant is described in Chapter 8.

A typical leaf of a broad-leaved plant is shown in Figure 6.18(a). (Figure 6.18(b) shows a transverse section through the leaf.) It is attached to the stem by a **leaf stalk**, which continues into the leaf as a **midrib**. Branching from the midrib is a network of veins that deliver water and salts to the leaf cells and carry away the food made by them.

As well as carrying food and water, the network of veins forms a kind of skeleton that supports the softer tissues of the leaf blade.

The **leaf blade** (or **lamina**) is broad. A vertical section through a small part of a leaf blade is shown in Figure 6.18(c) and Figure 6.19 is a photograph of a leaf section under the microscope.

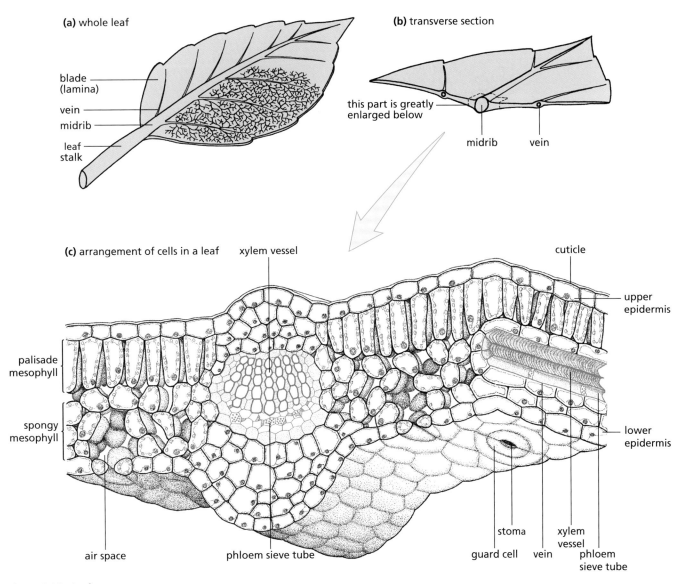

Figure 6.18 Leaf structure

6 PLANT NUTRITION

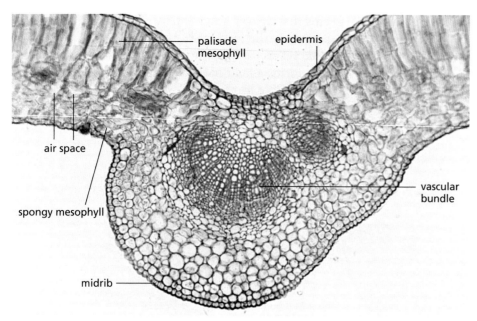

Figure 6.19 Transverse section through a leaf (×30)

Epidermis

The epidermis is a single layer of cells on the upper and lower surfaces of the leaf. There is a thin waxy layer called the **cuticle** over the epidermis.

Stomata

In the leaf epidermis there are structures called **stomata** (singular = stoma). A stoma consists of a pair of **guard cells** (Figure 6.20) surrounding an opening or stomatal pore. In most dicotyledons (i.e. the broad-leaved plants; see 'Features of organisms' in Chapter 1), the stomata occur only in the lower epidermis. In monocotyledons (i.e. narrow-leaved plants such as grasses) the stomata are equally distributed on both sides of the leaf.

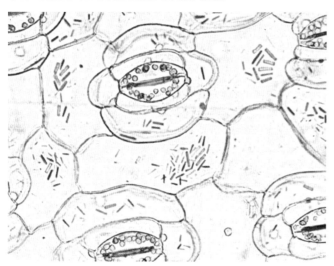

Figure 6.20 Stomata in the lower epidermis of a leaf (×350)

Mesophyll

The tissue between the upper and lower epidermis is called **mesophyll** (Figure 6.18(c)). It consists of two zones: the upper **palisade mesophyll** and the lower **spongy mesophyll** (Figure 6.23). The palisade cells are usually long and contain many **chloroplasts**. Chloroplasts are green organelles, due to the presence of the pigment chlorophyll, found in the cytoplasm of the photosynthesising cells. The spongy mesophyll cells vary in shape and fit loosely together, leaving many air spaces between them. They also contain chloroplasts.

Veins (vascular bundles)

The main **vein** of the leaf is called the midrib. Other veins branch off from this and form a network throughout the leaf. Vascular bundles consist of two different types of tissues, called **xylem** and **phloem**. The xylem vessels are long thin tubes with no cell contents when mature. They have thickened cell walls, impregnated with a material called **lignin**, which can form distinct patterns in the vessel walls, e.g. spirals (see Chapter 8). Xylem carries water and salts to cells in the leaf. The phloem is in the form of sieve tubes. The ends of each elongated cell are perforated to form sieve plates and the cells retain their contents. Phloem transports food substances such as sugars away from the leaf to other parts of the plant.

Leaf structure

Table 6.1 Summary of parts of a leaf

Part of leaf	Details
cuticle	Made of wax, waterproofing the leaf. It is secreted by cells of the upper epidermis.
upper epidermis	These cells are thin and transparent to allow light to pass through. No chloroplasts are present. They act as a barrier to disease organisms.
palisade mesophyll	The main region for photosynthesis. Cells are columnar (quite long) and packed with chloroplasts to trap light energy. They receive carbon dioxide by diffusion from air spaces in the spongy mesophyll.
spongy mesophyll	These cells are more spherical and loosely packed. They contain chloroplasts, but not as many as in palisade cells. Air spaces between cells allow gaseous exchange – carbon dioxide to the cells, oxygen from the cells during photosynthesis.
vascular bundle	This is a leaf vein, made up of xylem and phloem. Xylem vessels bring water and minerals to the leaf. Phloem vessels transport sugars and amino acids away (this is called translocation).
lower epidermis	This acts as a protective layer. Stomata are present to regulate the loss of water vapour (this is called transpiration). It is the site of gaseous exchange into and out of the leaf.
stomata	Each stoma is surrounded by a pair of guard cells. These can control whether the stoma is open or closed. Water vapour passes out during transpiration. Carbon dioxide diffuses in and oxygen diffuses out during photosynthesis.

Functions of parts of the leaf

Epidermis

The epidermis helps to keep the leaf's shape. The closely fitting cells (Figure 6.18(c)) reduce evaporation from the leaf and prevent bacteria and fungi from getting in. The cuticle is a waxy layer lying over the epidermis, which helps to reduce water loss. It is produced by the epidermal cells.

Stomata

Changes in the turgor (see 'Osmosis' in Chapter 3) and shape of the guard cells can open or close the stomatal pore. In very general terms, stomata are open during the hours of daylight but closed during the evening and most of the night (Figure 6.21). This pattern, however, varies greatly with the plant species. A satisfactory explanation of stomatal rhythm has not been worked out, but when the stomata are open (i.e. mostly during daylight), they allow carbon dioxide to diffuse into the leaf where it is used for photosynthesis.

If the stomata close, the carbon dioxide supply to the leaf cells is virtually cut off and photosynthesis stops. However, in many species, the stomata are closed during the hours of darkness, when photosynthesis is not taking place anyway.

It seems, therefore, that stomata allow carbon dioxide into the leaf when photosynthesis is taking place and prevent excessive loss of water vapour (see 'Transpiration' in Chapter 8) when photosynthesis stops, but the story is likely to be more complicated than this.

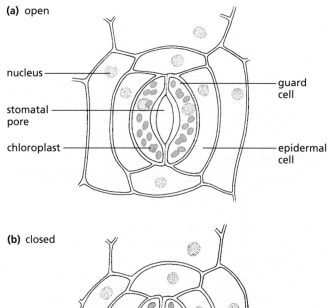

Figure 6.21 Stoma

The detailed mechanism by which stomata open and close is not fully understood, but it is known that in the light, the potassium concentration in the guard cell vacuoles increases. This lowers the water potential (see 'Osmosis' in Chapter 3) of the cell sap and water enters the guard cells by osmosis from their neighbouring epidermal cells. This inflow of water raises the turgor pressure inside the guard cells.

The cell wall next to the stomatal pore is thicker than elsewhere in the cell and is less able to stretch (Figure 6.22). So, although the increased turgor tends to expand the whole guard cell, the thick inner wall cannot expand. This causes the guard cells to curve in such a way that the stomatal pore between them is opened.

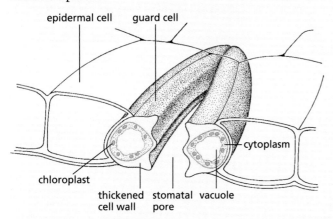

Figure 6.22 Structure of guard cells

When potassium ions leave the guard cell, the water potential rises, water passes out of the cells by osmosis, the turgor pressure falls and the guard cells straighten up and close the stoma.

Where the potassium ions come from and what triggers their movement into or out of the guard cells is still under active investigation.

You will notice from Figures 6.21 and 6.22 that the guard cells are the only epidermal cells containing chloroplasts. At one time it was thought that the chloroplasts built up sugar by photosynthesis during daylight, that the sugars made the cell sap more concentrated and so caused the increase in turgor. In fact, little or no photosynthesis takes place in these chloroplasts and their function has not been explained, though it is known that starch accumulates in them during the hours of darkness. In some species of plants, the guard cells have no chloroplasts.

Mesophyll
The function of the palisade cells and – to a lesser extent – of the spongy mesophyll cells is to make food by photosynthesis. Their chloroplasts absorb sunlight and use its energy to join carbon dioxide and water molecules to make sugar molecules as described earlier in this chapter.

In daylight, when photosynthesis is rapid, the mesophyll cells are using up carbon dioxide. As a result, the concentration of carbon dioxide in the air spaces falls to a low level and more carbon dioxide diffuses in (Chapter 3) from the outside air, through the stomata (Figure 6.23). This diffusion continues through the air spaces, up to the cells which are using carbon dioxide. These cells are also producing oxygen as a by-product of photosynthesis. When the concentration of oxygen in the air spaces rises, it diffuses out through the stomata.

Vascular bundles
The water needed for making sugar by photosynthesis is brought to the mesophyll cells by the veins. The mesophyll cells take in the water by osmosis (Chapter 3) because the concentration of free water molecules in a leaf cell, which contains sugars, will be less than the concentration of water in the water vessels of a vein. The branching network of leaf veins means that no cell is very far from a water supply.

The sugars made in the mesophyll cells are passed to the phloem cells (Chapter 8) of the veins, and these cells carry the sugars away from the leaf into the stem.

The ways in which a leaf is thought to be well adapted to its function of photosynthesis are listed in the next paragraph.

Adaptation of leaves for photosynthesis

When biologists say that something is **adapted**, they mean that its structure is well suited to its function. The detailed structure of the leaf is described in the first section of this chapter and although there are wide variations in leaf shape, the following general statements apply to a great many leaves, and are illustrated in Figures 6.18(b) and (c).

- Their broad, flat shape offers a large surface area for absorption of sunlight and carbon dioxide.
- Most leaves are thin and the carbon dioxide only has to diffuse across short distances to reach the inner cells.

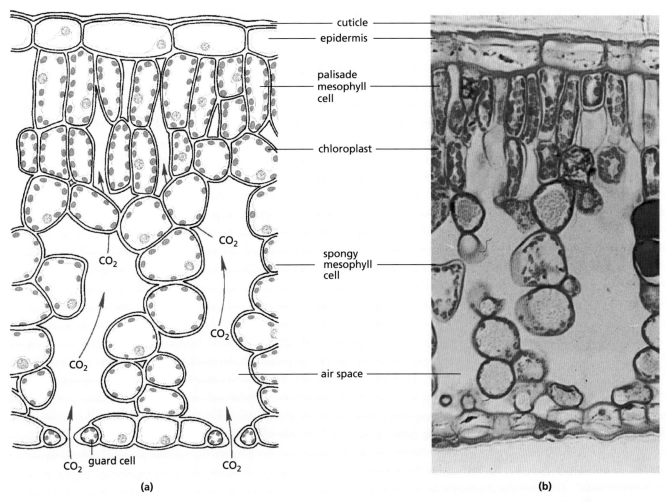

Figure 6.23 Vertical section through a leaf blade (×300)

- The large spaces between cells inside the leaf provide an easy passage through which carbon dioxide can diffuse.
- There are many stomata (pores) in the lower surface of the leaf. These allow the exchange of carbon dioxide and oxygen with the air outside.
- There are more chloroplasts in the upper (palisade) cells than in the lower (spongy mesophyll) cells. The palisade cells, being on the upper surface, will receive most sunlight and this will reach the chloroplasts without being absorbed by too many cell walls.
- The branching network of veins provides a good water supply to the photosynthesising cells. No cell is very far from a water-conducting vessel in one of these veins.

Although photosynthesis takes place mainly in the leaves, any part of the plant that contains chlorophyll will photosynthesise. Many plants have green stems in which photosynthesis takes place.

● Mineral requirements

Plants need a source of nitrate ions (NO_3^-) for making amino acids (Chapter 4). Amino acids are important because they are joined together to make proteins, needed to form the enzymes and cytoplasm of the cell. Nitrates are absorbed from the soil by the roots.

Magnesium ions (Mg^{2+}) are needed to form chlorophyll, the photosynthetic pigment in chloroplasts. This metallic element is also obtained in salts from the soil (see the salts listed under 'Water cultures' on page 82).

6 PLANT NUTRITION

Sources of mineral elements and effects of their deficiency

The substances mentioned previously (nitrates, magnesium) are often referred to as 'mineral salts' or 'mineral elements'. If any mineral element is lacking, or deficient, in the soil then the plants may show visible deficiency symptoms.

Many slow-growing wild plants will show no deficiency symptoms even on poor soils. Fast-growing crop plants, on the other hand, will show distinct deficiency symptoms though these will vary according to the species of plant. If nitrate ions are in short supply, the plant will show stunted growth. The stem becomes weak. The lower leaves become yellow and die, while the upper leaves turn pale green. If the plant is deficient in magnesium, it will not be able to make magnesium. The leaves turn yellow from the bottom of the stem upwards (a process called **chlorosis**). Farmers and gardeners can recognise these symptoms and take steps to replace the missing minerals.

The mineral elements needed by plants are absorbed from the soil in the form of salts. For example, a plant's needs for potassium (K) and nitrogen (N) might be met by absorbing the ions of the salt **potassium nitrate** (KNO_3). Salts like this come originally from rocks, which have been broken down to form the soil. They are continually being taken up from the soil by plants or washed out of the soil by rain. They are replaced partly from the dead remains of plants and animals. When these organisms die and their bodies decay, the salts they contain are released back into the soil. This process is explained in some detail, for nitrates, in Chapter 19 'Nutrient cycles'.

In arable farming, the ground is ploughed and whatever is grown is removed. There are no dead plants left to decay and replace the mineral salts. The farmer must replace them by spreading animal manure, sewage sludge or artificial fertilisers in measured quantities over the land.

Three manufactured fertilisers in common use are ammonium nitrate, superphosphate and compound NPK.

Ammonium nitrate (NH_4NO_3)

The formula shows that ammonium nitrate is a rich source of nitrogen but no other plant nutrients. It is sometimes mixed with calcium carbonate to form a compound fertiliser such as 'Nitro-chalk'.

Superphosphates

These fertilisers are mixtures of minerals. They all contain calcium and phosphate and some have sulfate as well.

Compound NPK fertiliser

'N' is the chemical symbol for nitrogen, 'P' for phosphorus and 'K' for potassium. NPK fertilisers are made by mixing ammonium sulfate, ammonium phosphate and potassium chloride in varying proportions. They provide the ions of nitrate, phosphate and potassium, which are the ones most likely to be below the optimum level in an agricultural soil.

Water cultures

It is possible to demonstrate the importance of the various mineral elements by growing plants in **water cultures**. A full water culture is a solution containing the salts that provide all the necessary elements for healthy growth, such as

- potassium nitrate for potassium and nitrogen
- magnesium sulfate for magnesium and sulfur
- potassium phosphate for potassium and phosphorus
- calcium nitrate for calcium and nitrogen.

From these elements, plus the carbon dioxide, water and sunlight needed for photosynthesis, a green plant can make all the substances it needs for a healthy existence.

Some branches of horticulture, e.g. growing of glasshouse crops, make use of water cultures on a large scale. Sage plants may be grown with their roots in flat polythene tubes. The appropriate water culture solution is pumped along these tubes (Figure 6.24). This method has the advantage that the yield is increased and the need to sterilise the soil each year, to destroy pests, is eliminated. This kind of technique is sometimes described as **hydroponics** or soil-less culture.

Mineral requirements

Figure 6.24 Soil-less culture. The sage plants are growing in a nutrient solution circulated through troughs of polythene.

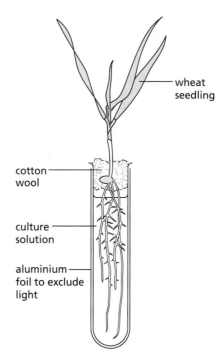

Figure 6.25 Apparatus for a water culture to investigate plant mineral requirements

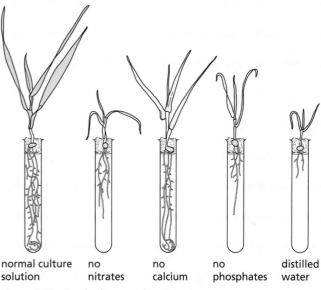

Figure 6.26 Result of water culture experiment

Practical work

The importance of different mineral elements

- Place wheat seedlings in test-tubes containing water cultures as shown in Figure 6.25.
- Cover the tubes with aluminium foil to keep out light and so stop green algae from growing in the solution.
- Some of the solutions have one of the elements missing. For example, magnesium chloride is used instead of magnesium sulfate and so the solution will lack sulfur. In a similar way, solutions lacking nitrogen, potassium and phosphorus can be prepared.
- Leave the seedlings to grow in these solutions for a few weeks, keeping the tubes topped up with distilled water.

Result

The kind of result that might be expected from wheat seedlings is shown in Figure 6.26. Generally, the plants in a complete culture will be tall and sturdy, with large, dark green leaves. The plants lacking nitrogen will usually be stunted and have small, pale leaves. In the absence of magnesium, chlorophyll cannot be made, and these plants will be small with yellow leaves.

Interpretation

The healthy plant in the full culture is the control and shows that this method of raising plants does not affect them. The other, less healthy plants show that a full range of mineral elements is necessary for normal growth.

Quantitative results

Although the effects of mineral deficiency can usually be seen simply by looking at the wheat seedlings, it is better if actual measurements are made.

The height of the shoot, or the total length of all the leaves on one plant, can be measured. The total root length can also be measured, though this is difficult if root growth is profuse.

Alternatively, the **dry weight** of the shoots and roots can be measured. In this case, it is best to pool the results of several experiments. All the shoots from the complete culture are placed in a labelled container; all those from the 'no nitrate' culture solution are placed in another container; and so on for all the plants from the different solutions. The shoots are then dried at 110°C for 24 hours and weighed. The same procedure can be carried out for the roots.

You would expect the roots and shoots from the complete culture to weigh more than those from the nutrient-deficient cultures.

Questions

Core

1. a What substances must a plant take in, in order to carry on photosynthesis?
   b Where does it get each of these substances from?
2. Look at Figure 6.23(a). Identify the palisade cells, the spongy mesophyll cells and the cells of the epidermis. In which of these would you expect photosynthesis to occur:
   a most rapidly
   b least rapidly
   c not at all?
   Explain your answers.
3. a What provides a plant with energy for photosynthesis?
   b What chemical process provides a plant with energy to carry on all other living activities?
4. Look at Figure 6.23. Why do you think that photosynthesis does not take place in the cells of the epidermis?
5. During bright sunlight, what gases are:
   a passing out of the leaf through the stomata
   b entering the leaf through the stomata?

Extended

6. a What substances does a green plant need to take in, to make:
      i sugar
      ii proteins?
   b What must be present in the cells to make reactions i and ii work?
7. A molecule of carbon dioxide enters a leaf cell at 4 p.m. and leaves the same cell at 6 p.m. What is likely to have happened to the carbon dioxide molecule during the 2 hours it was in the leaf cell?
8. In a partially controlled environment such as a greenhouse:
   a how could you alter the external factors to obtain maximum photosynthesis
   b which of these alterations might not be cost effective?
9. Figure 6.27 is a graph showing the average daily change in the carbon dioxide concentration, 1 metre above an agricultural crop in July. From what you have learned about photosynthesis and respiration, try to explain the changes in the carbon dioxide concentration.

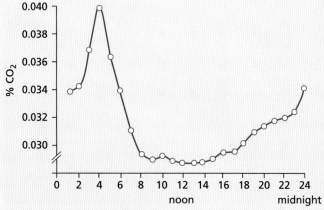

Figure 6.27 Daily changes in concentration of carbon dioxide 1 metre above a plant crop

10. What gases would you expect a leaf to be (i) taking in and (ii) giving out:
    a in bright sunlight
    b in darkness?
11. Measurements on a leaf show that it is giving out carbon dioxide and taking in oxygen. Does this prove that photosynthesis is *not* going on in the leaf? Explain your answer.
12. How could you adapt the experiment with hydrogencarbonate indicator on page 74 to find the light intensity that corresponded to the compensation point?
13. How would you expect the compensation points to differ between plants growing in a wood and those growing in a field?
14. What are the functions of:
    a the epidermis
    b the mesophyll of a leaf?
15. In some plants, the stomata close for a period at about midday. Suggest some possible advantages and disadvantages of this to the plant.
16. What salts would you put in a water culture which is to contain *no* nitrogen?
17. How can a floating pond plant, such as duckweed, survive without having its roots in soil?
18. In the water culture experiment, why should a lack of nitrate cause reduced growth?

19 Figure 6.28 shows the increased yield of winter wheat in response to adding more nitrogenous fertiliser.
   a If the applied nitrogen is doubled from 50 to 100 kg per hectare, how much extra wheat does the farmer get?
   b If the applied nitrogen is doubled from 100 to 200 kg per hectare, how much extra wheat is obtained?
   c What sort of calculations will a farmer need to make before deciding to increase the applied nitrogen from 150 to 200 kg per hectare?

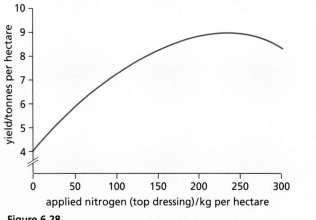

Figure 6.28

Checklist

After studying Chapter 6 you should know and understand the following:

- Photosynthesis is the way plants make their food.
- They combine carbon dioxide and water to make sugar.
- To do this, they need energy from sunlight, which is absorbed by chlorophyll.
- Chlorophyll converts light energy to chemical energy.
- The word equation to represent photosynthesis is

   carbon dioxide + water $\xrightarrow[\text{chlorophyll}]{\text{light energy}}$ glucose + oxygen

- Plant leaves are adapted for the process of photosynthesis by being broad and thin, with many chloroplasts in their cells.
- From the sugar made by photosynthesis, a plant can make all the other substances it needs, provided it has a supply of mineral salts like nitrates.
- In daylight, respiration and photosynthesis will be taking place in a leaf; in darkness, only respiration will be taking place.
- In daylight, a plant will be taking in carbon dioxide and giving out oxygen.
- In darkness, a plant will be taking in oxygen and giving out carbon dioxide.
- Experiments to test photosynthesis are designed to exclude light, or carbon dioxide, or chlorophyll, to see if the plant can still produce starch.
- A starch test can be carried out to test if photosynthesis has occurred in a leaf.
- Leaves have a structure which adapts them for photosynthesis.
- Plants need a supply of nitrate ions to make protein and magnesium ions to make chlorophyll.
- The balanced chemical equation for photosynthesis is

   $6CO_2 + 6H_2O \xrightarrow[\text{chlorophyll}]{\text{light energy}} C_6H_{12}O_6 + 6O_2$

- The rate of photosynthesis may be restricted by light intensity and temperature. These are 'limiting factors'.
- Glasshouses can be used to create optimal conditions for photosynthesis.
- Nitrate ions are needed to make proteins; magnesium ions are needed to make chlorophyll.

7 Human nutrition

Diet
Balanced diet
Sources and importance of food groups
Malnutrition
 	
  Kwashiorkor and marasmus
 	
Alimentary canal
Definitions of digestion, absorption, assimilation, egestion
Regions of the alimentary canal and their functions
Diarrhoea
Cholera
 	
  How cholera affects osmosis in the gut
	Mechanical digestion
	Teeth
	Dental decay
	Tooth care
	
	Chemical digestion
	Importance
	Sites of enzyme secretion
	Functions of enzymes and hydrochloric acid
	
	  Roles of bile and enzymes
	
	Absorption
	Role of small intestine
	Absorption of water
	
	  Significance of villi

The need for food

All living organisms need food. An important difference between plants and animals is that green plants can make food in their leaves but animals have to take it in 'ready-made' by eating plants or the bodies of other animals. In all plants and animals, food is used as follows:

For growth

It provides the substances needed for making new cells and tissues.

As a source of energy

Energy is required for the chemical reactions that take place in living organisms to keep them alive. When food is broken down during respiration (see Chapter 12), the energy from the food is used for chemical reactions such as building complex molecules (Chapter 4). In animals the energy is also used for activities such as movement, the heart beat and nerve impulses. Mammals and birds use energy to maintain their body temperature.

For replacement of worn and damaged tissues

The substances provided by food are needed to replace the millions of our red blood cells that break down each day, to replace the skin that is worn away and to repair wounds.

● Diet

Balanced diets

A **balanced diet** must contain enough carbohydrates and fats to meet our energy needs. It must also contain enough protein of the right kind to provide the essential amino acids to make new cells and tissues for growth or repair. The diet must also contain vitamins and mineral salts, plant fibre and water. The composition of four food samples is shown in Figure 7.1.

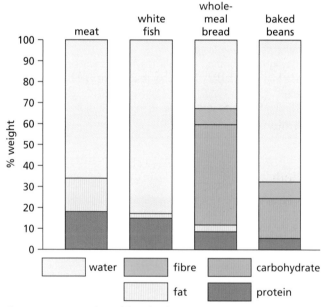

Figure 7.1 An analysis of four food samples
Note: The percentage of water includes any salts and vitamins. There are wide variations in the composition of any given food sample according to its source and the method of preservation and cooking. 'White fish' (e.g. cod, haddock, plaice) contains only 0.5% fat whereas herring and mackerel contain up to 14%. White bread contains only 2–3% fibre. Frying the food greatly adds to its fat content.

Energy requirements

Energy can be obtained from carbohydrates, fats and proteins. The cheapest energy-giving food is usually carbohydrate; the greatest amount of energy is available in fats; proteins give about the same energy as carbohydrates but are expensive. Whatever mixture of carbohydrate, fat and protein makes up the diet, the total energy must be sufficient:

- to keep our internal body processes working (e.g. heart beating, breathing action)
- to keep up our body temperature, and
- to meet the needs of work and other activities.

The amount of energy that can be obtained from food is measured in calories or joules. One gram of carbohydrate or protein can provide us with 16 or 17 kJ (kilojoules). A gram of fat can give 37 kJ. We need to obtain about 12 000 kJ of energy each day from our food. Table 7.1 shows how this figure is obtained. However, the figure will vary greatly according to our age, occupation and activity (Figure 7.2). It is fairly obvious that a person who does hard manual work, such as digging, will use more energy than someone who sits in an office. Similarly, someone who takes part in a lot of sport will need more energy input than someone who doesn't do much physical exercise.

Females tend to have lower energy requirements than males. Two reasons for this are that females have, on average, a lower body mass than males, which has a lower demand on energy intake, and there are also different physical demands made on boys and girls. However, an active female may well have a higher energy requirement than an inactive male of the same age.

As children grow, the energy requirement increases because of the energy demands of the growth process and the extra energy associated with maintaining their body temperature. However, metabolism, and therefore energy demands, tends to slow down with age once we become adults due to a progressive loss of muscle tissue.

Table 7.1 Energy requirements in kJ

8 hours asleep	2 400
8 hours awake; relatively inactive physically	3 000
8 hours physically active	6 600
Total	12 000

The 2400 kJ used during 8 hours' sleep represents the energy needed for **basal metabolism**, which maintains the circulation, breathing, body temperature, brain function and essential chemical processes in the liver and other organs.

If the diet includes more food than is needed to supply the energy demands of the body, the surplus food is stored either as glycogen in the liver or as fat below the skin and in the abdomen.

In 2006, the Food Standards Agency in Britain recommended that, for a balanced diet, 50% of our energy intake should be made up of carbohydrate, 35% of fat (with not more than 11% saturated fat) and the remaining percentage made up of fibre.

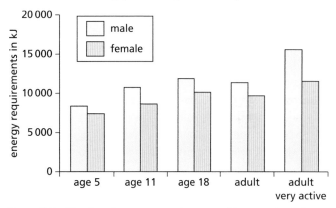

Figure 7.2 The changing energy requirements with age and activity

Protein requirements

Proteins are an essential part of the diet because they supply the amino acids needed to build up our own body structures. Estimates of how much protein we need have changed over the last few years. A recent WHO/FAO/UNU report recommended that an average person needs 0.57 g protein for every kilogram of body weight. That is, a 70 kg person would need 70 × 0.57 = 39.9, i.e. about 40 g protein per day.

This could be supplied by about 200 g (7 ounces) lean meat or 500 g bread but 2 kg potatoes would be needed to supply this much protein and even this will not contain all the essential amino acids.

Vegetarian and vegan diets

There is relatively less protein in food derived from plants than there is in animal products. **Vegetarians** and semi-vegetarians, who include dairy products, eggs and possibly fish in their diets, will obtain sufficient protein to meet their needs (Table 7.2). However, some vegetarian foods now contain relatively high proportions of protein: *Quorn* products (made from mycoprotein – derived from fungi) typically contain 14.5 g protein per 100 g, compared with 18.0 g protein per 100 g for beef sausage, and they do not contain animal fats. **Vegans**, who eat no animal products, need to ensure that their diets include a good variety of cereals, peas, beans and nuts in order to obtain all the essential amino acids to build their body proteins.

Special needs

Pregnancy

A pregnant woman who is already receiving an adequate diet needs no extra food. Her body's metabolism will adapt to the demands of the growing baby although the demand for energy and protein does increase. If, however, her diet is deficient in protein, calcium, iron, vitamin D or folic acid, she will need to increase her intake of these substances to meet the needs of the baby. The baby needs protein for making its tissues, calcium and vitamin D are needed for bone development, and iron is used to make the haemoglobin in its blood.

Lactation

'Lactation' means the production of breast milk for feeding the baby. The production of milk, rich in proteins and minerals, makes a large demand on the mother's resources. If her diet is already adequate, her metabolism will adjust to these demands. Otherwise, she may need to increase her intake of proteins, vitamins and calcium to produce milk of adequate quality and quantity.

Growing children

Most children up to the age of about 12 years need less food than adults, but they need more in proportion to their body weight. For example, an adult may need 0.57 g protein per kg body weight, but a 6–11-month baby needs 1.85 g per kg and a 10-year-old child needs 1.0 g per kg for growth. In addition, children need extra calcium for growing bones, iron for their red blood cells, vitamin D to help calcify their bones and vitamin A for disease resistance.

Malnutrition

Malnutrition is often taken to mean simply not getting enough food, but it has a much wider meaning than this, including getting too much food or the wrong sort of food.

If the total intake of food is not sufficient to meet the body's need for energy, the body tissues themselves are broken down to provide the energy to stay alive. This leads to loss of weight, muscle wastage, weakness and ultimately **starvation**. Extreme slimming diets, such as those that avoid carbohydrate foods, can result in the disease anorexia nervosa.

Coronary heart disease can occur when the diet contains too much fat (see 'Heart' in Chapter 9). Deposits of a fatty substance build up in the arteries, reducing the diameter of these blood vessels, including the coronary artery. Blood clots are then more likely to form. Blood supply to the heart can be reduced resulting in **angina** (chest pains when exercising or climbing stairs, for example) and eventually a coronary **heart attack**.

If food intake is drastically inadequate, it is likely that the diet will also be deficient in proteins, minerals and vitamins so that deficiency diseases such as anaemia, rickets and scurvy also make an appearance. **Scurvy** is caused by a lack of vitamin C (ascorbic acid) in the diet. Vitamin C is present in citrus fruit such as lemons, blackcurrants, tomatoes, fresh green vegetables and potatoes. It is not unusual for people in developed countries who rely on processed food such as tinned products, rather than eating fresh produce, to suffer from scurvy. Symptoms of scurvy include bleeding under the skin, swollen and bleeding gums and poor healing of wounds. The victims of malnutrition due to food deficiencies such as those mentioned above will also have reduced resistance to infectious diseases such as malaria or measles. Thus, the symptoms of malnutrition are usually the outcome of a variety of causes, but all resulting from an inadequate diet.

The **causes of malnutrition** can be famine due to drought or flood, soil erosion, wars, too little land for too many people, ignorance of proper dietary needs but, above all, poverty. Malnourished populations are often poor and cannot afford to buy enough nutritious food.

World food

The world population doubled in the last 30 years but food production, globally, rose even faster. The 'Green Revolution' of the 1960s greatly increased global food production by introducing high-yielding varieties of crops. However, these varieties needed a high input of fertiliser and the use of pesticides, so only the wealthy farmers could afford to use them. Moreover, since 1984, the yields are no longer rising fast enough to feed the growing population or keep pace with the loss of farmland due to erosion and urbanisation.

It is estimated that, despite the global increase in food production, 15% of the world population is undernourished and 180 million children are underweight (Figure 7.4).

There are no obvious, easy or universal solutions to this situation. Genetically modified crops (see 'Genetic engineering' in Chapter 20) may hold out some hope but they are some way off. There is resistance to their introduction in some countries because of concerns about their safety, gene transfer to wild plants or animals, the creation of allergies, the cost of seed and, with some GM seed, the necessity to buy particular pesticides to support them. Redistribution of food from the wealthy to the poorer countries is not a practical proposition except in emergencies, and the process can undermine local economies.

The strategies adopted need to be tailored to the needs and climate of individual countries. Crops suited to the region should be grown. Millet and sorghum grow far better in dry regions than do rice or wheat and need little or no irrigation. Cash crops such as coffee, tea or cotton can earn foreign currency but have no food value and do not feed the local population. There has been a surge in the production of palm oil (Figure 7.3) due to world demand for the product as a biofuel as well as for food manufacture. This has resulted in deforestation to provide land to grow the crop and is putting endangered species at risk of extinction. Countries such as Indonesia and Malaysia have been particularly affected. Where cash crops are grown, it might be better to use the land, where suitable, to cultivate food crops.

Figure 7.3 A new palm oil plantation, replacing a rainforest

The agricultural practices need to be sustainable and not result in erosion. Nearly one-third of the world's crop-growing land has had to be abandoned in the last 40 years because erosion has made it unproductive. Over-irrigation can also cause a build-up in soil salinity, making the land effectively sterile due to the osmotic problems the salt creates (see 'Osmosis' in Chapter 3). Conservation of land, water and energy is essential for sustainable agriculture. A reduction in the growth of the world's population, if it could be achieved, would have a profound effect in reducing malnutrition.

Apart from the measures outlined above, lives could be saved by such simple and inexpensive steps as provision of regular vitamin and mineral supplements. It is estimated that about 30 million children are deficient in vitamin A. This deficiency leads to blindness and death if untreated.

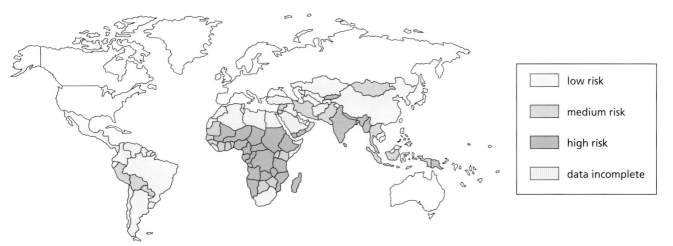

Figure 7.4 Countries with populations at risk of inadequate nutrition

Western diets

In the affluent societies, e.g. USA and Europe, there is no general shortage of food and most people can afford a diet with an adequate energy and protein content. So, few people are undernourished. Eating too much food or food of the 'wrong' sort, however, leads to malnutrition of a different kind.

Refined sugar (sucrose)

This is a very concentrated source of energy. You can absorb a lot of sugar from biscuits, ice-cream, sweets, soft drinks, tinned fruits and sweet tea without ever feeling 'full up'. So you tend to take in more sugar than your body needs, which may lead to you becoming overweight or obese. The food industry has been urged to reduce the sugar content of its products to help curb the increase in obesity in countries like Great Britain and America.

Sugar is also a major cause of tooth decay (see 'Mechanical digestion').

Fats

Fatty deposits, called 'plaques', in the arteries can lead to coronary heart disease and strokes (see 'Heart' in Chapter 9). These plaques are formed from lipids and **cholesterol** combined with proteins (**low density lipoproteins** or **LDLs**). Although the liver makes LDLs, there is evidence to suggest that a high intake of fats, particularly animal fats, helps raise the level of LDLs in the blood and increase the risk of plaque formation.

Most animal fats are formed from **saturated fatty acids**, so called because of their molecular structure. Plant oils are formed from **unsaturated fatty acids** (polyunsaturates) and are thought less likely to cause fatty plaques in the arteries. For this reason, vegetable fats and certain margarines are considered, by some nutritionists, to be healthier than butter and cream. However, there is still much debate about the evidence for this.

Fibre

Many of the processed foods in Western diets contain too little fibre. White bread, for example, has had the fibre (bran) removed. A lack of fibre can result in **constipation** (see 'Classes of food'). Unprocessed foods, such as unskinned potatoes, vegetables and fruit, contain plenty of fibre. Food rich in fibre is usually bulky and makes you feel 'full up' so that you are unlikely to overeat. Fibre enables the process of peristalsis (Figure 7.14) to move food through the gut more efficiently and may also protect the intestines from cancer and other disorders. As explained later, fibre helps prevent constipation.

Overweight and obesity

These are different degrees of the same disorder. If you take in more food than your body needs for energy, growth and replacement, the excess is converted to fat and stored in fat deposits under the skin or in the abdomen.

Obese people are more likely to suffer from high blood pressure, coronary heart disease (see the previous section on malnutrition) and diabetes (Chapter 14). Having extra weight to carry also makes you reluctant to take exercise. By measuring a person's height and body mass, it is possible to use a chart to predict whether or not they have an ideal body mass (Figure 7.5).

Why some people should be prone to obesity is unclear. There may be a genetic predisposition, in which the brain centre that responds to food intake may not signal when sufficient food has been taken in; in some cases it may be the outcome of an infectious disease. Whatever the cause, the remedy is to reduce food intake to a level that matches but does not exceed the body's needs. Taking exercise helps, but it takes a great deal of exercise to 'burn off' even a small amount of surplus fat.

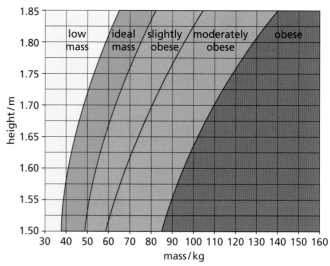

Figure 7.5 Ideal body mass chart

Classes of food

There are three classes of food: carbohydrates, proteins and fats. The chemical structure of these substances is described in Chapter 4. In addition to proteins, carbohydrates and fats, the diet must include salts, vitamins, water and vegetable fibre (roughage). These substances are present in a balanced diet and do not normally have to be taken in separately. A summary of the three classes of food and their sources is shown in Table 7.3.

Carbohydrates

Sugar and **starch** are important carbohydrates in our diet. Starch is abundant in potatoes, bread, maize, rice and other cereals. Sugar appears in our diet mainly as **sucrose** (table sugar) which is added to drinks and many prepared foods such as jam, biscuits and cakes. Glucose and fructose are sugars that occur naturally in many fruits and some vegetables.

Although all foods provide us with energy, carbohydrates are the cheapest and most readily available source of energy. They contain the elements carbon, hydrogen and oxygen (e.g. glucose is $C_6H_{12}O_6$). When carbohydrates are oxidised to provide energy by respiration they are broken down to carbon dioxide and water (Chapter 12). One gram of carbohydrate can provide, on average, 16 kilojoules (kJ) of energy (see practical work 'Energy from food', p. 95).

If we eat more carbohydrates than we need for our energy requirements, the excess is converted in the liver to either glycogen or fat. The glycogen is stored in the liver and muscles; the fat is stored in fat deposits in the abdomen, round the kidneys or under the skin (Figure 7.6).

The **cellulose** in the cell walls of all plant tissues is a carbohydrate. We probably derive relatively little nourishment from cellulose but it is important in the diet as **fibre**, which helps to maintain a healthy digestive system.

Fats

Animal fats are found in meat, milk, cheese, butter and egg-yolk. Plant fats occur as oils in fruits (e.g. palm oil) and seeds (e.g. sunflower seed oil), and are used for cooking and making margarine. Fats and oils are sometimes collectively called **lipids**.

Lipids are used in the cells of the body to form part of the cell membrane and other membrane systems. Lipids can also be oxidised in respiration, to carbon dioxide and water. When used to provide energy in this way, 1 g fat gives 37 kJ of energy. This is more than twice as much energy as can be obtained from the same weight of carbohydrate or protein.

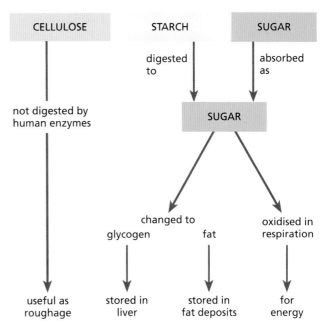

Figure 7.6 Digestion and use of carbohydrate

Fats can be stored in the body, so providing a means of long-term storage of energy in fat deposits. The fatty tissue, **adipose tissue**, under the skin forms a layer that, if its blood supply is restricted, can reduce heat losses from the body.

Proteins

Lean meat, fish, eggs, milk and cheese are important sources of animal protein. All plants contain some protein, but soybeans, seeds such as pumpkin, and nuts are the best sources (see Table 7.2).

Table 7.2 Comparing the protein content of foods (source: USDA database)

Food	Protein content/g per 100 g
soybeans	35
pumpkin seeds	30
beef, lean	27
peanuts	26
fish, e.g. salmon	25
cheese, e.g. cheddar	25
bacon	20
Tofu	18
beef sausage	18
chicken breast	17
Quorn sausage	14
eggs	13
wheat flour	13
yoghurt	4

7 HUMAN NUTRITION

Proteins, when digested, provide the chemical substances needed to build cells and tissues, e.g. skin, muscle, blood and bones. Neither carbohydrates nor fats can do this so it is essential to include some proteins in the diet.

Protein molecules consist of long chains of **amino acids** (see Chapter 4). When proteins are digested, the molecules are broken up into the constituent amino acids. The amino acids are absorbed into the bloodstream and used to build up different proteins. These proteins form part of the cytoplasm and enzymes of cells and tissues. Such a rearrangement of amino acids is shown in Figure 7.7.

The amino acids that are not used for making new tissues cannot be stored, but the liver removes their amino (—NH_2) groups and changes the residue to glycogen. The glycogen can be stored or oxidised to provide energy (Chapter 12). One gram of protein can provide 17 kJ of energy.

Chemically, proteins differ from both carbohydrates and fats because they contain nitrogen and sometimes sulfur as well as carbon, hydrogen and oxygen.

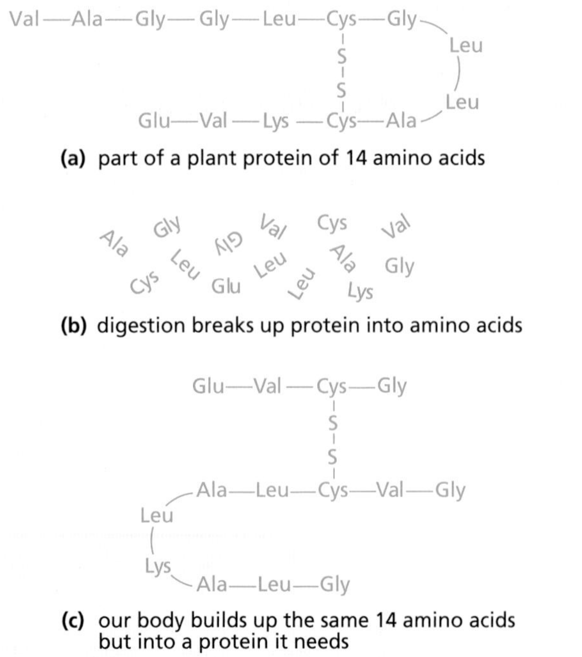

(a) part of a plant protein of 14 amino acids

(b) digestion breaks up protein into amino acids

(c) our body builds up the same 14 amino acids but into a protein it needs

key Ala = alanine, Gly = glycine, Leu = leucine
Cys = cysteine, Glu = glutamine, Lys = lysine,
Val = valine, S = sulfur atom

Figure 7.7 A model for digestion and use of a protein molecule

Table 7.3 Summary table for food classes

Nutrient	Good food sources	Use in the body
carbohydrate	rice, potato, yam, cassava, bread, millet, sugary foods (cake, jam, honey)	storage; source of energy
fat/oil (oils are liquid at room temperature, but fats are solid)	butter, milk, cheese, egg-yolk, animal fat, groundnuts (peanuts)	source of energy (twice as much as carbohydrate); insulation against heat loss; some hormones; cell membranes; insulation of nerve fibres
protein	meat, fish, eggs, soya, groundnuts, milk, Quorn, cowpeas	growth; tissue repair; enzymes; some hormones; cell membranes; hair; nails; can be broken down to provide energy

Vitamins

All proteins are similar to each other in their chemical structure, as are all carbohydrates. Vitamins, on the other hand, are a group of organic substances quite unrelated to each other in their chemical structure.

The features shared by all vitamins are:

- They are not digested or broken down for energy.
- Mostly, they are not built into the body structures.
- They are essential in small quantities for health.
- They are needed for chemical reactions in the cells, working in association with enzymes.

Plants can make these vitamins in their leaves, but animals have to obtain many of them ready-made either from plants or from other animals.

If any one of the vitamins is missing or deficient in the diet, a vitamin-deficiency disease may develop. Such a disease can be cured, at least in the early stages, simply by adding the vitamin to the diet.

Fifteen or more vitamins have been identified and they are sometimes grouped into two classes: water-soluble and fat-soluble. The fat-soluble vitamins are found mostly in animal fats or vegetable oils, which is one reason why our diet should include some of these fats. The water-soluble vitamins are present in green leaves, fruits and cereal grains.

See Table 7.4 for details of vitamins C and D.

Salts

These are sometimes called 'mineral salts' or just 'minerals'. Proteins, carbohydrates and fats provide the body with carbon, hydrogen, oxygen, nitrogen, sulfur and phosphorus but there are several more elements that the body needs and which occur as salts in the food we eat.

Iron

Red blood cells contain the pigment haemoglobin (see 'Blood' in Chapter 9). Part of the haemoglobin molecule contains iron and this plays an important role in carrying oxygen around the body. Millions of red cells break down each day and their iron is stored by the liver and used to make more haemoglobin. However, some iron is lost and needs to be replaced through dietary intake.

Red meat, especially liver and kidney, is the richest source of iron in the diet, but eggs, groundnuts, wholegrains such as brown rice, spinach and other green vegetables are also important sources.

If the diet is deficient in iron, a person may suffer from some form of **anaemia**. Insufficient haemoglobin is made and the oxygen-carrying capacity of the blood is reduced.

Calcium

Calcium, in the form of calcium phosphate, is deposited in the bones and the teeth and makes them hard. It is present in blood plasma and plays an essential part in normal blood clotting (see 'Blood' in Chapter 9). Calcium is also needed for the chemical changes that make muscles contract and for the transmission of nerve impulses.

The richest sources of calcium are milk (liquid, skimmed or dried) and cheese, but calcium is present in most foods in small quantities and also in 'hard' water.

Many calcium salts are not soluble in water and may pass through the alimentary canal without being absorbed. Simply increasing the calcium in the diet may not have much effect unless the calcium is in the right form, the diet is balanced and the intestine is healthy. Vitamin D and bile salts are needed for efficient absorption of calcium.

Dietary fibre (roughage)

When we eat vegetables and other fresh plant material, we take in a large quantity of plant cells. The cell walls of plants consist mainly of cellulose, but we do not have enzymes for digesting this substance. The result is that the plant cell walls reach the large intestine (colon) without being digested. This undigested part of the diet is called fibre or roughage. The colon contains many bacteria that can digest some of the substances in the plant cell walls to form fatty acids (Chapter 4). Vegetable fibre, therefore, may supply some useful food material, but it has other important functions.

The fibre itself and the bacteria, which multiply from feeding on it, add bulk to the contents of the colon and help it to retain water. This softens the faeces and reduces the time needed for the undigested residues to pass out of the body. Both effects help to prevent constipation and keep the colon healthy.

Most vegetables and whole cereal grains contain fibre, but white flour and white bread do not contain much. Good sources of dietary fibre are vegetables, fruit and wholemeal bread.

Water

About 70% of most tissue consists of water; it is an essential part of cytoplasm. The body fluids, blood, lymph and tissue fluid (Chapter 9) are composed mainly of water.

Digested food, salts and vitamins are carried around the body as a watery solution in the blood (Chapter 9) and excretory products such as excess salt and urea are removed from the body in solution by the kidneys (Chapter 13). Water thus acts as a solvent and as a transport medium for these substances.

Digestion is a process that uses water in a chemical reaction to break down insoluble substances to soluble ones. These products then pass, in solution, into the bloodstream. In all cells there are many reactions in which water plays an essential part as a reactant and a solvent.

Table 7.4 Vitamins

Name and source of vitamin	Importance of vitamin	Diseases and symptoms caused by lack of vitamin	Notes
vitamin C (ascorbic acid); water-soluble: oranges, lemons, grapefruit, tomatoes, fresh green vegetables, potatoes	prevents scurvy	Fibres in connective tissue of skin and blood vessels do not form properly, leading to bleeding under the skin, particularly at the joints, swollen, bleeding gums and poor healing of wounds. These are all symptoms of scurvy (Figure 7.8).	Possibly acts as a catalyst in cell respiration. Scurvy is only likely to occur when fresh food is not available. Cows' milk and milk powders contain little ascorbic acid so babies may need additional sources. Cannot be stored in the body; daily intake needed.
vitamin D (calciferol); fat-soluble: butter, milk, cheese, egg-yolk, liver, fish-liver oil	prevents rickets	Calcium is not deposited properly in the bones, causing **rickets** in young children. The bones remain soft and are deformed by the child's weight (Figure 7.9). Deficiency in adults causes **osteo-malacia**; fractures are likely.	Vitamin D helps the absorption of calcium from the intestine and the deposition of calcium salts in the bones. Natural fats in the skin are converted to a form of vitamin D by sunlight.

Since we lose water by evaporation, sweating, urinating and breathing, we have to make good this loss by taking in water with the diet.

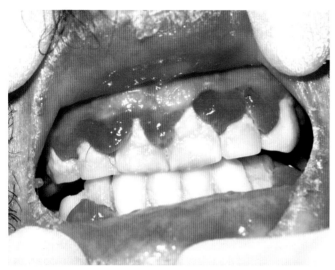

Figure 7.8 Symptoms of scurvy

Kwashiorkor
Kwashiorkor (roughly = 'deposed child') is an example of protein–energy malnutrition (PEM) in the developing world. When a mother has her second baby, the first baby is weaned on to a starchy diet of yam, cassava or sweet potato, all of which have inadequate protein. The first baby then develops symptoms of kwashiorkor (dry skin, pot-belly, changes to hair colour, weakness and irritability). Protein deficiency is not the only cause of kwashiorkor. Infection, plant toxins, digestive failure or even psychological effects may be involved. The good news, however, is that it can often be cured or prevented by an intake of protein in the form of dried skimmed milk.

Marasmus
The term 'marasmus' is derived from a Greek word, meaning decay. It is an acute form of malnutrition. The condition is due to a very poor diet with inadequate carbohydrate intake as well as a lack of protein. The incidence of marasmus increases in babies until they reach the age of 12 months. Sufferers are extremely emaciated with reduced fat and muscle tissue. Their skin is thin and hangs in folds. Marasmus is distinguished from kwashiorkor because kwashiorkor is due to lack of protein intake, while energy intake is adequate. Treatment involves provision of an energy-rich, balanced diet, but the complications of the disorder, which may include infections and dehydration, also need attention to increase chances of survival and recovery.

Causes and effects of mineral and vitamin deficiencies

Iron
Iron is present in red meat, eggs, nuts, brown rice, shellfish, soybean flour, dried fruit such as apricots, spinach and other dark-green leafy vegetables. Lack of iron in the diet can lead to iron-deficiency anaemia, which is a decrease in the number of red blood cells. Red blood cells, when mature, have no nucleus and this limits their life to about 3 months, after which they are broken down in the liver and replaced. Most of the iron is recycled, but some is lost as a chemical called bilirubin in the faeces and needs to be replaced. Adults need to take in about 15 mg each day. Without sufficient iron, your body is unable to produce enough haemoglobin, the protein in red blood cells responsible for transporting oxygen to respiring tissues. Iron is also needed by the muscles and for enzyme systems in all the body cells. The symptoms of anaemia are feeling weak, tired and irritable.

Vitamin D
Vitamin D is the only vitamin that the body can manufacture, when the skin is exposed to sunlight. However, for 6 months of the year (October to April), much of western Europe does not receive enough UV rays in sunlight to make vitamin D in the skin. So, many people living there are at risk of not getting enough vitamin D unless they get it in their diet. Also, people who have darker skin, such as people of African, African-Caribbean and South Asian origin, are at risk because their skin reduces UV light absorption.

Foods that provide vitamin D include oily fish such as sardines and mackerel, fish liver oil, butter, milk, cheese and egg-yolk. In addition, many manufactured food products contain vitamin D supplements.

Vitamin D helps in the absorption of calcium and phosphorus through the gut wall. Bone is made of the mineral calcium phosphate. A lack of the vitamin therefore results in poor calcium and phosphorus deposition in bones, leading to softening. The weight of the body can deform bones in the legs, causing the condition called rickets in children (Figure 7.9). Adults deficient in vitamin D can suffer from **osteo-malacia**; they are very vulnerable to fracturing bones if they fall.

Figure 7.9 A child with rickets

Practical work

Energy from food

- Set up the apparatus as shown in Figure 7.10.
- Use a measuring cylinder to place 20 cm³ cold water in the boiling tube.
- With a thermometer, find the temperature of the water and make a note of it.
- Weigh a peanut (or other piece of dried food), secure it onto a mounted needle and heat it with the Bunsen flame until it begins to burn. **Note**: make sure that no students have nut allergies.
- As soon as it starts burning, hold the nut under the boiling tube so that the flames heat the water.
- If the flame goes out, do not apply the Bunsen burner to the food while it is under the boiling tube, but return the nut to the Bunsen flame to start the nut burning again and replace it beneath the boiling tube as soon as the nut catches alight.
- When the nut has finished burning and cannot be ignited again, gently stir the water in the boiling tube with the thermometer and record its new temperature.

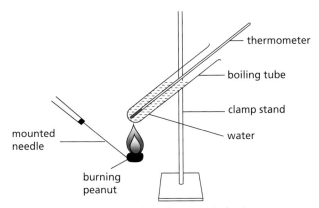

Figure 7.10 Experiment to show the energy in food

- Calculate the rise in temperature by subtracting the first from the second temperature.
- Work out the quantity of energy transferred to the water from the burning peanut as follows:

  4.2 J raise 1 g water by 1 °C
  20 cm³ cold water weighs 20 g
  The energy (in joules) released by the burning nut =
  rise in temperature × mass of water × 4.2

Note: The value 4.2 in the equation is used to convert the answer from calories to joules, as the calorie is an obsolete unit.

- To calculate the energy from 1 g of nut, divide your answer by the mass of nut you used. This gives a value in $J g^{-1}$.
- The experiment can now be repeated using different sizes of nut, or different varieties of nut, or other types of food. Remember to replace the warm water in the boiling tube with 20 cm³ cold water each time.
- The experiment is quite inaccurate: compare the value you obtained with an official value (2385 kJ per 100 g). There are plenty of websites with this sort of information if you use different nuts or other food. To make the comparison you may need to convert your energy value from joules to kilojoules (divide by 1000) and to 100 g of the food (multiply by 100).
- Try to list some of the faults in the design of the experiment to account for the difference you find. Where do you think some of the heat is going? Can you suggest ways of reducing this loss to make the results more accurate?

● Alimentary canal

Key definitions
Ingestion is the taking of substances such as food and drink into the body through the mouth.
Mechanical digestion is the breakdown of food into smaller pieces without chemical change to the food molecules.
Chemical digestion is the breakdown of large insoluble molecules into small soluble molecules.
Absorption is the movement of small food molecules and ions through the wall of the intestine into the blood.
Assimilation is the movement of digested food molecules into the cells of the body where they are used, becoming part of the cells.
Egestion is the passing out of food that has not been digested or absorbed, as faeces, through the anus.

Feeding involves taking food into the mouth, chewing it and swallowing it down into the stomach. This satisfies our hunger, but for food to be of any use to the whole body it has first to be **digested**. This means that the solid food is dissolved and the molecules reduced in size. The soluble products then have to be **absorbed** into the bloodstream and carried by the blood all around the body. In this way, the blood delivers dissolved food to the living cells in all parts of the body such as the muscles, brain, heart and kidneys. This section describes how the food is digested and absorbed. Chapter 9 describes how the blood carries it around the body.

Regions of the alimentary canal and their functions

The **alimentary canal** is a tube running through the body. Food is digested in the alimentary canal. The soluble products are absorbed and the indigestible residues expelled (egested). A simplified diagram of an alimentary canal is shown in Figure 7.11.

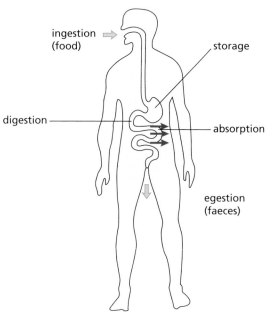

Figure 7.11 The alimentary canal (generalised)

The inside of the alimentary canal is lined with layers of cells forming what is called an **epithelium**. New cells in the epithelium are being produced all the time to replace the cells worn away by the movement of the food. There are also cells in the lining that produce **mucus**. Mucus is a slimy liquid that lubricates the lining of the canal and protects it from wear and tear. Mucus may also protect the lining from attack by the **digestive enzymes** which are released into the alimentary canal.

Some of the digestive enzymes are produced by cells in the lining of the alimentary canal, as in the stomach lining. Others are produced by **glands** that are outside the alimentary canal but pour their enzymes through tubes (called **ducts**) into the alimentary canal (Figure 7.12). The **salivary glands** and the **pancreas** (see Figure 7.13) are examples of such digestive glands.

The alimentary canal has a great many blood vessels in its walls, close to the lining. These bring oxygen needed by the cells and take away the carbon dioxide they produce. They also absorb the digested food from the alimentary canal.

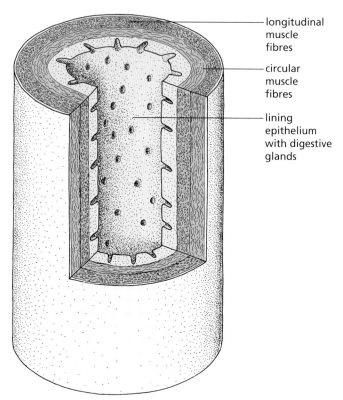

Figure 7.12 The general structure of the alimentary canal

Five main processes associated with digestion occur in the alimentary canal. These are ingestion, digestion, absorption, assimilation and egestion. The main parts of the alimentary canal are shown in Figure 7.13. An outline of the functions of its main parts is given in Table 7.5.

Peristalsis

The alimentary canal has layers of muscle in its walls (Figure 7.12). The fibres of one layer of muscles run around the canal (**circular muscle**) and the others run along its length (**longitudinal muscle**). When the circular muscles in one region contract, they make the alimentary canal narrow in that region.

A contraction in one region of the alimentary canal is followed by another contraction just below it so that a wave of contraction passes along the canal, pushing food in front of it. The wave of contraction, called **peristalsis**, is illustrated in Figure 7.14.

Table 7.5 Functions of main parts of the alimentary canal

Region of alimentary canal	Function
mouth	**ingestion** of food; **mechanical digestion** by teeth; **chemical digestion** of starch by amylase; formation of a bolus for swallowing
salivary glands	saliva contains amylase for **chemical digestion** of starch in food; also liquid to lubricate food and make small pieces stick together
oesophagus (gullet)	transfers food from the mouth to the stomach, by peristalsis
stomach	produces gastric juice containing pepsin, for **chemical digestion** of protein; also hydrochloric acid to kill bacteria; peristalsis churns food up into a liquid
duodenum	first part of the small intestine; receives pancreatic juice for **chemical digestion** of proteins, fats and starch as well as neutralising the acid from the stomach; receives bile to emulsify fats (a form of **physical digestion**)
ileum	second part of the small intestine; enzymes in the epithelial lining carry out **chemical digestion** of maltose and peptides; very long and has villi (see Figures 7.22 and 7.23) to increase surface area for **absorption** of digested food molecules
pancreas	secretes pancreatic juice into the duodenum via pancreatic duct (see Figure 7.21) for **chemical digestion** of proteins, fats and starch
liver	makes bile, containing salts to emulsify fats (**physical digestion**); **assimilation** of digested food such as glucose; **deamination** of excess amino acids (see Chapter 13)
gall bladder	stores bile, made in the liver, to be secreted into the duodenum via the bile duct (see Figure 7.21)
colon	first part of the large intestine; **absorption** of water from undigested food; **absorption** of bile salts to pass back to the liver
rectum	second part of the large intestine; stores faeces
anus	**egestion** of faeces

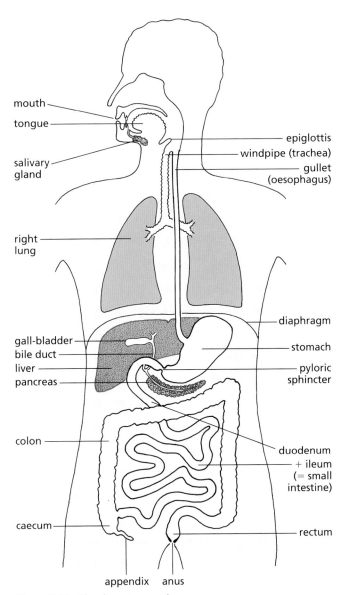

Figure 7.13 The alimentary canal

Diarrhoea

Diarrhoea is the loss of watery faeces. It is sometimes caused by bacterial or viral infection, for example from food or water. Once infected, the lining of the digestive system is damaged by the pathogens, resulting in the intestines being unable to absorb fluid from the contents of the colon or too much fluid being secreted into the colon. Undigested food then moves through the large intestine too quickly, resulting in insufficient time to absorb water from it. Unless the condition is treated, dehydration can occur.

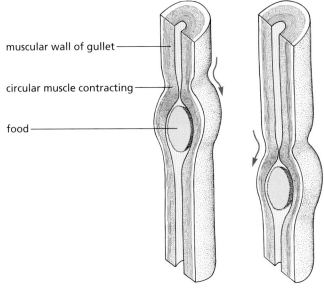

Figure 7.14 Diagram to illustrate peristalsis

Treatment is known as **oral hydration therapy**. This involves drinking plenty of fluids – sipping small amounts of water at a time to rehydrate the body.

Other possible causes of diarrhoea include anxiety, food allergies, lactose intolerance, a side-effect of antibiotics and bowel cancer.

Cholera

This disease is caused by the bacterium *Vibrio cholera* which causes acute diarrhoea. Treatment involves rehydration and restoration of the salts lost (administered by injecting a carefully controlled solution into the bloodstream) and use of an antibiotic such as tetracycline to kill the bacteria. The bacteria thrive in dirty water (often that contaminated by sewage) and are transmitted when the water is drunk or used to wash food. Long-term methods of control are to dispose of human sewage safely, ensuring that drinking water is free from bacteria and preventing food from being contaminated.

How cholera causes diarrhoea

When the *Vibrio cholera* bacteria are ingested, they multiply in the small intestine and invade its epithelial cells. As the bacteria become embedded, they release toxins (poisons) which irritate the intestinal lining and lead to the secretion of large amounts of water and salts, including chloride ions. The salts decrease the osmotic potential of the gut contents, drawing more water from surrounding tissues and blood by osmosis (see 'Osmosis' in Chapter 3). This makes the undigested food much more watery, leading to acute diarrhoea, and the loss of body fluids and salt leads to dehydration and kidney failure.

● Mechanical digestion

The process of mechanical digestion mainly occurs in the mouth by means of the teeth, through a process called **mastication**.

Humans are omnivores (organisms that eat animal and plant material). Broadly, we have the same types of teeth as carnivores, but human teeth are not used for catching, holding, killing or tearing up prey, and we cannot cope with bones. Thus, although we have incisors, canines, premolars and molars, they do not show such big variations in size and shape as, for example, a wolf's. Figure 7.15 shows the position of teeth in the upper jaw and Figure 7.16 shows how they appear in both jaws when seen from the side.

Table 7.6 gives a summary of the types of human teeth and their functions.

Our top incisors pass in front of our bottom incisors and cut pieces off the food, such as when biting into an apple or taking a bite out of a piece of toast.

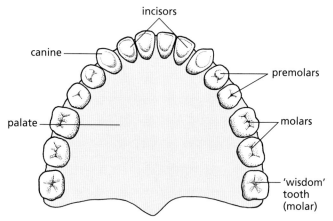

Figure 7.15 Teeth in human upper jaw

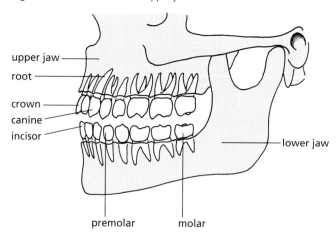

Figure 7.16 Human jaws and teeth

Our canines are more pointed than the incisors but are not much larger. They function like extra incisors.

Our premolars and molars are similar in shape and function. Their knobbly surfaces, called cusps, meet when the jaws are closed, and crush the food into small pieces. Small particles of food are easier to digest than large chunks.

Table 7.6 Summary of types of human teeth and their functions

Type	Incisor	Canine	Premolar	Molar
Diagram				
Position in mouth	front	either side of incisors	behind canines	back
Description	chisel-shaped (sharp edge)	slightly more pointed than incisors	have two points (cusps); have one or two roots	have four or five cusps; have two or three roots
Function	biting off pieces of food	similar function to incisors	tearing and grinding food	chewing and grinding food

Tooth structure

The part of a tooth that is visible above the gum line is called the **crown**. The **gum** is tissue that overlays the jaws. The rest, embedded in the jaw bone, is called the **root** (Figure 7.17). The surface of the crown is covered by a very hard layer of **enamel**. This layer is replaced by **cement** in the root, which enables the tooth to grip to its bony socket in the jaw. Below the enamel is a layer of **dentine**. Dentine is softer than enamel. Inside the dentine is a **pulp cavity**, containing nerves and blood vessels. These enter the tooth through a small hole at the base of the root.

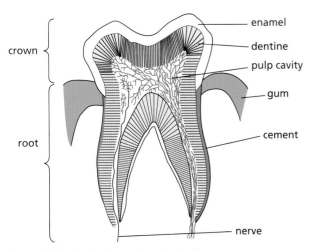

Figure 7.17 Section through a molar tooth

Dental decay (dental caries)

Decay begins when small holes (cavities) appear in the enamel. The cavities are caused by bacteria on the tooth surface. The bacteria feed on the sugars deposited on the teeth, respiring them and producing acid, which dissolves the calcium salts in the tooth enamel. The enamel is dissolved away in patches, exposing the dentine to the acids. Dentine is softer than enamel and dissolves more quickly so cavities are formed. The cavities reduce the distance between the outside of the tooth and the nerve endings. The acids produced by the bacteria irritate the nerve endings and cause toothache. If the cavity is not cleaned and filled by a dentist, the bacteria will get into the pulp cavity and cause a painful abscess at the root. Often, the only way to treat this is to have the tooth pulled out.

Although some people's teeth are more resistant to decay than others, it seems that it is the presence of refined sugar (sucrose) in the diet that contributes to decay.

Western diets contain a good deal of refined sugar and children suck sweets between one meal and the next. The high level of dental decay in Western society is thought to be caused mainly by keeping sugar in the mouth for long periods of time.

The graph in Figure 7.18(a) shows how the pH in the mouth falls (i.e. becomes more acid) when a single sweet is sucked. The pH below which the enamel is attacked is called the **critical pH** (between 5.5 and 6). In this case, the enamel is under acid attack for about 10 minutes.

The graph in Figure 7.18(b) shows the effect of sucking sweets at the rate of four an hour. In this case the teeth are exposed to acid attack almost continually.

The best way to prevent tooth decay, therefore, is to avoid eating sugar at frequent intervals either in the form of sweets or in sweet drinks such as orange squash or soft (fizzy) drinks.

It is advisable also to visit the dentist every 6 months or so for a 'check-up' so that any caries or gum disease can be treated at an early stage.

7 HUMAN NUTRITION

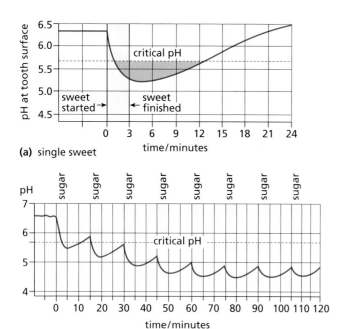

Figure 7.18 pH in the mouth when sweets are sucked

Brushing the teeth is very important in the prevention of gum disease. It may not be so effective in preventing caries, although the use of fluoride toothpaste does help to reduce the bacterial population on the teeth and to increase their resistance to decay (see below).

● Extension work

Gum disease (periodontal disease)

There is usually a layer of saliva and mucus over the teeth. This layer contains bacteria that live on the food residues in the mouth, building up a coating on the teeth called **plaque**. If the plaque is not removed, mineral salts of calcium and magnesium are deposited on it, forming a hard layer of 'tartar' or **calculus**. If the bacterial plaque that forms on teeth is not removed regularly, it spreads down the tooth into the narrow gap between the gum and enamel. Here it causes inflammation, called **gingivitis**, which leads to redness and bleeding of the gums and to bad breath. It also causes the gums to recede and expose the cement. If gingivitis is not treated, it progresses to **periodontitis**; the fibres holding the tooth in the jaw are destroyed, so the tooth becomes loose and falls out or has to be pulled out.

There is evidence that cleaning the teeth does help to prevent gum disease. It is best to clean the teeth about twice a day using a toothbrush. No one method of cleaning has proved to be any better than any other, but the cleaning should attempt to remove all the plaque from the narrow crevice between the gums and the teeth. Rinsing the mouth regularly with mouthwashes helps reduce the number of bacteria residing in the mouth.

Drawing a waxed thread ('dental floss') between the teeth, or using interdental brushes, helps to remove plaque in these regions.

● Chemical digestion

Digestion is mainly a chemical process and consists of breaking down large molecules to small molecules. The large molecules are usually not soluble in water, while the smaller ones are. The small molecules can be absorbed through the epithelium of the alimentary canal, through the walls of the blood vessels and into the blood.

Some food can be absorbed without digestion. The **glucose** in fruit juice, for example, could pass through the walls of the alimentary canal and enter the blood vessels without further change. Most food, however, is solid and cannot get into blood vessels. Digestion is the process by which solid food is dissolved to make a solution.

The chemicals that dissolve the food are **enzymes**, described in Chapter 5. A protein might take 50 years to dissolve if just placed in water but is completely digested by enzymes in a few hours. All the solid starch in foods such as bread and potatoes is digested to glucose, which is soluble in water. The solid proteins in meat, eggs and beans are digested to soluble substances called amino acids. Fats are digested to two soluble products called **glycerol** and **fatty acids** (see Chapter 4).

The chemical breakdown usually takes place in stages. For example, the starch molecule is made up of hundreds of carbon, hydrogen and oxygen atoms. The first stage of digestion breaks it down to a 12-carbon sugar molecule called **maltose**. The last stage of digestion breaks the maltose molecule into two 6-carbon sugar molecules called glucose (Figure 7.19). Protein molecules are digested first to smaller molecules called **peptides** and finally into completely soluble molecules called amino acids.

starch → maltose → glucose
protein → peptide → amino acid

These stages take place in different parts of the alimentary canal. The progress of food through the canal and the stages of digestion will now be described.

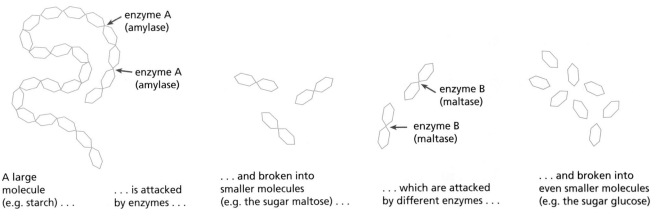

Figure 7.19 Enzymes acting on starch

The mouth

The act of taking food into the mouth is called **ingestion**. In the mouth, the food is chewed and mixed with **saliva**. The chewing breaks the food into pieces that can be swallowed and it also increases the surface area for the enzymes to work on later. Saliva is a digestive juice produced by three pairs of glands whose ducts lead into the mouth. It helps to lubricate the food and make the small pieces stick together. Saliva contains one enzyme, **salivary amylase** (sometimes called **ptyalin**), which acts on cooked starch and begins to break it down into maltose.

Strictly speaking, the 'mouth' is the aperture between the lips. The space inside, containing the tongue and teeth, is called the **buccal cavity**. Beyond the buccal cavity is the 'throat' or **pharynx**.

Swallowing

For food to enter the **gullet** (oesophagus), it has to pass over the windpipe. To ensure that food does not enter the windpipe and cause choking during swallowing, the **epiglottis** (a flap of cartilage) guides the food into the gullet.

The beginning of the swallowing action is voluntary, but once the food reaches the back of the mouth, swallowing becomes an automatic or reflex action. The food is forced into and down the gullet by peristalsis. This takes about 6 seconds with relatively solid food; the food is then admitted to the stomach. Liquid travels more rapidly down the gullet.

The stomach

The stomach has elastic walls, which stretch as the food collects in it. The **pyloric sphincter** is a circular band of muscle at the lower end of the stomach that stops solid pieces of food from passing through. The main function of the stomach is to store the food from a meal, turn it into a liquid and release it in small quantities at a time to the rest of the alimentary canal. An example of physical digestion is the peristaltic action of muscles in the wall of the stomach. These muscles alternately contract and relax, churning and squeezing the food in the stomach and mixing it with gastric juice, turning the mixture into a creamy liquid called **chyme**. This action gives the food a greater surface area so that it can be digested more efficiently.

Glands in the lining of the stomach (Figure 7.20) produce **gastric juice** containing the **protease** enzyme. It helps in the process of breaking down large protein molecules into small, soluble amino acids. The stomach lining also produces hydrochloric acid, which makes a weak solution in the gastric juice. This acid provides the best degree of acidity for stomach protease to work in (Chapter 4) and kills many of the bacteria taken in with the food.

The regular, peristaltic movements of the stomach, about once every 20 seconds, mix up the food and gastric juice into a creamy liquid. How long food remains in the stomach depends on its nature. Water may pass through in a few minutes; a meal of carbohydrate such as porridge may be held in the stomach for less than an hour, but a mixed meal containing protein and fat may be in the stomach for 1 or 2 hours.

The pyloric sphincter lets the liquid products of digestion pass, a little at a time, into the first part of the small intestine called the **duodenum**.

7 HUMAN NUTRITION

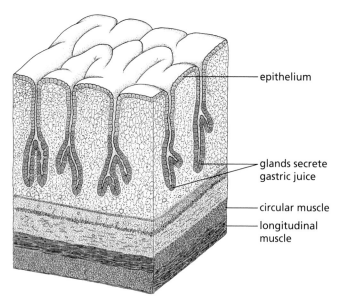

Figure 7.20 Diagram of section through stomach wall

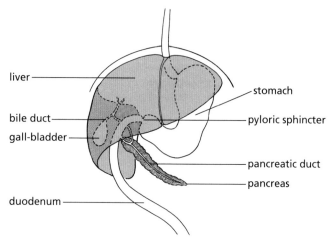

Figure 7.21 Relationship between stomach, liver and pancreas

The small intestine

A digestive juice from the pancreas (**pancreatic juice**) and bile from the liver are poured into the duodenum to act on food there. The pancreas is a digestive gland lying below the stomach (Figure 7.21). It makes a number of enzymes, which act on all classes of food. **Protease** breaks down proteins into amino acids. **Pancreatic amylase** attacks starch and converts it to maltose. **Lipase** digests fats (lipids) to fatty acids and glycerol.

Pancreatic juice contains sodium hydrogencarbonate, which partly neutralises the acidic liquid from the stomach. This is necessary because the enzymes of the pancreas do not work well in acid conditions.

All the digestible material is thus changed to soluble compounds, which can pass through the lining of the intestine and into the bloodstream. The final products of digestion are:

Food	Final products
starch →	glucose (a simple sugar)
proteins →	amino acids
fats (lipids) →	fatty acids and glycerol

Bile

Bile is a green, watery fluid made in the liver, stored in the gall-bladder and delivered to the duodenum by the bile duct (Figure 7.21). It contains no enzymes, but its green colour is caused by bile pigments, which are formed from the breakdown of haemoglobin in the liver. Bile also contains bile salts, which act on fats rather like a detergent. The bile salts **emulsify** the fats. That is, they break them up into small droplets with a large surface area, which are more efficiently digested by lipase.

Bile is slightly alkaline as it contains sodium hydrogencarbonate and, along with pancreatic juice, has the function of neutralising the acidic mixture of food and gastric juices as it enters the duodenum. This is important because enzymes secreted into the duodenum need alkaline conditions to work at their optimum rate.

Digestion of protein

There are actually several proteases (or proteinases) which break down proteins. One protease is **pepsin** and is secreted in the stomach. Pepsin acts on proteins and breaks them down into soluble compounds called peptides. These are shorter chains of amino acids than proteins. Another protease is called **trypsin**. Trypsin is secreted by the pancreas in an inactive form, which is changed to an active enzyme in the duodenum. It has a similar role to pepsin, breaking down proteins to peptides.

The small intestine itself does not appear to produce digestive enzymes. The structure labelled 'crypt' in Figure 7.23 is not a digestive gland, though some of its cells do produce mucus and other secretions. The main function of the crypts is to produce new epithelial cells (see 'Absorption') to replace those lost from the tips of the villi.

The epithelial cells of the villi contain enzymes in their cell membranes that complete the breakdown of sugars and peptides, before they pass through the cells on their way to the bloodstream. For example, **peptidase** breaks down polypeptides and peptides into amino acids.

Digestion of starch

Starch is digested in two places in the alimentary canal: by salivary amylase in the mouth and by pancreatic amylase in the duodenum. Amylase works best in a neutral or slightly alkaline pH and converts large, insoluble starch molecules into smaller, soluble maltose molecules. Maltose is a disaccharide sugar and is still too big to be absorbed through the wall of the intestine. Maltose is broken down to glucose by the enzyme **maltase**, which is present in the membranes of the epithelial cells of the villi.

Functions of hydrochloric acid in gastric juice

The hydrochloric acid, secreted by cells in the wall of the stomach, creates a very acid pH of 2. This pH is important because it denatures enzymes in harmful organisms in food, such as bacteria (which may otherwise cause food poisoning) and it provides the optimum pH for the protein-digesting enzyme pepsin to work.

Table 7.7 Principal substances produced by digestion

Region of alimentary canal	Digestive gland	Digestive juice produced	Enzymes in the juice/cells	Class of food acted upon	Substances produced
mouth	salivary glands	saliva	salivary amylase	starch	maltose
stomach	glands in stomach lining	gastric juice	pepsin	proteins	peptides
duodenum	pancreas	pancreatic juice	proteases, such as trypsin amylase lipase	proteins and peptides starch fats	peptides and amino acids maltose fatty acids and glycerol
ileum	epithelial cells	(none)	maltase peptidase	maltose peptides	glucose amino acids

(Note: details of peptidase and peptides are **not** a syllabus requirement)

● Extension work

Prevention of self-digestion

The gland cells of the stomach and pancreas make protein-digesting enzymes (proteases) and yet the proteins of the cells that make these enzymes are not digested. One reason for this is that the proteases are secreted in an inactive form. Pepsin is produced as **pepsinogen** and does not become the active enzyme until it encounters the hydrochloric acid in the stomach. The lining of the stomach is protected from the action of pepsin probably by the layer of mucus.

Similarly, trypsin, one of the proteases from the pancreas, is secreted as the inactive **trypsinogen** and is activated by **enterokinase**, an enzyme secreted by the lining of the duodenum.

● Absorption

The small intestine consists of the duodenum and the **ileum**. Nearly all the absorption of digested food takes place in the ileum, along with most of the water. Small molecules of the digested food such as glucose and amino acids pass into the bloodstream, while fatty acids and glycerol pass into the **lacteals** (Figure 7.23) connected to the **lymphatic system**.

The large intestine (colon and rectum)

The material passing into the large intestine consists of water with undigested matter, largely cellulose and vegetable fibres (roughage), mucus and dead cells from the lining of the alimentary canal. The large intestine secretes no enzymes but the bacteria in the colon digest part of the fibre to form fatty acids, which the colon can absorb. Bile salts are absorbed and returned to the liver by the blood circulation. The colon also absorbs much of the water from the undigested residues. About 7 litres of digestive juices are poured into the alimentary canal each day. If the water from these was not absorbed by the ileum and colon, the body would soon become dehydrated.

The semi-solid waste, the **faeces** or 'stool', is passed into the rectum by peristalsis and is expelled at intervals through the anus. The residues may spend from 12 to 24 hours in the intestine. The act of expelling the faeces is called **egestion** or **defecation**.

7 HUMAN NUTRITION

The ileum is efficient in the absorption of digested food for the following reasons:

- It is fairly long and presents a large absorbing surface to the digested food.
- Its internal surface is greatly increased by circular folds (Figure 7.22) bearing thousands of tiny projections called **villi** (singular = villus) (Figures 7.23 and 7.24). These villi are about 0.5 mm long and may be finger-like or flattened in shape.
- The lining epithelium is very thin and the fluids can pass rapidly through it. The outer membrane of each epithelial cell has **microvilli**, which increase by 20 times the exposed surface of the cell.
- There is a dense network of blood capillaries (tiny blood vessels, see 'Blood and lymphatic vessels' in Chapter 9) in each villus (Figure 7.22).

The small molecules of digested food, for example glucose and amino acids, pass into the epithelial cells and then through the wall of the capillaries in the villus and into the bloodstream. They are then carried away in the capillaries, which join up to form veins. These veins unite to form one large vein, the hepatic portal vein (see Chapter 9). This vein carries all the blood from the intestines to the liver, which may store or alter any of the digestion products. When these products are released from the liver, they enter the general blood circulation.

Some of the fatty acids and glycerol from the digestion of fats enter the blood capillaries of the villi. However, a large proportion of the fatty acids and glycerol may be combined to form fats again in the intestinal epithelium. These fats then pass into the lacteals (Figure 7.23). The fluid in the lacteals flows into the lymphatic system, which forms a network all over the body and eventually empties its contents into the bloodstream (see 'Blood and lymphatic vessels' in Chapter 9).

Water-soluble vitamins may diffuse into the epithelium but fat-soluble vitamins are carried in the microscopic fat droplets that enter the cells. The ions of mineral salts are probably absorbed by active transport. Calcium ions need vitamin D for their effective absorption.

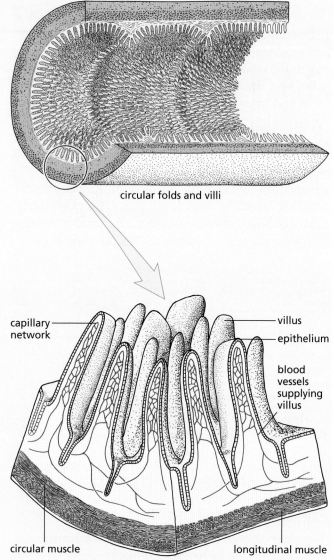

Figure 7.22 The absorbing surface of the ileum

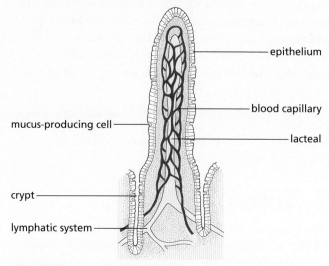

Figure 7.23 Structure of a single villus

Absorption

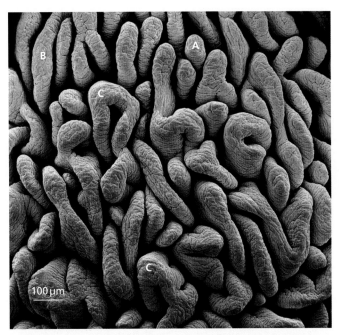

Figure 7.24 Scanning electron micrograph of the human intestinal lining (×60). The villi are about 0.5 mm long. In the duodenum they are mostly leaf-like (C), but further towards the ileum they become narrower (B), and in the ileum they are mostly finger-like (A). This micrograph is of a region in the duodenum.

Absorption of the products of digestion and other dietary items is not just a matter of simple diffusion, except perhaps for alcohol and, sometimes, water. Although the mechanisms for transport across the intestinal epithelium have not been fully worked out, it seems likely that various forms of active transport are involved. Even water can cross the epithelium against an osmotic gradient (Chapter 3). Amino acids, sugars and salts are, almost certainly, taken up by active transport. Glucose, for example, crosses the epithelium faster than fructose (another monosaccharide sugar) although their rates of diffusion would be about the same.

The epithelial cells of the villi are constantly being shed into the intestine. Rapid cell division in the epithelium of the crypts (Figure 7.23) replaces these lost cells. In effect there is a steady procession of epithelial cells moving up from the crypts to the villi.

Use of digested food

The products of digestion are carried around the body in the blood. From the blood, cells absorb and use glucose, fats and amino acids. This uptake and use of food is called **assimilation**.

Glucose

During respiration in the cells, glucose is oxidised to carbon dioxide and water (see 'Aerobic respiration' in Chapter 12). This reaction provides energy to drive the many chemical processes in the cells, which result in, for example, the building-up of proteins, contraction of muscles or electrical changes in nerves.

Fats

These are built into cell membranes and other cell structures. Fats also form an important source of energy for cell metabolism. Fatty acids produced from stored fats or taken in with the food, are oxidised in the cells to carbon dioxide and water. This releases energy for processes such as muscle contraction. Fats can provide twice as much energy as sugars.

Amino acids

These are absorbed by the cells and built up, with the aid of enzymes, into proteins. Some of the proteins will become plasma proteins in the blood (see 'Blood' in Chapter 9). Others may form structures such as cell membranes or they may become enzymes that control the chemical activity within the cell. Amino acids not needed for making cell proteins are converted by the liver into glycogen, which can then be used for energy.

Practical work

Experiments on digestion

1 The action of salivary amylase on starch

- Rinse the mouth with water to remove traces of food.
- Collect saliva* in two test-tubes, labelled A and B, to a depth of about 15 mm (see Figure 7.25).
- Heat the saliva in tube B over a small flame, or in a water bath of boiling water, until it boils for about 30 seconds and then cool the tube under the tap.
- Add about 2 cm³ of a 2% starch solution to each tube; shake each tube and leave them for 5 minutes.
- Share the contents of tube A between two clean test-tubes.
- To one of these add some iodine solution. To the other add some Benedict's solution and heat in a water bath as described in Chapter 4.
- Test the contents of tube B in exactly the same way.

*If there is some objection to using your own saliva, use a 5 per cent solution of commercially prepared amylase instead.

7 HUMAN NUTRITION

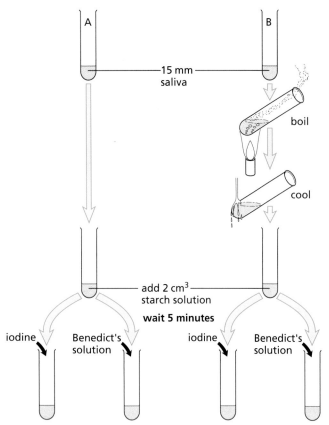

Figure 7.25 Experiment to show the action of salivary amylase on starch

Results
The contents of tube A fail to give a blue colour with iodine, showing that the starch has gone. The other half of the contents, however, gives a red or orange precipitate with Benedict's solution, showing that sugar is present.

The contents of tube B still give a blue colour with iodine but do not form a red precipitate on heating with Benedict's solution.

Interpretation
The results with tube A suggest that something in saliva has converted starch into sugar. The fact that the boiled saliva in tube B fails to do this suggests that it was an enzyme in saliva that brought about the change (see Chapter 5), because enzymes are proteins and are destroyed by boiling. If the boiled saliva had changed starch to sugar, it would have ruled out the possibility of an enzyme being responsible.

This interpretation assumes that it is something in saliva that changes starch into sugar. However, the results could equally well support the claim that starch can turn unboiled saliva into sugar. Our knowledge of (1) the chemical composition of starch and saliva and (2) the effect of heat on enzymes, makes the first interpretation more plausible.

2 Modelling the action of amylase on starch

- Collect a 15 cm length of Visking tubing which has been softened in water.
- Tie one end tightly. Use a syringe to introduce 2% starch solution into the Visking tubing, to about two thirds full.
- Add 2 cm³ of 5% amylase solution (or saliva if it is permissible).
- Pinch the top of the Visking tubing to keep it closed, before carefully mixing its contents by squeezing the tubing.
- Rinse the outside of the Visking tubing thoroughly with tap water, then place it in a boiling tube, trapping the top of the tubing with an elastic band (see Figure 7.26).
- Add enough distilled water to cover the Visking tubing.
- Test a small sample of the distilled water and the contents of the Visking tubing for starch and reducing sugar, using iodine solution and Benedict's solution (see page 58 for methods).
- Place the boiling tube in a beaker of water or a water bath at 37 °C.
- After 20 minutes, use clean teat pipettes to remove a sample of the water surrounding the Visking tubing and from inside the Visking tubing.
- Test some of each sample for starch, using iodine solution, and for reducing sugar, using Benedict's solution (see Chapter 4 for methods). Also test some of the original starch solution for reducing sugar, to make sure it is not contaminated with glucose.

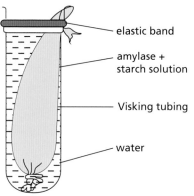

Figure 7.26 Experiment to model the digestion of starch

Result
At the start of the investigation the distilled water tests negative for starch (stays brown) and reducing sugar (stays turquoise). The contents of the Visking tubing are positive for starch (blue-black), but negative for reducing sugars (stays turquoise).

After 20 minutes, the contents of the Visking tubing are yellow/brown with iodine solution, but turn orange or brick red with Benedict's solution. The water sample stays yellow/brown with iodine solution, but turns orange or brick red with Benedict's solution.

Interpretation
The amylase digests the starch in the Visking tubing, producing reducing sugar. The complete digestion of starch results in a negative colour change with iodine solution. The presence of reducing sugar (maltose or glucose) causes the Benedict's solution to turn orange or brick red. The reducing sugar molecules can diffuse through the Visking tubing into the surrounding water, so the water gives a positive result with Benedict's solution. Starch is a large molecule, so it cannot diffuse through the tubing: the water gives a negative result with iodine solution.

This model can be used to represent digestion in the gut. The starch solution and amylase are the contents of the mouth or duodenum. The Visking tubing represents the duodenum wall and the distilled water represents the bloodstream, into which the products of digestion are absorbed.

3 The action of pepsin on egg-white protein

A cloudy suspension of egg-white is prepared by stirring the white of one egg into 500 cm³ tap water, heating it to boiling point and filtering it through glass wool to remove the larger particles.

- Label four test-tubes A, B, C and D and place 2 cm³ egg-white suspension in each of them. Then add pepsin solution and/or dilute hydrochloric acid (HCl) to the tubes as follows (Figure 7.27):

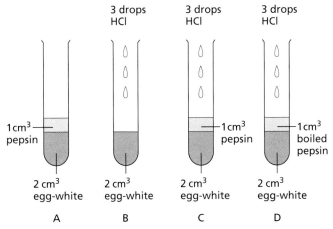

Figure 7.27 Experiment to show the action of pepsin on egg-white

- A egg-white suspension + 1 cm³ pepsin solution (1%)
- B egg-white suspension + 3 drops dilute HCl
- C egg-white suspension + 1 cm³ pepsin + 3 drops HCl
- D egg-white suspension + 1 cm³ boiled pepsin + 3 drops HCl

- Place all four tubes in a beaker of warm water at 35 °C for 10–15 minutes.

Result
The contents of tube C go clear. The rest remain cloudy.

Interpretation
The change from a cloudy suspension to a clear solution shows that the solid particles of egg protein have been digested to soluble products. The failure of the other three tubes to give clear solutions shows that:

- pepsin will only work in acid solutions
- it is the pepsin and not the hydrochloric acid that does the digestion
- pepsin is an enzyme, because its activity is destroyed by boiling.

4 The action of lipase

- Place 5 cm³ milk and 7 cm³ dilute (0.05 mol dm⁻³) sodium carbonate solution into each of three test-tubes labelled 1 to 3 (Figure 7.28).

- Add six drops of phenolphthalein to each to turn the contents pink.
- Add 1 cm³ of 3% bile salts solution to tubes 2 and 3.
- Add 1 cm³ of 5% lipase solution to tubes 1 and 3, and an equal volume of boiled lipase to tube 2.

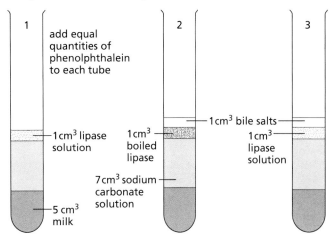

Figure 7.28 Experiment to show the action of lipase

Result
In 10 minutes or less, the colour of the liquids in tubes 1 and 3 will change to white, with tube 3 changing first. The liquid in tube 2 will remain pink.

Interpretation
Lipase is an enzyme that digests fats to fatty acids and glycerol. When lipase acts on milk fats, the fatty acids that have been produced react with the alkaline sodium carbonate and make the solution more acid. In acid conditions the pH indicator, phenolphthalein, changes from pink to colourless. The presence of bile salts in tube 3 seems to speed up the reaction, although bile salts with the denatured enzyme in tube 2 cannot bring about the change on their own.

For experiments investigating the effect of temperature and pH on enzyme action see Chapter 5.

Questions

1. In Experiment 2, why does some reducing sugar remain inside the Visking tubing?
2. In Experiment 3, why does the change from cloudy to clear suggest that digestion has occurred?
3. How would you modify Experiment 3 if you wanted to find the optimum temperature for the action of pepsin on egg-white?
4. Experiment 3 is really two experiments combined because there are two variables.
   a Identify the variables.
   b Which of the tubes could be the control?
5. It was suggested that an alternative interpretation of the result in Experiment 1 might be that starch has turned saliva into sugar. From what you know about starch, saliva and the design of the experiment, explain why this is a less acceptable interpretation.

7 HUMAN NUTRITION

Questions

Core

1. What sources of protein-rich foods are available to a vegetarian who:
   a. will eat animal products but not meat itself
   b. will eat only plants and their products?
2. Why must all diets contain some protein?
3. Could you survive on a diet that contained no carbohydrate? Justify your answer.
4. In what sense can the fats in your diet be said to contribute to 'keeping you warm'?
5. How do proteins differ from fats (lipids) in:
   a. their chemical composition (Chapter 4)
   b. their energy value
   c. their role in the body?
6. Construct a flowchart for the digestion and use of proteins, similar to the one for carbohydrates in Figure 7.6.
7. Which tissues of the body need:
   a. iron
   b. glucose
   c. calcium
   d. protein?
8. Some examples of the food that would give a balanced diet are shown in Figure 7.29. Consider the picture and say what class of food or item of diet is mainly present. For example, the meat is mainly protein but will also contain some iron.

Figure 7.29 Examples of types of food in a balanced diet (see question 8)

9. What is the value of leafy vegetables, such as cabbage and lettuce, in the diet?
10. Why is a diet consisting mainly of one type of food, e.g. rice or potatoes, likely to be unsatisfactory even if it is sufficient to meet our energy needs?
11. A zoologist is trying to find out whether rabbits need vitamin C in their diet. Assuming that a sufficiently large number of rabbits is used and adequate controls are applied, the best design of experiment would be to give the rabbits:
    a. an artificial diet of pure protein, carbohydrate, fats, minerals and vitamins but lacking vitamin C
    b. an artificial diet as above but with extra vitamin C
    c. a natural diet of grass, carrots, etc. but with added vitamin C
    d. natural food but of one kind only, e.g. exclusively grass or exclusively carrots?

    Justify your choice and say why you excluded the other alternatives.
12. Name three functions of the alimentary canal shown in Figure 7.11.
13. Into what parts of the alimentary canal do the following pour their digestive juices?
    a. the pancreas
    b. the salivary glands
14. Starting from the inside, name the layers of tissue that make up the alimentary canal.
15. a. Why is it necessary for our food to be digested?
    b. Why do plants not need a digestive system? (See 'Photosynthesis' in Chapter 6.)
16. In which parts of the alimentary canal are the following digested?
    a. starch
    b. protein
17. Study the characteristics of enzymes in Chapter 5. In what ways does pepsin show the characteristics of an enzyme?
18. In experiments with enzymes, the control often involves the boiled enzyme. Suggest why this type of control is used.
19. a. What process in the body enables the *majority* of the reducing sugar in the ileum to be absorbed by the bloodstream?
    b. What is needed to achieve this process?
20. Write down the menu for your breakfast and lunch (or supper). State the main food substances present in each item of the meal. State the final digestion product of each.

Extended

21. What are the products of digestion of the following, which are absorbed by the ileum?
    a. starch
    b. protein
    c. fats
22. What characteristics of the small intestine enable it to absorb digested food efficiently?
23. State briefly what happens to a protein molecule in food, from the time it is swallowed, to the time its products are built up into the cytoplasm of a muscle cell.
24. List the chemical changes that a starch molecule undergoes from the time it reaches the duodenum to the time its carbon atoms become part of carbon dioxide molecules. Say where in the body these changes occur.

Checklist

After studying Chapter 7 you should know and understand the following:

- A balanced diet must contain proteins, carbohydrates, fats, minerals, vitamins, fibre and water, in the correct proportions. Dietary needs are affected by the age, gender and activity of humans.
- Growing children and pregnant women have special dietary needs.
- Malnutrition is the result of taking in food that does not match the energy needs of the body, or is lacking in proteins, vitamins or minerals.
- The effects of malnutrition include starvation, coronary heart disease, constipation and scurvy.
- Western diets often contain too much sugar and fat and too little fibre.
- Obesity results from taking in more food than the body needs for energy, growth or replacement.
- Examples of good food sources for the components of a balanced diet.
- Fats, carbohydrates and proteins provide energy.
- Proteins provide amino acids for the growth and replacement of the tissues.
- Mineral salts like calcium and iron are needed in tissues such as bone and blood.
- Vegetable fibre helps to maintain a healthy intestine.
- Vitamins are essential in small quantities for chemical reactions in cells.
- Shortage of vitamin C causes scurvy; inadequate vitamin D causes rickets.
- Mechanical digestion breaks down food into smaller pieces, without any chemical change of the food molecules. This process involves teeth, which can become decayed if not cared for properly.
- Chemical digestion is the process that changes large, insoluble food molecules into small, soluble molecules.
- Digestion takes place in the alimentary canal.
- The changes are brought about by chemicals called digestive enzymes.
- The stomach produces gastric juice, which contains hydrochloric acid as well as pepsin.
- The ileum absorbs amino acids, glucose and fats.
- These are carried in the bloodstream first to the liver and then to all parts of the body.
- The small intestine and the colon both absorb water.
- Undigested food is egested through the anus as faeces.
- Diarrhoea is the loss of watery faeces.
- Cholera is a disease caused by a bacterium.

- Malnutrition includes kwashiorkor and marasmus.
- Cholera bacteria produce a toxin that affects osmosis in the gut.
- Internal folds, villi and microvilli greatly increase the absorbing surface of the small intestine.
- The villi have a special structure to enable efficient absorption of digested food.

8 Transport in plants

> **Transport in plants**
> Structure and function of xylem and phloem
>
> **Water uptake**
> Pathway taken by water into and through the plant
>
> Root hairs and surface area, linked to osmosis and active transport
>
> **Transpiration**
> Transport of water through the plant
> Loss by evaporation through plant leaves
> Causes of changes in transpiration rate
>
> Explanation of the mechanism of water uptake and movement
> Wilting
>
> **Translocation**
> (no details needed for the Core syllabus)
>
> Structure and function of phloem
> Pathway taken by sucrose and amino acids from sources to sinks

● Extension work

Before looking in detail at leaf, stem and root structure, it is useful to consider the relationship between these parts and the whole plant.

A young sycamore plant is shown in Figure 8.1. It is typical of many flowering plants in having a **root system** below the ground and a **shoot system** above ground. The shoot consists of an upright stem, with leaves and buds. The buds on the side of the stem are called **lateral buds**. When they grow, they will produce branches. The bud at the tip of the shoot is the **terminal bud** and when it grows, it will continue the upward growth of the stem. The lateral buds and the terminal buds may also produce flowers.

The region of stem from which leaves and buds arise is called a **node**. The region of stem between two nodes is the **internode**.

The **leaves** make food by photosynthesis (Chapter 6) and pass it back to the stem.

The **stem** carries this food to all parts of the plant that need it and also carries water and dissolved salts from the roots to the leaves and flowers.

In addition, the stem supports and spaces out the leaves so that they can receive sunlight and absorb carbon dioxide, which they need for photosynthesis.

An upright stem also holds the flowers above the ground, helping the pollination by insects or the wind (see 'Sexual reproduction in plants' in Chapter 16). A tall stem may help in seed dispersal later on.

The **roots** anchor the plant in the soil and prevent it from falling over or being blown over by the wind. They also absorb the water and salts that the plant needs for making food in the leaves. A third function is sometimes the storage of food made by the leaves.

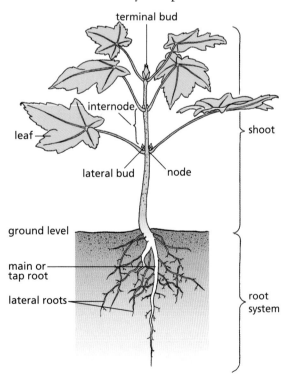

Figure 8.1 Structure of a typical flowering plant

● Transport in plants

Plant structure and function

Leaf

The structure of a leaf has already been described in Chapter 6. Xylem and phloem appear in the midrib of the leaf, as well as in the leaf veins. These features are identified in Chapter 6, Figures 6.18 and 6.19.

Stem

Figure 8.2 shows a stem cut across (transversely) and down its length (longitudinally) to show its internal structure.

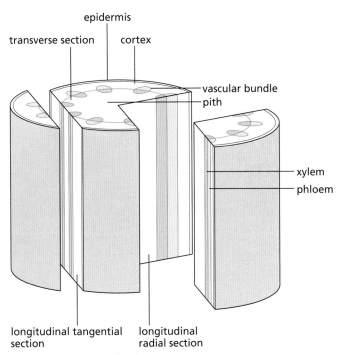

Figure 8.2 Structure of a plant stem

these cells are digested away to form a continuous, fine tube (Figure 8.4(c)). At the same time, the cell walls are thickened and impregnated with a substance called **lignin**, which makes the cell wall very strong and impermeable. Since these lignified cell walls prevent the free passage of water and nutrients, the cytoplasm dies. This does not affect the passage of water in the vessels. Xylem also contains many elongated, lignified supporting cells called **fibres**.

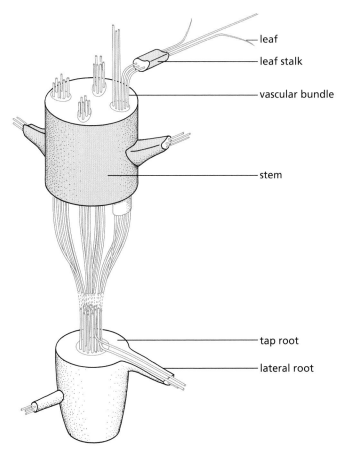

Figure 8.3 Distribution of veins from root to leaf

Epidermis
Like the leaf epidermis, this is a single layer of cells that helps to keep the shape of the stem and cuts down the loss of water vapour. Stomata in the epidermis allow the tissues inside to take up oxygen and get rid of carbon dioxide. In woody stems, the epidermis is replaced by bark, which consists of many layers of dead cells.

Vascular bundles
These are made up of groups of specialised cells that conduct water, dissolved salts and food up or down the stem. The vascular bundles in the roots, stem, leaf stalks and leaf veins all connect up to form a transport system throughout the entire plant (Figure 8.3). The two main tissues in the vascular bundles are called **xylem** and **phloem** (Figure 8.4). Food substances travel in the phloem; water and salts travel mainly in the xylem. The cells in each tissue form elongated tubes called **vessels** (in the xylem) or **sieve tubes** (in the phloem) and they are surrounded and supported by other cells.

Vessels
The cells in the xylem that carry water become vessels. A vessel is made up of a series of long cells joined end to end (Figure 8.5(a)). Once a region of the plant has ceased growing, the end walls of

Sieve tubes
The conducting cells in the phloem remain alive and form sieve tubes. Like vessels, they are formed by vertical columns of cells (Figure 8.5(b)). Perforations appear in the end walls, allowing substances to pass from cell to cell, but the cell walls are not lignified and the cell contents do not die, although they do lose their nuclei. The perforated end walls are called **sieve plates**.

Phloem contains supporting cells as well as sieve tubes.

8 TRANSPORT IN PLANTS

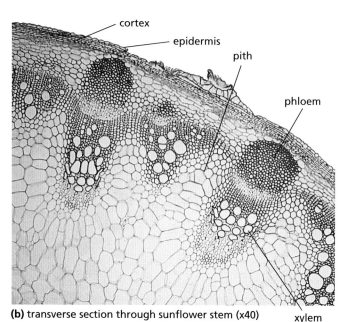
(b) transverse section through sunflower stem (×40)

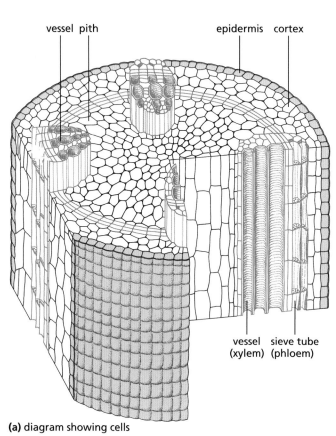
(a) diagram showing cells
Figure 8.4 Structure of plant stem

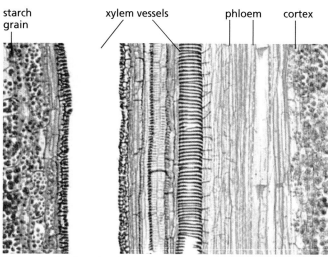
(c) longitudinal section through sunflower stem (×200)

Functions of vascular bundles
In general, water travels up the stem in the xylem from the roots to the leaves. Food may travel either up or down the stem in the phloem, from the leaves where it is made (the '**source**'), to any part of the plant that is using or storing it (the '**sink**').

Vascular bundles have a supporting function as well as a transport function, because they contain vessels, fibres and other thick-walled, lignified, elongated cells. In many stems, the vascular bundles are arranged in a cylinder, a little way in from the epidermis. This pattern of distribution helps the stem to resist the sideways bending forces caused by the wind. In a root, the vascular bundles are in the centre (Figure 8.6) where they resist the pulling forces that the root is likely to experience when the shoot is being blown about by the wind.

The network of veins in many leaves supports the soft mesophyll tissues and resists stresses that could lead to tearing.

The methods by which water, salts and food are moved through the vessels and sieve tubes are discussed in 'Transpiration' and 'Translocation' later in this chapter.

Transport in plants

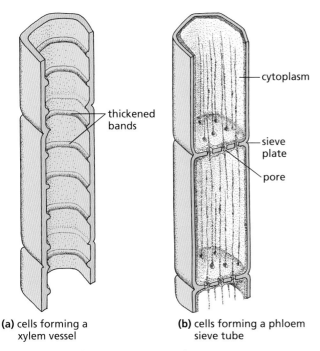

(a) cells forming a xylem vessel

(b) cells forming a phloem sieve tube

Figure 8.5 Conducting structures in a plant

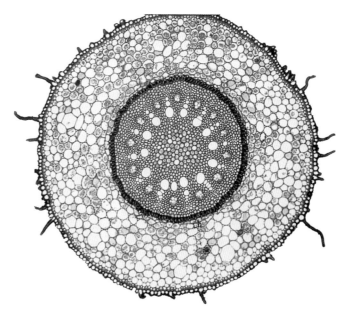

Figure 8.6 Transverse section through a root (×40). Notice that the vascular tissue is in the centre. Some root hairs can be seen in the outer layer of cells.

Cortex and pith

The tissue between the vascular bundles and the epidermis is called the **cortex**. Its cells often store starch. In green stems, the outer cortex cells contain chloroplasts and make food by photosynthesis. The central tissue of the stem is called **pith**. The cells of the pith and cortex act as packing tissues and help to support the stem in the same way that a lot of blown-up balloons packed tightly into a plastic bag would form quite a rigid structure.

Root

The internal structure of a typical root is shown in Figure 8.7. The vascular bundle is in the centre of the root (Figure 8.6), unlike the stem where the bundles form a cylinder in the cortex.

The xylem carries water and salts from the root to the stem. The phloem brings food from the stem to the root, to provide the root cells with substances for their energy and growth.

Outer layer and root hairs

There is no distinct epidermis in a root. At the root tip are several layers of cells forming the **root cap**. These cells are continually replaced as fast as they are worn away when the root tip is pushed through the soil.

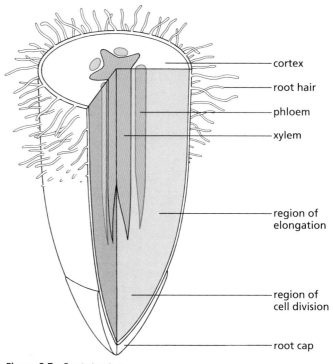

Figure 8.7 Root structure

In a region above the root tip, where the root has just stopped growing, the cells of the outer layer produce tiny, tube-like outgrowths called **root hairs** (Figure 8.11, page 115). These can just be seen as a white furry layer on the roots of seedlings grown in moist air (Figure 8.8). In the soil, the root hairs grow between the soil particles and stick closely to them. The root hairs take up water from the soil by osmosis and absorb mineral salts (as ions) by active transport (Chapter 3).

Figure 8.8 Root hairs (×5) as they appear on a root grown in moist air

Root hairs remain alive for only a short time. The region of root just below a root hair zone is producing new root hairs while the root hairs at the top of the zone are shrivelling (Figure 8.9). Above the root hair zone, the cell walls of the outer layer become less permeable. This means that water cannot get in so easily.

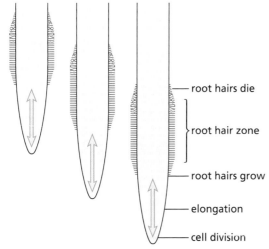

Figure 8.9 The root hair zone changes as the root grows.

● Extension work

Tap root
When a seed germinates, a single root grows vertically down into the soil. Later, lateral roots grow from this at an acute angle outwards and downwards, and from these laterals other branches may arise. Where a main root is recognisable the arrangement is called a **tap-root system** (Figure 8.10(a)).

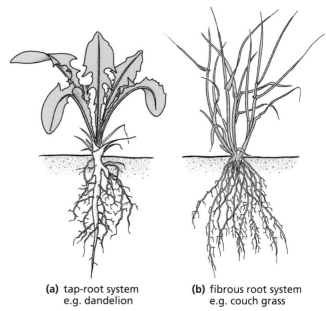

(a) tap-root system e.g. dandelion
(b) fibrous root system e.g. couch grass

Figure 8.10 Types of root system

Fibrous root
When a seed of the grass and cereal group germinates, several roots grow out at the same time and laterals grow from them. There is no distinguishable main root and it is called a **fibrous root system** (Figure 8.10(b)).

Adventitious root
Where roots grow not from a main root, but directly from the stem as they do in bulbs, corms, rhizomes or ivy, they are called **adventitious roots**, but such a system may also be described as a fibrous rooting system.

● Water uptake

Pathway taken by water
The water tension developed in the vessels by a rapidly transpiring plant (see next section) is thought to be sufficient to draw water through the root from the soil. The water enters the root hair cells and is then passed on to cells in the root cortex. It enters the xylem vessels to be transported up the stem and into the leaves, arriving at the leaf mesophyll cells.

Practical work

Transport in the vascular bundles

- Place the shoots of several leafy plants in a solution of 1% methylene blue. 'Busy Lizzie' (*Impatiens*) or celery stalks with leaves are usually effective.
- Leave the shoots in the light for up to 24 hours.

Result
If some of the stems are cut across, the dye will be seen in the vascular bundles (see Figure 2.2). In some cases the blue dye will also appear in the leaf veins.

Interpretation
These results show that the dye and, therefore, probably also the water, travel up the stem in the vascular bundles. Closer study would show that they travel in the xylem vessels.

Transport of water in the xylem

- Cut three leafy shoots from a deciduous tree or shrub. Each shoot should have about the same number of leaves.
- On one twig remove a ring of bark about 5 mm wide, about 100 mm up from the cut base.
- With the second shoot, smear a layer of Vaseline over the cut base so that it blocks the vessels. The third twig is a control.
- Place all three twigs in a jar with a little water. The water level must be below the region from which you removed the ring of bark.
- Leave the twigs where they can receive direct sunlight.

Result
After an hour or two, you will probably find that the twig with blocked vessels shows signs of wilting. The other two twigs should still have turgid leaves.

Interpretation
Removal of the bark (including the phloem) has not prevented water from reaching the leaves, but blocking the xylem vessels has. The vessels of the xylem, therefore, offer the most likely route for water passing up the stem.

As Figures 8.7 and 8.8 illustrate, the large number of tiny root hairs greatly increases the absorbing surface of a root system. The surface area of the root system of a mature rye plant has been estimated at about 200 m². The additional surface provided by the root hairs was calculated to be 400 m². The water in the surrounding soil is absorbed by osmosis (see Chapter 3). The precise pathway taken by the water is the subject of some debate, but the path of least resistance seems to be in or between the cell walls rather than through the cells.

When water loss through transpiration is slow, e.g. at night-time or just before bud burst in a deciduous tree, then osmosis may play a more important part in the uptake of water than water tension developed in the vessels. In Figure 8.11, showing a root hair in the soil, the cytoplasm of the root hair is partially permeable to water. The soil water is more dilute than the cell sap and so water passes by osmosis from the soil into the cell sap of the root hair cell. This flow of water into the root hair cell raises the cell's turgor pressure. So water is forced out through the cell wall into the next cell and so on, right through the cortex of the root to the xylem vessels (Figure 8.12).

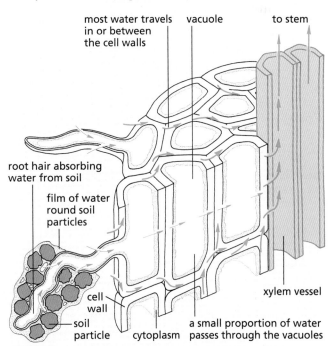

Figure 8.11 The probable pathways of water through a root

One problem for this explanation is that it has not been possible to demonstrate that there is an osmotic gradient across the root cortex that could produce this flow of water from cell to cell. Nevertheless, root pressure developed probably by osmosis does force water up the root system and into the stem.

Uptake of salts

The methods by which roots take up salts from the soil are not fully understood. Some salts may be carried in with the water drawn up by transpiration and pass mainly along the cell walls in the root cortex and into the xylem.

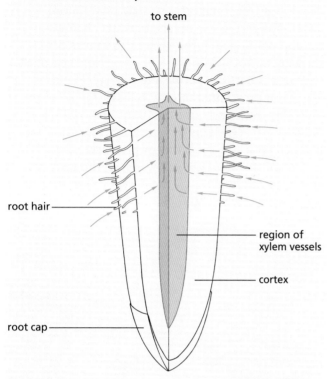

Figure 8.12 Diagrammatic section of root to show passage of water from the soil

It may be that diffusion from a relatively high concentration in the soil to a lower concentration in the root cells accounts for uptake of some individual salts, but it has been shown: (a) that salts can be taken from the soil even when their concentration is below that in the roots, and (b) that anything which interferes with respiration impairs the uptake of salts. This suggests that active transport (Chapter 3) plays an important part in the uptake of salts.

The growing regions of the root and the root hair zone (Figure 8.9) seem to be most active in taking up salts. Most of the salts appear to be carried at first in the xylem vessels, though they soon appear in the phloem as well.

The salts are used by the plant's cells to build up essential molecules. Nitrates, for example, are combined with carbohydrates to make amino acids in the roots. These amino acids are used later to make proteins.

Transpiration

The main force that draws water from the soil and through the plant is caused by a process called **transpiration**. Water evaporates from the leaves and causes a kind of 'suction', which pulls water up the stem (Figure 8.13). The water travels up the xylem vessels in the vascular bundles (see Figure 8.3, page 111) and this flow of water is called the **transpiration stream**.

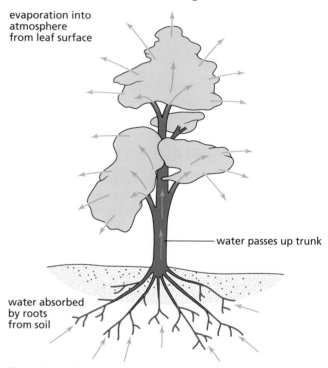

Figure 8.13 The transpiration stream

> **Key definition**
> **Transpiration** is the loss of water vapour from plant leaves by evaporation of water at the surfaces of the mesophyll cells followed by the diffusion of water vapour through the stomata.

Practical work

To demonstrate water loss by a plant

The apparatus shown in Figure 8.14 is called a weight **potometer**. A well-watered potted plant is prepared by surrounding the pot with a plastic bag, sealed around the stem of the plant with an elastic band or string. The plant is then placed on a top-pan balance and its mass is recorded. After a measured time period e.g. 24 hours, the plant is re-weighed and the difference in mass calculated. Knowing the time which has elapsed, the rate of mass loss per hour can be calculated. The process can be repeated, exposing the plant to different environmental conditions, such as higher temperature, wind speed, humidity or light intensity.

Transpiration

Results
The plant loses mass over the measured time period. Increases in temperature, wind speed and light intensity result in larger rates of loss of mass. An increase in humidity would be expected to reduce the rate of loss of mass.

Interpretation
As the roots and soil surrounding the plant have been sealed in a plastic bag, it can be assumed that any mass lost must be due to the evaporation of water vapour from the stem or leaves (transpiration). Increases in temperature, wind speed and light intensity all cause the rate of transpiration to get higher, so the rate of loss of mass from the plant increases. An increase in humidity reduces transpiration, so the rate of loss of mass slows down.

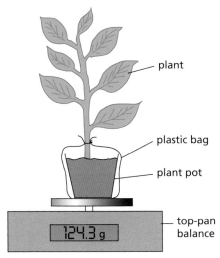

Figure 8.14 A weight potometer

Rates of water uptake in different conditions

The apparatus shown in Figure 8.15 is called a potometer. It is designed to measure the rate of uptake of water in a cut shoot.

- Fill the syringe with water and attach it to the side arm of the 3-way tap.
- Turn the tap downwards (i) and press the syringe until water comes out of the rubber tubing at the top.
- Collect a leafy shoot and push its stem into the rubber tubing as far as possible. Set up the apparatus in a part of the laboratory that is not receiving direct sunlight.
- Turn the tap up (ii) and press the syringe until water comes out of the bottom of the capillary tube. Turn the tap horizontally (iii).
- As the shoot transpires, it will draw water from the capillary tube and the level can be seen to rise. Record the distance moved by the water column in 30 seconds or a minute.
- Turn the tap up and send the water column back to the bottom of the capillary. Turn the tap horizontally and make another measurement of the rate of uptake. In this way obtain the average of three readings.

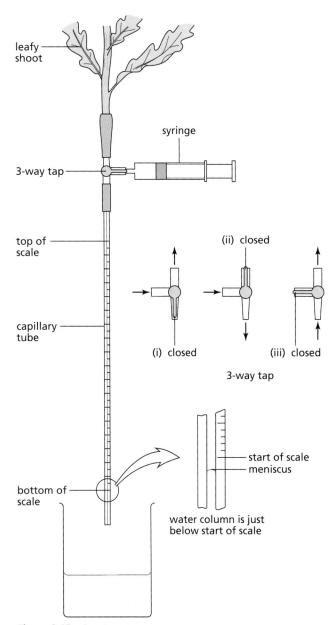

Figure 8.15 A potometer

- The conditions can now be changed in one of the following ways:
  1. Move the apparatus into sunlight or under a fluorescent lamp.
  2. Blow air past the shoot with an electric fan or merely fan it with an exercise book.
  3. Cover the shoot with a plastic bag.
- After each change of conditions, take three more readings of the rate of uptake and notice whether they represent an increase or a decrease in the rate of transpiration.

Results
1. An increase in light intensity should make the stomata open and allow more rapid transpiration.
2. Moving air should increase the rate of evaporation and, therefore, the rate of uptake.
3. The plastic bag will cause a rise in humidity round the leaves and suppress transpiration.

Interpretation

Ideally, you should change only one condition at a time. If you took the experiment outside, you would be changing the light intensity, the temperature and the air movement. When the rate of uptake increased, you would not know which of these three changes was mainly responsible.

To obtain reliable results, you should really keep taking readings until three of them are nearly the same. A change in conditions may take 10 or 15 minutes before it produces a new, steady rate of uptake. In practice, you may not have time to do this, but even your first three readings should indicate a trend towards increased or decreased uptake.

Note: a simpler version of potometer can be used effectively. This does not include the syringe or scaled capillary tubing shown in Figure 8.15.

- The plant stem can be attached directly to a length of capillary tubing with a short section of rubber tubing. This is best carried out in a bowl of water.
- While still in the water, squeeze the rubber tubing to force out any air bubbles.
- Remove the potometer from the water and rub a piece of filter paper against the end of the capillary tubing to introduce an air bubble. The capillary tubing does not need to have a scale: a ruler can be clamped next to the tubing.
- Record the distance moved by the bubble over a measured period of time. Then place the end of the capillary tubing in a beaker of water and squeeze out the air bubble.
- Introduce a new air bubble as previously described and take further readings.

Limitations of the potometer

Although we use the potometer to compare rates of transpiration, it is really the rates of uptake that we are observing. Not all the water taken up will be transpired; some will be used in photosynthesis; some may be absorbed by cells to increase their turgor. However, these quantities are very small compared with the volume of water transpired and they can be disregarded.

The rate of uptake of a cut shoot may not reflect the rate in the intact plant. If the root system were present, it might offer resistance to the flow of water or it could be helping the flow by means of its root pressure.

To find which surface of a leaf loses more water vapour

- Cut four leaves of about the same size from a plant (do not use an evergreen plant). Protect the bench with newspaper and then treat each leaf as follows:
  a Smear a thin layer of Vaseline (petroleum jelly) on the lower surface.
  b Smear Vaseline on the upper surface.
  c Smear Vaseline on both surfaces.
  d Leave both surfaces free of Vaseline.
- Place a little Vaseline on the cut end of the leaf stalk and then suspend the four leaves from a retort stand with cotton threads for several days.

Result

All the leaves will have shrivelled and curled up to some extent but the ones that lost most water will be the most shrivelled (Figure 8.16).

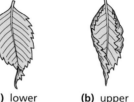

(a) lower surface (b) upper surface (c) both surfaces (d) neither surface

Figure 8.16 The results of evaporation from leaves subjected to different treatments

Interpretation

The Vaseline prevents evaporation. The untreated leaf and the leaf with its upper surface sealed show the greatest degree of shrivelling, so it is from the lower surface that leaves lose most water by evaporation.

More accurate results may be obtained by weighing the leaves at the start and the end of the experiment. It is best to group the leaves from the whole class into their respective batches and weigh each batch. Ideally, the weight loss should be expressed as a percentage of the initial weight.

More rapid results can be obtained by sticking small squares of blue cobalt chloride paper to the upper and lower surface of the same leaf using transparent adhesive tape (Figure 8.17). Cobalt chloride paper changes from blue to pink as it takes up moisture. By comparing the time taken for each square to go pink, the relative rates of evaporation from each surface can be compared.

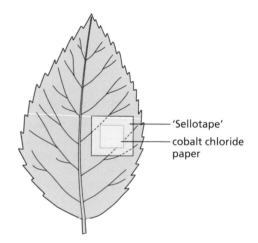

Figure 8.17 To find which surface of a leaf loses more water vapour

The results of either experiment can be correlated with the numbers of stomata on the upper and lower epidermis. This can be done by painting clear nail varnish or 'Germoline New-skin' over each surface and allowing it to dry. The varnish is then peeled off and examined under the microscope. The outlines of the guard cells can be seen and counted.

The cells in part of a leaf blade are shown in Figure 8.18. As explained in 'Osmosis' in Chapter 3, the cell sap in each cell is exerting a turgor pressure outwards on the cell wall. This pressure forces some water out of the cell wall, evaporating into the air space between the cells. The water vapour passes by diffusion through the air spaces in the mesophyll and out of the stomata. It is this loss of water vapour from the leaves that is called 'transpiration'. Each leaf contains many air spaces in the spongy mesophyll and the air becomes saturated with water vapour. There are hundreds of stomata, particularly on the lower epidermis of the leaf, enabling water vapour to diffuse from a high concentration in the air spaces into the atmosphere (representing a lower concentration of water vapour, unless the humidity is high).

The cell walls that are losing water in this way replace it by drawing water from the nearest vein. Most of this water travels along the cell walls without actually going inside the cells (Figure 8.19). Thousands of leaf cells are evaporating water like this: their surfaces represent a very large surface area. More water is drawn up to replace the evaporated water, from the xylem vessels in the veins. As a result, water is pulled through the xylem vessels and up the stem from the roots. This transpiration pull is strong enough to draw up water 50 metres or more in trees (Figure 8.20).

In addition to the water passing along the cell walls, a small amount will pass right through the cells. When leaf cell A in Figure 8.19 loses water, its turgor pressure will fall. This fall in pressure allows the water in the cell wall to enter the vacuole and so restore the turgor pressure. In conditions of water shortage, cell A may be able to get water by osmosis from cell B more easily than B can get it from the xylem vessels. In this case, all the mesophyll cells will be losing water faster than they can absorb it from the vessels, and the leaf will wilt (see 'Osmosis' in Chapter 3). Water loss from the cell vacuoles results in the cells losing their turgor and becoming **flaccid**. A leaf with flaccid cells will be limp and the stem will droop. A plant that loses water to this extent is said to be 'wilting' (see Figure 3.11).

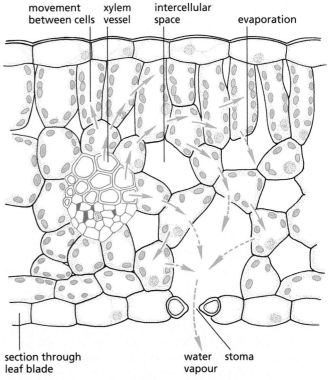

Figure 8.18 Movement of water through a leaf

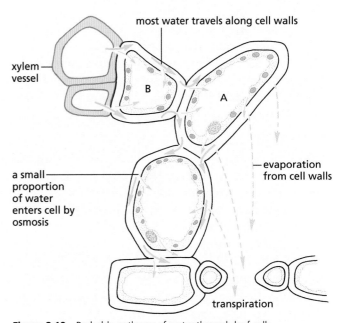

Figure 8.19 Probable pathway of water through leaf cells

Importance of transpiration

A tree, on a hot day, may draw up hundreds of litres of water from the soil (Figure 8.20). Most of this water evaporates from the leaves; only a tiny fraction is retained for photosynthesis and to maintain the turgor of the cells. The advantage to the plant of this excessive evaporation is not clear. A rapid water flow may be needed to obtain sufficient mineral salts, which are in very dilute solution in the soil. Evaporation may also help to cool the leaf when it is exposed to intense sunlight.

Against the first possibility, it has to be pointed out that, in some cases, an increased transpiration rate does not increase the uptake of minerals.

The second possibility, the cooling effect, might be very important. A leaf exposed to direct sunlight will absorb heat and its temperature may rise to a level that could kill the cytoplasm. Water evaporating from a leaf absorbs its latent heat and cools the leaf down. This is probably one value of transpiration. However, there are plants whose stomata close at around midday, greatly reducing transpiration. How do these plants avoid overheating?

Many biologists regard transpiration as an inevitable consequence of photosynthesis. In order to photosynthesise, a leaf has to take in carbon dioxide from the air. The pathway that allows carbon dioxide in will also let water vapour out whether the plant needs to lose water or not. In all probability, plants have to maintain a careful balance between the optimum intake of carbon dioxide and a damaging loss of water. Plants achieve this balance in different ways, some of which are described in 'Adaptive features' in Chapter 18.

The role of stomata

The opening and closing of stomata can be triggered by a variety of factors, principally light intensity, carbon dioxide concentration and humidity. These factors interact with each other. For example, a rise in light intensity will increase the rate of photosynthesis and so lower the carbon dioxide concentration in the leaf. These are the conditions you would expect to influence stomatal aperture if the stomata are to control the balance between loss of water vapour and uptake of carbon dioxide.

The stomata also react to water stress, i.e. if the leaf is losing water by transpiration faster than it is being taken up by the roots. Before wilting sets in, the stomata start to close. Although they do not prevent wilting, the stomata do seem to delay its onset.

Figure 8.20 Californian redwoods. Some of these trees are over 100 metres tall. Transpiration from their leaves pulls hundreds of litres of water up the trunk.

Rate of transpiration

Transpiration is the evaporation of water from the leaves, so any change that increases or reduces evaporation will have the same effect on transpiration.

Light intensity

Light itself does not affect evaporation, but in daylight the stomata of the leaves are open (see 'Leaf structure' in Chapter 6). This allows the water vapour in the leaves to diffuse out into the atmosphere. At night, when the stomata close, transpiration is greatly reduced.

Generally speaking, then, transpiration speeds up when light intensity increases because the stomata respond to changes in light intensity.

Sunlight may also warm up the leaves and increase evaporation (see below).

Humidity

If the air is very humid, i.e. contains a great deal of water vapour, it can accept very little more from the plants and so transpiration slows down. In dry air, the diffusion of water vapour from the leaf to the atmosphere will be rapid.

Air movements

In still air, the region round a transpiring leaf will become saturated with water vapour so that no more can escape from the leaf. In these conditions, transpiration would slow down. In moving air, the water vapour will be swept away from the leaf as fast as it diffuses out. This will speed up transpiration.

Temperature

Warm air can hold more water vapour than cold air. Thus evaporation or transpiration will take place more rapidly into warm air.

Furthermore, when the Sun shines on the leaves, they will absorb heat as well as light. This warms them up and increases the rate of evaporation of water.

Investigations into the effect of some of these conditions on the rate of transpiration are described earlier in this chapter.

Water movement in the xylem

You may have learned that you cannot draw water up by 'suction' to a height of more than about 10 metres. Many trees are taller than this yet they can draw up water effectively. The explanation offered is that, in long vertical columns of water in very thin tubes, the attractive forces between the water molecules result in **cohesion** (the molecules stick together). The attractive forces are greater than the forces trying to separate them. So, in effect, the transpiration stream is pulling up thin threads of water, which resist the tendency to break.

There are still problems, however. It is likely that the water columns in some of the vessels do have air breaks in them and yet the total water flow is not affected.

Evidence for the pathway of water

The experiment on page 115 uses a dye to show that in a cut stem, the dye and, therefore, presumably the water, travels in the vascular bundles. Closer examination with a microscope would show that it travels in the xylem vessels.

Removal of a ring of bark (which includes the phloem) does not affect the passage of water along a branch. Killing parts of a branch by heat or poisons does not interrupt the flow of water, but anything that blocks the vessels does stop the flow.

The evidence all points to the non-living xylem vessels as the main route by which water passes from the soil to the leaves.

● Translocation

> **Key definition**
> **Translocation** is the movement of sucrose and amino acids in the phloem, from regions of production (the 'source') to regions of storage or to regions where they are used in respiration or growth (the 'sink').

The xylem sap is always a very dilute solution, but the phloem sap may contain up to 25% of dissolved solids, the bulk of which consists of sucrose and amino acids. There is a good deal of evidence to support the view that sucrose, amino acids and many other substances are transported in the phloem. This is called **translocation**.

The movement of water and salts in the xylem is always upwards, from soil to leaf, but in the phloem the solutes may be travelling up or down the stem. The carbohydrates made in the leaf during photosynthesis are converted to sucrose and carried out of the leaf (the source) to the stem. From here, the sucrose may pass upwards to growing buds and fruits or downwards to the roots and storage organs (sink). All parts of a plant that cannot photosynthesise will need a supply of nutrients brought by the phloem. It is quite possible for substances to be travelling upwards and downwards at the same time in the phloem.

Some insects feed using syringe-like mouthparts, piercing the stems of plants to extract liquid from the phloem vessels. Figure 8.21 shows aphids feeding on a rose plant. The pressure of sucrose solution in the phloem can be so great that it is forced through the gut of the aphid and droplets of the sticky liquid exude from its anus.

8 TRANSPORT IN PLANTS

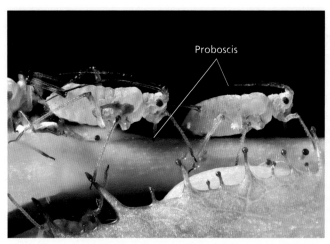

Figure 8.21 Aphids feeding on a rose plant

Some parts of a plant can act as a source and a sink at different times during the life of a plant. For example, while a bud containing new leaves is forming it would require nutrients and therefore act as a sink. However, once the bud has burst and the leaves are photosynthesising, the region would act as a source, sending newly synthesised sugars and amino acids to other parts of the plant. Similarly, the new tuber of a potato plant would act as a sink while it was growing, storing sugars as starch. (Starch is a good storage molecule because it is insoluble and quite compact.) However, once the buds on the tubers start to grow, the stored starch is converted to sucrose, a soluble nutrient, which will be passed to these buds from the tuber. So the tuber becomes the source. The shoots will eventually become sources, once they break through the soil and produce new leaves that can photosynthesise. Bulbs, such as those of the daffodil and snowdrop (see 'Asexual reproduction' in Chapter 16), act in the same way, although they tend to store sugars as well as starch.

There is no doubt that substances travel in the sieve tubes of the phloem, but the mechanism by which they are moved is not fully understood. We do know that translocation depends on living processes because anything that inhibits cell metabolism, e.g. poisons or high temperatures, also arrests translocation.

Questions

Core

1. Make a list of the types of cells or tissues you would expect to find in a vascular bundle.
2. What structures help to keep the stem's shape and upright position?
3. What are the differences between xylem and phloem:
   a in structure
   b in function?
4. State briefly the functions of the following: xylem, root hair, root cap, epidermis.
5. If you were given a cylindrical structure cut from part of a plant, how could you tell whether it was a piece of stem or a piece of root:
   a with the naked eye
   b with the aid of a microscope or hand lens?
6. Describe the path taken by a water molecule from the soil until it reaches a mesophyll cell of a leaf to be made into sugar.
7. Why do you think that root hairs are produced only on the parts of the root system that have stopped growing?
8. Discuss whether you would expect to find a vascular bundle in a flower petal.

Extended

9. If root hairs take up water from the soil by osmosis, what would you expect to happen if so much nitrate fertiliser was put on the soil that the soil water became a stronger solution than the cell sap of the root hairs?
10. A plant's roots may take up water and salts less efficiently from a waterlogged soil than from a fairly dry soil. Revise 'Active transport' (Chapter 3) and suggest reasons for this.
11. Why do you think that, in a deciduous tree in spring, transpiration is negligible before bud burst?
12. Describe the pathway followed by a water molecule from the time it enters a plant root to the time it escapes into the atmosphere from a leaf.
13. What kind of climate and weather conditions do you think will cause a high rate of transpiration?
14. What would happen to the leaves of a plant that was losing water by transpiration faster than it was taking it up from the roots?
15. In what two ways does sunlight increase the rate of transpiration?
16. Apart from drawing water through the plant, what else may be drawn up by the transpiration stream?
17. Transpiration has been described in this chapter as if it takes place only in leaves. In what other parts of a plant might transpiration occur?
18. How do sieve tubes and vessels differ:
    a in the substances they transport
    b in the directions these substances are carried?
19. A complete ring of bark cut from around the circumference of a tree-trunk causes the tree to die. The xylem continues to carry water and salts to the leaves, which can make all the substances needed by the tree. So why does the tree die?
20. Make a list of all the non-photosynthetic parts of a plant that need a supply of sucrose and amino acids.

Checklist

After studying Chapter 8 you should know and understand the following:

- The shoot of a plant consists of the stem, leaves, buds and flowers.
- The roots hold the plant in the soil, absorb the water and mineral salts needed by the plant for making sugars and proteins and, in some cases, store food for the plant.
- The root hairs make very close contact with soil particles and are the main route by which water and mineral salts enter the plant.
- The stem supports the leaves and flowers.
- The stem contains vascular bundles (veins).
- The leaves carry out photosynthesis and allow gaseous exchange of carbon dioxide, oxygen and water vapour.
- Closure of the stomata stops the entry of carbon dioxide into a leaf but also reduces water loss.
- The xylem vessels in the veins carry water up the stem to the leaves.
- The phloem in the veins carries food up or down the stem to wherever it is needed.
- The position of vascular bundles helps the stem to withstand sideways bending and the root to resist pulling forces.
- Transpiration is the evaporation of water vapour from the leaves of a plant.
- The water travelling in the transpiration stream will contain dissolved salts.
- Closure of stomata and shedding of leaves may help to regulate the transpiration rate.
- The rate of transpiration is increased by sunlight, high temperature and low humidity.
- Salts are taken up from the soil by roots, and are carried in the xylem vessels.

- Transpiration produces the force that draws water up the stem.
- Root pressure forces water up the stem as a result of osmosis in the roots.
- The large surface area provided by root hairs increases the rate of absorption of water (osmosis) and mineral ions (active transport).
- The large surface area provided by cell surfaces, interconnecting air spaces and stomata in the leaf encourages water loss.
- Wilting occurs when the volume of water vapour lost by leaves is greater than that absorbed by roots.
- Translocation is the movement of sucrose and amino acids in phloem.
- The point where food is made is called a source.
- The place where food is taken to and used is called a sink.

9 Transport in animals

Transport in animals
Single circulation in fish
Double circulation and its advantages

Heart
Structures of the heart
Monitoring heart activity
Coronary heart disease

Heart valves
Explanation of heart features
Functioning of the heart
Explanation of the effect of exercise
Treatment and prevention of coronary heart disease

Blood and lymphatic vessels
Arteries, veins, capillaries
Main blood vessels of the heart and lungs

Adaptations of blood vessels
Lymphatic system

Blood
Components of blood – appearance and functions

Lymphocyes
Phagocytes
Blood clotting
Transfer of materials between capillaries and tissue fluid

● Transport in animals

The blood, pumped by the heart, travels all around the body in blood vessels. It leaves the heart in arteries and returns in veins. Valves, present in the heart and veins, ensure a one-way flow for the blood. As blood enters an organ, the arteries divide into smaller arterioles, which supply capillaries. In these vessels the blood moves much more slowly, allowing the exchange of materials such as oxygen and glucose, carbon dioxide and other wastes. Blood leaving an organ is collected in venules, which transfer it on to larger veins.

Single circulation of fish

Fish have the simplest circulatory system of all the vertebrates. A heart, consisting of one blood-collecting chamber (the atrium) and one blood-ejection chamber (the ventricle), sends blood to the gills where it is oxygenated. The blood then flows to all the parts of the body before returning to the heart (Figure 9.1). This is known as a **single circulation** because the blood goes through the heart once for each complete circulation of the body. However, as the blood passes through capillaries in the gills, blood pressure is lost, but the blood still needs to circulate through other organs of the body before returning to the heart to increase blood pressure. This makes the fish circulatory system inefficient.

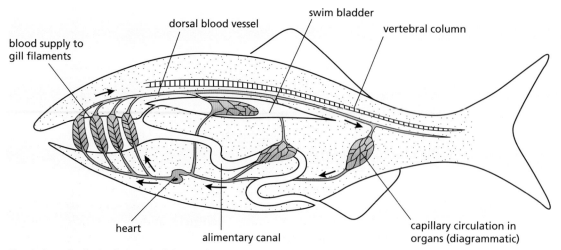

Figure 9.1 Single circulation of a fish

Double circulation of mammals

The route of the circulation of blood in a mammal is shown in Figure 9.2.

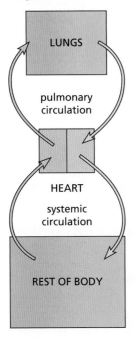

Figure 9.2 Double circulation of a mammal

The blood passes twice through the heart during one complete circuit: once on its way to the body and again on its way to the lungs. The circulation through the lungs is called the **pulmonary circulation**; the circulation around the rest of the body is called the **systemic circulation**. On average, a red blood cell would go around the whole circulation in 45 seconds. A more detailed diagram of the circulation is shown in Figure 9.20.

A **double circulation** has the advantage of maintaining a high blood pressure to all the major organs of the body. The right side of the heart collects blood from the body, builds up the blood pressure and sends it to the lungs to be oxygenated, but the pressure drops during the process. The left side of the heart receives oxygenated blood from the lungs, builds up the blood pressure again and pumps the oxygenated blood to the body.

● Heart

The heart pumps blood through the circulatory system to all the major organs of the body. The appearance of the heart from the outside is shown in Figure 9.3. Figure 9.4 shows the left side cut open, while Figure 9.5 is a diagram of a vertical section to show its internal structure. Since the heart is seen as if in a dissection of a person facing you, the left side is drawn on the right.

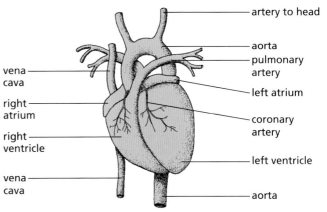

Figure 9.3 External view of the heart

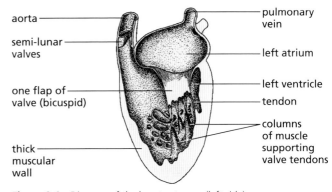

Figure 9.4 Diagram of the heart cut open (left side)

If you study Figure 9.5 you will see that there are four chambers. The upper, thin-walled chambers are the **atria** (singular = atrium) and each of these opens into a thick-walled chamber, the **ventricle**, below.

Blood enters the atria from large veins. The **pulmonary vein** brings oxygenated blood from the lungs into the left atrium. The **vena cava** brings deoxygenated blood from the body tissues into the right atrium. The blood passes from each atrium to its corresponding ventricle, and the ventricle pumps it out into the arteries. The left chambers are separated from the right chambers by a wall of muscle called a **septum**.

9 TRANSPORT IN ANIMALS

The artery carrying oxygenated blood to the body from the left ventricle is the **aorta**. The **pulmonary artery** carries deoxygenated blood from the right ventricle to the lungs.

In pumping the blood, the muscle in the walls of the atria and ventricles contracts and relaxes (Figure 9.6). The walls of the atria contract first and force blood into the two ventricles. Then the ventricles contract and send blood into the arteries. Valves prevent blood flowing backwards during or after heart contractions.

The heart muscle is supplied with food and oxygen by the **coronary arteries** (Figure 9.3).

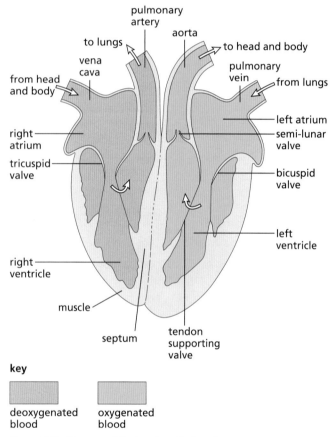

Figure 9.5 Diagram of the heart, vertical section

There are a number of ways by which the activity of the heart can be monitored. These include measuring pulse rate, listening to heart sounds and the use of electrocardiograms (ECGs).

Pulse rate

The ripple of pressure that passes down an artery as a result of the heart beat can be felt as a **'pulse'** when the artery is near the surface of the body. You can feel the pulse in your radial artery by pressing the fingertips of one hand on the wrist of the other (Figure 9.7). It is important that the thumb is *not* used because it has its own pulse. There is also a detectable pulse in the carotid artery in the neck. Digital pulse rate monitors are also available. These can be applied to a finger, wrist or earlobe depending on the type and provide a very accurate reading.

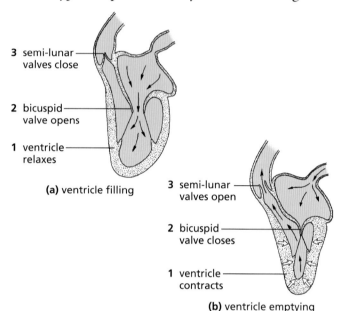

Figure 9.6 Diagram of heartbeat (only the left side is shown)

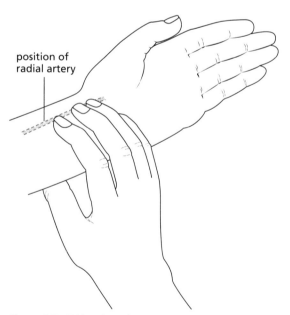

Figure 9.7 Taking the pulse

Heart sounds

These can be heard using a **stethoscope**. This instrument amplifies the sounds of the heart valves opening and closing. A healthy heart produces a regular 'lub-dub' sound. The first ('lub') sound is caused by the closure of the valves separating the atria from the ventricles. The second ('dub') sound represents the closure of the valves at the entrance of the pulmonary artery and aorta. Observation of irregular sounds may indicate an irregular heartbeat. If the 'lub' or 'dub' sounds are not clear then this may point to a problem with faulty valves.

ECGs

An ECG is an **electrocardiogram**. To obtain an ECG, electrodes, attached to an ECG recording machine, are stuck onto the surface of the skin on the arms, legs and chest (Figure 9.8). Electrical activity associated with heartbeat is then monitored and viewed on a computer screen or printed out (Figure 9.9). Any irregularity in the trace can be used to diagnose heart problems.

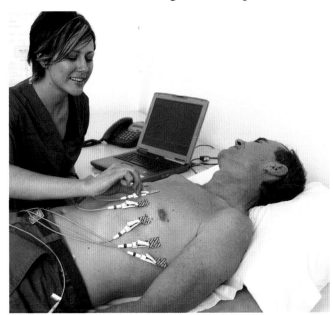

Figure 9.8 A patient undergoing an ECG

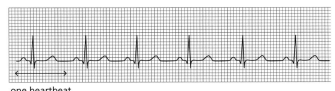

one heartbeat

Figure 9.9 ECG trace

The effect of physical activity on the pulse rate

A heartbeat is a contraction. Each contraction squeezes blood to the lungs and body. The pulse is a pressure wave passing through the arteries as a result of the heartbeat. At rest, the heart beats about 70 times a minute, but this varies according to a person's age, gender and fitness: higher if you are younger, higher if you are female and lower if you are fit. An increase in physical activity increases the pulse rate, which can rise to 200 beats per minute. After exercise has stopped, the pulse rate gradually drops to its resting state. How quickly this happens depends on the fitness of the individual (an unfit person's pulse rate will take longer to return to normal).

Coronary heart disease

In the lining of the large and medium arteries, deposits of a fatty substance, called **atheroma**, are laid down in patches. This happens to everyone and the patches get more numerous and extensive with age, but until one of them actually blocks an important artery the effects are not noticed. It is not known how or why the deposits form. Some doctors think that fatty substances in the blood pass into the lining. Others believe that small blood clots form on damaged areas of the lining and are covered over by the atheroma patches. The patches may join up to form a continuous layer, which reduces the internal diameter of the vessel (Figure 9.10).

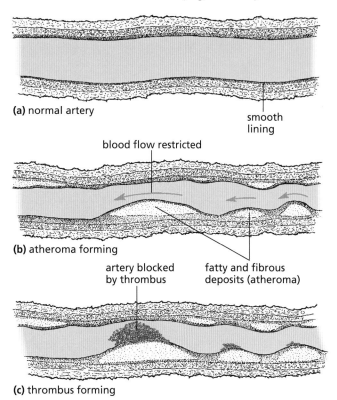

Figure 9.10 Atheroma and thrombus formation

The surface of a patch of atheroma sometimes becomes rough and causes fibrinogen in the plasma to deposit fibrin on it, causing a blood clot (a **thrombus**) to form. If the blood clot blocks the coronary artery (Figure 9.3), which supplies the muscles of the ventricles with blood, it starves the muscles of oxygenated blood and the heart may stop beating. This is a severe heart attack from **coronary thrombosis**. A thrombus might form anywhere in the arterial system, but its effects in the coronary artery and in parts of the brain (strokes) are the most drastic.

In the early stages of coronary heart disease, the atheroma may partially block the coronary artery and reduce the blood supply to the heart (Figure 9.11). This can lead to **angina**, i.e. a pain in the chest that occurs during exercise or exertion. This is a warning to the person that he or she is at risk and should take precautions to avoid a heart attack.

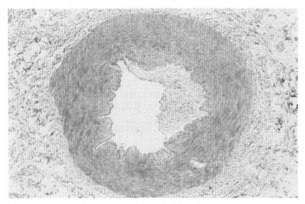

Figure 9.11 Atheroma partially blocking the coronary artery

Possible causes of coronary heart disease

Atheroma and thrombus formation are the immediate causes of a heart attack but the long-term causes that give rise to these conditions are not well understood.

There is an inherited tendency towards the disease but incidences of the disease have increased very significantly in affluent countries in recent years. This makes us think that some features of 'Western' diets or lifestyles might be causing it. The main risk factors are thought to be an unbalanced diet with too much fat, stress, smoking, genetic disposition, age, gender and lack of exercise.

Diet
The atheroma deposits contain **cholesterol**, which is present, combined with lipids and proteins, in the blood. Cholesterol plays an essential part in our physiology, but it is known that people with high levels of blood cholesterol are more likely to suffer from heart attacks than people with low cholesterol levels.

Blood cholesterol can be influenced, to some extent, by the amount and type of fat in the diet. Many doctors and dieticians believe that animal fats (milk, cream, butter, cheese, egg-yolk, fatty meat) are more likely to raise the blood cholesterol than are the vegetable oils, which contain a high proportion of unsaturated fatty acids (see 'Diet' in Chapter 7).

An unbalanced diet with too many calories can lead to obesity. Being overweight puts extra strain on the heart and makes it more difficult for the person to exercise.

Stress
Emotional stress often leads to raised blood pressure. High blood pressure may increase the rate at which atheroma are formed in the arteries.

Smoking
Statistical studies suggest that smokers are two to three times more likely to die from a heart attack than are non-smokers of a similar age (Figure 9.12). The carbon monoxide and other chemicals in cigarette smoke may damage the lining of the arteries, allowing atheroma to form, but there is very little direct evidence for this.

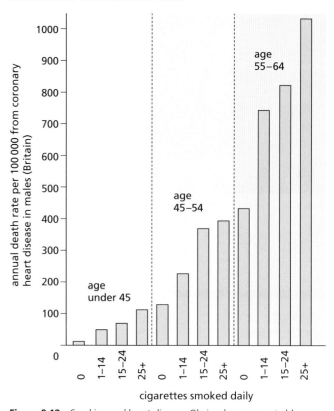

Figure 9.12 Smoking and heart disease. Obviously, as you get older you are more likely to die from a heart attack, but notice that, in any age group, the more you smoke the higher your chances of dying from heart disease.

Genetic predisposition

Coronary heart disease appears to be passed from one generation to the next in some families. This is not something we have any control over, but we can be aware of this risk and reduce some of the other risk factors to compensate.

Age and gender

As we get older our risk of suffering from coronary heart disease increases. Males are more at risk of a heart attack than females: it may be that males tend to have less healthy lifestyles than females.

Lack of exercise

Heart muscle loses its tone and becomes less efficient at pumping blood when exercise is not untaken. A sluggish blood flow, resulting from lack of exercise, may allow atheroma to form in the arterial lining but, once again, the direct evidence for this is slim.

Control of blood flow through the heart

The blood is stopped from flowing backwards by four sets of valves. Valves that separate each atrium from the ventricle below it are known as **atrioventricular valves**. Between the right atrium and the right ventricle is the **tricuspid** (= three flaps) valve. Between the left atrium and left ventricle is the **bicuspid** (= two flaps) valve. The flaps of these valves are shaped rather like parachutes, with 'strings' called **tendons** or **cords** to prevent them from being turned inside out.

In the pulmonary artery and aorta are the **semi-lunar** (= half-moon) valves. These each consist of three 'pockets', which are pushed flat against the artery walls when blood flows one way. If blood tries to flow the other way, the pockets fill up and meet in the middle to stop the flow of blood (Figure 9.13).

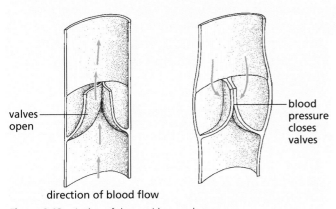

Figure 9.13 Action of the semi-lunar valves

When the ventricles contract, blood pressure closes the bicuspid and tricuspid valves and these prevent blood returning to the atria. When the ventricles relax, the blood pressure in the arteries closes the semi-lunar valves, preventing the return of blood to the ventricles.

From the description above, it may seem that the ventricles are filled with blood as a result of the contraction of the atria. However, the atria have much thinner muscle walls than the ventricles. In fact, when the ventricles relax, their internal volume increases and they draw in blood from the pulmonary vein or vena cava through the relaxed atria. Atrial contraction then forces the final amount of blood into the ventricles just before ventricular contraction.

The left ventricle (sometimes referred to as the 'large left ventricle') has a wall made of cardiac muscle that is about three times thicker than the wall of the right ventricle. This is because the right ventricle only needs to create enough pressure to pump blood to one organ, the lungs, which are next to the heart. However, the left ventricle has to pump blood to all the major organs of the body, as shown in Figure 9.20. It should be noted that the left and right ventricles pump the same volume of blood: the left ventricle does *not* have a thicker wall to pump *more* blood!

● Extension work

Blood circulation in the fetus

The septum separating the left and right heart chambers prevents the oxygenated blood in the left chambers from mixing with the deoxygenated blood in the right chambers. When a fetus is developing, there is a hole (the **foramen ovale**) between the right atrium and the left atrium, allowing blood to bypass the lungs. This is because the fetal blood is oxygenated by the placenta rather than the lungs. During the birth sequence, the foramen ovale closes, so all blood in the right atrium passes into the right ventricle and on to the lungs for oxygenation. Occasionally, the foramen ovale does not seal completely and the baby suffers from a 'hole in the heart'. Babies suffering from this condition tend to look blue because their blood is not being adequately oxygenated: some of it bypasses the lungs.

Control of the heartbeat

Heart muscle has a natural rhythmic contraction of its own, about 40 contractions per minute. However, it is supplied by nerves, which maintain a faster rate that can be adjusted to meet the body's needs for oxygen. At rest, the normal heart rate may lie between 50 and 100 beats per minute, according to age, gender and other factors. During exercise, the rate may increase to 200 beats per minute.

The heart beat is initiated by the '**pacemaker**', a small group of specialised muscle cells at the top of the right atrium. The pacemaker receives two sets of nerves from the brain. One group of nerves speeds up the heart rate and the other group slows it down. These nerves originate from a centre in the brain that receives an input from receptors (See 'Nervous control in humans' in Chapter 14) in the circulatory system that are sensitive to blood pressure and levels of oxygen and carbon dioxide in the blood.

If blood pressure rises, nervous impulses reduce the heart rate. A fall in blood pressure causes a rise in the rate. Reduced oxygen concentration or increased carbon dioxide in the blood also contributes to a faster rate. By this means, the heart rate is adjusted to meet the needs of the body at times of rest, exertion and excitement.

The hormone **adrenaline** (see 'Hormones in humans' in Chapter 14) also affects the heart rate. In conditions of excitement, activity or stress, adrenaline is released into the blood circulation from the adrenal glands. On reaching the heart it causes an increase in the rate and strength of the heartbeat.

Physical activity and heart rate

During periods of physical activity, active parts of the body (mainly skeletal muscle) respire faster, demanding more oxygen and glucose. Increased respiration also produces more carbon dioxide, which needs to be removed. Blood carries the oxygen and glucose, so the heart rate needs to increase to satisfy demand. If the muscle does not get enough oxygen, it will start to respire anaerobically, producing lactic acid (lactate). Lactic acid build-up causes muscle fatigue, leading to cramp. An 'oxygen debt' is created, which needs to be repaid after exercise by continued rapid breathing and higher than normal heart rate (see 'Anaerobic respiration' in Chapter 12).

Correlation and cause

It is not possible or desirable to conduct experiments on humans to find out, more precisely, the causes of heart attacks. The evidence has to be collected from long-term studies on populations of individuals, e.g. smokers and non-smokers. Statistical analysis of these studies will often show a correlation, e.g. more smokers, within a given age band, suffer heart attacks than do non-smokers of the same age. This correlation does not prove that smoking causes heart attacks. It could be argued that people who are already prone to heart attacks for other reasons (e.g. high blood pressure) are more likely to take up smoking. This may strike you as implausible, but until it can be shown that substances in tobacco smoke do cause an increase in atheroma, the correlation cannot be used on its own to claim a cause and effect.

Nevertheless, there are so many other correlations between smoking and ill-health (e.g. bronchitis, emphysema, lung cancer) that the circumstantial evidence against smoking is very strong.

Another example of a positive correlation is between the possession of a television set and heart disease. Nobody would seriously claim that television sets cause heart attacks. The correlation probably reflects an affluent way of life, associated with over-eating, fatty diets, lack of exercise and other factors that may contribute to coronary heart disease.

Prevention of coronary heart disease

Maintaining a healthy, balanced diet will result in less chance of a person becoming obese. There will also be a low intake of saturated fats, so the chances of atheroma and thrombus formation are reduced.

There is some evidence that regular, vigorous exercise reduces the chances of a heart attack. This may be because it increases muscle tone – not only of skeletal muscle, but also of cardiac muscle. Good heart muscle tone leads to an improved coronary blood flow and the heart requires less effort to keep pumping.

Treatment of coronary heart disease

The simplest treatment for a patient who suffers from coronary heart disease is to be given a regular dose of aspirin (salicylic acid). Aspirin prevents the formation of blood clots in the arteries, which can lead to a heart attack. It has been found that long-term use of low-dose aspirin also reduces the risk of coronary heart disease.

Methods of removing or treating atheroma and thrombus formations include the use of **angioplasty**, a **stent** and, in the most severe cases, by-pass surgery.

Angioplasty and stent

Angioplasty involves the insertion of a long, thin tube called a **catheter** into the blocked or narrowed blood vessel. A wire attached to a deflated balloon is then fed through the catheter to the damaged artery. Once in place, the balloon is inflated to widen the artery wall, effectively freeing the blockage. In some cases a stent is also applied. This is a wire-mesh tube that can be expanded and left in place (Figure 9.14). It then acts as scaffolding, keeping the blood vessel open and maintaining the free flow of blood. Some stents are designed to give a slow release of chemicals to prevent further blockage of the artery.

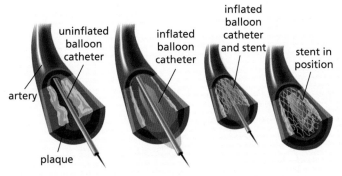

Figure 9.14 Application of a stent to overcome a blockage in an artery

By-pass surgery

The surgeon removes a section of blood vessel from a different part of the body, such as the leg. The blood vessel is then attached around the blocked region of artery to by-pass it, allowing blood to pass freely. This is a major, invasive operation because it involves open-heart surgery.

Practical work

Heart dissection

- Obtain an intact heart (sheep or goat for example) from a butcher's shop or abattoir.
- Rinse it under a tap to remove excess blood.
- Observe the surface of the heart, identifying the main visible features (shown in Figure 9.3). The blood vessels may have been cut off, but it is possible to identify where these would have been attached later in the dissection.
- Gently squeeze the ventricles. They can be distinguished because the wall of the right ventricle is much thinner than that of the left ventricle.
- Using a pair of sharp scissors or a scalpel, make an incision from the base of the left ventricle, up through the left atrium.
- Using a pair of forceps, remove any blood clots lying in the exposed chambers.
- Identify the main features as shown in Figure 9.4.
- If you have not cut open the aorta, gently push the handle of a blunt seeker or an old pencil, behind the bicuspid valve. It should find its way into the aorta. Note how thick the wall of this blood vessel is.
- Compare the semi-lunar valves in the base of the aorta with the bicuspid valve between the atrium and ventricle. Note that the latter has tendons to prevent it turning inside-out.
- Now repeat the procedure on the right side of the heart to expose the right atrium and ventricle.
- Pushing the handle of the seeker behind the tricuspid valve should allow it to enter the pulmonary artery. Cut open the artery to expose semi-lunar valves. Note the relative thinness of the wall, compared to that of the aorta.
- Also compare the thickness of the left ventricle wall to that of the right ventricle.

Investigating the effect of exercise on pulse rate

- Find your pulse in your wrist or neck – see Figure 9.7.
- Count the number of beats in 15 seconds, then multiply the result by four to provide a pulse rate in beats per minute. This is your resting pulse rate.
- Repeat the process two more times and then calculate an average resting pulse rate.
- Carry out 2 minutes of exercise, e.g. running on the spot, then sit down and immediately start a stopwatch and measure your pulse rate over 15 seconds as before.
- Allow the stopwatch to keep timing. Measure your pulse rate every minute for 10 minutes.
- Convert all the readings to beats per minute. Plot a graph of pulse rate after exercise against time, with the first reading being 0 minutes.
- Finally, draw a line across the graph representing your average resting pulse rate.

9 TRANSPORT IN ANIMALS

Result
The pulse rate immediately after exercise should be much higher than the average resting pulse rate. With time the pulse rate gradually falls back to the average resting pulse rate.

Interpretation
During exercise the muscles need more oxygen and glucose for aerobic respiration to provide the energy needed for the increased movement. The heart rate increases to provide these materials. After exercise, demand for oxygen and glucose decreases, so the pulse rate gradually returns to normal.

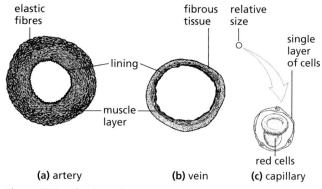

Figure 9.15 Blood vessels, transverse section

● Blood and lymphatic vessels

Arteries

These are fairly wide vessels (Figure 9.15) which carry blood from the heart to the limbs and organs of the body (Figure 9.20). The blood in the arteries, except for the pulmonary arteries, is oxygenated.

Arteries have elastic tissue and muscle fibres in their thick walls. The arteries divide into smaller vessels called **arterioles**.

The arterioles divide repeatedly to form a branching network of microscopic vessels passing between the cells of every living tissue. These final branches are called **capillaries**.

Capillaries

These are tiny vessels, often as little as 0.001 mm in diameter and with walls only one cell thick (Figures 9.15(c) and 9.17). Although the blood as a whole cannot escape from the capillary, the thin capillary walls allow some liquid to pass through, i.e. they are permeable. Blood pressure in the capillaries forces part of the plasma out through the walls.

The capillary network is so dense that no living cell is far from a supply of oxygen and food. The capillaries join up into larger vessels, called **venules**, which then combine to form **veins**.

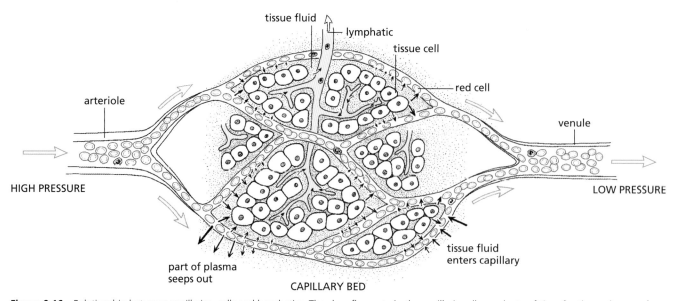

Figure 9.16 Relationship between capillaries, cells and lymphatics. The slow flow rate in the capillaries allows plenty of time for the exchange of oxygen, food, carbon dioxide and waste products.

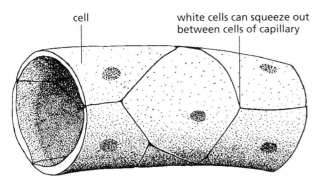

Figure 9.17 Diagram of blood capillary

Veins

Veins return blood from the tissues to the heart (Figure 9.20). The blood pressure in them is steady and is less than that in the arteries. They are wider and their walls are thinner, less elastic and less muscular than those of the arteries (Figures 9.15(b) and 9.18). They also have valves in them similar to the semi-lunar valves (Figure 9.13, page 129).

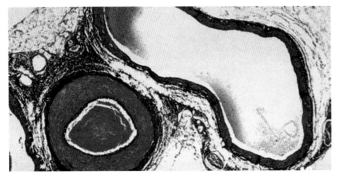

Figure 9.18 Transverse section through a vein and artery. The vein is on the right, the artery on the left. Notice that the wall of the artery is much thicker than that of the vein. The material filling the artery is formed from coagulated red blood cells. These are also visible in two regions of the vein.

The blood in most veins is deoxygenated and contains less food but more carbon dioxide than the blood in most arteries. This is because respiring cells have used the oxygen and food and produced carbon dioxide (Figure 9.19). The pulmonary veins, which return blood from the lungs to the heart, are an exception. They contain oxygenated blood and a reduced level of carbon dioxide.

The main blood vessels associated with the heart, lungs and kidneys are shown in Figure 9.20. The right side of the heart is supplied by the vena cava (the main vein of the body) and sends blood to the lungs along the pulmonary artery. The left side of the heart receives blood from the lungs in the pulmonary vein and sends it to the body in the aorta, the main artery (see Chapter 11). In reality there are two pulmonary arteries and two pulmonary veins, because there are two lungs. There are also two vena cavae: one returns blood from the lower body; the other from the upper body. Each kidney receives blood from a renal artery. Once the blood has been filtered it is returned to the vena cava through a renal vein (see Chapter 13).

Blood pressure

The pumping action of the heart produces a pressure that drives blood around the circulatory system (Figure 9.20). In the arteries, the pressure fluctuates with the heartbeat, and the pressure wave can be felt as a pulse. The millions of tiny capillaries offer resistance to the blood flow and, by the time the blood enters the veins, the surges due to the heartbeat are lost and the blood pressure is greatly reduced.

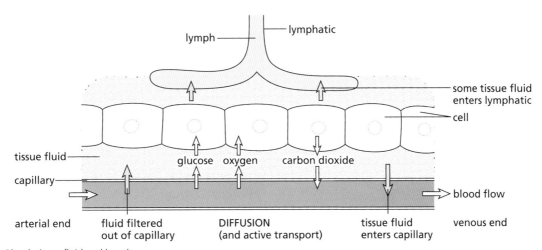

Figure 9.19 Blood, tissue fluid and lymph

9 TRANSPORT IN ANIMALS

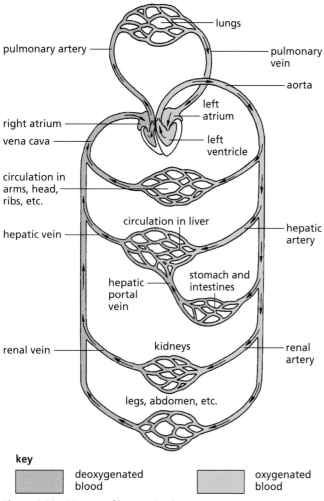

Figure 9.20 Diagram of human circulation

Table 9.1 compares the structure of arteries, veins and capillaries and provides an explanation of how their structures are related to their functions.

Table 9.1 Comparing arteries, veins and capillaries

Blood vessel	Structure	Explanation of how structure is related to function
artery	thick, tough wall with muscles, elastic fibres and fibrous tissue	Carries blood at high pressure – prevents bursting and maintains pressure wave. The large arteries, near the heart, have a greater proportion of elastic tissue, which allows these vessels to stand up to the surges of high pressure caused by the heartbeat.
	lumen quite narrow, but increases as a pulse of blood passes through	This helps to maintain blood pressure.
	valves absent	High pressure prevents blood flowing backwards.
vein	thin wall – mainly fibrous tissue, with little muscle or elastic fibres	Carries blood at low pressure.
	lumen large	To reduce resistance to blood flow
	valves present	To prevent backflow of blood. Contraction of body muscles, particularly in the limbs, compresses the thin-walled veins. The valves in the veins prevent the blood flowing backwards when the vessels are compressed in this way. This assists the return of venous blood to the heart.
capillary	permeable wall, one cell thick, with no muscle or elastic tissue	This allows diffusion of materials between the capillary and surrounding tissues.
	lumen approximately one red blood cell wide	White blood cells can squeeze between cells of the wall. Blood cells pass through slowly to allow diffusion of materials and tissue fluid.
	valves absent	Blood is still under pressure.

Although blood pressure varies with age and activity, it is normally kept within specific limits by negative feedback (see 'Homeostasis' in Chapter 14). The filtration process in the kidneys (Chapter 13) needs a fairly consistent blood pressure. If blood pressure falls significantly because, for example, of loss of blood or shock, then the kidneys may fail. Blood pressure consistently higher than normal increases the risk of heart disease or stroke.

Arterioles, shunt vessels and venules

Arterioles and shunt vessels

The small arteries and the arterioles have proportionately less elastic tissue and more muscle fibres than the great arteries. When the muscle fibres of the arterioles contract, they make the vessels narrower and restrict the blood flow (a process called **vasoconstriction**). In this way, the distribution of blood to different parts of the body can be regulated. One example is in the skin. If the body temperature drops below normal, arterioles in the skin constrict to reduce the amount of blood flowing through capillaries near the skin surface. **Shunt vessels**, linking the arterioles with venules, dilate to allow the blood to bypass the capillaries (Figure 9.21). This helps to reduce further heat loss. (See also 'Homeostasis' in Chapter 14.)

At certain points in the lymphatic vessels there are swellings called **lymph nodes** (Figure 9.22). Lymphocytes are stored in the lymph nodes and released into the lymph to eventually reach the blood system. There are also phagocytes in the lymph nodes. If bacteria enter a wound and are not ingested by the white cells of the blood or lymph, they will be carried in the lymph to a lymph node and white cells there will ingest them. The lymph nodes thus form part of the body's defence system against infection.

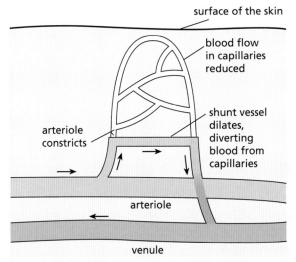

Figure 9.21 Shunt vessels in the skin in cold conditions

The lymphatic system

Not all the tissue fluid returns to the capillaries. Some of it enters blind-ended, thin-walled vessels called **lymphatics** (Figure 9.16). The lymphatics from all parts of the body join up to make two large vessels, which empty their contents into the blood system as shown in Figure 9.22.

The lacteals from the villi in the small intestine (Figure 7.24) join up with the lymphatic system, so most of the fats absorbed in the intestine reach the circulation by this route. The fluid in the lymphatic vessels is called **lymph** and is similar in composition to tissue fluid.

Some of the larger lymphatics can contract, but most of the lymph flow results from the vessels being compressed from time to time when the body muscles contract in movements such as walking or breathing. There are valves in the lymphatics (Figure 9.23) like those in the veins and the pulmonary artery (Figure 9.13), so that when the lymphatics are squashed, the fluid in them is forced in one direction only: towards the heart.

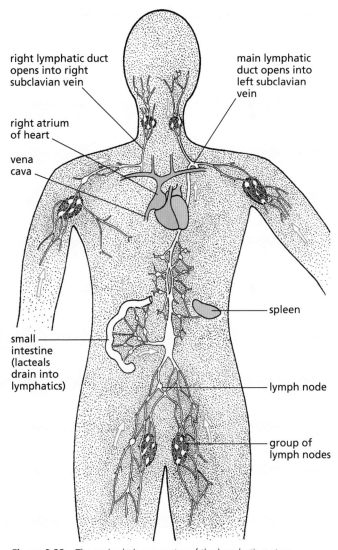

Figure 9.22 The main drainage routes of the lymphatic system

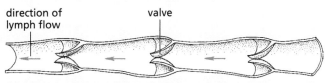

Figure 9.23 Lymphatic vessel cut open to show valves

9 TRANSPORT IN ANIMALS

● Blood

Blood consists of red cells, white cells and platelets floating in a liquid called plasma. There are between 5 and 6 litres of blood in the body of an adult, and each cubic centimetre contains about 5 billion red cells.

Red cells

These are tiny, disc-like cells (Figures 9.24(a) and 9.26) which do not have nuclei. They are made of spongy cytoplasm enclosed in an elastic cell membrane. In their cytoplasm is the red pigment haemoglobin, a protein combined with iron. Haemoglobin combines with oxygen in places where there is a high concentration of oxygen, to form **oxyhaemoglobin**. Oxyhaemoglobin is an unstable compound. It breaks down and releases its oxygen in places where the oxygen concentration is low (Figure 9.25). This makes haemoglobin very useful in carrying oxygen from the lungs to the tissues.

Blood that contains mainly oxyhaemoglobin is said to be **oxygenated**. Blood with little oxyhaemoglobin is **deoxygenated**.

Each red cell lives for about 4 months, after which it breaks down. The red haemoglobin changes to a yellow pigment, bilirubin, which is excreted in the bile. The iron from the haemoglobin is stored in the liver. About 200 000 million red cells wear out and are replaced each day. This is about 1% of the total. Red cells are made by the red bone marrow of certain bones in the skeleton – in the ribs, vertebrae and breastbone for example.

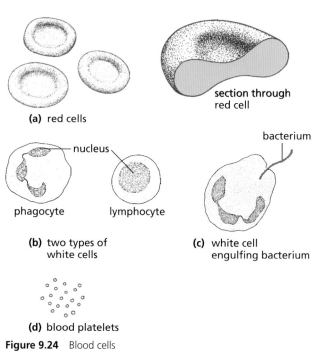

Figure 9.24 Blood cells

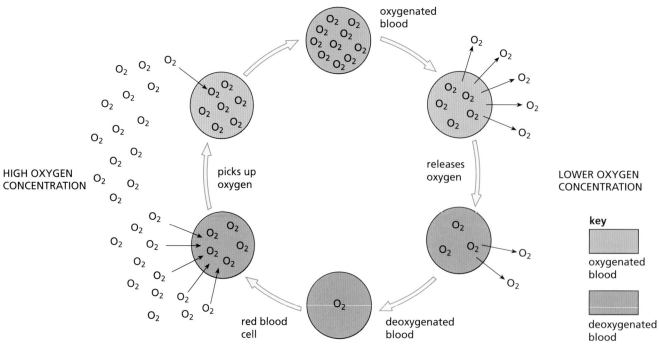

Figure 9.25 The function of the red cells

White cells

There are several different kinds of white cell (Figures 9.24(b) and 9.26). Most are larger than the red cells and they all have a nucleus. There is one white cell to every 600 red cells and they are made in the same bone marrow that makes red cells. Many of them undergo a process of maturation and development in the thymus gland, lymph nodes or spleen. White blood cells are involved with phagocytosis and antibody production.

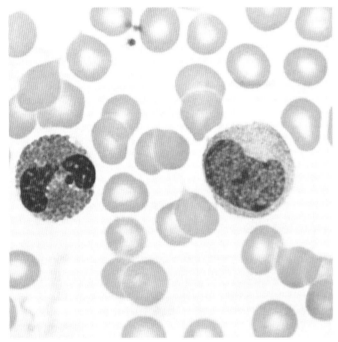

Figure 9.26 Red and white cells from human blood (×2500). The large nucleus can be seen clearly in the white cells.

Platelets

These are pieces of special blood cells budded off in the red bone marrow. They help to clot the blood at wounds and so stop the bleeding.

White blood cells

The two most numerous types of white cells are **phagocytes** and **lymphocytes**.

The phagocytes can move about by a flowing action of their cytoplasm and can escape from the blood capillaries into the tissues by squeezing between the cells of the capillary walls. They collect at the site of an infection, engulfing (**ingesting**) and digesting harmful bacteria and cell debris – a process called **phagocytosis** (Figure 9.24(c)). In this way they prevent the spread of infection through the body. One of the functions of lymphocytes is to produce antibodies.

Plasma

The liquid part of the blood is called plasma. It is water with a large number of substances dissolved in it. The ions of sodium, potassium, calcium, chloride and hydrogen carbonate, for example, are present. Proteins such as fibrinogen, albumin and globulins make up an important part of the plasma. Fibrinogen is needed for clotting (see below), and the globulin proteins include antibodies, which combat bacteria and other foreign matter (page 149). The plasma will also contain varying amounts of food substances such as amino acids, glucose and lipids (fats). There may also be hormones (Chapter 14) present, depending on the activities taking place in the body. The excretory product, urea, is dissolved in the plasma, along with carbon dioxide.

The liver and kidneys keep the composition of the plasma more or less constant, but the amount of digested food, salts and water will vary within narrow limits according to food intake and body activities.

Table 9.2 summarises the role of transport by the blood system

Table 9.2 Transport by the blood system

Substance	From	To
oxygen	lungs	whole body
carbon dioxide	whole body	lungs
urea	liver	kidneys
hormones	glands	target organs
digested food	intestine	whole body
heat	abdomen and muscles	whole body

Note that the blood is not directed to a particular organ. A molecule of urea may go round the circulation many times before it enters the renal artery, by chance, and is removed by the kidneys.

Clotting

When tissues are damaged and blood vessels cut, platelets clump together and block the smaller capillaries. The platelets and damaged cells at the wound also produce a substance that acts, through a series of enzymes, on the soluble plasma protein called fibrinogen. As a result of this action, the fibrinogen is changed into insoluble **fibrin**, which forms a network of fibres across the wound. Red cells become trapped in this network and so form a blood clot. The clot not only stops further loss of blood, but also prevents the entry of harmful bacteria into the wound (Figures 9.27 and 9.28).

9 TRANSPORT IN ANIMALS

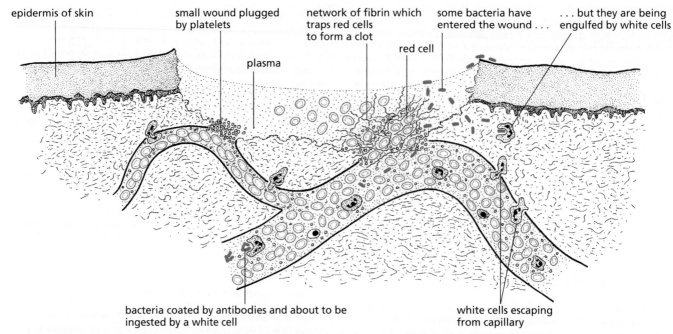

Figure 9.27 The defence against infection by pathogens. An area of skin has been damaged and two capillaries broken open.

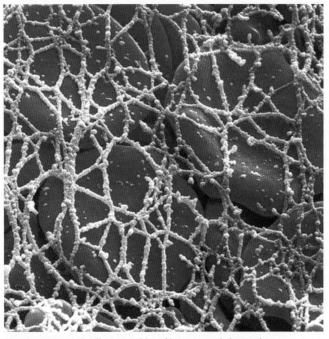

Figure 9.28 Red cells trapped in a fibrin network (×6500)

The transfer of materials between capillaries and tissue fluid

The fluid that escapes from capillaries is not blood, nor plasma, but tissue fluid. Tissue fluid is similar to plasma but contains less protein, because protein molecules are too large to pass through the walls of the capillaries. This fluid bathes all the living cells of the body and, since it contains dissolved food and oxygen from the blood, it supplies the cells with their needs (Figures 9.16 and 9.19). Some of the tissue fluid eventually seeps back into the capillaries, having given up its oxygen and dissolved food to the cells, but it has now received the waste products of the cells, such as carbon dioxide, which are carried away by the bloodstream. The tissue fluid that doesn't return to the capillaries joins the lymphatic system.

● Extension work

Ideas about the circulatory system

There must have been knowledge of human internal anatomy thousands of years ago. This might have come, for example, from the practice of removing internal organs before the process of mummification in Ancient Egypt. However, there seems to have been little or no systematic study of human anatomy in the sense that the parts were named, described or illustrated.

Some of the earliest records of anatomical study come from the Greek physician, Galen.

Galen (AD 130–200)

Galen dissected goats, monkeys and other animals and produced detailed and accurate records. He was not allowed to dissect human bodies, so his descriptions were often not applicable to human anatomy.

The anatomical knowledge was important but the functions of the various parts could only be guessed at. It was known that the veins contained blood but arteries at death are usually empty and it was assumed that they carried air or, more obscurely, 'animal spirit'. Galen observed the pulse, but thought that it was caused by surges of blood into the veins.

William Harvey (1578–1657)

In the 15th and 16th centuries, vague ideas about the movement of blood began to emerge, but it was William Harvey, an English physician, who produced evidence to support the circulation theory.

Harvey's predecessors had made informed guesses, but Harvey conducted experiments to support his ideas. He noted that the valves in the heart would permit blood to pass in one direction only. So the notion that blood shunted back and forth was false. When he restricted the blood flow in an artery he observed that it bulged on the side nearest the heart, whereas a vein bulged on the side away from the heart.

Figure 9.29 shows a simple experiment that reveals the presence of valves in the veins and supports the idea of a one-way flow.

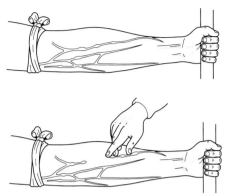

Figure 9.29 Harvey's demonstration of valves and one-way flow in a vein. The vein is compressed and the blood expelled by running a finger up the arm. The vein refills, but only as far as the valve. (Compare with Figure 9.13, page 129.)

Harvey published his results in 1628. They were at first rejected and ridiculed, not because anyone tried his experiments or tested his observations, but simply because his conclusions contradicted the writings of Galen 1500 years previously.

By 1654, Harvey's theory of circulation was widely accepted but it was still not known how blood passed from the arteries to the veins. Harvey observed that arteries and veins branched and re-branched until the vessels were too small to be seen and suggested that the connection was made through these tiny vessels. This was confirmed after the microscope had been invented in 1660 and the vessels were called 'capillaries'.

The significance of this history is that, although it is reasonable to make an informed guess at the function of a structure or organ, it is only by testing these guesses by experiment that they can be supported or disproved.

Questions

Core

1. Starting from the left atrium, put the following in the correct order for circulation of the blood:
left atrium, vena cava, aorta, lungs, pulmonary artery, right atrium, pulmonary vein, right ventricle, left ventricle
2. Why is it incorrect to say 'all arteries carry oxygenated blood and all veins carry deoxygenated blood'?
3. How do veins differ from arteries in:
    a their function
    b their structure?
4. How do capillaries differ from other blood vessels in:
    a their structure
    b their function?
5. Why is it misleading to say that a person 'suffers from blood pressure'?
6. Which important veins are not labelled in Figure 9.3?
7. In what ways are white cells different from red cells in:
    a their structure
    b their function?
8. Where, in the body, would you expect haemoglobin to be combining with oxygen to form oxyhaemoglobin?
9. In what parts of the body would you expect oxyhaemoglobin to be breaking down to oxygen and haemoglobin?
10. a Why is it important for oxyhaemoglobin to be an unstable compound, i.e. easily changed to oxygen and haemoglobin?
    b What might be the effect on a person whose diet contained too little iron?

Extended

11. Which parts of the heart:
    a pump blood into the arteries
    b stop blood flowing the wrong way?

12 Put the following in the correct order:
   a blood enters arteries
   b ventricles contract
   c atria contract
   d ventricles relax
   e blood enters ventricles
   f semi-lunar valves close
   g tri- and bicuspid valves close.
13 Why do you think that:
   a the walls of the ventricles are more muscular than the walls of the atria
   b the muscle of the left ventricle is thicker than that of the right ventricle?
   (Hint: look back at Figure 9.20.)
14 Why is a person whose heart valves are damaged by disease unable to take part in active sport?
15 a What positive steps could you take, and
   b what things should you avoid, to reduce your risk of coronary heart disease in later life?
16 About 95% of patients with disease of the leg arteries are cigarette smokers. Arterial disease of the leg is the most frequent cause of leg amputation.
   a Is there a correlation between smoking and leg amputation?
   b Does smoking cause leg amputation?
   c In what way could smoking be a possible cause of leg amputation?

17 Figure 9.30 shows the relative increase in the rates of four body processes in response to vigorous exercise.
   a How are the changes related physiologically to one another?
   b What other physiological changes are likely to occur during exercise?
   c Why do you think that the increase in blood flow in muscle is less than the total increase in the blood flow?

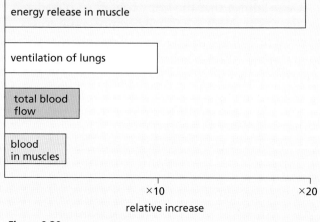

Figure 9.30

18 List the things you would expect to find if you analysed a sample of lymph.

Checklist

After studying Chapter 9 you should know and understand the following:

- The circulatory system is made up of blood vessels with a heart and valves to ensure one-way flow of blood.
- The heart is a muscular pump with valves, which sends blood around the circulatory system.
- The left side of the heart pumps oxygenated blood around the body.
- The right side of the heart pumps deoxygenated blood to the lungs.
- The atria are thin walled and receive blood from veins.
- The ventricles have thick muscular walls to pump blood through arteries.
- Blood pressure is essential in order to pump blood around the body.
- Arteries carry blood from the heart to the tissues.
- Veins return blood to the heart from the tissues.
- Capillaries form a network of tiny vessels in all tissues. Their thin walls allow dissolved food and oxygen to pass from the blood into the tissues, and carbon dioxide and other waste substances to pass back into the blood.
- The main blood vessels to and from the heart are: vena cavae, pulmonary veins, pulmonary arteries and aorta.
- The lungs are supplied by the pulmonary arteries and veins.
- The kidneys are supplied by the renal arteries and veins.
- Heart activity can be monitored by ECG, pulse rate and stethoscope, which transmits the sound of valves closing.
- Blockage of the coronary arteries in the heart leads to a heart attack.
- Smoking, fatty diets, stress, lack of exercise, genetic disposition and age may contribute to heart disease.
- Blood consists of red cells, white cells and platelets suspended in plasma.
- Plasma transports blood cells, ions, soluble nutrients, e.g. glucose, hormones and carbon dioxide.
- The red cells carry oxygen. The white cells attack bacteria by phagocytosis and production of antibodies. Platelets are needed to clot blood.

- Fish have a single circulation; mammals have a double circulation, with advantages over a single circulation.
- The heart contains atrioventricular and semi-lunar valves, preventing backflow of blood.
- The left and right sides of the heart are divided by a septum, keeping oxygenated and deoxygenated blood separate.

- The risk of coronary heart disease can be reduced by an appropriate diet and exercise regime.
- Coronary heart disease can be treated by the use of drugs (aspirin), stents, angioplasty and by-pass.
- Lymphocytes and phagocytes have distinctive shapes and features.
- Antibodies are chemicals made by white cells in the blood. They attack any micro-organisms or foreign proteins that get into the body.
- Blood clotting involves the conversion of the soluble blood protein fibrinogen to insoluble fibrin, which traps blood cells.
- Blood clotting prevents loss of blood and entry of pathogens into the body.
- Materials are transferred between capillaries and tissue fluid.
- All cells in the body are bathed in tissue fluid, which is derived from plasma.
- Lymph vessels return tissue fluid to the lymphatic system and finally into the blood system.
- Lymph nodes are important immunological organs.

10 Diseases and immunity

Pathogens and transmission
Definitions
Transmissible diseases

Defences against diseases
Defences of the body against pathogens
Vaccination
Controlling the spread of disease

How antibodies work
Active immunity, including definition
Vaccination
Passive immunity
Type 1 diabetes

● Pathogens and transmission

Key definitions
A **pathogen** is a disease-causing organism.
A **transmissible disease** is a disease in which the pathogen can be passed from one host to another.

Pathogens

Pathogens include many bacteria, viruses and some fungi, as well as a number of protoctista and other organisms. Pathogenic bacteria may cause diseases because of the damage they do to the host's cells, but most bacteria also produce poisonous waste products called **toxins**. Toxins damage the cells in which the bacteria are growing. They also upset some of the systems in the body. This gives rise to a raised temperature, headache, tiredness and weakness, and sometimes diarrhoea and vomiting. The toxin produced by the *Clostridium* bacteria (which causes tetanus) is so poisonous that as little as 0.000 23 g is fatal.

Many viruses cause diseases in plants and animals. Human virus diseases include the common cold, poliomyelitis, measles, mumps, chickenpox, herpes, rubella, influenza and AIDS (See 'Sexually transmitted infections' in Chapter 16). Tobacco mosaic virus affects tomato plants as well as tobacco. It causes mottling and discolouration of the leaves, eventually stunting the growth of the plant.

While most fungi are saprophytic (feeding on dead organic matter) some are parasitic, obtaining their nutrients from living organisms. The hyphae of parasitic fungi penetrate the tissues of their host plant and digest the cells and their contents. If the mycelium spreads extensively through the host, it usually causes the death of the plant. The bracket fungus shown in Chapter 1, Figure 1.27, is the fruiting body of a mycelium that is spreading through the tree and will eventually kill it.

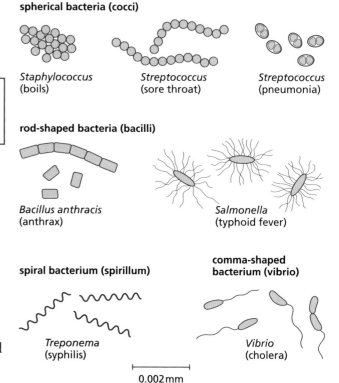

Figure 10.1 Some pathogenic bacteria

Fungus diseases such as blight, mildews or rusts (see Chapter 1, Figure 1.28) are responsible for causing considerable losses to arable farmers, and there is a constant search for new varieties of crop plants that are resistant to fungus disease, and for new chemicals (fungicides) to kill parasitic fungi without harming the host.

A few parasitic fungi cause diseases in animals, including humans. One group of these fungi cause tinea or ringworm. The fungus grows in the epidermis of the skin and causes irritation and inflammation. One form of tinea is athlete's foot, in which the skin between the toes becomes infected. Tinea is very easily spread by contact with infected towels or clothing, but can usually be cured quickly with a fungicidal ointment.

Transmission

Pathogens responsible for transmissible diseases can be spread either through direct contact or indirectly.

Direct contact

This may involve transfer through blood or other body fluids. HIV is commonly passed on by drug addicts who inject the drug into their bloodstream, sharing needles with other drug users. If one user injects himself, the pathogens in his blood will contaminate the syringe needle. If this is then used by a second drug user, the pathogens are passed on. Anyone cleaning up dirty needles is at risk of infection if they accidently stab themselves. Surgeons carrying out operations have to be especially careful not to be in direct contact with the patient's blood, for example by cutting themselves while conducting an operation. A person with HIV or another sexually transmitted disease (see Chapters 15 and 16) who has unprotected sex, can pass on the pathogen to their partner through body fluids. It used to be said that HIV could be transferred from one person to another through saliva, but this is now considered to be a very low risk.

● Extension work

Malaria

About 219 million people suffer from malaria in over 100 countries (Figure 10.2). In 2010 there were an estimated 660 000 malaria deaths according to the World Health Organization.

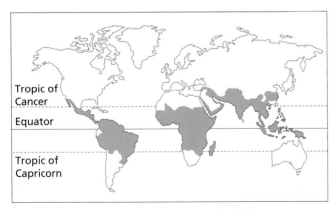

Figure 10.2 The worldwide distribution of malaria

The disease is caused by a protozoan parasite called *Plasmodium* which is transmitted from person to person by the bites of infected mosquitoes of the genus *Anopheles*. The mosquito is said to be the **vector** of the disease. When a mosquito 'bites' a human, it inserts its sharp, pointed mouthparts through the skin till they reach a capillary (Figure 10.3). The mosquito then injects saliva, which stops the blood from clotting. If the mosquito is infected, it will also inject hundreds of malarial parasites.

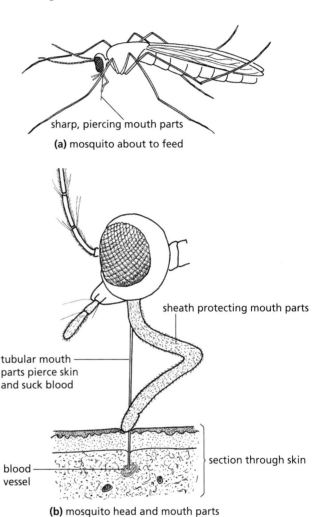

Figure 10.3 Mosquito feeding on blood

The parasites reach the liver via the circulation and burrow into the liver cells where they reproduce. A week or two later, the daughter cells break out of the liver cells and invade the red blood cells. Here they reproduce rapidly and then escape from the original red cells to invade others (Figure 10.4).

The cycle of reproduction in the red cells takes 2 or 3 days (depending on the species of *Plasmodium*). Each time the daughter plasmodia are released simultaneously from thousands of red cells the patient experiences the symptoms of malaria. These are chills accompanied by violent shivering, followed by a fever and profuse sweating. With so many red cells being destroyed, the patient will also become anaemic (see 'Diet' in Chapter 7).

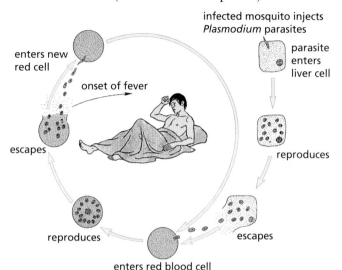

Figure 10.4 *Plasmodium*, the malarial parasite

If a mosquito sucks blood from an infected person, it will take up the parasites in the red cells. The parasites reproduce in the mosquito and finally invade the salivary glands, ready to infect the next human.

Control

There are drugs which kill the parasites in the bloodstream but they do not reach those in the liver. The parasites in the liver may emerge at any time and start the cycle again. If these drugs are taken by a healthy person before entering a malarious country, they kill any parasites as soon as they are injected. This is a protective or **prophylactic** use of the drug.

Unfortunately there are now many mutant forms of *Plasmodium* that have developed resistance to these drugs.

A great deal of work has been devoted to finding an effective vaccine, without much success. Trials are currently taking place of a vaccine that may offer at least partial protection against the disease.

The most far-reaching form of malarial control is based on the elimination of the mosquito. It is known that mosquitoes lay their eggs in stagnant water and that the larvae hatch, feed and grow in the water, but have to come to the surface to breathe air.

Spraying stagnant water with oil and insecticides suffocates or poisons the larvae and pupae. Spraying must include not only lakes and ponds but any accumulation of fresh water that mosquitoes can reach, e.g. drains, gutters, tanks, tin cans and old car tyres. By draining swamps and turning sluggish rivers into swifter streams, the breeding grounds of the mosquito are destroyed.

Spraying the walls of dwellings with chemicals like DDT was once very effective because the insecticide remained active for several months and the mosquito picked up a lethal dose merely by settling on the wall. See page 324 for further details about the use of DDT and its effects on the environment.

However, in at least 60 countries, many species of *Anopheles* have developed resistance to these insecticides and this method of control is now far less effective. The emphasis has changed back to the removal of the mosquito's breeding grounds or the destruction of the larvae and pupae.

Indirect contact

This may involve infection from pathogens on contaminated surfaces, for example during food preparation. Raw meat carries bacteria, which are killed if the meat is adequately cooked. However, if the raw meat is prepared on a surface that is then used for other food preparation, such as cutting up fruit or vegetables that are later eaten raw, then the pathogens from meat can be transferred to the fresh food. The person handling the food is also a potential vector of disease if he or she does not wash their hands after using the toilet, moving rubbish or handling raw produce. In Britain there have been serious cases where customers in butchers' shops have been infected with the bacterium *Escherichia coli* (*E. coli*), because germs from raw meat were transferred to cooked meat unwittingly by shop assistants using poor hygiene practices. For example, in 1996, 21 people died after eating contaminated meat supplied by a butcher's shop in Scotland.

Salmonella food poisoning

One of the commonest causes of food poisoning is the toxin produced by the bacteria *Salmonella typhimurium* and *S. enteritidis*. These bacteria live in the intestines of cattle, chickens and ducks without causing disease symptoms. Humans, however, may develop food poisoning if they drink milk or eat meat or eggs that are contaminated with *Salmonella* bacteria from the alimentary canal of an infected animal.

Intensive methods of animal rearing may contribute to a spread of infection unless care is taken to reduce the exposure of animals to infected faeces.

The symptoms of food poisoning are diarrhoea, vomiting and abdominal pain. They occur from 12 to 24 hours after eating the contaminated food. Although these symptoms are unpleasant, the disease is not usually serious and does not need treatment with drugs. Elderly people and very young children, however, may be made very ill by food poisoning.

The *Salmonella* bacteria are killed when meat is cooked or milk is pasteurised. Infection is most likely if untreated milk is drunk, meat is not properly cooked, or cooked meat is contaminated with bacteria transferred from raw meat (Figure 10.5). Frozen poultry must be thoroughly defrosted before cooking, otherwise the inside of the bird may not get hot enough during cooking to kill the *Salmonella*.

It follows that, to avoid the disease, all milk should be pasteurised and meat should be thoroughly cooked. People such as shop assistants and cooks should not handle cooked food at the same time as they handle raw meat. If they must do so, they should wash their hands thoroughly between the two activities.

The liquid that escapes when a frozen chicken is defrosted may contain *Salmonella* bacteria. The dishes and utensils used while the bird is defrosting must not be allowed to come into contact with any other food.

Uncooked meat or poultry should not be kept alongside any food that is likely to be eaten without cooking. Previously cooked meat should never be warmed up; the raised temperature accelerates the reproduction of any bacteria present. The meat should be eaten cold or cooked at a high temperature.

In the past few years there has been an increase in outbreaks of *Salmonella* food poisoning in which the bacteria are resistant to antibiotics. Some scientists suspect that this results from the practice of feeding antibiotics to farm animals to increase their growth rate. This could allow populations of drug-resistant salmonellae to develop.

Salmonella bacteria, and also bacteria that cause typhoid, are present in the faeces of infected people and may reach food from the unwashed hands of the sufferer.

People recovering from one of these diseases may feel quite well, but bacteria may still be present in their faeces. If they don't wash their hands thoroughly after going to the lavatory, they may have small numbers of bacteria on their fingers. If they then handle food, the bacteria may be transferred to the food. When this food is eaten by healthy people, the bacteria will multiply in their bodies and give them the disease.

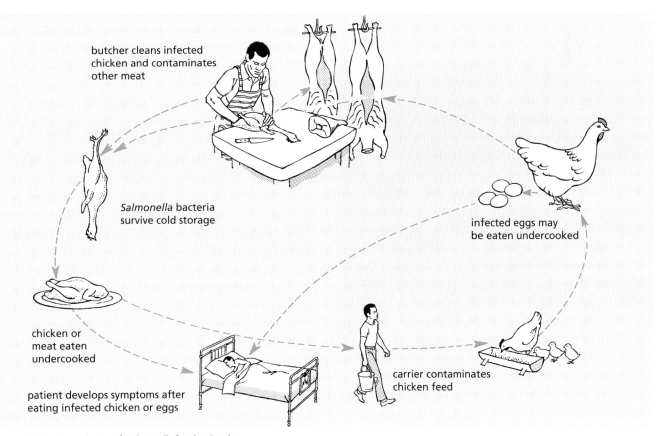

Figure 10.5 Transmission of *Salmonella* food poisoning

People working in food shops, kitchens and food-processing factories could infect thousands of other people in this way if they were careless about their personal cleanliness.

Some forms of food poisoning result from poisons (toxins) that are produced by bacteria that get into food. Cooking kills the bacteria in the food but does not destroy the toxins that cause the illness. Only one form of this kind of food poisoning, called **botulism**, is dangerous. It is also very rare.

In the 1970s another genus of bacteria, *Campylobacter*, was identified as a cause of food poisoning. This bacterium causes acute abdominal pains and diarrhoea for about 24 hours. The sources of infection are thought to be undercooked meat, particularly 'burgers'.

In summary, people who handle and prepare food need to be extremely careful about their personal hygiene. It is essential that they wash their hands before touching food, particularly after they have visited the lavatory (Figure 10.6). Hand-washing is also important after handling raw meat, particularly poultry (see Figure 10.5). Food on display in shops needs to be protected (Figure 10.7).

Some people carry intestinal pathogens without showing any symptoms of disease. These people are called 'carriers'. Once identified, they should not be allowed to work in canteens or food-processing factories.

Figure 10.6 Hygienic handling of food. Shop assistants avoid handling meat and shellfish with their fingers by using disposable gloves.

Figure 10.7 Protection of food on display. The glass barrier stops customers from touching the products, keeps flies off the food and helps stop droplets from coughs and sneezes falling on the food.

Contamination of water

If disease bacteria get into water supplies used for drinking, hundreds of people can become infected. Diseases of the alimentary canal, like typhoid and cholera (see 'Alimentary canal' in Chapter 7), are especially dangerous. Millions of bacteria infest the intestinal lining of a sick person.

Some of these bacteria will pass out with the faeces. If the faeces get into streams or rivers, the bacteria may be carried into reservoirs of water used for drinking. Even if faeces are left on the soil or buried, rainwater may wash the bacteria into a nearby stream.

To prevent this method of infection, drinking water needs to be purified and faeces must be made harmless, a process involving sewage treatment (see 'Conservation' in Chapter 21).

Water treatment

On a small scale, simply boiling the water used for drinking will destroy any pathogens. On a large scale, water supplies are protected by (a) ensuring that untreated human sewage cannot reach them and (b) treating the water to make it safe.

The treatment needed to make water safe for drinking depends on the source of the water. Some sources, e.g. mountain streams, may be almost pure; others, e.g. sluggish rivers, may be contaminated.

The object of the treatment is to remove all micro-organisms that might cause disease. This is done by filtration and chlorination. The water is passed through beds of sand in which harmless bacteria and protozoa are growing. These produce a gelatinous film which acts as a fine filter and removes pathogens.

Finally, chlorine gas is added to the filtered water and remains in contact with it for long enough to kill any bacteria that have passed through the filter. How much chlorine is added and the length of the contact time both depend on how contaminated the water source is likely to be. Most of the chlorine disappears before the water reaches the consumers.

The purified water is pumped to a high-level reservoir or water tower. These are enclosed to ensure that no pathogens can get into the water. The height of the reservoir provides the pressure needed to deliver the water to the consumer.

Waste disposal

Waste from domestic or commercial premises should be stored in dustbins or garbage cans made of galvanised steel or strong plastic, with a closely fitted lid to exclude flies and keep out scavenging animals. If this is not done, pathogens will breed in the waste and become a source of disease organisms. The waste is taken away and disposed of by burning, or burying deep enough to prevent rats using it as food, or (less effectively) tightly packed to keep out flies and vermin.

Contamination by houseflies

Flies walk about on food. They place their mouthparts on it and pump saliva onto the food. Then they suck up the digested food as a liquid.

This would not matter much if flies fed only on clean food, but they also visit decaying food or human faeces. Here they may pick up bacteria on their feet or their mouthparts. They then alight on our food and the bacteria on their bodies are transferred to the food. Figure 10.8 shows the many ways in which this can happen.

Food poisoning, amoebic dysentery and polio can be spread by houseflies.

Tinea ('ringworm') – a fungal parasite

Several species of fungus give rise to the various forms of this disease. The fungus attacks the epidermis (see 'Homeostasis' in Chapter 14) and produces a patch of inflamed tissue. On the skin the infected patch spreads outwards and heals in the centre, giving a ring-like appearance ('ringworm').

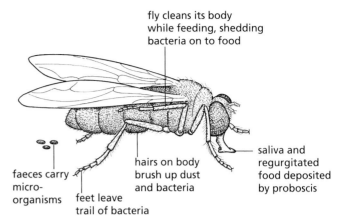

Figure 10.8 Transmission of bacteria by houseflies

The different species of tinea fungi may live on the skin of humans or domestic animals, or in the soil. The region of the body affected will depend on the species of fungus.

One kind affects the scalp and causes circular bald patches. The hair usually grows again when the patient recovers from the disease.

The species of fungus that affects the feet usually causes cracks in the skin between the toes. This is known as 'athlete's foot'.

Tinea of the crutch is a fungus infection, occurring usually in males, which affects the inner part of the thighs on each side of the scrotum. It causes a spreading, inflamed area of skin with an itching or burning sensation.

All forms of the disease are very contagious. That means, they are spread by contact with an infected person or their personal property. Tinea of the scalp is spread by using infected hairbrushes, combs or pillows. Tinea of the crutch can be caught by using towels or bedclothes contaminated by the fungus or its spores, and 'athlete's foot' by wearing infected socks or shoes, or from the floors of showers and swimming pools.

When an infection is diagnosed, the clothing, bed linen, infected hairbrushes, combs or towels must be boiled to destroy the fungus. It is best, anyway, to avoid sharing these items as their owners may be carrying the infection without knowing or admitting it.

In young people, tinea infections often clear up without treatment. Where treatment is needed, a fungicide cream or dusting powder is applied to the affected areas of skin. Infected feet may be dipped in a solution of potassium permanganate (potassium manganate(VII)).

10 DISEASES AND IMMUNITY

Amoebic dysentery

Entamoeba histolytica is a species of small amoebae that normally live harmlessly in the human intestine, feeding on food particles or bacteria. In certain conditions, however, *Entamoeba* invades the lining of the intestine causing ulceration and bleeding, with pain, vomiting and diarrhoea: the symptoms of amoebic dysentery.

The diarrhoea and vomiting lead to a loss of water and salts from the body and if they persist for very long can cause **dehydration**. Dehydration, if untreated, can lead to kidney failure and death. The treatment for dehydration is to give the patient a carefully prepared mixture of water, salts and sugar. The intestine absorbs this solution more readily than water and it restores the volume and concentration of the body fluids. This simple, effective and inexpensive treatment is called **oral rehydration therapy** and has probably saved thousands of lives since it was first discovered. There are also drugs that attack *Entamoeba*.

The faeces of infected people contain *Entamoeba* amoebae which, if they reach food or drinking water, can infect other people. The disease is prevalent in tropical, sub-tropical and, to some extent, temperate countries and is associated with low standards of hygiene and sanitation.

Airborne, 'droplet' or aerosol infection

When we sneeze, cough, laugh, speak or just breathe out, we send a fine spray of liquid drops into the air. These droplets are so tiny that they remain floating in the air for a long time. They may be breathed in by other people or fall on to exposed food (Figure 10.9). If the droplets contain viruses or bacteria, they may cause disease when they are eaten with food or inhaled.

Virus diseases such as colds, 'flu, measles and chickenpox are spread in this way. So are the bacteria (*Streptococci*) that cause sore throats. When the water in the droplets evaporates, the bacteria often die as they dry out. The viruses remain infectious, however, floating in the air for a long time.

In buses, trains, cinemas and night clubs the air is warm and moist, and full of floating droplets. These are places where you are likely to pick up one of these infections.

Figure 10.9 Droplet infection. The visible drops expelled by this sneeze will soon sink to the floor, but smaller droplets will remain suspended in the air.

● Defences against diseases

The body has three main lines of defence against disease. These involve mechanical barriers, chemical barriers and cells.

Mechanical barriers

Although many bacteria live on the surface of the skin, the outer layer of the epidermis (see 'Homeostasis' in Chapter 14) seems to act as a barrier that stops them getting into the body. But if the skin is cut or damaged, the bacteria may get into the deeper tissues and cause infection.

Hairs in the nose help to filter out bacteria that are breathed in. However, if air is breathed in through the mouth, this defence is by-passed.

Chemical barriers

The acid conditions in the stomach destroy most of the bacteria that may be taken in with food. The moist lining of the nasal passages traps many bacteria, as does the mucus produced by the lining of the trachea and bronchi. The ciliated cells of these organs carry the trapped bacteria away from the lungs.

Tears contain an enzyme called **lysozyme**. This dissolves the cell walls of some bacteria and so protects the eyes from infection.

Cells

When bacteria get through the mechanical and chemical barriers, the body has two more lines of defence – white blood cells and antibodies, produced by white blood cells. One type of white blood cells fights infection by engulfing bacteria (a process called phagocytosis) and digesting them. Further details of the way these work is also described in 'Blood' in Chapter 9. Another type produce antibodies that attach themselves to bacteria, making it easier for other white blood cells to engulf them.

Antibodies and immunity

> **Key definition**
> **Active immunity** is the defence against a pathogen by antibody production in the body.

On the surface of all cells there are chemical substances called **antigens**. Lymphocytes produce proteins called **antibodies** which attack the antigens of bacteria or any alien cells or proteins that invade the body. The antibodies may attach to the surface of the bacteria to mark them, making it easier for the phagocytes to find and ingest them, they may clump the bacteria together or they may neutralise the poisonous proteins (**toxins**) that the bacteria produce.

Each antibody is very **specific**. This means that an antibody that attacks a typhoid bacterium will not affect a pneumonia bacterium. This is illustrated in the form of a diagram in Figure 10.10.

Some of the lymphocytes that produced the specific antibodies remain in the lymph nodes for some time and divide rapidly and make more antibodies if the same antigen gets into the body again. This means that the body has become immune to the disease caused by the antigen and explains why, once you have recovered from measles or chickenpox, for example, you are very unlikely to catch the same disease again. This is called **active immunity**. Active immunity can also be gained by vaccination. You may also inherit some forms of immunity or acquire antibodies from your mother's milk (see 'Sexual reproduction in humans' in Chapter 16). This is **innate immunity**.

Vaccination

The body's defences can be enhanced by **vaccination**. This involves a harmless form of the pathogen (bacteria or virus) being introduced into the body by injection or swallowing. The presence of the pathogen triggers white blood cells to make specific antibodies to combat possible infection. If the person is exposed to the disease later, defences are already in place to prevent it developing (the person is **immune** to that disease). Without vaccination, white blood cells need to be exposed to the disease organism before they make the appropriate antibody. If the disease is potentially lethal, the patient could die before the white blood cells have time to act.

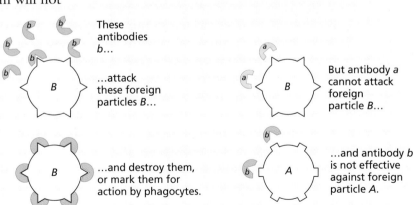

Figure 10.10 Antibodies are specific

Vaccination

When you are **inoculated** (vaccinated) against a disease, a harmless form of the bacteria or viruses is introduced into your body (Figure 10.11). The white cells make the correct antibodies, so that if the real micro-organisms get into the blood, the antibody is already present or very quickly made by the blood.

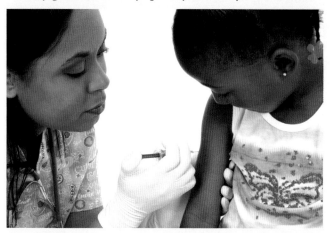

Figure 10.11 Vaccination. The girl is being vaccinated against rubella (German measles).

The material that is injected or swallowed is called a **vaccine** and is one of the following:

- a harmless form of the micro-organism, e.g. the BCG inoculation against tuberculosis and the Sabin oral vaccine against polio (oral, in this context, means 'taken by mouth')
- the killed micro-organisms, e.g. the Salk anti-polio vaccine and the whooping cough vaccine
- a **toxoid**, i.e. the inactivated toxin from the bacteria, e.g. the diphtheria and tetanus vaccines. (A toxin is the poisonous substance produced by certain bacteria, which causes the symptoms of the disease.)

B and T lymphocytes

There are two main types of lymphocyte. Both types undergo rapid cell division in response to the presence of specific antigens but their functions are different (though interdependent). The **B cells** (from **B**one marrow) become short-lived **plasma** cells and produce antibodies that are released into the blood. These antibodies may attack antigens directly or stick to the surface membrane of infected or alien cells, e.g. cells carrying a virus, bacteria, cancer cells or transplanted cells.

'Killer' T cells (from the **T**hymus gland) have receptor molecules on their surface, which attach them to these surface antibodies. The T cells then kill the cell by damaging its cell membrane.

'Helper' T cells stimulate the B cells to divide and produce antibodies. They also stimulate the phagocytes to ingest any cells carrying antibodies on their surface.

Some of the B cells remain in the lymph nodes as **memory cells**. These can reproduce swiftly and produce antibodies in response to any subsequent invasion of the body by the same foreign organism.

When mass vaccination fails, the population is at risk of infection with potential epidemics resulting. An example of this was with the MMR vaccine in Britain. MMR is a combination of vaccines protecting against measles, mumps and rubella (German measles). A researcher and surgeon called Andrew Wakefield claimed (incorrectly) to have found a link between the MMR vaccine and the incidence of autism and bowel disease in children. The story got into the national press and many parents reacted by refusing to allow their children to have the MMR vaccination, leaving them vulnerable to the three potentially life-threatening diseases. The drop in MMR vaccination rates left whole populations more susceptible to the spread of measles, mumps and rubella. There needs to be a significant proportion of a population immunised to prevent an epidemic of a disease, ideally over 90%. The percentage of people protected against measles, mumps and rubella dropped well below this figure in some areas after the MMR vaccine scare. It has taken years for doctors to restore parents' faith in the safety of the MMR vaccine.

There is a small risk of serious side-effects from vaccines, just as there is with all medicines. These risks are always far lower than the risk of catching the disease itself. For example, the measles vaccine carries a risk of 1 in 87 000 of causing encephalitis (inflammation of the brain). This is much less than the risk of getting encephalitis as a result of catching measles. Also, the vaccines themselves are becoming much safer, and the risk of side-effects is now almost nil.

Routine vaccination not only protects the individual but also prevents the spread of infectious disease. Diseases like diphtheria and whooping cough were once common, and are now quite rare. This is the result of improved social conditions and routine vaccination. Smallpox was completely wiped out throughout the world by a World Health Organization programme of vaccination between 1959 and 1980.

Global travel

In the 18th and 19th centuries, explorers, traders and missionaries carried European diseases to countries where the population had no natural immunity. It is thought that devastating epidemics of smallpox and measles in, for example, North American Indians and Australian aborigines resulted from contact with infected Europeans.

Today, the ease with which we can travel around the world raises the possibility that travellers may catch a disease in a region where it is **endemic** and subsequently introduce it into a region where the incidence of disease is low or non-existent.

An 'endemic' disease is one that is constantly present in a population. Figure 10.2 shows areas in which malaria is endemic. Small numbers of travellers returning to Britain from such a region may have become infected during their stay. Fortunately, British mosquitoes do not transmit malaria, but global warming might change this.

If you plan to visit a country where an infectious disease is endemic, you are likely to be offered advice on vaccination. There is no vaccine against malaria but, if you are travelling to a malarious country, you will probably be advised to take a drug (e.g. chloroquine) that kills malarial parasites, starting a week or more before your departure, throughout your stay and for a few weeks after your return. Drugs such as this, which help to *prevent* you getting a disease are called **prophylactics**.

Also, you may find your aircraft cabin being sprayed with insecticide to kill any malaria-carrying mosquitoes that might have entered.

If you visit a country where a disease, e.g. yellow fever, is endemic, you may be required to produce a certificate of vaccination (Figure 10.12) before being allowed into a country where the disease does not occur.

Passive immunity

Some diseases can be prevented or cured by injecting the patient with serum from a person who has recovered from the disease. Serum is plasma with

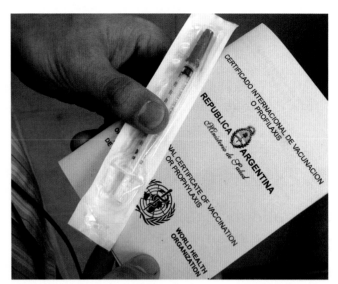

Figure 10.12 International certificate of vaccination

the fibrinogen removed. A serum is prepared from the plasma given by blood donors. People who have recently received an anti-tetanus inoculation will have made anti-tetanus antibodies in their blood. Some of these people volunteer to donate their blood, but their plasma is separated at once and the red cells returned to their circulation. The anti-tetanus antibodies are then extracted from the plasma and used to treat patients who are at risk of contracting tetanus, as a result of an accident, for example. Antibodies against chickenpox and rabies can be produced in a similar way.

The temporary immunity conferred by these methods is called **passive immunity** because the antibodies have not been produced by the patient. It is only temporary because it does not result in the formation of memory cells.

When a mother breastfeeds her baby, the milk contains some of the mother's white blood cells, which produce antibodies. These antibodies provide the baby with protection against infection at a vulnerable time: the baby's immune responses are not yet fully developed. However, this is another case of passive immunity as it is only short-term protection: memory cells are not produced.

Type 1 diabetes

This type of diabetes, also known as juvenile-onset diabetes, mainly affects young people. It is due to the inability of islet cells in the pancreas to produce sufficient insulin. There is a slight inherited tendency towards the disease, but it may be triggered by some event, possibly a virus infection, which causes the body's immune system to attack the islet cells that produce insulin. It is therefore classed as an **autoimmune** disease. The outcome is that the patient's blood is deficient in insulin and he or she needs regular injections of the hormone in order to control blood sugar levels and so lead a normal life. This form of the disease is, therefore, sometimes called 'insulin-dependent' diabetes (see 'Homeostasis' in Chapter 14).

● Extension work

Ideas about disease transmission and micro-organisms

Edward Jenner (1749–1823)

The history of immunisation centres on the disease **smallpox**, which is caused by a virus. Only a few years ago it was a serious, worldwide disease causing hundreds of thousands of deaths.

It had long been noticed that people who had recovered from smallpox never caught the disease again. In the late 1600s this observation was exploited in countries such as Greece, Turkey, China and India. Fluid from the blisters, which characterised the disease, was introduced into healthy people through cuts in the skin. The patient suffered a mild form of smallpox but was, thereafter, immune to the disease. It was a risky practice, however, and some people developed smallpox and died as a result of the vaccination.

In the 1750s, a Suffolk surgeon, Robert Sutton, refined the technique with considerable success. Edward Jenner is usually given the credit for smallpox vaccination. While using Sutton's technique he noticed that milkmaids who had caught 'cowpox' from infected cows did not develop the mild symptoms of illness after vaccination.

In 1796, Jenner conducted a crucial, if somewhat risky, experiment. He took fluid from a cowpox blister on a milkmaid's hand and injected it into a young boy. Two months later, he inoculated the boy with smallpox and demonstrated that the boy was immune. After publication of the results, the practice spread widely throughout Europe, reducing deaths from smallpox by about two-thirds.

Jenner called his technique 'vaccination' to distinguish it from inoculation with smallpox. 'Vacca' is Latin for 'cow' and 'vaccinia' is the medical name for cowpox. We now know that viruses and bacteria often lose much of their virulence if they are allowed to pass through different animals or are cultured in a particular way. Such non-virulent microbes are said to be **attenuated**. Jenner and his contemporaries, of course, knew nothing about viruses or attenuation but their shrewd observations, logical deductions and bold experiments led to a massive reduction in suffering.

In 1967, the World Health Organization embarked on a programme to eradicate smallpox from the whole world. The strategy was to trace all cases of smallpox and isolate the patients so that they could not pass on the disease. Everyone at risk was then vaccinated. By 1987 the disease had been eradicated.

Louis Pasteur (1822–95)

Pasteur made outstanding contributions to chemistry, biology and medicine. In 1854, as professor of chemistry at the University of Lille, he was called in by the French wine industry to investigate the problem of wines going sour.

Under the microscope he observed the yeast cells that were present and proposed that these were responsible for the fermentation. Thus, he claimed, fermentation was the outcome of a living process in yeast and not caused solely by a chemical change in the grape juice. In time, Pasteur observed that the yeast cells were supplanted by microbes (which we now call 'bacteria'), which appeared to change the alcohol into acetic and lactic acids.

Pasteur showed that souring was prevented by heating the wine to 120°F (49°C). He reasoned that this was because the microbes responsible for souring had been killed by the heat and, if the wine was promptly bottled, they could not return. This process is now called 'pasteurisation'.

Spontaneous generation

The micro-organisms in decaying products could be seen under the microscope, but where did they come from? Many scientists claimed that they were the *result* of decay rather than the *cause*; they had arisen 'spontaneously' in the decaying fluids.

In the 17th century, it was believed that organisms could be generated from decaying matter. The organisms were usually 'vermin' such as insects, worms and mice. To contest this notion, an experiment was conducted in 1668, comparing meat freely exposed to the air with meat protected from blowflies by a gauze lid on the container. Maggots appeared only in the meat to which blowflies had access.

This, and other experiments, laid to rest theories about spontaneous generation, as far as visible organisms were concerned, but the controversy about the origin of microbes continued into the 1870s.

It was already known that prolonged boiling, followed by enclosure, prevented liquids from putrefying. Exponents of spontaneous generation claimed that this was because the heat had affected some property of the air in the vessel. Pasteur designed experiments to put this to the test.

He made a variety of flasks, two of which are shown in Figure 10.13, and boiled meat broth in each of them. Fresh air was not excluded from the flask but could enter only through a tube, which was designed to prevent 'dust' (and microbes) from reaching the liquid. The broths remained sterile until either the flask was opened or until it was tilted to allow some broth to reach the U-bend and then tipped back again.

Figure 10.13 Two of Pasteur's flask shapes. The thin tubes admitted air but microbes were trapped in the U-bend.

This series of experiments, and many others, supported the theory that micro-organisms *caused* decay and did not arise spontaneously in the liquids.

The germ theory of disease

In 1865, Pasteur was asked to investigate the cause of a disease of silkworms (silk-moth caterpillars) that was devastating the commercial production of silk. He observed that particular micro-organisms were present in the diseased caterpillars but not in the healthy ones. He demonstrated that, by removing all of the diseased caterpillars and moths, the disease could be controlled. This evidence supported the idea that the microbes passed from diseased caterpillars to healthy ones, thus causing the disease to spread.

He extended this observation to include many forms of transmissible disease, including anthrax. He also persuaded doctors to sterilise their instruments by boiling, and to steam-heat their bandages. In this way, the number of infections that followed surgery was much reduced.

Pasteur's discoveries led to the introduction of antiseptic surgery and also to the production of a rabies vaccine.

10 DISEASES AND IMMUNITY

Questions

Core

1. a What are the two main lines of attack on malaria?
   b What is the connection between stagnant water and malaria?
   c What are the principal 'set-backs' in the battle against malaria?
2. Study the cartoon shown in Figure 10.14. Identify the potential hygiene risks in Sid's Store.
3. In what ways might improved sanitation and hygiene help to reduce the spread of amoebic dysentery?
4. How might a medical officer try to control an outbreak of amoebic dysentery?
5. Why should people who sell, handle and cook food be particularly careful about their personal hygiene?
6. Coughing or sneezing without covering the mouth and nose with a handkerchief is thought to be inconsiderate behaviour. Why is this?
7. Inhaling cigarette smoke can stop the action of cilia in the trachea and bronchi for about 20 minutes. Why should this increase a smoker's chance of catching a respiratory infection?

Figure 10.14 An unhygienic shop

Extended

8. Figure 10.15 shows the changes in the levels of antibody in response to an inoculation of a vaccine, followed by a booster injection 3 weeks later. Use your knowledge of the immune reaction to explain these changes.

9. How might a harmful bacterium be destroyed or removed by the body if it arrived:
   a on the hand
   b in a bronchus
   c in the stomach?
10. After a disaster such as an earthquake, the survivors are urged to boil all drinking water. Why do you think this is so?
11. Explain why vaccination against diphtheria does not protect you against polio as well.
12. Even if there have been no cases of diphtheria in a country for many years, children may still be vaccinated against it. What do you think is the point of this?

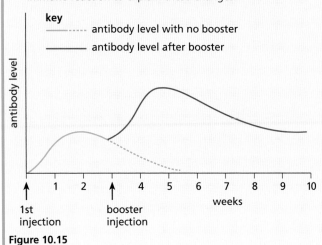

Figure 10.15

Checklist

After studying Chapter 10 you should know and understand the following:

- Transmissible diseases are infections caused by viruses, bacteria, fungi or protoctista.
- Infectious diseases may be transmitted by air, water, food or contact.
- The body has defences against pathogens, including mechanical and chemical barriers and white blood cells.
- A vaccine stimulates the blood system to produce antibodies against a disease, without causing the disease itself.
- The presence of antibodies in the blood, or the ability to produce them rapidly, gives immunity to a disease.
- Water-borne diseases are controlled by sewage treatment and water purification.
- Food-borne diseases can be controlled by hygienic food preparation, hygienic handling and good personal hygiene.
- The spread of disease can be controlled by waste disposal and sewage treatment.
- Antibodies, produced by lymphocytes, work by locking on to antigens.
- Antigens have specific shapes, so each type of antigen needs a different antibody.
- Active immunity is a defence against a pathogen by antibody production in the body.
- Vaccination involves the administration of a dead or inactive form of the pathogen to a patient to stimulate antibody production.
- Memory cells provide long-term immunity.
- Systematic immunisation can protect whole populations.
- Passive immunity only provides short-term protection because memory cells are not produced.
- Type 1 diabetes is caused by the immune system targeting and destroying cells in the pancreas.

Gas exchange in humans

Gas exchange in humans
Features of human gas exchange surfaces
Parts of the breathing system
Composition of inspired and expired air
Test for carbon dioxide

Identification of muscles associated with breathing
Roles of parts of the breathing system in ventilation
Explaining differences between inspired and expired air
Role of brain in monitoring carbon dioxide
Protection of the gas exchange system against pathogens

● Gas exchange in humans

All the processes carried out by the body, such as movement, growth and reproduction, require energy. In animals, this energy can be obtained only from the food they eat. Before the energy can be used by the cells of the body, it must be set free from the chemicals of the food by a process called 'respiration' (see Chapter 12). Aerobic respiration needs a supply of oxygen and produces carbon dioxide as a waste product. All cells, therefore, must be supplied with oxygen and must be able to get rid of carbon dioxide.

In humans and other mammals, the oxygen is obtained from the air by means of the lungs. In the lungs, the oxygen dissolves in the blood and is carried to the tissues by the circulatory system (Chapter 9).

Characteristics of respiratory surfaces

The exchange of oxygen and carbon dioxide across a respiratory surface, as in the lungs, depends on the diffusion of these two gases. Diffusion occurs more rapidly if:

- there is a large surface area exposed to the gas
- the distance across which diffusion has to take place is small
- there is a good blood supply, and
- there is a big difference in the concentrations of the gas at two points brought about by **ventilation**.

Large surface area

The presence of millions of alveoli in the lungs provides a very large surface for gaseous exchange. The many branching filaments in a fish's gills have the same effect.

Thin epithelium

There is only a two-cell layer, at the most, separating the air in the alveoli from the blood in the capillaries (Figure 11.4). One layer is the alveolus wall; the other is the capillary wall. Thus, the distance for diffusion is very short.

Good blood supply

The alveoli are surrounded by networks of blood capillaries. The continual removal of oxygen by the blood in the capillaries lining the alveoli keeps its concentration low. In this way, a steep diffusion gradient is maintained, which favours the rapid diffusion of oxygen from the air passages to the alveolar lining.

The continual delivery of carbon dioxide from the blood into the alveoli, and its removal from the air passages by ventilation, similarly maintains a diffusion gradient that promotes the diffusion of carbon dioxide from the alveolar lining into the bronchioles.

Ventilation

Ventilation of the lungs helps to maintain a steep diffusion gradient (see 'Diffusion' in Chapter 3) between the air at the end of the air passages and the alveolar air. The concentration of the oxygen in the air at the end of the air passages is high, because the air is constantly replaced by the breathing actions.

The respiratory surfaces of land-dwelling mammals are invariably moist. Oxygen has to dissolve in the thin film of moisture before passing across the epithelium.

Lung structure

The lungs are enclosed in the thorax (chest region) (see Figure 7.13). They have a spongy texture and can be expanded and compressed by movements of the thorax in such a way that air is sucked in and

Gas exchange in humans

blown out. The lungs are joined to the back of the mouth by the windpipe or **trachea** (Figure 11.1). The trachea divides into two smaller tubes, called **bronchi** (singular = bronchus), which enter the lungs and divide into even smaller branches. When these branches are only about 0.2 mm in diameter, they are called **bronchioles** (Figure 11.3(a)). These fine branches end in a mass of little, thin-walled, pouch-like air sacs called **alveoli** (Figures 11.3(b), (c) and 11.4).

The **epiglottis** and other structures at the top of the trachea stop food and drink from entering the air passages when we swallow.

Figure 11.2 shows a section through the thorax. The ribs, shown in cross section, form a cage, which has two main functions:

- to protect the lungs and heart
- to move to ventilate the lungs.

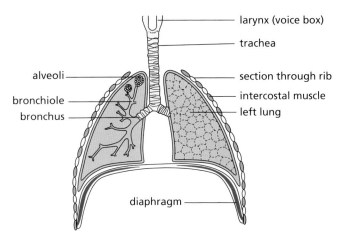

Figure 11.2 Section through the thorax

The alveoli have thin elastic walls, formed from a single-cell layer or **epithelium**. Beneath the epithelium is a dense network of capillaries (Figure 11.3(c)) supplied with deoxygenated blood (see 'Blood' in Chapter 9). This blood, from which the body has taken oxygen, is pumped from the right ventricle, through the pulmonary artery (see Figure 9.20). In humans, there are about 350 million alveoli, with a total absorbing surface of about 90 m². This large absorbing surface makes it possible to take in oxygen and give out carbon dioxide at a rate to meet the body's needs.

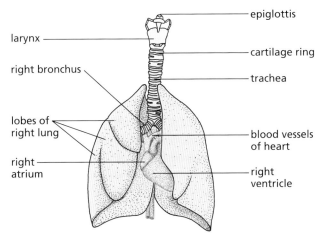

Figure 11.1 Diagram of lungs, showing position of heart

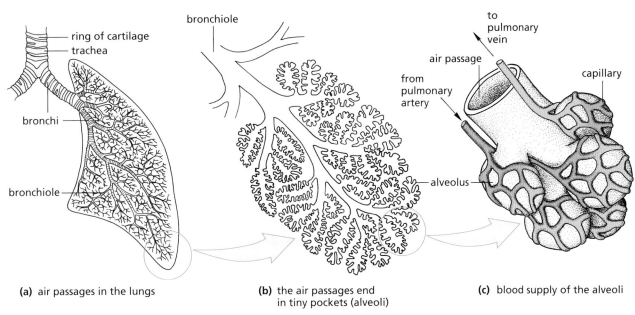

(a) air passages in the lungs

(b) the air passages end in tiny pockets (alveoli)

(c) blood supply of the alveoli

Figure 11.3 Lung structure

11 GAS EXCHANGE IN HUMANS

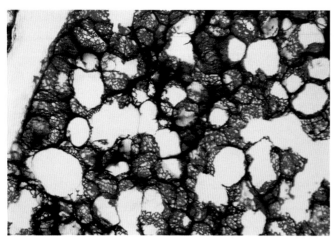

Figure 11.4 Small piece of lung tissue (×40). The capillaries have been injected with red and blue dye. The networks surrounding the alveoli can be seen.

Gaseous exchange

Ventilation refers to the movement of air into and out of the lungs. Gaseous exchange refers to the exchange of oxygen and carbon dioxide, which takes place between the air and the blood vessels in the lungs (Figure 11.5).

The 1.5 litres of residual air in the alveoli is not exchanged during ventilation and oxygen has to reach the capillaries by the slower process of diffusion. Figure 11.5 shows how oxygen reaches the red blood cells and how carbon dioxide escapes from the blood.

The oxygen combines with the haemoglobin in the red blood cells, forming **oxyhaemoglobin** (see 'Blood' in Chapter 9). The carbon dioxide in the plasma is released when the hydrogencarbonate ions ($-HCO_3$) break down to CO_2 and H_2O.

The capillaries carrying oxygenated blood from the alveoli join up to form the pulmonary vein (see Figure 9.20), which returns blood to the left atrium of the heart. From here it enters the left ventricle and is pumped all around the body, so supplying the tissues with oxygen.

Table 11.1 shows changes in the composition of air as it is breathed in and out.

Table 11.1 Changes in the composition of breathed air

	Inhaled/%	Exhaled/%
oxygen	21	16
carbon dioxide	0.04	4
water vapour	variable	saturated

Sometimes the word **respiration** or **respiratory** is used in connection with breathing. The lungs, trachea and bronchi are called the **respiratory system**; a person's rate of breathing may be called his or her **respiration rate**. This use of the word should not be confused with the biological meaning of respiration, namely the release of energy in cells (Chapter 12). This chemical process is sometimes called **tissue respiration** or **internal respiration** to distinguish it from breathing.

Lung capacity and breathing rate

The total volume of the lungs when fully inflated is about 5 litres in an adult. However, in quiet breathing, when asleep or at rest, you normally exchange only about 500 cm³. During exercise you can take in and expel an extra 3 litres. There is a **residual volume** of 1.5 litres, which cannot be expelled no matter how hard you breathe out.

At rest, you normally inhale and exhale about 12 times per minute. During exercise, the breathing rate may rise to over 20 breaths per minute and the depth also increases.

Breathing rate and exercise

The increased rate and depth of breathing during exercise allows more oxygen to dissolve in the blood and supply the active muscles. The extra carbon dioxide that the muscles put into the blood is detected by the brain, which instructs the intercostal muscles and diaphragm muscles to contract and relax more rapidly, increasing the breathing rate. Carbon dioxide will be removed by the faster, deeper breathing.

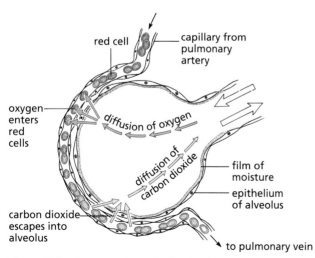

Figure 11.5 Gaseous exchange in the alveolus

Gas exchange in humans

Practical work

Oxygen in exhaled air

- Place a large screw-top jar on its side in a bowl of water (Figure 11.6(a)).
- Put a rubber tube in the mouth of the jar and then turn the jar upside-down, still full of water and with the rubber tube still in it.
- Start breathing out and when you feel your lungs must be about half empty, breathe the last part of the air down the rubber tubing so that the air collects in the upturned jar and fills it (Figure 11.6(b)).
- Put the screw top back on the jar under water, remove the jar from the bowl and place it upright on the bench.
- Light the candle on the special wire holder (Figure 11.6(c)), remove the lid of the jar, lower the burning candle into the jar and count the number of seconds the candle stays alight.
- Now take a fresh jar, with ordinary air, and see how long the candle stays alight in this.

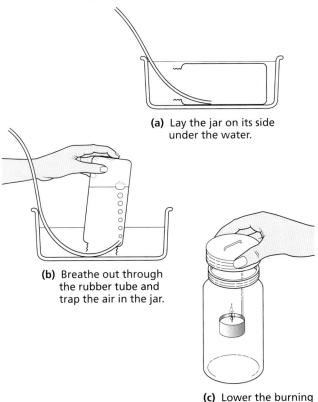

(a) Lay the jar on its side under the water.

(b) Breathe out through the rubber tube and trap the air in the jar.

(c) Lower the burning candle into the jar until the lid is resting on the rim.

Figure 11.6 Experiment to test exhaled air for oxygen

Results
The candle will burn for about 15–20 seconds in a large jar of ordinary air. In exhaled air it will go out in about 5 seconds.

Interpretation
Burning needs oxygen. When the oxygen is used up, the flame goes out. It looks as if exhaled air contains much less oxygen than atmospheric air.

Carbon dioxide in exhaled air

- Prepare two large test-tubes, A and B, as shown in Figure 11.7, each containing a little clear limewater.
- Put the mouthpiece in your mouth and breathe in and out *gently* through it for about 15 seconds. Notice which tube is bubbling when you breathe out and which one bubbles when you breathe in.

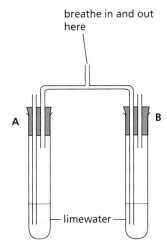

Figure 11.7 Experiment to compare the carbon dioxide content of inhaled and exhaled air

If after 15 seconds there is no difference in the appearance of the limewater in the two tubes, continue breathing through them for another 15 seconds.

Results
The limewater in tube B goes milky. The limewater in tube A stays clear.

Interpretation
Carbon dioxide turns limewater milky. Exhaled air passes through tube B. Inhaled air passes through tube A. Exhaled air must, therefore, contain more carbon dioxide than inhaled air.

Note 1: if the breathing process is carried out for too long, the limewater that had turned milky will revert to being colourless. This is because the calcium carbonate formed (milky precipitate) reacts in water with carbon dioxide to form calcium hydrogencarbonate, which is soluble and colourless.

Note 2: Hydrogencarbonate indicator is an alternative to limewater. It changes from red to yellow when carbon dioxide is bubbled through it.

Volume of air in the lungs

- Calibrate a large (5 litre) plastic bottle by filling it with water, half a litre at a time, and marking the water levels on the outside.
- Fill the bottle with water and put on the stopper.
- Put about 50 mm depth of water in a large plastic bowl.
- Hold the bottle upside-down with its neck under water and remove the screw top. Some of the water will run out but this does not matter.

11 GAS EXCHANGE IN HUMANS

- Push a rubber tube into the mouth of the bottle to position A, shown on the diagram (Figure 11.8).
- Take a deep breath and then exhale as much air as possible down the tubing into the bottle. The final water level inside the bottle will tell you how much air you can exchange in one deep breath.
- Now push the rubber tubing further into the bottle, to position B (Figure 11.8), and blow out any water left in the tube.
- Support the bottle with your hand and breathe gently in and out through the tube, keeping the water level inside and outside the bottle the same. This will give you an idea of how much air you exchange when breathing normally.

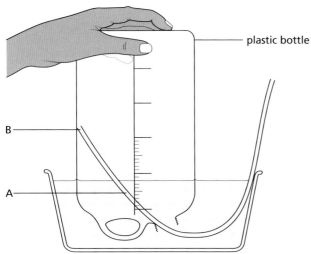

Figure 11.8 Experiment to measure the volume of air exhaled from the lungs. (A) shows the position of the tube when measuring the maximum usable lung volume. (B) is the position for measuring the volume exchanged in gentle breathing.

Investigating the effect of exercise on carbon dioxide production

- Half fill two clean boiling tubes with limewater.
- Place a drinking straw in one of the boiling tubes and gently blow into it, with normal, relaxed breaths.
- Count how many breaths are needed to turn the limewater milky.
- Now exercise for 1 to 2 minutes, e.g. running on the spot.
- Place a drinking straw in the second boiling tube, blowing into it as before.
- Count the number of breaths needed to turn the limewater milky.

Results
The number of breaths needed after exercise will be less than before exercise.

Interpretation
Cells in the body are constantly respiring, even when we are not doing physical work. They produce carbon dioxide, which is expired by the lungs. The carbon dioxide turns limewater milky. During exercise, cells (particularly in the skeletal muscles) respire more rapidly producing more carbon dioxide. This turns the limewater milky more rapidly.

Investigating the effect of exercise on rate and depth of breathing

This investigation makes use of an instrument called a spirometer. It may be one as illustrated in Figure 11.9, or a digital version, connected to a computer. A traditional spirometer has a hinged chamber, which rises and falls as a person breathes through the mouthpiece. The chamber is filled with medical oxygen from a cylinder. There is a filter containing soda lime, which removes any carbon dioxide in the user's breath, so that it is not re-breathed. The hinged chamber has a pen attached (shown in red in Figure 11.9), which rests against the paper-covered drum of a kymograph. This can be set to revolve at a fixed rate so that the trace produced by the user progresses across the paper.

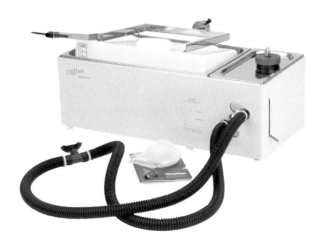

Figure 11.9 A spirometer. This instrument measures the volume of air breathed in and out of the lungs and can be used to measure oxygen consumption.

- A volunteer is asked to breathe in and out through the mouthpiece and the kymograph is set to revolve slowly. This will generate a trace, which will provide information about the volunteer's tidal volume and breathing rate (each peak on the trace represents one breath and the depth between a peak and trough can be used to calculate the tidal volume).
- Next, the volunteer is asked to take a deep breath with the mouthpiece removed, then breathe out through the mouthpiece for one long continuous breath. The depth between the peak and trough produced can be used to calculate the vital capacity.
- Finally, the volunteer is asked insert the mouthpiece, then run on the spot or pedal an exercise bicycle, while breathing through the spirometer. The trace produced (Figure 11.10) can be used to compare the breathing rate and depth during exercise with that at rest. A study of the trace would also show a drop in the trace with time. This can be used to calculate the volume of oxygen consumed over time.

Results
Tidal volume is about 500 cm^3, but tends to appear higher if the person is nervous or influenced by the trace being created.

Vital capacity can be between 2.5 and 5.0 litres, depending on the sex, physical size and fitness of the person.

Gas exchange in humans

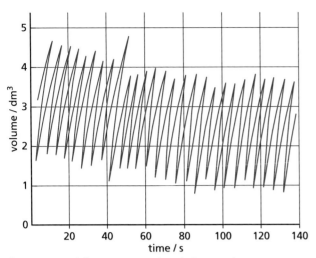

Figure 11.10 Spirometer trace taken during exercise

The breathing rate at rest is around 12 breaths per minute. During exercise this increases and may reach 20 or more breaths per minute.

Note: this experiment makes use of medical oxygen. This has a high purity and is toxic if inhaled for a prolonged period of time. If the volunteer starts to feel dizzy while using the spirometer, he or she should remove the mouthpiece immediately and rest.

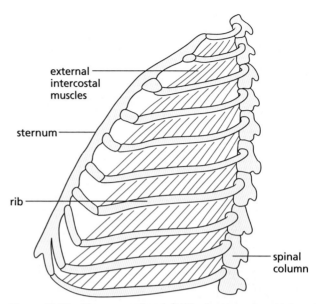

Figure 11.11 Ribcage seen from left side, showing external intercostal muscles

Ventilation of the lungs

The movement of air into and out of the lungs, called **ventilation**, renews the oxygen supply in the lungs and removes the surplus carbon dioxide. Horseshoe-shaped hoops of cartilage are present in the trachea and bronchi to prevent them collapsing when we breathe in. The lungs contain no muscle fibres and are made to expand and contract by movements of the ribs and diaphragm.

The **diaphragm** is a sheet of tissue that separates the thorax from the abdomen (see Figure 7.13). When relaxed, it is domed slightly upwards. The ribs are moved by the **intercostal muscles**. The external intercostals (Figure 11.11) contract to pull the ribs upwards and outwards. The internal intercostals contract to pull them downwards and inwards. Figure 11.12 shows the contraction of the external intercostals making the ribs move upwards.

Inhaling

1 The diaphragm muscles contract and pull it down (Figure 11.13(a)).
2 The internal intercostal muscles relax, while the external intercostal muscles contract and pull the ribcage upwards and outwards (Figure 11.14(a)).

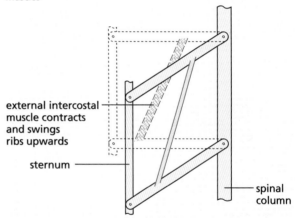

Figure 11.12 Model to show action of intercostal muscles

These two movements make the volume in the thorax bigger, so forcing the lungs to expand. The reduction in air pressure in the lungs results in air being drawn in through the nose and trachea. This movement of air into the lungs is known as ventilation.

Exhaling

1 The diaphragm muscles relax, allowing the diaphragm to return to its domed shape (Figure 11.13(b)).
2 The external intercostal muscles relax, while the internal intercostal muscles contract, pulling the ribs downwards to bring about a forced expiration (Figure 11.14(b)).

The lungs are elastic and shrink back to their relaxed volume, increasing the air pressure inside them. This results in air being forced out again.

11 GAS EXCHANGE IN HUMANS

The outside of the lungs and the inside of the thorax are lined with a smooth membrane called the **pleural membrane**. This produces a thin layer of liquid called **pleural fluid**, which reduces the friction between the lungs and the inside of the thorax.

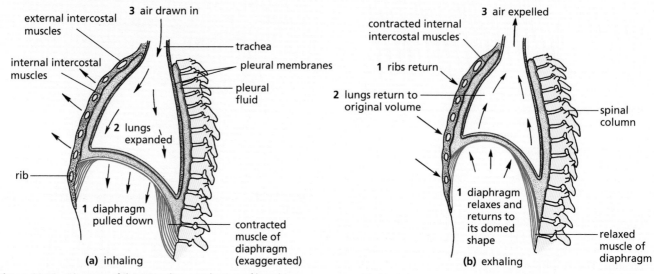

Figure 11.13 Diagrams of thorax to show mechanism of breathing

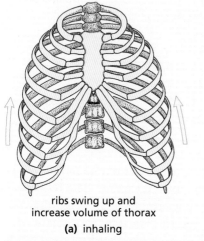
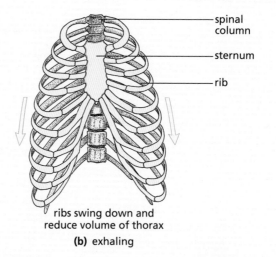

Figure 11.14 Movement of ribcage during breathing

A piece of apparatus known as the 'bell-jar model' (Figure 11.15) can be used to show the way in which movement of the diaphragm results in inspiration and expiration. The balloons start off deflated. When the handle attached to the rubber sheet is pulled down, the balloons inflate. If the handle is released, the balloons deflate again.

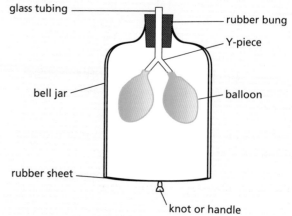

Figure 11.15 Bell-jar model

When the rubber sheet is pulled down, the volume inside the bell jar increases. This reduces the air pressure inside, making it lower than outside. The air rushes in, through the glass tubing, to equalise the air pressure, causing the balloons to inflate.

Differences in composition of inspired and expired air

Air in the atmosphere (which is breathed in) contains about 21% oxygen (see Table 11.1). Some of this is absorbed into the bloodstream when it enters the alveoli, resulting in a reduction of oxygen in exhaled air to 16% (the process of gaseous exchange in the alveoli does not remove all the oxygen from the air). Gas exchange relies on diffusion to transfer the oxygen into red blood cells and the air breathed in mixes with air that has not all been breathed out from the previous breath, so the process of gas exchange is not very efficient.

The remaining 79% of the air consists mainly of nitrogen, the percentage composition of which does not change significantly during breathing.

Inspired air contains 0.04% carbon dioxide. Cells of the body produce carbon dioxide as a waste product during aerobic respiration (see 'Aerobic respiration' in Chapter 12). The bloodstream carries carbon dioxide to the lungs for excretion. It diffuses across the walls of the alveoli to be expired. The percentage breathed out is 4%, 100 times greater than the percentage breathed in.

The lining of the alveoli is coated with a film of moisture in which the oxygen dissolves. Some of this moisture evaporates into the alveoli and saturates the air with water vapour. The air you breathe out, therefore, always contains a great deal more water vapour than the air you breathe in. The presence of water vapour in expired air is easily demonstrated by breathing onto a cold mirror: condensation quickly builds up on the glass surface. The exhaled air is warmer as well, so in cold and temperate climates you lose heat to the atmosphere by breathing.

The relationship between physical activity and the rate and depth of breathing

It has already been stated that the rate and depth of breathing increase during exercise. In order for the limbs to move faster, aerobic respiration in the skeletal muscles increases. Carbon dioxide is a waste product of aerobic respiration. As a result, CO_2 builds up in the muscle cells and diffuses into the plasma in the bloodstream more rapidly. The brain detects increases in carbon dioxide concentration in the blood and stimulates the breathing mechanism to speed up, increasing the rate of expiration of the gas. An increase in the breathing rate also has the advantage of making more oxygen available to the more rapidly respiring muscle cells.

Protection of the gas exchange system from pathogens and particles

Pathogens are disease-causing organisms (see Chapter 10). Pathogens, such as bacteria, and dust particles are present in the air we breathe in and are potentially dangerous if not actively removed. There are two types of cells that provide mechanisms to help achieve this.

Goblet cells are found in the epithelial lining of the trachea, bronchi and some bronchioles of the respiratory tract (Figure 11.16). Their role is to secrete **mucus**. The mucus forms a thin film over the internal lining. This sticky liquid traps pathogens and small particles, preventing them from entering the alveoli where they could cause infection or physical damage.

Ciliated cells are also present in the epithelial lining of the respiratory tract (Figure 11.16; see also 'Levels of organisation' in Chapter 2). They are in a continually flicking motion to move the mucus, secreted by the goblet cells, upwards and away from the lungs. When the mucus reaches the top of the trachea, it passes down the gullet during normal swallowing.

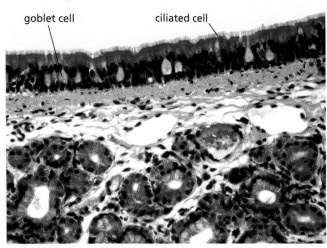

Figure 11.16 Goblet cells and ciliated cells in the trachea

11 GAS EXCHANGE IN HUMANS

Questions

Core
1. Place the following structures in the order in which air will reach them when breathing in: bronchus, trachea, nasal cavity, alveolus.
2. One function of the small intestine is to absorb food (see 'Absorption' in Chapter 7). One function of the lungs is to absorb oxygen. Point out the basic similarities in these two structures, which help to speed up the process of absorption.

Extended
3. a Compare the bell-jar model in Figure 11.15 with the diagram of the lungs (Figure 11.1). What do the following parts represent on the model?
   i glass tubing
   ii Y-piece
   iii balloons
   iv bell jar
   v rubber sheet
   b Explain why this model does not give a complete simulation of the process of breathing.
4. What are the two principal muscular contractions that cause air to be inhaled?
5. Place the following in the correct order:
   lungs expand, ribs rise, air enters lungs, external intercostal muscles contract, thorax expands.
6. During inhalation, which parts of the lung structure would you expect to expand the most?

Checklist

After studying Chapter 11 you should know and understand the following:

- Alveoli in the lungs are very numerous, provide a large surface area, have a thin, moist surface and are well-ventilated for efficient gas exchange.
- Alveoli have a good blood supply.
- Exchange of oxygen and carbon dioxide in the alveoli takes place by diffusion.
- The blood in the capillaries picks up oxygen from the air in the alveoli and gives out carbon dioxide. This is called gaseous exchange.
- The oxygen is carried around the body by the blood and used by the cells for their respiration.
- The ribs, rib muscles and diaphragm make the lungs expand and contract. This causes inhaling and exhaling.
- Air is drawn into the lungs through the trachea, bronchi and bronchioles.
- Inhaled air contains a higher percentage of oxygen and a lower percentage of carbon dioxide and (usually) water vapour than exhaled air.
- Limewater is used as a test for the presence of carbon dioxide. It turns milky.
- During exercise, the rate and depth of breathing increase.

- Cartilage, present in the trachea, keeps the airway open and unrestricted.
- The diaphragm, internal and external intercostal muscles play a part in ventilation of the lungs.
- During exercise, the rate and depth of breathing increase. This supplies extra oxygen to the muscles and removes their excess carbon dioxide.
- Movement of the ribcage and diaphragm results in volume and pressure changes in the thorax, leading to ventilation of the lungs.
- During physical activity, increases in levels of carbon dioxide in the blood are detected in the brain, causing an increased rate of breathing.
- Goblet cells make mucus to trap pathogens and particles to protect the gas exchange system.
- Ciliated cells move mucus away from the alveoli.

12 Respiration

Respiration
Use of energy in humans
Role of enzymes

Aerobic respiration
Define aerobic respiration
Word equation
Investigating uptake of oxygen in respiring organisms

Balanced chemical equation
Investigating the effect of temperature on respiration

Anaerobic respiration
Define anaerobic respiration
Word equations
Energy output compared with aerobic respiration

Balanced chemical equation
Effects of lactic acid
Oxygen debt

● Respiration

Most of the processes taking place in cells need energy to make them happen. Examples of energy-consuming processes in living organisms are:

- the contraction of muscle cells – to create movement of the organism, or peristalsis to move food along the alimentary canal, or contraction of the uterus wall during childbirth
- building up proteins from amino acids
- the process of cell division (Chapter 17) to create more cells, or replace damaged or worn out cells, or to make reproductive cells
- the process of active transport (Chapter 3), involving the movement of molecules across a cell membrane against a concentration gradient
- growth of an organism through the formation of new cells or a permanent increase in cell size
- the conduction of electrical impulses by nerve cells (Chapter 14)
- maintaining a constant body temperature in homoiothermic (warm-blooded) animals ('Homeostasis' in Chapter 14) to ensure that vital chemical reactions continue at a predictable rate and do not slow down or speed up as the surrounding temperature varies.

This energy comes from the food that cells take in. The food mainly used for energy in cells is glucose.

The process by which energy is produced from food is called **respiration**.

Respiration is a chemical process that takes place in cells and involves the action of enzymes. It must not be confused with the process of breathing, which is also sometimes called 'respiration'. To make the difference quite clear, the chemical process in cells is sometimes called **cellular respiration**, **internal respiration** or **tissue respiration**. The use of the word 'respiration' for breathing is best avoided altogether.

● Aerobic respiration

> **Key definition**
> **Aerobic respiration** is the term for the chemical reactions in cells that use oxygen to break down nutrient molecules to release energy.

The word **aerobic** means that oxygen is needed for this chemical reaction. The food molecules are combined with oxygen. The process is called **oxidation** and the food is said to be **oxidised**. All food molecules contain carbon, hydrogen and oxygen atoms. The process of oxidation converts the carbon to carbon dioxide (CO_2) and the hydrogen to water (H_2O) and, at the same time, sets free energy, which the cell can use to drive other reactions.

Aerobic respiration can be summed up by the equation

$$\text{glucose} + \text{oxygen} \xrightarrow{\text{enzymes}} \text{carbon dioxide} + \text{water} + 2830\,\text{kJ energy}$$

The amount of energy you would get by completely oxidising 180 grams (g) of glucose to carbon dioxide and water is 2830 kilojoules (kJ). In the cells, the energy is not released all at once. The oxidation takes place in a series of small steps and not in one jump as the equation suggests. Each small step needs its own enzyme and at each stage a little energy is released (Figure 12.1).

Although the energy is used for the processes mentioned above, some of it always appears as heat. In 'warm-blooded' animals (birds and mammals) some of this heat is retained to maintain their body temperature.

12 RESPIRATION

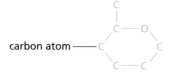

(a) molecule of glucose (H and O atoms not all shown)

(b) the enzyme attacks and breaks the glucose molecule into two 3-carbon molecules

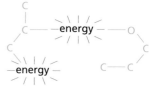

(c) this breakdown sets free energy

(d) each 3-carbon molecule is broken down to carbon dioxide

(e) more energy is released and CO_2 is produced

(f) the glucose has been completely oxidised to carbon dioxide (and water), and all the energy released

Figure 12.1 Aerobic respiration

In 'cold-blooded' animals (e.g. reptiles and fish) the heat may build up for a time in the body and allow the animal to move about more quickly. In plants the heat is lost to the surroundings (by conduction, convection and evaporation) as fast as it is produced.

Practical work

Experiments on respiration and energy

If you look below at the chemical equation that represents aerobic respiration you will see that a tissue or an organism that is respiring should be (a) using up food, (b) using up oxygen, (c) giving off carbon dioxide, (d) giving out water and (e) releasing energy, which can be used for other processes.

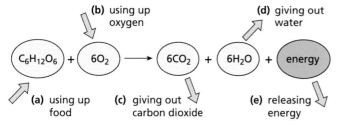

If we wish to test whether aerobic respiration is taking place:

- '(d) giving out water' is not a good test because non-living material will give off water vapour if it is wet to start with.
- '(a) using up food' can be tested by seeing if an organism loses weight. This is not as easy as it seems because most organisms lose weight as a result of evaporation of water and this may have nothing to do with respiration. It is the decrease in 'dry weight' that must be measured.

We will focus on the uptake of oxygen and the production of carbon dioxide as indications that respiration is taking place.

Seeds are often used as the living organisms because when they start to grow (germinate) there is a high level of chemical activity in the cells. Seeds are easy to obtain and to handle and they fit into small-scale apparatus. In some cases blowfly maggots or woodlice can be used as animal material. Yeast is useful when studying anaerobic respiration.

1 Using up oxygen during respiration

The apparatus in Figure 12.2 is a **respirometer** (a 'respire meter'), which can measure the rate of respiration by seeing how quickly oxygen is taken up. Germinating seeds, or blowfly larvae or woodlice are placed in the test-tube and, as they use up the oxygen for respiration, the level of liquid in the delivery tubing will go up.

There are two drawbacks to this. One is that the organisms usually give out as much carbon dioxide as they take in oxygen. So there may be no change in the total amount of air in the test-tube and the liquid level will not move. This drawback is overcome by placing **soda-lime** in the test-tube. Soda-lime will absorb carbon dioxide as fast as the organisms give it out. So only the uptake of oxygen will affect the amount of air in the tube. The second drawback is that quite small changes in temperature will make the air in the test-tube expand or contract and so cause the liquid to rise or fall whether or not respiration is taking place. To overcome this, the test-tube is kept in a beaker of water (a water bath). The temperature of water changes far more slowly than that of air, so there will not be much change during a 30-minute experiment.

Control
To show that it is a living process that uses up oxygen, a similar respirometer is prepared but containing an equal quantity of germinating seeds that have been killed by boiling. (If blowfly larvae or woodlice are used, the control can consist of an equivalent volume of glass beads. This is not a very good control but is probably more acceptable than killing an equivalent number of animals.)

The apparatus is finally set up as shown in Figure 12.2 and left for 30 minutes (10 minutes if blowfly larvae or woodlice are used).

The capillary tube and reservoir of liquid are called a **manometer**.

Result
The level of liquid in the experiment goes up more than in the control. The level in the control may not move at all.

Interpretation
The rise of liquid in the delivery tubing shows that the living seedlings have taken up part of the air. It does not prove that it is oxygen that has been taken up. Oxygen seems the most likely gas, however, because (1) there is only 0.03% carbon dioxide in the air to start with and (2) the other gas, nitrogen, is known to be less active than oxygen.

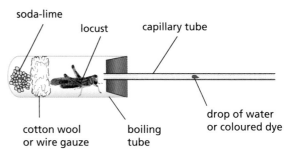

Figure 12.3 A simple respirometer

- A drop of water or coloured dye is introduced to the capillary tube by touching it against the liquid.
- The capillary tube is rested against a ruler and the position of the water drop is noted.
- After 1 minute (or longer if the drop moves very slowly) the new position of the water drop is recorded.

Note: Care must be taken when handling living organisms. Wash hands thoroughly with water if they come into contact with caustic soda.

Results
The water drop moves towards the organism. If the diameter of the bore of the capillary tube is measured, the volume of air taken in by the organism can be calculated:

$$\text{volume} = \pi r^2 l$$

where r = radius of the capillary tube bore

l = distance travelled by the water drop

This value can be converted into a rate if the volume is divided by the time taken.

Interpretation
The movement of the water drop towards the organism shows that it is taking in air. By using a range of organisms (locust, woodlice, blowfly larvae, germinating seeds) the rates of uptake can be compared to see which is respiring most actively.

A control could be set up using the same apparatus, but with glass beads instead of the organism(s). The bubble may still move because the soda-lime will absorb any carbon dioxide in the air in the boiling tube, but the movement should be less than that for living organisms.

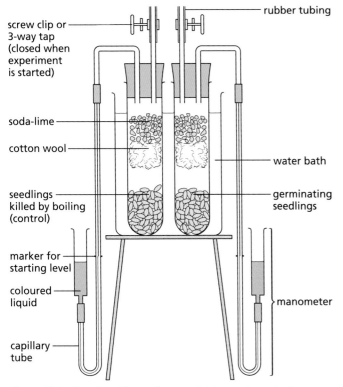

Figure 12.2 Experiment to see if oxygen is taken up in respiration

If the experiment is allowed to run for a long time, the uptake of oxygen could be checked at the end by placing a lighted splint in each test-tube in turn. If some of the oxygen has been removed by the living seedlings, the flame should go out more quickly than it does in the tube with dead seedlings.

2 Using up oxygen during respiration (alternative method)

A respirometer such as the one illustrated in Figure 12.2 is not an easy piece of apparatus to set up and collect data from. An alternative way of showing that oxygen is used up during respiration can be achieved using a simple respirometer (Figure 12.3).

- A larger invertebrate such as a locust, or a group of woodlice or blowfly maggots, is placed in the boiling tube (an alternative is a large plastic syringe, linked to the capillary tube with a short section of rubber or silicone tubing). The organisms are protected from the soda-lime by means of cotton wool or a wire gauze (soda-lime is caustic).

If you are following the extended curriculum you need to be able to state the balanced chemical equation for aerobic respiration:

$$C_6H_{12}O_6 + 6O_2 \longrightarrow 6CO_2 + 6H_2O + 2830\,\text{kJ}$$

glucose   oxygen   carbon   water   energy
                   dioxide

Mitochondria

It is in the mitochondria that the chemistry of aerobic respiration takes place (Chapter 2). The mitochondria generate a compound called **ATP**, which is used by the cell as the source of energy for driving other chemical reactions in the cytoplasm and nucleus.

Practical work

More experiments on respiration and energy

3 Investigating the effect of temperature on the rate of respiration of germinating seeds

- Use the same apparatus as shown in Experiment 2, but set up the boiling tube so it is vertical and supported in a water bath such as a beaker (Figure 12.4).
- Use wheat grains or pea seeds that have been soaked for 24 hours and rinsed in 1% formaldehyde (or domestic bleach diluted 1:4) for 5 minutes. These solutions will kill any bacteria or fungi on the surface of the seeds.
- Kill an equal quantity of soaked seeds by boiling them for 5 minutes.
- Cool the boiled seeds in cold tap water; rinse them in bleach or formaldehyde for 5 minutes as before. These can be used as the control (or, alternatively, use an equivalent volume of glass beads).
- Start with a water bath at about 20°C and allow the seeds to acclimatise to that temperature for a few minutes before taking any readings. The initial and final positions of the water drop could be recorded on the capillary tube with a permanent marker or chinagraph pencil, or by sticking a small label onto the glass. The distance travelled can then be measured with a ruler.
- Repeat the procedure (introducing a new bubble each time) at a range of different temperatures, remembering to allow time for the seeds to acclimatise to the new conditions before taking further readings.

Results
As the temperature is increased the rate of movement of the water bubble towards the seeds increases. The movement may stop at higher temperatures.

Interpretation
As the temperature increases, the rate of respiration in the germinating seeds increases. This is because the enzymes controlling respiration are more active at higher temperatures. However, respiration may stop above around 40°C because the enzymes become denatured if they get too hot.

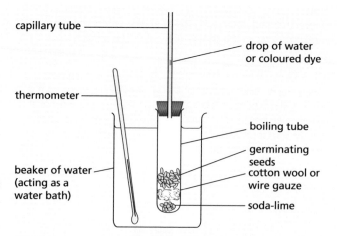

Figure 12.4 Simple respirometer for investigating the effect of temperature on germinating seeds

Controlled experiments

In most biological experiments, a second experiment called a **control** is set up. This is to make sure that the results of the first experiment are due to the conditions being studied and not to some other cause that has been overlooked.

In the experiment in Figure 12.2, the liquid rising up the capillary tube could have been the result of the test-tube cooling down, so making the air inside it contract. The identical experiment with dead seeds – the control – showed that the result was not due to a temperature change, because the level of liquid in the control did not move.

The term 'controlled experiment' refers to the fact that the experimenter (1) sets up a control and (2) controls the conditions in the experiment. In the experiment shown in Figure 12.2 the seeds are enclosed in a test-tube and soda-lime is added. This makes sure that any uptake or output of oxygen will make the liquid go up or down, and that the output of carbon dioxide will not affect the results. The experimenter had controlled both the amount and the composition of the air available to the germinating seeds.

If you did an experiment to compare the growth of plants in the house and in a greenhouse, you could not be sure whether it was the extra light or the high temperature of the greenhouse that caused better growth. This would not, therefore, be a properly controlled experiment. You must alter only one condition (called a **variable**) at a time, either the light or the temperature, and then you can compare the results with the control experiment.

A properly controlled experiment, therefore, alters only one variable at a time and includes a control, which shows that it is this condition and nothing else that gave the result.

● Anaerobic respiration

Key definition
Anaerobic respiration is the term for the chemical reactions in cells that break down nutrient molecules to release energy without using oxygen.

The word **anaerobic** means 'in the absence of oxygen'. In this process, energy is still released from food by breaking it down chemically but the reactions do not use oxygen though they do often produce carbon dioxide. A common example is the action of yeast on sugar solution to produce alcohol. The sugar is not completely oxidised to carbon dioxide and water but converted to carbon dioxide and alcohol. This process is called **fermentation** and is shown by the following equation:

$$\text{glucose} \xrightarrow{\text{enzymes}} \text{alcohol} + \text{carbon dioxide} + 118\,\text{kJ energy}$$

The processes of brewing and bread-making rely on anaerobic respiration by yeast. As with aerobic respiration, the reaction takes place in small steps and needs several different enzymes. The yeast uses the energy for its growth and living activities, but you can see from the equation that less energy is produced by anaerobic respiration than in aerobic respiration. This is because the alcohol still contains a great deal of energy that the yeast is unable to use.

Anaerobic respiration also occurs in muscles during vigorous exercise, because oxygen cannot be delivered fast enough to satisfy the needs of the respiring muscle cells. The products are different to those produced by anaerobic respiration in yeast. The process is shown by the following equation:

$$\text{glucose} \longrightarrow \text{lactic acid}$$

The lactic acid builds up in the muscles and causes muscle fatigue (cramp).

Anaerobic respiration is much less efficient than aerobic respiration because it releases much less energy per glucose molecule broken down (respired).

Practical work

More experiments on respiration and energy

4 **Releasing energy in respiration**

- Fill a small vacuum flask with wheat grains or pea seeds that have been soaked for 24 hours and rinsed in 1% formaldehyde (or domestic bleach diluted 1:4) for 5 minutes. These solutions will kill any bacteria or fungi on the surface of the seeds.
- Kill an equal quantity of soaked seeds by boiling them for 5 minutes.
- Cool the boiled seeds in cold tap water, rinse them in bleach or formaldehyde for 5 minutes as before and then put them in a vacuum flask of the same size as the first one. This flask is the control.
- Place a thermometer in each flask so that its bulb is in the middle of the seeds (Figure 12.5).
- Plug the mouth of each flask with cotton wool and leave both flasks for 2 days, noting the thermometer readings whenever possible.

Result
The temperature in the flask with the living seeds will be 5–10 °C higher than that of the dead seeds.

Interpretation
Provided there are no signs of the living seeds going mouldy, the heat produced must have come from living processes in the seeds, because the dead seeds in the control did not give out any heat. There is no evidence that this process is respiration rather than any other chemical change but the result is what you would expect if respiration does produce energy.

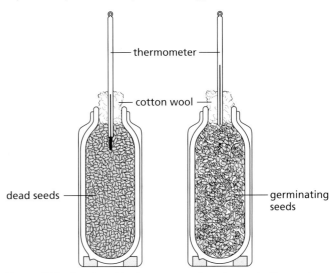

Figure 12.5 Experiment to show energy release in germinating seeds

5 Anaerobic respiration in yeast

- Boil some water to expel all the dissolved oxygen.
- When cool, use the boiled water to make up a 5% solution of glucose and a 10% suspension of dried yeast.
- Place 5 cm³ of the glucose solution and 1 cm³ of the yeast suspension in a test-tube and cover the mixture with a thin layer of liquid paraffin to exclude atmospheric oxygen.
- Fit a delivery tube as shown in Figure 12.6 and allow it to dip into clear limewater.

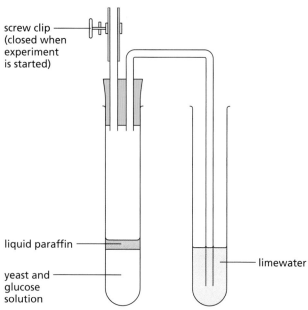

Figure 12.6 Experiment to show anaerobic respiration in yeast

Result
After 10–15 minutes, with gentle warming if necessary, there should be signs of fermentation in the yeast–glucose mixture and the bubbles of gas escaping through the limewater should turn it milky.

Interpretation
The fact that the limewater goes milky shows that the yeast–glucose mixture is producing carbon dioxide. If we assume that the production of carbon dioxide is evidence of respiration, then it looks as if the yeast is respiring. In setting up the experiment, you took care to see that oxygen was removed from the glucose solution and the yeast suspension, and the liquid paraffin excluded air (including oxygen) from the mixture. Any respiration taking place must, therefore, be anaerobic (i.e. without oxygen).

Control
It might be suggested that the carbon dioxide came from a chemical reaction between yeast and glucose (as between chalk and acid), which had nothing to do with respiration or any other living process. A control should, therefore, be set up using the same procedure as before but with yeast that has been killed by boiling. The failure, in this case, to produce carbon dioxide supports the claim that it was a living process in the yeast in the first experiment that produced the carbon dioxide.

The balanced chemical equation for anaerobic respiration in organisms such as yeast is shown below:

$$C_6H_{12}O_6 \xrightarrow{\text{enzymes}} 2C_2H_5OH + 2CO_2 + 118\,\text{kJ}$$

glucose alcohol carbon dioxide energy

This amount of energy released per mole of glucose respired is much less than that released in aerobic respiration (2830 kJ per mole).

During vigorous exercise, **lactic acid** may build up in a muscle. In this case it is removed in the bloodstream. The blood needs to move more quickly during and after exercise to maintain this lactic acid removal process, so the heart rate is rapid. On reaching the liver, some of the lactic acid is oxidised to carbon dioxide and water, using up oxygen in the process. After exercise has stopped, a high level of oxygen consumption may persist until the excess of lactic acid has been oxidised. This is characterised by deeper breathing (an athlete pants for breath). The build-up of lactic acid that is oxidised later is said to create an **oxygen debt**.

Accumulation of lactic acid in the muscles results in muscular fatigue, leading to cramp.

Athletes and climbers who are used to working at low altitude (normal air pressure) have problems if they then perform at high altitude (low air pressure). High-altitude air has a lower percentage of oxygen, so an oxygen debt can be experienced much more easily than at low altitude. The problem can be resolved if the person spends time at high altitude before performing to allow the body to acclimatise (making more red blood cells and increasing blood volume).

● Extension work

Metabolism
All the chemical changes taking place inside a cell or a living organism are called its **metabolism**. The minimum turnover of energy needed simply to keep an organism alive, without movement or growth, is called the **basal metabolism**. Our basal metabolism maintains vital processes such as breathing, heartbeat, digestion and excretion.

The processes that break substances down are sometimes called **catabolism**. Respiration is an example of catabolism in which carbohydrates are broken down to carbon dioxide and water. Chemical reactions that build up substances are called **anabolism**. Building up a protein from amino acids is an example of anabolism. The energy released by the **catabolic** process of respiration is used to drive the **anabolic** reactions that build up proteins.

You may have heard of anabolic steroids in connection with drug taking by athletes. These chemicals reduce the rate of protein breakdown and may enhance the build-up of certain proteins. However, their effects are complicated and not fully understood, they have undesirable side-effects and their use contravenes athletics codes (see 'Misused drugs' in Chapter 15).

Practical work

More experiments on respiration and energy

6 The effect of temperature on yeast respiration

- Make some bread dough using flour, water and activated yeast (yeast in a warm sugar solution).
- Rub the inside of a boiling tube with oil (this makes it easier to remove the dough after the experiment).
- Use a glass rod or the end of an old pencil to push a piece of dough into the bottom of the boiling tube, so that the tube is about a quarter full of dough.
- Mark the height of the top of the dough on the boiling tube, using a chinagraph pencil or permanent marker pen.
- Place the boiling tube into a beaker of water set to a preselected temperature, e.g. 20 °C.
- Leave the dough for 20 minutes, checking to make sure the temperature of the water bath remains constant (adding warm or cold water to maintain this).
- Record the new height of the dough.
- Repeat the procedure at different temperatures and compare the rate of rising of the bread dough.

Results
The dough rises faster as the temperature is increased to 35 or 40 °C. Higher temperatures slow down the rate. Low temperatures may result in no change in height of the dough.

Explanation
Yeast respires anaerobically, producing carbon dioxide. This causes the dough to rise. The process is controlled by enzymes, which work faster as the temperature is increased to the optimum (around 35–40 °C). Higher temperatures cause the enzymes to denature (Chapter 5).

Extension work

Hypothesis testing

You will have noticed that none of the experiments described above claim to have *proved* that respiration is taking place. The most we can claim is that they have not disproved the proposal that energy is produced from respiration. There are many reactions taking place in living organisms and, for all we know at this stage, some of them may be using oxygen or giving out carbon dioxide without releasing energy, i.e. they would not fit our definition of respiration.

This inability to 'prove' that a particular proposal is 'true' is not restricted to experiments on respiration. It is a feature of many scientific experiments. One way in which science makes progress is by putting forward a **hypothesis**, making predictions from the hypothesis, and then testing these predictions by experiments.

A hypothesis is an attempt to explain some event or observation using the information currently available. If an experiment's results do not confirm the predictions, the hypothesis must be abandoned or altered.

For example, biologists observing that living organisms take up oxygen might put forward the hypothesis that 'oxygen is used to convert food to carbon dioxide, so producing energy for movement, growth, reproduction, etc.' This hypothesis can be tested by predicting that, '*if* the oxygen is used to oxidise food *then* an organism that takes up oxygen will also produce carbon dioxide'. Experiment 1 on page 166 tests this and fulfils this prediction and, therefore, supports the hypothesis. Looking at the equation for respiration, we might also predict that an organism that is respiring will produce carbon dioxide and take up oxygen. Experiment 5 with yeast, however, does not fulfil this prediction and so does not support the hypothesis as it stands, because here is an organism producing carbon dioxide without taking up oxygen. The hypothesis will have to be modified, e.g. 'energy is released from food by breaking it down to carbon dioxide; some organisms use oxygen for this process, others do not'.

There are still plenty of tests that we have not done. For example, we have not attempted to see whether it is food that is the source of energy and carbon dioxide. One way of doing this is to provide the organism with food, e.g. glucose, in which the carbon atoms are radioactive. Carbon-14 (^{14}C) is a radioactive form of carbon and can be detected by using a Geiger counter. If the organism produces radioactive carbon dioxide, it is reasonable to suppose that the carbon dioxide comes from the glucose.

$$C_6H_{12}O_6 + 6O_2 \longrightarrow 6CO_2 + 6H_2O + \text{energy}$$

This is **direct evidence** in support of the hypothesis. All the previous experiments have provided only **indirect evidence**.

Criteria for a good hypothesis

A good hypothesis must:

- explain *all* aspects of the observation
- be the simplest possible explanation
- be expressed in such a way that predictions can be made from it
- be testable by experiment.

Questions

Core

1. a If, in one word, you had to say what respiration was about, which word would you choose from this list: breathing, energy, oxygen, cells, food?
   b In which parts of a living organism does respiration take place?
2. What are the main differences between aerobic and anaerobic respiration?
3. What chemical substances must be provided for aerobic respiration to take place:
   a from outside the cell
   b from inside the cell?
   c What are the products of aerobic respiration?
4. Which of the following statements are true? If an organism is respiring you would expect it to be:
   a giving out carbon dioxide
   b losing heat
   c breaking down food
   d using up oxygen
   e gaining weight
   f moving about.
5. What was the purpose of:
   a the soda-lime in the respirometer in Figure 12.2
   b the limewater in Figure 12.6?

Extended

6. What is the difference between aerobic and anaerobic respiration in the amount of energy released from one molecule of glucose?
7. Victims of drowning who have stopped breathing are sometimes revived by a process called 'artificial respiration'. Why would a biologist object to the use of this expression? ('Resuscitation' is a better word to use.)
8. Why do you think your breathing rate and heart rate stay high for some time after completing a spell of vigorous exercise?
9. In an experiment like the one shown in Figure 12.2, the growing seeds took in 5 cm³ oxygen and gave out 7 cm³ carbon dioxide. How does the volume change:
   a if no soda-lime is present
   b if soda-lime is present?
10. The germinating seeds in Figure 12.5 will release the same amount of heat whether they are in a beaker or a vacuum flask. Why then is it necessary to use a vacuum flask for this experiment?
11. Experiment 5 with yeast supported the claim that anaerobic respiration was taking place. The experiment was repeated using unboiled water and without the liquid paraffin. Fermentation still took place and carbon dioxide was produced. Does this mean that the design or the interpretation of the first experiment was wrong? Explain your answer.
12. Twenty seeds are placed on soaked cotton wool in a closed glass dish. After 5 days in the light 15 of the seeds had germinated. If the experiment is intended to see if light is needed for germination, which of the following would be a suitable control:
    a exactly the same set-up but with dead seeds
    b the same set-up but with 50 seeds
    c an identical experiment but with 20 seeds of a different species
    d an identical experiment but left in darkness for 5 days?
13. Certain bacteria that live in sulfurous springs in areas of volcanic activity take up hydrogen sulfide (H_2S) and produce sulfates ($-SO_4$). Put forward a hypothesis to account for this chemical activity. Suggest one way of testing your hypothesis.

14 The table below shows the energy used up each day either as kilojoules per kilogram of body mass or as kilojoules per square metre of body surface.

Animal	Mass/kg	kJ per day	
		per kg body mass	per m² body surface
man	64.3	134	4360
mouse	0.018	2736	4971

Reprinted from *Textbook of Physiology*, Emslie-Smith, Paterson, Scratcherd and Read, by permission of the publisher Churchill Livingstone, 1988

a According to the table, what is the total amount of energy used each day by
  i a man
  ii a mouse?
b Which of these two shows a greater rate of respiration in its body cells?
c Why, do you think, is there so little difference in the energy expenditure per square metre of body surface?

Checklist

After studying Chapter 12 you should know and understand the following:

- The word equation for aerobic respiration is

  glucose + oxygen $\xrightarrow{\text{enzymes}}$ carbon dioxide + water + energy

- Aerobic respiration is the term for the chemical reactions in cells that convert energy in nutrient molecules using oxygen so that cells can use this energy.
- The word equation for anaerobic respiration in muscles is

  glucose $\xrightarrow{\text{enzymes}}$ lactic acid + energy

- The word equation for anaerobic respiration in yeast is

  glucose $\xrightarrow{\text{enzymes}}$ alcohol + carbon dioxide + energy

- Anaerobic respiration is the term for the chemical reactions in cells that convert energy in nutrient molecules without the use of oxygen so that cells can use this energy.
- Respiration is the process in cells that releases energy from food.
- Aerobic respiration needs oxygen; anaerobic respiration does not.
- Aerobic respiration releases much more energy per glucose molecule than anaerobic respiration.
- The oxidation of food produces carbon dioxide as well as releasing energy.
- Experiments to investigate respiration try to detect uptake of oxygen, production of carbon dioxide, release of energy as heat or a reduction in dry weight.
- The balanced chemical equation for aerobic respiration is

  $C_6H_{12}O_6 + 6O_2 \longrightarrow 6CO_2 + 6H_2O + 2830\,kJ$

- Experiments to investigate the effect of temperature on the rate of respiration of germinating seeds.
- The balanced chemical equation for anaerobic respiration in yeast is

  $C_6H_{12}O_6 \longrightarrow 2C_2H_5OH + 2CO_2 + 118\,kJ$

- Lactic acid builds up in muscles due to anaerobic respiration, causing an oxygen debt.
- An outline of how oxygen is removed during recovery.
- In a controlled experiment, the scientist tries to alter only one condition at a time, and sets up a control to check this.
- A control is a second experiment, identical to the first experiment except for the one condition being investigated.
- The control is designed to show that only the condition under investigation is responsible for the results.
- Experiments are designed to test predictions made from hypotheses; they cannot 'prove' a hypothesis.

13 Excretion in humans

Excretion
Excretory products: urea, carbon dioxide
Contents of urine
Urine output
Parts of urinary system

Role of liver in conversion of amino acids to proteins
Define deamination
Explain the need for excretion
Structure and function of kidney tubule
Dialysis
Compare dialysis with kidney transplant

● Excretion

Excretion is the removal from organisms of toxic materials and substances in excess of requirements. These include:

- the waste products of its chemical reactions
- the excess water and salts taken in with the diet
- spent hormones.

Excretion also includes the removal of drugs or other foreign substances taken into the alimentary canal and absorbed by the blood.

Many chemical reactions take place inside the cells of an organism in order to keep it alive. Some products of these reactions are poisonous and must be removed from the body. For example, the breakdown of glucose during respiration (see 'Aerobic respiration' in Chapter 12) produces carbon dioxide. This is carried away by the blood and removed in the lungs. Excess amino acids are deaminated in the liver to form glycogen and **urea**. The urea is removed from the tissues by the blood and expelled by the kidneys.

Urea and similar waste products, like **uric acid**, from the breakdown of proteins, contain the element nitrogen. For this reason they are often called **nitrogenous waste products**.

During feeding, more water and salts are taken in with the food than are needed by the body. So these excess substances need to be removed as fast as they build up.

The hormones produced by the endocrine glands (Chapter 14) affect the rate at which various body systems work. Adrenaline, for example, speeds up the heartbeat. When hormones have done their job, they are modified in the liver and excreted by the kidneys.

The nitrogenous waste products, excess salts and spent hormones are excreted by the kidneys as a watery solution called **urine**.

Excretory organs

Liver

The liver breaks down excess amino acids and produces urea. The yellow/green bile pigment, **bilirubin**, is a breakdown product of haemoglobin (Chapter 9). Bilirubin is excreted with the bile into the small intestine and expelled with the faeces. The pigment undergoes changes in the intestine and is largely responsible for the brown colour of the faeces.

Lungs

The lungs supply the body with oxygen, but they are also excretory organs because they get rid of carbon dioxide. They also lose a great deal of water vapour but this loss is unavoidable and is not a method of controlling the water content of the body (Table 13.1).

Kidneys

The kidneys remove urea and other nitrogenous waste from the blood. They also expel excess water, salts, hormones (Chapter 14) and drugs (Chapter 15).

Skin

Sweat consists of water, with sodium chloride and traces of urea dissolved in it. When you sweat, you will expel these substances from your body so, in one sense, they are being excreted. However, sweating is a response to a rise in temperature and not to a change in the blood composition. In this sense, therefore, skin is not an excretory organ like the lungs and kidneys. See 'Homeostasis' in Chapter 14 for more details of skin structure and its functions.

Table 13.1 Excretory products and incidental losses

Excretory organ	Excretory products	Incidental losses
lungs	carbon dioxide	water
kidneys	nitrogenous waste, water, salts, toxins, hormones, drugs	
liver	bile pigments	
skin		water, salt, urea

Excretion

The kidneys

The two kidneys are fairly solid, oval structures. They are red-brown, enclosed in a transparent membrane and attached to the back of the abdominal cavity (Figure 13.1). The **renal artery** branches off from the aorta and brings oxygenated blood to them. The **renal vein** takes deoxygenated blood away from the kidneys to the vena cava (see Figure 9.20). A tube, called the **ureter**, runs from each kidney to the bladder in the lower part of the abdomen.

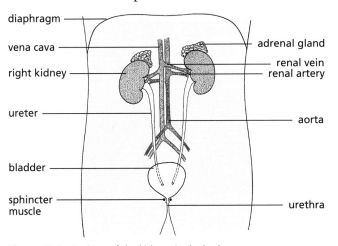

Figure 13.1 Position of the kidneys in the body

> **Key definition**
> **Deamination** is the removal of the nitrogen-containing part of amino acids to form urea.

The liver and its role in producing proteins

As well as being an excretory organ, the liver plays a very important role in **assimilating** amino acids. Assimilation means the absorption of substances, which are then built into other compounds in the organism. The liver removes amino acids from the plasma of the bloodstream and builds them up into proteins. Proteins are long chains of amino acids, joined together by peptide bonds (see Chapter 4 for details of protein structure). These include plasma proteins such as fibrinogen (Chapter 9), which have a role in blood clotting.

The need for excretion

Some of the compounds made in reactions in the body are potentially toxic (poisonous) if their

Water balance and osmoregulation

Your body gains water from food and drink. It loses water by evaporation, urination and defecation (Chapter 7). Evaporation from the skin takes place all the time but is particularly rapid when we sweat. Air from the lungs is saturated with water vapour, which is lost to the atmosphere every time we exhale. Despite these gains and losses of water, the concentration of body fluids is kept within very narrow limits by the kidneys, which adjust the concentration of the blood flowing through them. If it is too dilute (i.e. has too much water), less water is reabsorbed, leaving more to enter the bladder. After drinking a lot of fluid, a large volume of dilute urine is produced. On a cold day, sweating decreases so more water is removed from the blood by the kidneys, again increasing the volume of dilute urine.

If the blood is too concentrated, more water is absorbed back into the blood from the kidney tubules. So, if the body is short of water, e.g. after sweating profusely on a hot day, or through doing a lot of physical activity, or not having enough to drink, only a small quantity of concentrated urine is produced.

concentrations build up. Carbon dioxide dissolves in fluids such as tissue fluid and blood plasma to form carbonic acid. This increase in acidity can affect the actions of enzymes and can be fatal. Ammonia is made in the liver when excess amino acids are broken down. However, ammonia is very alkaline and toxic. It is converted to urea which is much less poisonous, making it a safe way of excreting excess nitrogen.

Microscopic structure of the kidneys

The kidney tissue consists of many capillaries and tiny tubes, called **renal tubules**, held together with connective tissue. If the kidney is cut down its length (sectioned), it is seen to have a dark, outer region called the **cortex** and a lighter, inner zone, the **medulla**. Where the ureter joins the kidney there is a space called the **pelvis** (Figure 13.2).

13 EXCRETION IN HUMANS

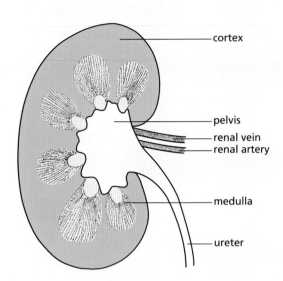

Figure 13.2 Section through the kidney to show regions

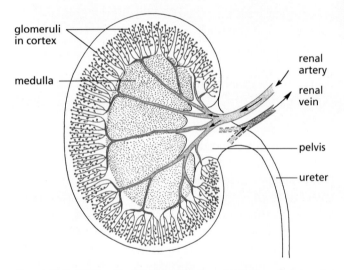

Figure 13.3 Section through kidney to show distribution of glomeruli

The renal artery divides up into a great many arterioles and capillaries, mostly in the cortex (Figure 13.3). Each arteriole leads to a **glomerulus**. This is a capillary repeatedly divided and coiled, making a knot of vessels (Figure 13.4). Each glomerulus is almost entirely surrounded by a cup-shaped organ called a **renal capsule**, which leads to a coiled **renal tubule**. This tubule, after a series of coils and loops, joins a **collecting duct**, which passes through the medulla to open into the pelvis (Figure 13.5). There are thousands of glomeruli in the kidney cortex and the total surface area of their capillaries is very great.

A **nephron** is a single glomerulus with its renal capsule, renal tubule and blood capillaries (see Figure 13.6).

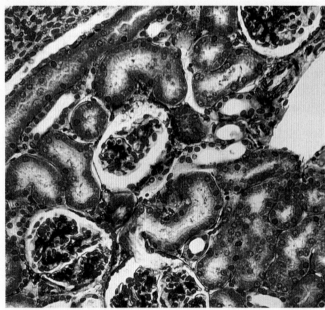

Figure 13.4 Glomeruli in the kidney cortex (×300). The three glomeruli are surrounded by kidney tubules sectioned at different angles. The light space around each glomerulus represents the renal capsule.

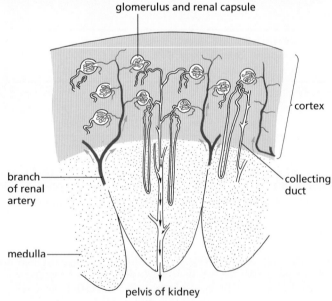

Figure 13.5 There are up to 4 million nephrons in a kidney. Only a few can be represented here, and not to scale.

Function of the kidneys

The blood pressure in a glomerulus causes part of the blood plasma to leak through the capillary walls.

The red blood cells and the plasma proteins are too big to pass out of the capillary, so the fluid that does filter through is plasma without the protein, i.e. similar to tissue fluid (Chapter 9). The fluid thus consists mainly of water with dissolved salts, glucose, urea and uric acid. The process by which the fluid is filtered out of the blood by the glomerulus is called **ultrafiltration**.

The filtrate from the glomerulus collects in the renal capsule and trickles down the renal tubule (Figure 13.6). As it does so, the capillaries that surround the tubule absorb back into the blood those substances which the body needs. First, all the glucose is reabsorbed, with much of the water. Then some of the salts are taken back to keep the correct concentration in the blood. The process of absorbing back the substances needed by the body is called **selective reabsorption**.

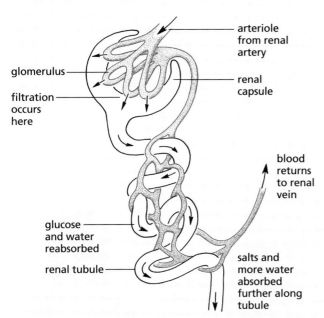

Figure 13.6 Part of a nephron (glomerulus, renal capsule and renal tubule)

Salts not needed by the body are left to pass on down the kidney tubule together with the urea and uric acid. So, these nitrogenous waste products, excess salts and water continue down the renal tube into the pelvis of the kidney. From here the fluid, now called **urine**, passes down the ureter to the bladder.

Table 13.2 shows some of the differences in composition between the blood plasma and the urine. The figures represent average values because the composition of the urine varies a great deal according to the diet, activity, temperature and intake of liquid.

Table 13.2 Composition of blood plasma and urine

	Plasma/%	Urine/%
water	90–93	95.0
urea	0.03	2.0
uric acid	0.003	0.05
ammonia	0.0001	0.05
sodium	0.3	0.6
potassium	0.02	0.15
chloride	0.37	0.6
phosphate	0.003	0.12

The **bladder** can expand to hold about 400 cm³ of urine. The urine cannot escape from the bladder because a band of circular muscle, called a **sphincter**, is contracted, so shutting off the exit. When this sphincter muscle relaxes, the muscular walls of the bladder expel the urine through the **urethra**. Adults can control this sphincter muscle and relax it only when they want to urinate. In babies, the sphincter relaxes by a reflex action (Chapter 14), set off by pressure in the bladder. By 3 years old, most children can control the sphincter voluntarily.

The dialysis machine ('artificial kidney')

Kidney failure may result from an accident involving a drop in blood pressure, or from a disease of the kidneys. In the former case, recovery is usually spontaneous, but if it takes longer than 2 weeks, the patient may die as a result of a potassium imbalance in the blood, which causes heart failure. In the case of kidney disease, the patient can survive with only one kidney, but if both fail, the patient's blood composition has to be regulated by a **dialysis** machine. Similarly, the accident victim can be kept alive on a dialysis machine until his or her blood pressure is restored.

In principle, a dialysis machine consists of a long cellulose tube coiled up in a water bath. The patient's blood is led from a vein in the arm and pumped through the cellulose (dialysis) tubing (Figures 13.7 and 13.8). The tiny pores in the dialysis tubing allow small molecules, such as those of salts, glucose and urea, to leak out into the water bath. Blood cells and protein molecules are too large to get through the pores (see Experiment 5, Chapter 4). This stage is similar to the filtration process in the glomerulus.

13 EXCRETION IN HUMANS

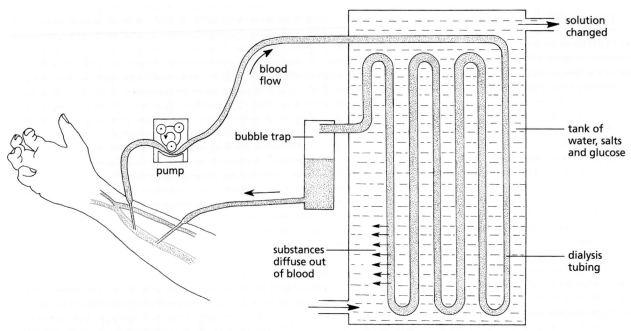

Figure 13.7 The principle of the kidney dialysis machine

To prevent a loss of glucose and essential salts from the blood, the liquid in the water bath consists of a solution of salts and sugar of the correct composition, so that only the substances above this concentration can diffuse out of the blood into the bathing solution. Thus, urea, uric acid and excess salts are removed.

The bathing solution is also kept at body temperature and is constantly changed as the unwanted blood solutes accumulate in it. The blood is then returned to the patient's arm vein.

A patient with total kidney failure has to spend 2 or 3 nights each week connected to the machine (Figure 13.8). With this treatment and a carefully controlled diet, the patient can lead a fairly normal life. A kidney transplant, however, is a better solution because the patient is not obliged to return to the dialysis machine.

The problem with kidney transplants is to find enough suitable donors of healthy kidneys and to prevent the transplanted kidney from being rejected.

The donor may be a close relative who is prepared to donate one of his or her kidneys (you can survive adequately with one kidney). Alternatively, the donated kidney may be taken from a healthy person who dies, for example, as a result of a road accident. People willing for their kidneys to be used after their death can carry a kidney donor card but the relatives must give their permission for the kidneys to be used.

The problem with rejection is that the body reacts to any transplanted cells or tissues as it does to all foreign proteins and produces lymphocytes, which attack and destroy them. This rejection can be overcome by:

- choosing a donor whose tissues are as similar as possible to those of the patient, e.g. a close relative
- using immunosuppressive drugs, which suppress the production of lymphocytes and their antibodies against the transplanted organ.

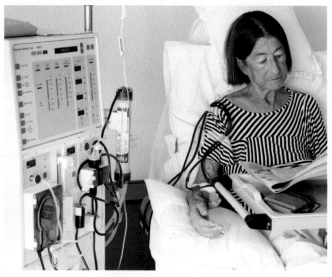

Figure 13.8 Kidney dialysis machine. The patient's blood is pumped to the dialyser, which removes urea and excess salts.

Excretion

The advantages and disadvantages of kidney transplants, compared with dialysis

Advantages
- The patient can return to a normal lifestyle – dialysis may require a lengthy session in hospital, three times a week, leaving the patient very tired after each session.
- The dialysis machine will be available for other patients to use.
- Dialysis machines are expensive to buy and maintain.

Disadvantages
- Transplants require a suitable donor – with a good tissue match. The donor may be from a dead person, or from a close living relative who is prepared to donate a healthy kidney (we can survive with one kidney).
- The operation is very expensive.
- There is a risk of rejection of the donated kidney – immunosuppressive drugs have to be used.
- Transplants are not accepted by some religions.

Questions

Core
1. Write a list of the substances that are likely to be excreted from the body during the day.
2. Why do you think that urine analysis is an important part of medical diagnosis?

Extended
3. How does the dialysis machine:
   a resemble and
   b differ from
   the nephron of a kidney in the way it functions?

Checklist

After studying Chapter 13 you should know and understand the following:

- Excretion is getting rid of toxic, surplus or unwanted substances produced by chemical reactions in the body or taken in with the diet.
- The lungs excrete carbon dioxide.
- The kidneys excrete urea, unwanted salts and excess water.
- Part of the blood plasma entering the kidneys is filtered out by the capillaries. Substances which the body needs, like glucose, are absorbed back into the blood. The unwanted substances are left to pass down the ureters into the bladder.
- The bladder stores urine, which is discharged at intervals.
- The kidneys help to keep the blood at a steady concentration by excreting excess salts and by adjusting the amounts of water (osmoregulation).
- The volume and concentration of urine produced is affected by water intake, temperature and exercise.
- The ureters, bladder and urethra on diagrams.

- The liver produces urea, formed from excess amino acids.
- Deamination is the removal of the nitrogen-containing part of amino acids to form urea.
- The liver has a role in the assimilation of amino acids by converting them to proteins, including plasma proteins.
- Outline of the structure and function of a kidney tubule.
- Explain the process of dialysis.
- Treatment, in response to damage to kidneys, may involve dialysis or transplant.
- The advantages and disadvantages of kidney transplants and dialysis.

14 Co-ordination and response

Nervous control in humans
Human nervous system
Structure of neurones
Nerve impulse
Reflex arc, spinal cord and reflexes
Define synapse
Structure of synapse

Voluntary and involuntary actions
Transfer of impulse across synapse
Effects of drugs on synapses

Sense organs
Define sense organ
Structure of eye
Pupil reflex

Explanation of pupil reflex
Accommodation
Function of rods and cones

Hormones in humans
Define hormone
Endocrine glands

Adrenaline
Functions of hormones

Role of adrenaline
Compare nervous and hormonal control systems

Homeostasis
Define homeostasis
Skin structure
Control of body temperature
Homeostasis

Negative feedback
Regulation of blood sugar
Type 1 diabetes
Vasodilation and vasoconstriction

Tropic responses
Define phototropism and gravitropism
Investigate tropic responses

Role of auxins in tropisms
Use of plant hormones in weedkillers

Co-ordination is the way all the organs and systems of the body are made to work efficiently together (Figure 14.1). If, for example, the leg muscles are being used for running, they will need extra supplies of glucose and oxygen. To meet this demand, the lungs breathe faster and deeper to obtain the extra oxygen and the heart pumps more rapidly to get the oxygen and glucose to the muscles more quickly.

The brain detects changes in the oxygen and carbon dioxide content of the blood and sends nervous impulses to the diaphragm, intercostal muscles and heart. In this example, the co-ordination of the systems is brought about by the **nervous system**.

The extra supplies of glucose needed for running come from the liver. Glycogen in the liver is changed to glucose, which is released into the bloodstream (see 'Homeostasis' on page 192). The conversion of glycogen to glucose is stimulated by, among other things, a chemical called adrenaline (see 'Hormones in humans' on page 190). Co-ordination by chemicals is brought about by the **endocrine system**.

The nervous system works by sending electrical impulses along nerves. The endocrine system depends on the release of chemicals, called **hormones**, from **endocrine glands**. Hormones are carried by the bloodstream. For example, insulin is carried from the pancreas to the liver by the circulatory system.

Figure 14.1 Co-ordination. The badminton player's brain is receiving sensory impulses from his eyes, ears (sound and balance) and muscle stretch receptors. Using this information, the brain co-ordinates the muscles of his limbs so that even while running or leaping he can control his stroke.

Nervous control in humans

The human nervous system is shown in Figure 14.2. The brain and spinal cord together form the **central nervous system**. Nerves carry electrical impulses from the central nervous system to all parts of the body, making muscles contract or glands produce enzymes or hormones. Electrical impulses are electrical signals that pass along nerve cells (neurones).

Glands and muscles are called **effectors** because they go into action when they receive nerve impulses or hormones. The biceps muscle is an effector that flexes the arm; the salivary gland (see 'Alimentary canal' in Chapter 7) is an effector that produces saliva when it receives a nerve impulse from the brain.

The nerves also carry impulses back to the central nervous system from receptors in the sense organs of the body. These impulses from the eyes, ears, skin, etc. make us aware of changes in our surroundings or in ourselves. Nerve impulses from the sense organs to the central nervous system are called **sensory impulses**; those from the central nervous system to the effectors, resulting in action, are called **motor impulses**.

The nerves that connect the body to the central nervous system make up the **peripheral** nervous system.

Nerve cells (neurones)

The central nervous system and the peripheral nerves are made up of nerve cells, called **neurones**. Three types of neurone are shown in Figure 14.3. **Motor neurones** carry impulses from the central nervous system to muscles and glands. **Sensory neurones** carry impulses from the sense organs to the central nervous system. **Relay neurones** (also called multi-polar or connector neurones) are neither sensory nor motor but make connections to other neurones inside the central nervous system.

Each neurone has a **cell body** consisting of a nucleus surrounded by a little cytoplasm. Branching fibres, called **dendrites**, from the cell body make contact with other neurones. A long filament of cytoplasm, surrounded by an insulating sheath, runs from the cell body of the neurone. This filament is called a **nerve fibre** (Figure 14.3(a) and (b)). The cell bodies of the neurones are mostly located in the brain or in the spinal cord and it is the nerve fibres that run in the nerves. A **nerve** is easily visible, white, tough and stringy and consists of hundreds of microscopic nerve fibres bundled together (Figure 14.4). Most nerves will contain a mixture of sensory and motor fibres. So a nerve can carry many different impulses. These impulses will travel in one direction in sensory fibres and in the opposite direction in motor fibres.

Some of the nerve fibres are very long. The nerve fibres to the foot have their cell bodies in the spinal cord and the fibres run inside the nerves, without a break, to the skin of the toes or the muscles of the foot. A single nerve cell may have a fibre 1 m long.

The nerve impulse

The nerve fibres do not carry sensations like pain or cold. These sensations are felt only when a nerve impulse reaches the brain. The impulse itself is a series of electrical pulses that travel down the fibre. Each pulse lasts about 0.001 s and travels at speeds of up to $100 \, \text{m s}^{-1}$. All nerve impulses are similar; there is no difference between nerve impulses from the eyes, ears or hands.

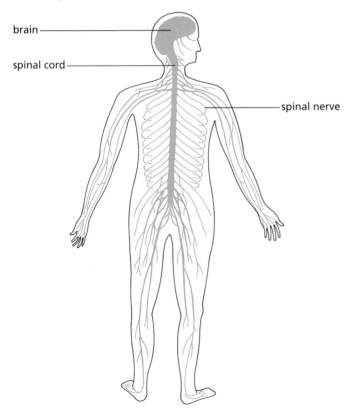

Figure 14.2 The human nervous system

14 CO-ORDINATION AND RESPONSE

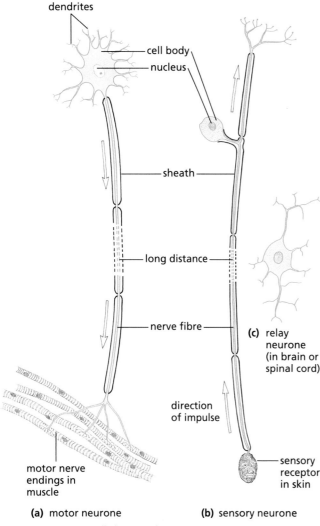

Figure 14.3 Nerve cells (neurones)

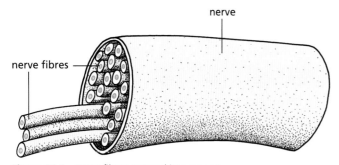

Figure 14.4 Nerve fibres grouped into a nerve

We are able to tell where the sensory impulses have come from and what caused them only because the impulses are sent to different parts of the brain. The nerves from the eye go to the part of the brain concerned with sight. So when impulses are received in this area, the brain recognises that they have come from the eyes and we 'see' something.

The reflex arc

One of the simplest situations where impulses cross synapses to produce action is in the reflex arc. A **reflex action** is an automatic response to a **stimulus**. (A stimulus is a change in the external or internal environment of an organism.) It provides a means of rapidly integrating and co-ordinating a stimulus with the response of an effector (a muscle or a gland) without the need for thought or a decision. When a particle of dust touches the cornea of the eye, you will blink; you cannot prevent yourself from blinking. A particle of food touching the lining of the windpipe will set off a coughing reflex that cannot be suppressed. When a bright light shines in the eye, the pupil contracts (see 'Sense organs' later in this chapter). You cannot stop this reflex and you are not even aware that it is happening.

The nervous pathway for such reflexes is called a **reflex arc**. In Figure 14.5 the nervous pathway for a well-known reflex called the 'knee-jerk' reflex is shown.

One leg is crossed over the other and the muscles are totally relaxed. If the tendon just below the kneecap of the upper leg is tapped sharply, a reflex arc makes the thigh muscle contract and the lower part of the leg swings forward.

The pathway of this reflex arc is traced in Figure 14.6. Hitting the tendon stretches the muscle and stimulates a stretch receptor. The receptor sends off impulses in a sensory fibre. These sensory impulses travel in the nerve to the spinal cord.

In the central region of the spinal cord, the sensory fibre passes the impulse across a synapse to a motor neurone, which conducts the impulse down the fibre, back to the thigh muscle (the effector). The arrival of the impulses at the muscle makes it contract and jerk the lower part of the limb forward. You are aware that this is happening (which means that sensory impulses must be reaching the brain), but there is nothing you can do to stop it.

Nervous control in humans

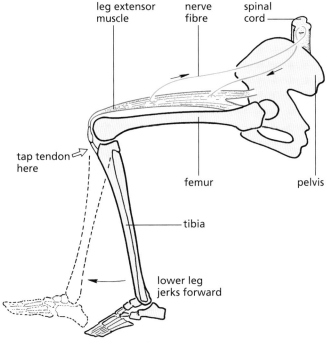

Figure 14.5 The reflex knee jerk

The sequence of events in a simple reflex arc is shown below.

stimulus (tapping the tendon below the kneecap)
↓
receptor (stretch receptor)
↓
sensory neurone
↓
co-ordinator (spinal cord)
↓
motor neurone
↓
effector (leg extensor muscle)
↓
response (leg extensor muscle contracts, making the leg kick forwards)

● Extension work

The spinal cord

Like all other parts of the nervous system, the spinal cord consists of thousands of nerve cells. The structure of the spinal cord is shown in Figures 14.6, 14.7 and 14.8.

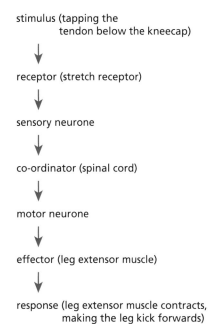

Figure 14.7 Section through spinal cord (×7). The light area is the white matter, consisting largely of nerve fibres running to and from the brain. The darker central area is the grey matter, consisting largely of nerve cell bodies.

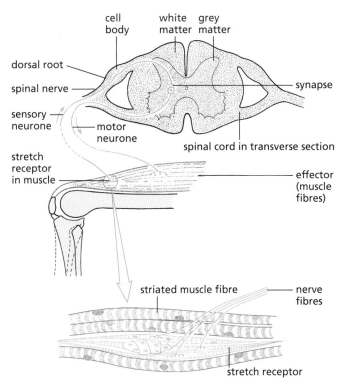

Figure 14.6 The reflex arc. This reflex arc needs only one synapse for making the response. Most reflex actions need many more synapses (i) to adjust other muscles in the body and (ii) to send impulses to the brain.

All the cell bodies, apart from those in the dorsal root ganglia, are concentrated in the central region called the **grey matter**. The **white matter** consists of nerve fibres. Some of these will be passing from the grey matter to the spinal nerves and others

14 CO-ORDINATION AND RESPONSE

will be running along the spinal cord connecting the spinal nerve fibres to the brain. The spinal cord is thus concerned with:

- reflex actions involving body structures below the neck
- conducting sensory impulses from the skin and muscles to the brain, and
- carrying motor impulses from the brain to the muscles of the trunk and limbs.

In Figure 14.6 the spinal cord is drawn in transverse section. The spinal nerve divides into two 'roots' at the point where it joins the spinal cord. All the sensory fibres enter through the **dorsal root** and the motor fibres all leave through the **ventral root**, but both kinds of fibre are contained in the same spinal nerve. This is like a group of insulated wires in the same electric cable. The cell bodies of all the sensory fibres are situated in the dorsal root and they make a bulge called a **ganglion** (Figure 14.9).

In even the simplest reflex action, many more nerve fibres, synapses and muscles are involved than are described here. Figure 14.8 illustrates the reflex arc that would result in the hand being removed from a painful stimulus. On the left side of the spinal cord, an incoming sensory fibre makes its first synapse with a relay neurone. This can pass the impulse on to many other motor neurones, although only one is shown in the diagram. On the right side of the spinal cord, some of the incoming sensory fibres are shown making synapses with neurones that send nerve fibres to the brain, thus keeping the brain informed about events in the body. Also, nerve fibres from the brain make synapses with motor neurones in the spinal cord so that 'commands' from the brain can be sent to muscles of the body.

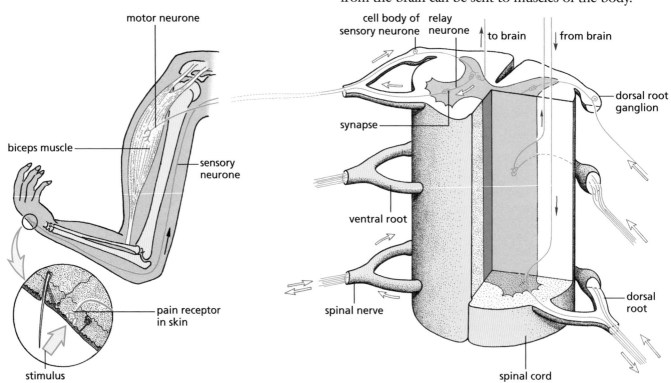

Figure 14.8 Reflex arc (withdrawal reflex)

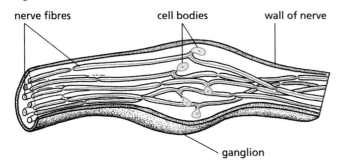

Figure 14.9 Cell bodies forming a ganglion

Reflexes

The reflex just described is a **spinal reflex**. The brain, theoretically, is not needed for it to happen. Responses that take place in the head, such as blinking, coughing and iris contraction, have their reflex arcs in the brain, but may still not be consciously controlled.

Bright light stimulates the light-sensitive cells of the retina. The nerve impulses in the sensory fibres from these receptors travel through the optic nerve to the brain. In the mid-brain the fibres synapse with relay and motor fibres, which carry impulses back through the optic nerve to the circular muscle of the iris and stimulate it to contract.

Synapses

> **Key definition**
> A **synapse** is a junction between two neurones.

Voluntary and involuntary actions

Voluntary actions

A **voluntary action** starts in the brain. It may be the result of external events, such as seeing a book on the floor, but any resulting action, such as picking up the book, is entirely voluntary. Unlike a reflex action it does not happen automatically; you can decide whether or not you carry out the action.

The brain sends motor impulses down the spinal cord in the nerve fibres. These make synapses with motor fibres, which enter spinal nerves and make connections to the sets of muscles needed to produce effective action. Many sets of muscles in the arms, legs and trunk would be brought into play in order to stoop and pick up the book, and impulses passing between the eyes, brain and arm would direct the hand to the right place and 'tell' the fingers when to close on the book.

One of the main functions of the brain is to co-ordinate these actions so that they happen in the right sequence and at the right time and place.

Involuntary actions

The reflex closure of the iris (see 'Sense organs' later in this chapter) protects the retina from bright light; the withdrawal reflex removes the hand from a dangerously hot object; the coughing reflex dislodges a foreign particle from the windpipe. Thus, these reflexes have a protective function and all are **involuntary actions**.

There are many other reflexes going on inside our bodies. We are usually unaware of these, but they maintain our blood pressure, breathing rate, heartbeat, etc. and so maintain the body processes.

How a synapse transmits an electrical impulse

At a synapse, a branch at the end of one fibre is in close contact with the cell body or dendrite of another neurone (Figure 14.10).

Although nerve fibres are insulated, it is necessary for impulses to pass from one neurone to another. An impulse from the fingertips has to pass through at least three neurones before reaching the brain and so produce a conscious sensation. The regions where impulses are able to cross from one neurone to the next are called **synapses**.

When an impulse arrives at the synapse, vesicles in the cytoplasm release a tiny amount of the neurotransmitter substance. It rapidly diffuses across the gap (also known as the **synaptic cleft**) and binds with neurotransmitter receptor molecules in the membrane of the neurone on the other side of the synapse. This then sets off an impulse in the neurone. Sometimes several impulses have to arrive at the synapse before enough transmitter substance is released to cause an impulse to be fired off in the next neurone.

Synapses control the direction of impulses because neurotransmitter substances are only synthesised on one side of the synapse, while receptor molecules are only present on the other side. They slow down the speed of nerve impulses slightly because of the time taken for the chemical to diffuse across the **synaptic gap**.

Many drugs produce their effects by interacting with receptor molecules at synapses. **Heroin**, for example, stimulates receptor molecules in synapses in the brain, triggering the release of dopamine (a neurotransmitter), which gives a short-lived 'high'.

Spider toxin, and also the toxin released by tetanus (an infection caused by *Clostridium* bacteria), breaks down vesicles, releasing massive amounts of transmitter substance and disrupting normal synaptic function. Symptoms caused by the tetanus toxin include muscle spasms, lock-jaw and heart failure.

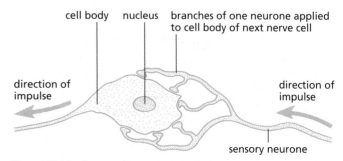

Figure 14.10 Synapses between nerve neurones

14 CO-ORDINATION AND RESPONSE

● Sense organs

> **Key definition**
> **Sense organs** are groups of sensory cells responding to specific stimuli, such as light, sound, touch, temperature and chemicals.

Our senses make us aware of changes in our surroundings and in our own bodies. We have sense cells that respond to stimuli (singular = stimulus). A **stimulus** is a change in light, temperature, pressure, etc., which produces a reaction in a living organism. Structures that detect stimuli are called **receptors**. Some of these receptors are scattered through the skin: this organ has a number of different types of receptor, as shown in Figure 14.21. Other receptors are concentrated into special **sense organs** such as the eye and the ear. Table 14.1 gives examples of these and their stimuli.

Table 14.1 Sense organs and their stimuli

Sense organ	Stimulus
ear	sound, body movement (balance)
eye	light
nose	chemicals (smells)
tongue	chemicals (taste)
skin	temperature, pressure, touch, pain

The special property of sensory cells and sense organs is that they are able to convert one form of energy to another. The eyes can convert light energy into the electrical energy of a nerve impulse. The ears convert the energy in sound vibrations into nerve impulses. The forms of energy that make up the stimuli may be very different, e.g. mechanical, chemical, light, but they are all transduced into pulses of electrical energy in the nerves.

When a receptor responds to a stimulus, it sends a nerve impulse to the brain, which makes us aware of the sensation.

The eye

Note: details of conjunctiva, humours, choroid and tear glands are **not** a syllabus requirement, but are included here to put parts seen in a diagram of the eye in context.

The structure of the eye is shown in Figures 14.11 and 14.12. The **sclera** is the tough, white outer coating. The front part of the sclera is clear and allows light to enter the eye. This part is called the **cornea**. The **conjunctiva** is a thin epithelium, which lines the inside of the eyelids and the front of the sclera and is continuous with the epithelium of the cornea.

The eye contains a clear liquid whose outward pressure on the sclera keeps the spherical shape of the eyeball. The liquid behind the lens is jelly-like and called **vitreous humour**. The **aqueous humour** in front of the lens is watery.

The **lens** is a transparent structure, held in place by a ring of fibres called the **suspensory ligament**. Unlike the lens of a camera or a telescope, the eye lens is flexible and can change its shape. In front of the lens is a disc of tissue called the **iris**. It is the iris we refer to when we describe the colour of the eye as brown or blue. The iris controls how much light enters the **pupil**, which is a hole in the centre of the iris. The pupil lets in light to the rest of the eye.

The pupil looks black because all the light entering the eye is absorbed by the black pigment in the **choroid**. The choroid layer, which contains many blood vessels, lies between the retina and the sclera. In the front of the eyeball, it forms the iris and the **ciliary body**. The ciliary body produces aqueous humour.

The internal lining at the back of the eye is the **retina** and it consists of many thousands of cells that respond to light. When light falls on these cells, they send off nervous impulses, which travel in nerve fibres, through the **optic nerve**, to the brain and so give rise to the sensation of sight. The part of the retina lying directly in front of the optic nerve contains no light-sensitive cells. This region is called the **blind spot**.

Tear glands under the top eyelid produce tear fluid. This is a dilute solution of sodium chloride and sodium hydrogencarbonate. The fluid is spread over the eye surface by the blinking of the eyelids, keeping the surface moist and washing away any dust particles or foreign bodies. Tear fluid also contains an enzyme, **lysozyme**, which attacks bacteria.

Table 14.2 gives the functions of the parts of the eye required for the Core section of the syllabus.

Table 14.2 Functions of parts of the eye

Part	Function
cornea	a transparent, curved layer at the front of the eye that refracts the light entering and helps to focus it
iris	a coloured ring of circular and radial muscle that controls the size of the pupil
lens	a transparent, convex, flexible, jelly-like structure that refracts light to focus it onto the retina
retina	a light-sensitive layer made up of rods, which detect light of low intensity, and cones, which detect different colours
optic nerve	transmits electrical impulses from the retina to the brain

Sense organs

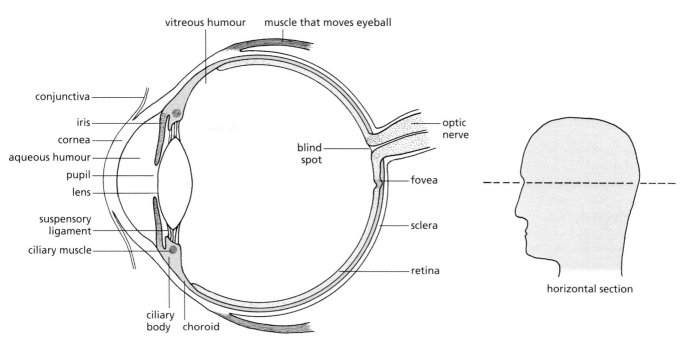

Figure 14.11 Horizontal section through left eye

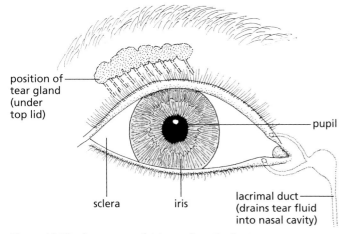

Figure 14.12 Appearance of right eye from the front

Vision

Light from an object produces a focused **image** on the retina (like a 'picture' on a cinema screen) (Figures 14.13 and 14.17). The curved surfaces of the cornea and lens both refract ('bend') the light rays that enter the eye, in such a way that each 'point of light' from the object forms a 'point of light' on the retina. These points of light will form an image, upside-down and smaller than the object.

The cornea and the aqueous and vitreous humours are mainly responsible for the refraction of light. The lens makes the final adjustments to the focus (Figure 14.13(b)).

The pattern of sensory cells stimulated by the image will produce a pattern of nerve impulses sent to the brain. The brain interprets this pattern, using past experience and learning, and forms an impression of the size, distance and upright nature of the object.

The pupil reflex

The change in size of the pupil is caused by exposure of the eye to different light intensities. It is an automatic reaction: you cannot control it. When bright light falls on the eye, the iris responds by making the diameter of the pupil smaller. This restricts the amount of light reaching the retina, which contains the light-sensitive cells. If dim light falls on the eye, the iris responds by making the diameter of the pupil larger, so that as much light as is available can reach the retina to stimulate the light-sensitive cells. Figure 14.12 shows an eye exposed to bright light: the pupil is small. It would become much larger if the light intensity was reduced.

14 CO-ORDINATION AND RESPONSE

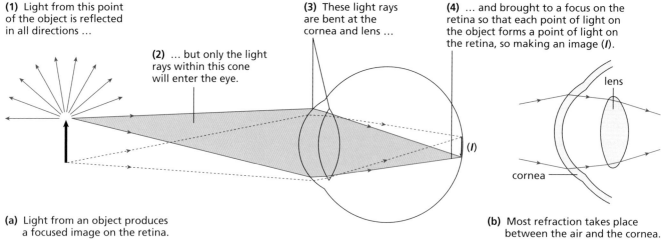

(a) Light from an object produces a focused image on the retina.

(b) Most refraction takes place between the air and the cornea.

Figure 14.13 Image formation on the retina

Control of light intensity

This section gives more detail about the roles of the iris and pupil in controlling light intensity falling on the retina, needed if you are following the extended syllabus.

The amount of light entering the eye is controlled by altering the size of the pupil (Figure 14.14). If the light intensity is high, it causes a contraction in a ring of muscle fibres (**circular muscle**) in the iris. This reduces the size of the pupil and cuts down the intensity of light entering the eye. High-intensity light can damage the retina, so this reaction has a protective function.

In low light intensities, the circular muscle of the iris relaxes and **radial muscle** fibres (which are arranged like the spokes of a bicycle wheel) contract. This makes the pupil enlarge and allows more light to enter. The circular and radial muscles act **antagonistically**. This means that they oppose each other in their actions – when the circular muscles contract they constrict the pupil and when the radial muscles contract the pupil dilates.

The change in size of the pupil is caused by an automatic reflex action; you cannot control it consciously.

Accommodation (focusing)

The eye can produce a focused image of either a near object or a distant object. To do this the lens changes its shape, becoming thinner for distant objects and fatter for near objects. This change in shape is caused by contracting or relaxing the **ciliary muscle**, which forms a circular band of muscle in the **ciliary body** (Figure 14.15). When the ciliary muscle is relaxed, the outward pressure of the humours on the sclera pulls on the suspensory ligament and stretches the lens to its thin shape. The eye is now accommodated (i.e. focused) for distant objects (Figures 14.15(a) and 14.16(a)). To focus a near object, the ciliary muscle contracts to a smaller circle and this takes the tension out of the suspensory ligament (Figures 14.15(b) and 14.16(b)). The lens is elastic and flexible and so is able to change to its fatter shape. This shape is better at bending the light rays from a close object.

Retina

The millions of light-sensitive cells in the retina are of two kinds, the **rods** and the **cones** (according to shape). The cones enable us to distinguish colours, but the rods are more sensitive to low intensities of light and therefore play an important part in night vision when the light intensity is not sufficient to stimulate the cone cells. Images formed at night appear as shades of grey, with no bright colours detected. There are thought to be three types of cone cell. One type responds best to red light, one to green and one to blue. If all three types are equally stimulated we get the sensation of white. The cone cells are concentrated in a central part of the retina, called the **fovea** (Figure 14.11); when you study an object closely you are making its image fall on the fovea.

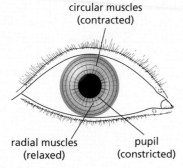

Figure 14.14 The iris reflex

Sense organs

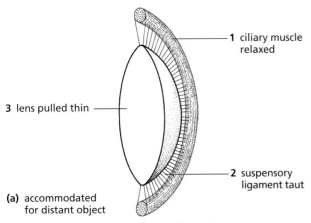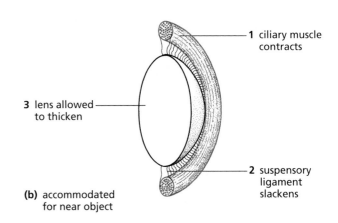

Figure 14.15 How accommodation is brought about

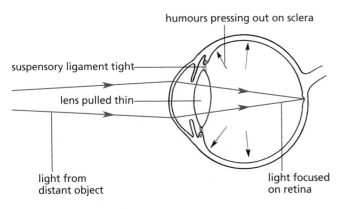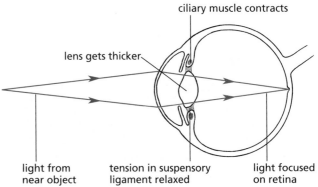

Figure 14.16 Accommodation

Fovea

It is in the fovea that the image on the retina is analysed in detail. Only objects within a 2° cone from the eye form an image on the fovea. This means that only about two letters in any word on this page can be seen in detail. It is the constant scanning movements of the eye that enable you to build up an accurate 'picture' of a scene. The centre of the fovea contains only cones: it is here that colour discrimination occurs.

Blind spot

At the point where the optic nerve leaves the retina, there are no sensory cells and so no information reaches the brain about that part of the image which falls on this blind spot (Figure 14.18).

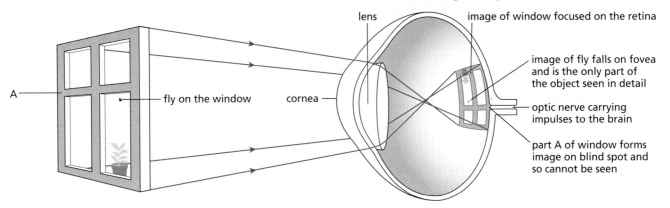

Figure 14.17 Image formation in the eye

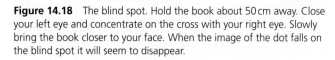

Figure 14.18 The blind spot. Hold the book about 50 cm away. Close your left eye and concentrate on the cross with your right eye. Slowly bring the book closer to your face. When the image of the dot falls on the blind spot it will seem to disappear.

● Hormones in humans

> **Key definition**
> A **hormone** is a chemical substance, produced by a gland and carried by the blood, which alters the activity of one or more specific target organs.

Co-ordination by the nervous system is usually rapid and precise. Nerve impulses, travelling at up to 100 metres per second, are delivered to specific parts of the body and produce an almost immediate response. A different kind of co-ordination is brought about by the **endocrine system**. This system depends on chemicals, called **hormones**, which are released from special glands, called **endocrine glands**, into the bloodstream. The hormones circulate around the body in the blood and eventually reach certain organs, called **target organs**. Hormones speed up, slow down or alter the activity of those organs. After being secreted, hormones do not remain permanently in the blood but are changed by the liver into inactive compounds and excreted by the kidneys. Insulin, for example, may stay in the bloodstream for just 4–8 hours before being broken down. Table 14.3 compares control by the endocrine and nervous systems.

Table 14.3 Endocrine and nervous control compared

Endocrine	Nervous
transmission of chemicals	transmission of electrical impulses
transmission via blood	transmission in nerves
slow transmission	rapid transmission
hormones dispersed throughout body	impulse sent directly to target organ
long-term effects	short-lived effects

Unlike the digestive glands, endocrine glands do not deliver their secretions through ducts (tubes). For this reason, the endocrine glands are sometimes called 'ductless glands'. The hormones are picked up directly from the glands by the blood circulation.

Responses of the body to hormones are much slower than responses to nerve impulses. They depend, in the first instance, on the speed of the circulatory system and then on the time it takes for the cells to change their chemical activities. Many hormones affect long-term changes such as growth rate, puberty and pregnancy. Nerve impulses often cause a response in a very limited area of the body, such as an eye-blink or a finger movement. Hormones often affect many organ systems at once.

Serious deficiencies or excesses of hormone production give rise to illnesses. Small differences in hormone activity between individuals probably contribute to differences of personality and temperament.

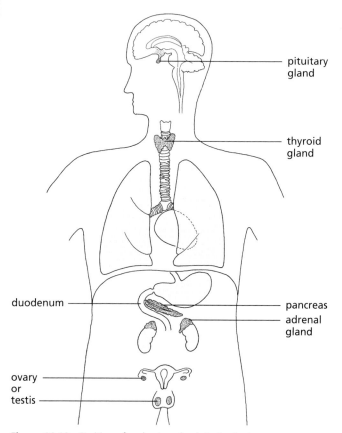

Figure 14.19 Position of endocrine glands in the body
Note: knowledge of the pituitary and thyroid glands is **not** a syllabus requirement

The position of the endocrine glands in the body is shown in Figure 14.19. Notice that the pancreas and the reproductive organs have a dual function.

● Extension work

Thyroid gland

The thyroid gland is situated in the front part of the neck and lies in front of the windpipe. It produces a hormone called **thyroxine**. This hormone has a stimulatory effect on the metabolic rate of nearly all the body cells, such as the speed or rate of cell respiration (Chapter 12) and other chemical reactions. It controls our level of activity, promotes skeletal growth and is essential for the normal development of the brain.

Pituitary gland

This gland is attached to the base of the brain. It produces many hormones. For example, the pituitary releases into the blood **follicle-stimulating hormone** (FSH) which, when it reaches the ovaries, makes one of the follicles start to mature and to produce oestrogen. **Luteinising hormone** (LH), also known as lutropin, is also produced from the pituitary and, together with FSH, induces ovulation (see 'Sex hormones in humans' in Chapter 16).

Adrenal glands

These glands are attached to the back of the abdominal cavity, one above each kidney (see also Figure 13.1). One part of the adrenal gland is a zone called the **adrenal medulla**. The medulla receives nerves from the brain and produces the hormone **adrenaline**.

Adrenaline has obvious effects on the body:

- In response to a stressful situation, nerve impulses are sent from the brain to the adrenal medulla, which releases adrenaline into the blood.
- Its presence causes breathing to become faster and deeper. This may be particularly apparent as we pant for breath.
- The heart beats faster, resulting in an increase in pulse rate. This increase in heart rate can be quite alarming, making us feel as if our heart is going to burst out of our chest.
- The pupils of our eyes dilate, making them look much blacker.

These effects all make us more able to react quickly and vigorously in dangerous situations (known as 'fight or flight situations') that might require us to run away or put up a struggle. However, in many stressful situations, such as taking examinations or giving a public performance, vigorous activity is not called for. So the extra adrenaline in our bodies just makes us feel tense and anxious.

The pancreas

The pancreas is a digestive gland that secretes enzymes into the duodenum through the pancreatic duct (Chapter 7). It is also an endocrine (ductless) gland. Most of the pancreas cells produce digestive enzymes but some of them produce hormones. The hormone-producing cells are arranged in small isolated groups called islets (Figure 14.20) and secrete their hormones directly into the bloodstream. One of the hormones is called **insulin**.

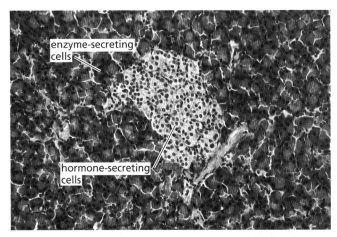

Figure 14.20 Section of pancreas tissue showing an islet (×250)

Insulin controls the levels of glucose in the blood by instructing the liver to remove the sugars and store them. This happens when levels get too high, such as after a meal rich in carbohydrate. (See page 196 for further details of the action of insulin.)

Reproductive organs

The ovaries and testes produce hormones as well as gametes (sperms and ova) and their effects are described in Chapter 16.

One of the hormones from the ovary, **oestrogen**, prepares the uterus for the implantation of the embryo, by making its lining thicker and increasing its blood supply.

The hormones **testosterone** (from the testes) and oestrogen (from the ovaries) play a part in the development of the secondary sexual characteristics.

14 CO-ORDINATION AND RESPONSE

The role of adrenaline

As adrenaline circulates around the body it affects a number of organs, as shown in Table 14.4.

You will recognise the sensations described in column four of Table 14.4 as characteristic of fear and anxiety.

Table 14.4 Responses to adrenaline

Target organ	Effects of adrenaline	Biological advantage	Effect or sensation
heart	beats faster	sends more glucose and oxygen to the muscles	thumping heart
breathing centre of the brain	faster and deeper breathing	increased oxygenation of the blood; rapid removal of carbon dioxide	panting
arterioles of the skin	constricts them (see 'Homeostasis')	less blood going to the skin means more is available to the muscles	person goes paler
arterioles of the digestive system	constricts them	less blood for the digestive system allows more to reach the muscles	dry mouth
muscles of alimentary canal	relax	peristalsis and digestion slow down; more energy available for action	'hollow' feeling in stomach
muscles of body	tenses them	ready for immediate action	tense feeling; shivering
liver	conversion of glycogen to glucose	more glucose available in blood for energy production, to allow metabolic activity to increase	no sensation
fat deposits	conversion of fats to fatty acids	fatty acids available in blood for muscle contraction	

Adrenaline is quickly converted by the liver to a less active compound, which is excreted by the kidneys. All hormones are similarly altered and excreted, some within minutes, others within days.

Thus their effects are not long-lasting. The long-term hormones, such as thyroxine, are secreted continuously to maintain a steady level.

● Homeostasis

> **Key definition**
> **Homeostasis** is the maintenance of a constant internal environment.

Homeostasis literally means 'staying similar'. It refers to the fact that the composition of the tissue fluid (see 'Blood' in Chapter 9) in the body is kept within narrow limits. The concentration, acidity and temperature of this fluid are being adjusted all the time to prevent any big changes.

The skin and temperature control
Skin structure

Figure 14.21 shows a section through skin. In the **basal layer** some of the cells are continually dividing and pushing the older cells nearer the surface. Here they die and are shed at the same rate as they are replaced. The basal layer and the cells above it constitute the **epidermis**. The basal layer also contributes to the hair follicles. The dividing cells give rise to the hair.

There are specialised pigment cells in the basal layer and epidermis. These produce a black pigment, **melanin**, which gives the skin its colour. The more melanin, the darker is the skin.

The thickness of the epidermis and the abundance of hairs vary in different parts of the body (Figure 14.22).

The **dermis** contains connective tissue with hair follicles, sebaceous glands, sweat glands, blood vessels and nerve endings. There is usually a layer of adipose tissue (a fat deposit) beneath the dermis.

Skin function
Protection
The outermost layer of dead cells of the epidermis helps to reduce water loss and provides a barrier against bacteria. The pigment cells protect the skin from damage by the ultraviolet rays in sunlight. In white-skinned people, more melanin is produced in response to exposure to sunlight, giving rise to a tan.

Sensitivity
Scattered throughout the skin are large numbers of tiny sense receptors, which give rise to sensations of touch, pressure, heat, cold and pain. These make us aware of changes in our surroundings and enable us to take action to avoid damage, to recognise objects by touch and to manipulate objects with our hands.

Homeostasis

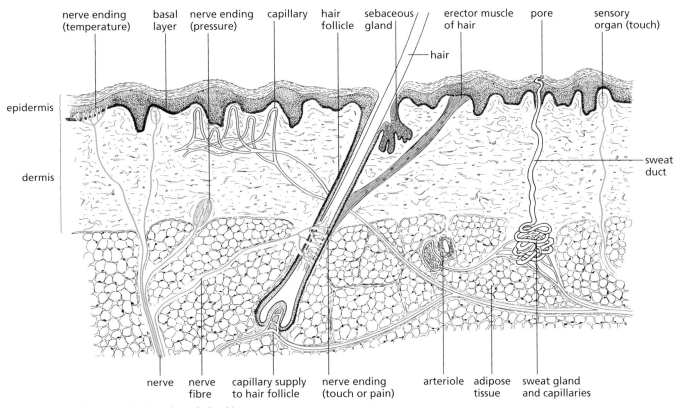

Figure 14.21 Generalised section through the skin

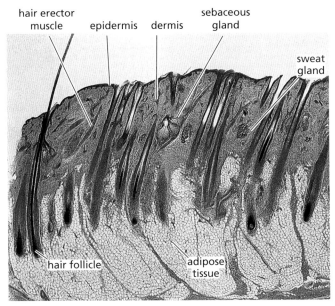

Figure 14.22 Section through hairy skin (×20)

Temperature regulation

The skin helps to keep the body temperature more or less constant. This is done by adjusting the flow of blood near the skin surface and by sweating. These processes are described more fully below.

Temperature control

Normal human body temperature varies between 35.8 °C and 37.7 °C. Temperatures below 34 °C or above 40 °C, if maintained for long, are considered dangerous. Different body regions, e.g. the hands, feet, head or internal organs, will be at different temperatures, but the **core** temperature, as measured with a thermometer under the tongue, will vary by only 1 or 2 degrees.

Heat is lost from the body surface by conduction, convection, radiation and evaporation. The amount of heat lost is reduced to an extent due to the insulating properties of adipose (fatty) tissue in the dermis. Some mammals living in extreme conditions, such as whales and seals, make much greater use of this: they have thick layers of blubber to reduce heat loss more effectively. Just how much insulation the blubber gives depends on the amount of water in the tissue: a smaller proportion of water and more fat provide better insulating properties.

Heat is gained, internally, from the process of respiration (Chapter 12) in the tissues and, externally, from the surroundings or from the Sun.

14 CO-ORDINATION AND RESPONSE

The two processes of heat gain and heat loss are normally in balance but any imbalance is corrected by a number of methods, including those described below.

Overheating
- More blood flows near the surface of the skin, allowing more heat to be exchanged with the surroundings.
- **Sweating** – the sweat glands secrete sweat on to the skin surface. When this layer of liquid evaporates, it takes heat (latent heat) from the body and cools it down (Figure 14.23).

Overcooling
- Less blood flows near the surface of the skin, reducing the amount of heat lost to the surroundings.
- Sweat production stops – thus the heat lost by evaporation is reduced.
- **Shivering** – uncontrollable bursts of rapid muscular contraction in the limbs release heat as a result of respiration in the muscles.

In these ways, the body temperature remains at about 37 °C. We also control our temperature by adding or removing clothing or deliberately taking exercise.

Whether we feel hot or cold depends on the sensory nerve endings in the skin, which respond to heat loss or gain. You cannot consciously detect changes in your core temperature. The brain plays a direct role in detecting any changes from normal by monitoring the temperature of the blood. A region called the **hypothalamus** contains a thermoregulatory centre in which temperature receptors detect temperature changes in the blood and co-ordinate a response to them. Temperature receptors are also present in the skin. They send information to the brain about temperature changes.

Figure 14.23 Sweating. During vigorous activity the sweat evaporates from the skin and helps to cool the body. When the activity stops, continued evaporation of sweat may overcool the body unless it is towelled off.

Homeostasis

It is vital that there are homeostatic mechanisms in the body to control internal conditions within set limits.

In Chapter 5 it was explained that, in living cells, all the chemical reactions are controlled by enzymes. The enzymes are very sensitive to the conditions in which they work. A slight fall in temperature or a rise in acidity may slow down or stop an enzyme from working and thus prevent an important reaction from taking place in the cell.

The cell membrane controls the substances that enter and leave the cell, but it is the tissue fluid that supplies or removes these substances, and it is therefore important to keep the composition of the tissue fluid as steady as possible. If the tissue fluid were to become too concentrated, it would withdraw water from the cells by osmosis (Chapter 3) and the body would be dehydrated. If the tissue fluid were to become too dilute, the cells would take up too much water from it by osmosis and the tissues would become waterlogged and swollen.

Many systems in the body contribute to homeostasis (Figure 14.24). The obvious example is the kidneys, which remove substances that might poison the enzymes. The kidneys also control the level of salts, water and acids in the blood. The composition of the blood affects the tissue fluid which, in turn, affects the cells.

Another example of a homeostatic organ is the liver, which regulates the level of glucose in the blood. The liver stores any excess glucose as glycogen, or turns glycogen back into glucose if the concentration in the blood gets too low. The brain cells are very sensitive to the glucose concentration in the blood and if the level drops too far, they stop working properly, and the person becomes unconscious and will die unless glucose is injected into the blood system. This shows how important homeostasis is to the body.

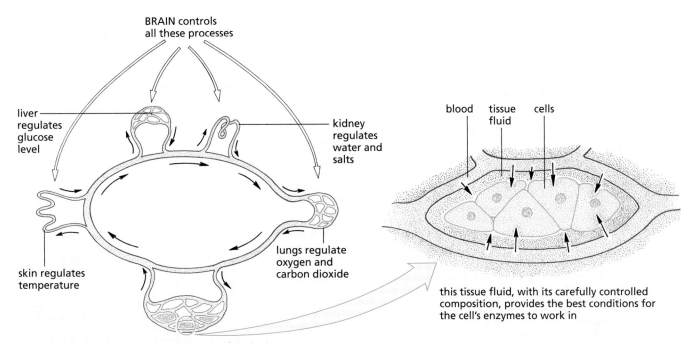

Figure 14.24 The homeostatic mechanisms of the body

The lungs (Chapter 11) play a part in homeostasis by keeping the concentrations of oxygen and carbon dioxide in the blood at the best level for the cells' chemical reactions, especially respiration.

The skin regulates the temperature of the blood. If the cells were to get too cold, the chemical reactions would become too slow to maintain life. If they became too hot, the enzymes would be destroyed.

The brain has overall control of the homeostatic processes in the body. It checks the composition of the blood flowing through it and if it is too warm, too cold, too concentrated or has too little glucose, nerve impulses or hormones are sent to the organs concerned, causing them to make the necessary adjustments.

Homeostasis and negative feedback

Temperature regulation is an example of homeostasis. Maintenance of a constant body temperature ensures that vital chemical reactions continue at a predictable rate and do not speed up or slow down when the surrounding temperature changes. The constant-temperature or **homoiothermic** ('warm-blooded') animals, the birds and mammals, therefore have an advantage over the variable-temperature or **poikilothermic** ('cold-blooded') animals. Poikilotherms such as reptiles and insects can regulate their body temperature to some extent by, for example, basking in the sun or seeking shade. Nevertheless, if their body temperature falls, their vital chemistry slows down and their reactions become more sluggish. They are then more vulnerable to predators.

The 'price' that homoiotherms have to pay is the intake of enough food to maintain their body temperature, usually above that of their surroundings.

In the hypothalamus of a homoiotherm's brain there is a thermoregulatory centre. This centre monitors the temperature of the blood passing through it and also receives sensory nerve impulses from temperature receptors in the skin. A rise in body temperature is detected by the thermoregulatory centre and it sends nerve impulses to the skin, which result in vasodilation and sweating. Similarly, a fall in body temperature will be detected and will promote impulses that produce vasoconstriction and shivering.

This system of control is called **negative feedback**. The outgoing impulses counteract the effects that produced the incoming impulses. For example, a rise in temperature triggers responses that counteract the rise.

Regulation of blood sugar

If the level of sugar in the blood falls, the islets release a hormone called **glucagon** into the bloodstream. Glucagon acts on the cells in the liver and causes them to convert some of their stored glycogen into glucose and so restore the blood sugar level.

Insulin has the opposite effect to glucagon. If the concentration of blood sugar increases (e.g. after a meal rich in carbohydrate), insulin is released from the islet cells. When the insulin reaches the liver it stimulates the liver cells to take up glucose from the blood and store it as glycogen.

Insulin has many other effects; it increases the uptake of glucose in all cells for use in respiration; it promotes the conversion of carbohydrates to fats and slows down the conversion of protein to carbohydrate.

All these changes have the effect of regulating the level of glucose in the blood to within narrow limits – a very important example of homeostasis.

blood glucose levels too high

$$\text{glucose} \underset{\text{glucagon}}{\overset{\text{insulin}}{\rightleftarrows}} \text{glycogen}$$

blood glucose levels too low

The concentration of glucose in the blood of a person who has not eaten for 8 hours is usually between 90 and 100 mg 100 cm^{-3} blood. After a meal containing carbohydrate, the blood sugar level may rise to 140 mg 100 cm^{-3} but 2 hours later, the level returns to about 95 mg as the liver has converted the excess glucose to glycogen.

About 100 g glycogen is stored in the liver of a healthy man. If the concentration of glucose in the blood falls below about 80 mg 100 cm^{-3} blood, some of the glycogen stored in the liver is converted by enzyme action into glucose, which enters the circulation. If the blood sugar level rises above 160 mg 100 cm^{-3}, glucose is excreted by the kidneys.

A blood glucose level below 40 mg 100 cm^{-3} affects the brain cells adversely, leading to convulsions and coma. By helping to keep the glucose concentration between 80 and 150 mg, the liver prevents these undesirable effects and so contributes to the homeostasis of the body.

If anything goes wrong with the production or function of insulin, the person will show the symptoms of **diabetes**.

Type 1 diabetes

There are two types of diabetes and type 1 is the less common form, the cause of which has been outlined in Chapter 10. It results from a failure of the islet cells to produce sufficient insulin. The outcome is that the patient's blood is deficient in insulin and he or she needs regular injections of the hormone in order to control blood sugar level and so lead a normal life. This form of the disease is, therefore, sometimes called 'insulin-dependent' diabetes. The patient is unable to regulate the level of glucose in the blood. It may rise to such a high level that it is excreted in the urine, or fall so low that the brain cells cannot work properly and the person goes into a coma.

The symptoms of type 1 diabetes include feeling tired, feeling very thirsty, frequent urination and weight loss. Weight loss is experienced because the body starts to break down muscle and fat.

Diabetics need a carefully regulated diet to keep the blood sugar within reasonable limits. They should have regular blood tests to monitor their blood sugar levels and take regular exercise.

Temperature control

In addition to the methods already described, the skin has another very important mechanism for maintaining a constant body temperature. This involves arterioles in the dermis of the skin, which can widen or narrow to allow more or less blood to flow near the skin surface through the blood capillaries. Further details of this process, involving the use of shunt vessels, are given in Chapter 9.

Vasodilation – the widening of the arterioles in the dermis allows more warm blood to flow through blood capillaries near the skin surface and so lose more heat (Figure 14.25(a)).

Vasoconstriction – narrowing (constriction) of the arterioles in the skin reduces the amount of warm blood flowing through blood capillaries near the surface (Figure 14.25(b)).

Tropic responses

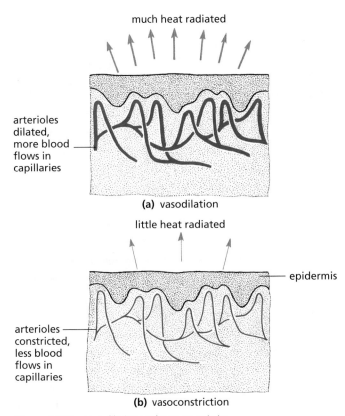

Figure 14.25 Vasodilation and vasoconstriction

● Tropic responses

Sensitivity is the ability of living organisms to respond to stimuli. Although plants do not respond by moving their whole bodies, parts of them do respond to stimuli. Some of these responses are described as tropic responses or **tropisms**.

Tropisms

Tropisms are growth movements related to directional stimuli, e.g. a shoot will grow towards a source of light but away from the direction of gravity. Growth movements of this kind are usually in response to the *direction* of light or gravity. Responses to light are called **phototropisms**; responses to gravity are **gravitropisms** (or **geotropisms**).

> **Key definitions**
> **Gravitropism** is a response in which a plant grows towards or away from gravity.
> **Phototropism** is a response in which a plant grows towards or away from the direction from which light is coming.

If the plant organ responds by growing towards the stimulus, the response is said to be 'positive'. If the response is growth away from the stimulus it is said to be 'negative'. For example, if a plant is placed horizontally, its stem will change its direction and grow upwards, away from gravity (Figure 14.26).

Figure 14.26 Negative gravitropism. The tomato plant has been left on its side for 24 hours.

The shoot is **negatively gravitropic**. The roots, however, will change their direction of growth to grow vertically downwards towards the pull of gravity (Experiment 1). Roots, therefore, are **positively gravitropic**.

Phototropism and gravitropism are best illustrated by some simple controlled experiments. Seedlings are good material for experiments on sensitivity because their growing roots (radicles) and shoots respond readily to the stimuli of light and gravity.

Practical work

Experiments on tropisms

1 Gravitropism in pea radicles

- Soak about 20 peas in water for a day and then let them germinate in a vertical roll of moist blotting-paper.
- After 3 days, choose 12 seedlings with straight radicles and pin six of these to the turntable of a clinostat so that the radicles are horizontal.
- Pin another six seedlings to a cork that will fit in a wide-mouthed jar. Leave the jar on its side.
- A **clinostat** is a clockwork or electric turntable, which rotates the seedlings slowly about four times an hour. Although gravity is pulling sideways on their roots, it will pull equally on all sides as they rotate.
- Place the jar and the clinostat in the same conditions of lighting or leave them in darkness for 2 days.

14 CO-ORDINATION AND RESPONSE

Result
The radicles in the clinostat will continue to grow horizontally but those in the jar will have changed their direction of growth, to grow vertically downwards (Figure 14.27).

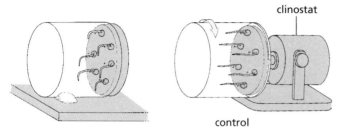

Figure 14.27 Results of an experiment to show gravitropism in roots

Interpretation
The stationary radicles have responded to the stimulus of one-sided gravity by growing towards it. The radicles are positively gravitropic.

The radicles in the clinostat are the controls. Rotation of the clinostat has allowed gravity to act on all sides equally and there is no one-sided stimulus, even though the radicles were horizontal.

2 Phototropism in shoots

- Select two potted seedlings, e.g. sunflower or runner bean, of similar size and water them both.
- Place one of them under a cardboard box with a window cut in one side so that light reaches the shoot from one direction only (Figure 14.28).
- Place the other plant in an identical situation but on a clinostat. This will rotate the plant about four times per hour and expose each side of the shoot equally to the source of light. This is the control.

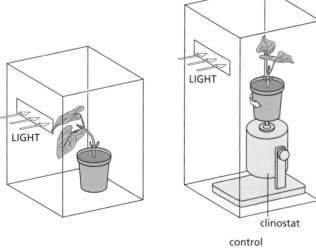

Figure 14.28 Experiment to show phototropism in a shoot

Result
After 1 or 2 days, the two plants are removed from the boxes and compared. It will be found that the stem of the plant with one-sided illumination has changed its direction of growth and is growing towards the light (Figure 14.29). The control shoot has continued to grow vertically.

Figure 14.29 Positive phototropism. The sunflower seedlings have received one-sided lighting for a day.

Interpretation
The results suggest that the young shoot has responded to one-sided lighting by growing towards the light. The shoot is said to be positively phototropic because it grows towards the direction of the stimulus.

However, the results of an experiment with a single plant cannot be used to draw conclusions that apply to green plants as a whole. The experiment described here is more of an illustration than a critical investigation. To investigate phototropisms thoroughly, a large number of plants from a wide variety of species would have to be used.

Advantages of tropic responses

Positive phototropism of shoots
By growing towards the source of light, a shoot brings its leaves into the best situation for photosynthesis. Similarly, the flowers are brought into an exposed position where they are most likely to be seen and pollinated by flying insects.

Negative gravitropism in shoots
Shoots that are negatively gravitropic grow vertically. This lifts the leaves and flowers above the ground and helps the plant to compete for light and carbon dioxide. The flowers are brought into an advantageous position for insect or wind pollination. Seed dispersal may be more effective from fruits on a long, vertical stem. However, these advantages are a product of a tall shoot rather than negative gravitropism.

Stems that form rhizomes (stems that grow underground) are not negatively gravitropic; they grow horizontally below the ground, though the shoots that grow up from them are negatively gravitropic.

Branches from upright stems are not negatively gravitropic; they grow at 90 degrees or, usually, at a more acute angle to the directional pull of gravity. The lower branches of a potato plant must be partially *positively* gravitropic when they grow down into the soil and produce potato tubers (see 'Asexual reproduction' in Chapter 16).

Positive gravitropism in roots

By growing towards gravity, roots penetrate the soil, which is their means of anchorage and their source of water and mineral salts. Lateral roots are not positively gravitropic; they grow at right angles or slightly downwards from the main root. This response enables a large volume of soil to be exploited and helps to anchor the plants securely.

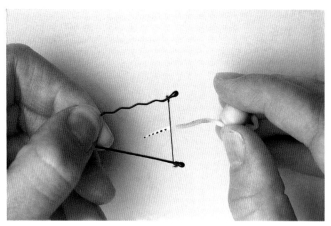

Figure 14.30 Marking a root. A piece of cotton is held by the hairpin and dipped into black ink.

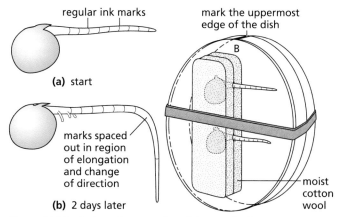

Figure 14.31 Region of response in radicles. Result of Experiment 3 on the B seedlings

Practical work

More experiments on tropisms

3 Region of response

- Grow pea seedlings in a vertical roll of blotting paper and select four with straight radicles about 25 mm long.
- Mark all the radicles with lines about 1 mm apart (Figures 14.30 and 14.31(a)).
- Use four strips of moist cotton wool to wedge two seedlings in each of two Petri dishes (Figure 14.31).
- Leave the dishes on their sides for 2 days, one (A) with the radicles vertical and the other (B) with the radicles horizontal.

Result

The ink marks will be more widely spaced in the region of greatest extension (Figure 14.31(b)). By comparing the seedlings in the two dishes, it can be seen that the region of curvature in the B seedlings corresponds to the region of extension in the A seedlings.

Interpretation

The response to the stimulus of one-sided gravity takes place in the region of extension. It does not necessarily mean that this is also the region which detects the stimulus.

Plant growth substances and tropisms

Control of growth

In animals and plants, the growth rate and extent of growth are controlled by chemicals: **hormones** in animals and **growth substances** in plants. Additionally, growth may be limited in animals by the availability of food, and in plants by light, water and minerals.

There are many different growth substances ('plant hormones') in plants. They are similar in some ways to animal hormones because they are produced in specific regions of the plant and transported to 'target' organs such as roots, shoots and buds. However, the sites of production are not specialised organs, as in animals, but regions of actively dividing cells such as the tips of shoots and roots. Also, plant growth substances are not transported in vessels.

One of the growth substances is **auxin**. Chemically it is indoleacetic acid (IAA). It is produced in the tips of actively growing roots and shoots and carried by active transport (Chapter 3) to the regions of extension where it promotes cell enlargement (Figure 14.32).

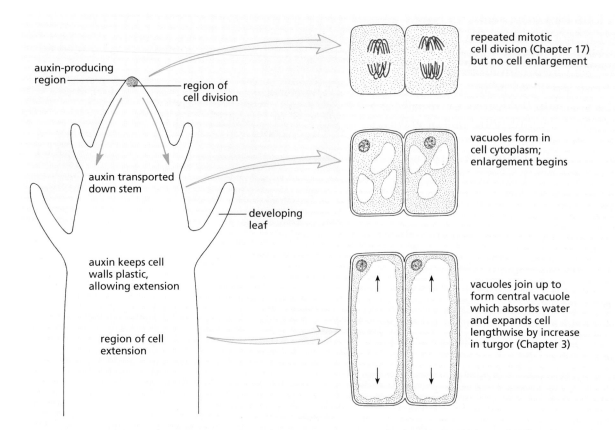

Figure 14.32 Extension growth at shoot tip

The responses made by shoots and roots to light and gravity are influenced by growth substances.

Growth substances also control seed germination, bud burst, leaf fall, initiation of lateral roots and many other processes.

It has already been explained that growth substances, e.g. auxin, are produced by the tips of roots and shoots and can stimulate or, in some cases, inhibit extension growth. Tropic responses could be explained if the one-sided stimuli produced a corresponding one-sided distribution of growth substance.

In the case of positive gravitropism in roots there is evidence that, in a horizontal root, more growth substance accumulates on the lower side. In this case the growth substance is presumed to inhibit extension growth, so that the root tip curves downwards (Figure 14.33).

In the case of phototropism, it is generally accepted that the distribution of growth substance causes reduced extension on the illuminated side and/or increased extension on the non-illuminated side.

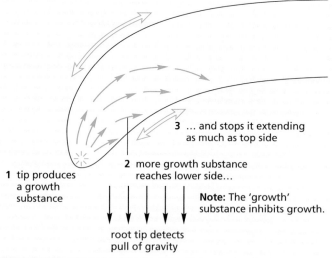

Figure 14.33 Possible explanation of positive gravitropism in roots

Summary of control of shoot growth by auxin

When a shoot is exposed to light from one side, auxins that have been produced by the tip move towards the shaded side of the shoot (or the auxins are destroyed on the light side, causing an unequal distribution). Cells on the shaded side are stimulated to absorb *more* water than those on the light side, so the unequal growth causes the stem to bend towards the light. Growth of a shoot towards light is called **positive phototropism**.

If a shoot is placed horizontally in the absence of light, auxins accumulate on the lower side of the shoot, due to gravity. This makes the cells on the lower side grow *faster* than those on the upper side, so the shoot bends upwards. This is called **negative gravitropism**.

The opposite applies to roots because root cell elongation appears to be slowed down by exposure to auxin.

Classic experiments to test how auxins work

Wheat and other grass species belong to the monocotyledon group of flowering plants (Chapter 1). When wheat seeds germinate (start to grow) they produce a shoot covered by a protective sheath called a **coleoptile**. This helps to prevent damage to the new leaves as they push through the soil. The coleoptile shows responses to light and gravity in a similar way to other plant parts. Wheat coleoptiles only take 2 or 3 days to grow and they show responses very quickly, so they are ideal for tropism experiments. The tip of the coleoptile, where it is expected that auxins would be produced, can be cut off without killing the plant, but effectively removing the source of the auxin. Figure 14.34 shows an investigation, treating coleoptiles in different ways.

Results

A No growth of the coleoptile occurs and there is no bending.

B The coleoptile grows taller and bends towards the light.

C The coleoptile grows taller, but there is no bending.

D The coleoptile grows taller and bends towards the light.

Interpretation

In **A**, the source of auxin has been removed. Auxin is needed to stimulate growth and stimulates a response to light. It could also be argued that the tip provides cells for growth and this source of cells has been removed.

In **B**, auxin is produced by the tip of the coleoptile. It diffuses down the coleoptile and collects on the shaded side of the coleoptile (or is destroyed by the light on the light side). Cells on the shaded side respond to the auxin by growing faster than on the light side causing the coleoptile to grow towards the light.

In **C**, auxin is produced by the tip and diffuses down, causing all cells on both sides of the coleoptile to grow at an equal rate, causing an increase in length. However, the black paper prevents the light influencing the auxin, so there is no response to the direction of light.

In **D**, auxin is produced by the tip of the coleoptile. It diffuses into the agar block. When the agar block is replaced on the cut coleoptile, the auxin diffuses down from the agar and collects on the shaded side of the coleoptile (or is destroyed by the light on the light side). Cells on the shaded side respond to the auxin by growing faster than on the light side causing the coleoptile to grow towards the light.

Use of plant growth substances

Chemicals can be manufactured which closely resemble natural growth substances and may be used to control various aspects of growth and development of crop plants.

The weedkiller, 2,4-D, is very similar to one of the auxins. When sprayed on a lawn, it affects the broad-leaved weeds (e.g. daisies and dandelions) but not the grasses. (It is called a 'selective weedkiller'.) Among other effects, it distorts the weeds' growth and speeds up their rate of respiration to the extent that they exhaust their food reserves and die.

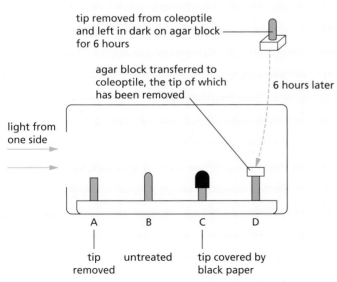

Figure 14.34 Investigation into how auxin works

14 CO-ORDINATION AND RESPONSE

Questions

Core

1. What is the difference between a nerve and a nerve fibre?
2. a In what ways are sensory neurones and motor neurones similar:
      i   in structure
      ii in function?
   b How do they differ?
3. Can a nerve fibre and a nerve carry both sensory and motor impulses? Explain your answers.
   a a nerve fibre
   b a nerve
4. Put the following in the correct order for a simple reflex arc
   a impulse travels in motor fibre
   b impulse travels in sensory fibre
   c effector organ stimulated
   d receptor organ stimulated
   e impulse crosses synapse.
5. Which receptors and effectors are involved in the reflex actions of:
   a sneezing
   b blinking
   c contraction of the iris?
6. Explain why the tongue may be considered to be both a receptor and an effector organ.
7. Discuss whether coughing is a voluntary or reflex action.
8. What sensation would you expect to feel if a warm pin-head was pressed on to a touch receptor in your skin? Explain your answer.
9. If a piece of ice is pressed on to the skin, which receptors are likely to send impulses to the brain?
10. Apart from the cells that detect chemicals, what other types of receptor must be present in the tongue?
11. a To what directional stimuli do:
       i   roots respond
       ii shoots respond?
    b Name the plant organs which are
       i   positively phototropic
       ii positively gravitropic
       iii negatively gravitropic.
12. Why is it incorrect to say:
    a 'Plants grow towards the light.'
    b 'If a root is placed horizontally, it will bend towards gravity'?
13. Explain why a clinostat is used for the controls in tropism experiments.
14. Look at Figure 14.26. What will the shoot look like in 24 hours after the pot has been stood upright again? (Just draw the outline of the stem.)
15. What do you think might happen if a potted plant were placed on its side and the shoot illuminated from below (i.e. light and gravity are acting from the same direction)?

Extended

16. Look at Figures 14.6 and Figure 14.8. For each diagram, state
    a how many cell bodies are drawn
    b how many synapses are shown.
17. If you could intercept and 'listen to' the nerve impulses travelling in the spinal cord, could you tell which ones came from pain receptors and which from temperature receptors? Explain your answer.
18. Would you expect synapses to occur in grey matter or in white matter? Explain your answer.
19. Study Figure 14.2. If the spinal cord were damaged at a point about one-third of the way up the vertebral column, what effect would you expect this to have on the bodily functions?
20. Study Table 14.3 and give one example for each point of comparison.
21. The pancreas has a dual function in producing digestive enzymes as well as hormones. Which other endocrine glands have a dual function and what are their other functions? (See also 'Sex hormones in humans' in Chapter 16.)
22. What are the effects on body functions of:
    a too much insulin
    b too little insulin?
23. Why do you think urine tests are carried out to see if a woman is pregnant?
24. What conscious actions do we take to reduce the heat lost from the body?
25. a What sort of chemical reaction in active muscle will produce heat?
    b How does this heat get to other parts of the body?
26. Draw up a balance sheet to show all the possible ways the human body can gain or lose heat. Make two columns, with 'Gains' on the left and 'Losses' on the right.
27. a Which structures in the skin of a furry mammal help to reduce heat loss?
    b What changes take place in the skin of humans to reduce heat loss?
28. Sweating cools you down only if the sweat can evaporate.
    a In what conditions might the sweat be unable to evaporate from your skin?
    b What conditions might speed up the evaporation of sweat and so make you feel very cold?
29. In Figure 14.35 the two sets of pea seedlings were sown at the same time, but the pot on the left was kept under a lightproof box. From the evidence in the picture:
    a what effects does light appear to have on growing seedlings
    b how might this explain positive phototropism?

Figure 14.35 Effect of light on shoots

30. It is suggested that it is the very tip of the radicle that detects the one-sided pull of gravity even though it is the region of extension that responds. How could you modify Experiment 3 to test this hypothesis?

Checklist

After studying Chapter 14 you should know and understand the following:

The nervous system
- The central nervous system consists of the brain and the spinal cord.
- The peripheral nervous system consists of the nerves.
- The nerves consist of bundles of nerve fibres.
- Each nerve fibre is a thin filament that grows out of a nerve cell body.
- The nerve cell bodies are mostly in the brain and spinal cord.
- Nerve fibres carry electrical impulses from sense organs to the brain or from the brain to muscles and glands.
- A reflex is an automatic nervous reaction that cannot be consciously controlled.
- A reflex arc is the nervous pathway that carries the impulses causing a reflex action.
- The simplest reflex involves a sensory nerve cell and a motor nerve cell, connected by synapses in the spinal cord.
- The brain and spinal cord contain millions of nerve cells.
- The millions of possible connections between the nerve cells in the brain allow complicated actions, learning, memory and intelligence.

- Voluntary actions start in the brain, while involuntary actions are automatic.
- Reflexes have a protective function.
- A synapse is a junction between two neurones consisting of a minute gap across which impulses pass by diffusion of a neurotransmitter.
- Identify parts of a synapse and describe how it transmits an impulse from one neurone to another.
- Drugs such as morphine and heroin can affect synapses.
- In reflex arcs, synapses ensure the movement of impulses in one direction.

Sense organs
- Sense organs are groups of receptor cells responding to specific stimuli: light, sound, touch, temperature and chemicals.
- Describe the structure of the eye.
- Describe the function of the parts of the eye.
- Describe the pupil reflex.

- Explain the pupil reflex.
- Explain accommodation to view near and distant objects.
- Describe the roles of parts of the eye in accommodation.
- State the distribution of rods and cones in the retina of a human.
- Describe the function of rods and cones.

Hormones in humans
- A hormone is a chemical substance, produced by a gland, carried by the blood, which alters the activity of one or more specific target organs
- The testes, ovaries and pancreas are also endocrine glands in addition to their other functions.

- The endocrine glands release hormones into the blood system.
- When the hormones reach certain organs they change the rate or kind of activity of the organ.
- Too much or too little of a hormone can cause a metabolic disorder.
- Adrenalin is secreted in 'fight or flight' situations.
- It causes an increased breathing and pulse rate and widened pupils.

- Adrenaline has a role in the chemical control of metabolic activity, including increasing the blood glucose concentration and pulse rate.
- The nervous system is much faster and its action tends to be over a shorter time span than hormonal control systems.

Homeostasis
- Homeostasis is the maintenance of a constant internal environment.
- Skin consists of an outer layer of epidermis and an inner dermis.
- The epidermis is growing all the time and has an outer layer of dead cells.
- The dermis contains the sweat glands, hair follicles, sense organs and capillaries.
- Skin (1) protects the body from bacteria and drying out, (2) contains sense organs which give us the sense of touch, warmth, cold and pain, and (3) controls the body temperature.
- Chemical activity in the body and muscular contractions produce heat.
- Heat is lost to the surroundings by conduction, convection, radiation and evaporation.
- If the body temperature rises too much, the skin cools it down by sweating and vasodilation.
- If the body loses too much heat, vasoconstriction and shivering help to keep it warm.

- Negative feedback provides a means of control: if levels of substances in the body change, the change is monitored and a response to adjust levels to normal is brought about.
- Glucose concentration in the blood is controlled using insulin and glucagon.
- Type 1 diabetes is the result of islet cells in the pancreas failing to produce enough insulin.
- Vasodilation and vasoconstriction of arterioles in the skin are mechanisms to control body temperature.

Tropic responses
- A response related to the direction of the stimulus is a tropism.
- The roots and shoots of plants may respond to the stimuli of light or gravity.
- Gravitropism is a response in which a plant grows towards or away from gravity.
- Phototropism is a response in which a plant grows towards or away from the direction from which light is coming.

14 CO-ORDINATION AND RESPONSE

- Growth towards the direction of the stimulus is called 'positive'; growth away from the stimulus is called 'negative'.
- Tropic responses bring shoots and roots into the most favourable positions for their life-supporting functions.
- Describe investigations into gravitropism and phototropism in shoots and roots.
- Explain phototropism and gravitropism of a shoot as examples of the chemical control of plant growth by auxin.
- Auxin is only made in the shoot tip and moves through the plant, dissolved in water.
- Auxin is unequally distributed in response to light and gravity.
- Auxin stimulates cell elongation.
- The synthetic plant hormone 2,4-D is used in weedkillers.

15 Drugs

Drugs
Define drug

Medicinal drugs
Use of antibiotics
Development of resistance in bacteria to antibiotics

Development of resistant bacteria
Antibiotics and viral diseases

Misused drugs
Effects of heroin, alcohol, tobacco
Role of liver in breaking down toxins

Effects of heroin on the nervous system
Link between smoking and cancer
Use of performance-enhancing drugs

Drugs

Key definition
A **drug** is any substance taken into the body that modifies or affects chemical reactions in the body.

The drug may be one taken legally to reduce a symptom such as a headache or to treat a bacterial infection (medicinal drugs), but it could also be one taken – often illegally – to provide stimulation or induce sleep or create hallucinations (recreational drugs). Drugs are present in many products such as: tea, coffee and 'energy drinks' (caffeine); tobacco (nicotine); and alcoholic drinks (alcohol) which, although legal, can cause serious effects when taken excessively or over extended periods of time.

Medicinal drugs

Any substance used in medicine to help our bodies fight illness or disease is called a drug.

Antibiotics

The ideal drug for curing disease would be a chemical that destroyed the pathogen without harming the tissues of the host. In practice, modern antibiotics such as penicillin come pretty close to this ideal for bacterial infections.

A tiny minority of bacteria are harmful (pathogenic). Figure 10.1 shows some examples and the diseases they cause.

Most of the antibiotics we use come from bacteria or fungi that live in the soil. The function of the antibiotics in this situation is not clear. One theory suggests that the chemicals help to suppress competition for limited food resources, but the evidence does not support this theory.

One of the most prolific sources of antibiotics is *Actinomycetes*. These are filamentous bacteria that resemble microscopic mould fungi. The actinomycete *Streptomyces* produces the antibiotic **streptomycin**.

Perhaps the best known antibiotic is **penicillin**, which is produced by the mould fungus *Penicillium* and was discovered by Sir Alexander Fleming in 1928. Penicillin is still an important antibiotic but it is produced by mutant forms of a different species of *Penicillium* from that studied by Fleming. The different mutant forms of the fungus produce different types of penicillin.

The penicillin types are chemically altered in the laboratory to make them more effective and to 'tailor' them for use with different diseases. 'Ampicillin', 'methicillin' and 'oxacillin' are examples.

Antibiotics attack bacteria in a variety of ways. Some of them disrupt the production of the cell wall and so prevent the bacteria from reproducing, or even cause them to burst open; some interfere with protein synthesis and thus arrest bacterial growth.

Animal cells do not have cell walls, and the cell structures involved in protein production are different. Consequently, antibiotics do not damage human cells although they may produce some side-effects such as allergic reactions.

Not all bacteria are killed by antibiotics. Some bacteria have a nasty habit of mutating to forms that are resistant to these drugs.

For this reason it is important not to use antibiotics in a diluted form, for too short a period or for trivial complaints. These practices lead to a build-up of a resistant population of bacteria. The drug resistance can be passed from harmless bacteria to pathogens.

It is important to note that antibiotics are ineffective in the treatment of viral diseases.

Development of resistant bacteria

If a course of antibiotics is not completed, some of the bacteria it is being used to destroy will not be killed, but will have been exposed to the drug. Some of the survivors may be drug-resistant mutants. When they reproduce, all their offspring will have the drug resistance, so the antibiotic will become less effective (Figure 15.1).

15 DRUGS

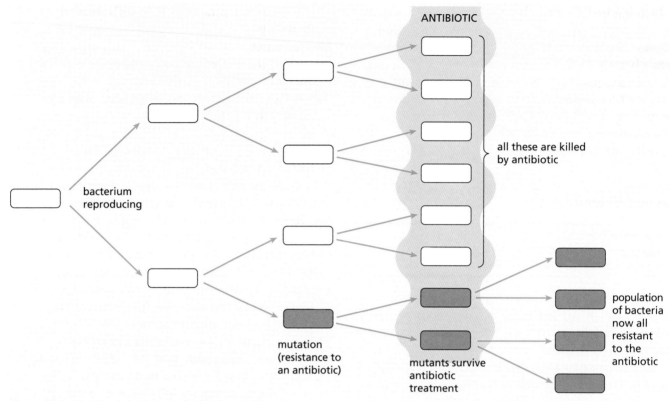

Figure 15.1 Mutation in bacteria can lead to drug resistance

One type of bacteria that has developed resistance to a number of widely used antibiotics is called MRSA (methicillin-resistant *Staphylococcus aureus*). These types of bacteria are sometime referred to as 'superbugs' because they are so difficult to treat. *Staphylococcus aureus* is very common and is found living harmlessly on the skin, the nose and throat, sometimes causing mild infections. It becomes dangerous if there is a break in the skin, allowing it to infect internal organs and causing blood poisoning. This can happen in hospitals with infection during operations, especially if hygiene precautions are not adequate.

Doctors now have to be much more cautious about prescribing antibiotics, to reduce the risk of resistant strains developing. Patients need to be aware of the importance of completing a course of antibiotics, again to reduce the risk of development of resistant strains.

Antibiotics and viral diseases

Antibiotics are not effective against viral diseases. This is because antibiotics work by disrupting structures in bacteria such as cell walls and membranes, or processes associated with protein synthesis and replication of DNA. Viruses have totally different characteristics to bacteria, so antibiotics do not affect them. Compare the image of a virus in Figure 1.34 with that of a bacterium in Figure 1.29.

● Extension work

Ideas about antibiotics

Alexander Fleming (1881–1955)

Before 1934 there were few effective drugs. Some herbal preparations may have been useful; after all, many of our present-day drugs are derived from or based on plant products. Quinine, for example, was used for the treatment of malaria and was extracted from a specific kind of tree bark.

In 1935, a group of chemicals called **sulfanilamides** were found to be effective against some bacterial diseases such as blood poisoning, pneumonia and septic wounds.

Fleming had discovered penicillin in 1928, 7 years before the use of sulfanilamides, but he had been unable to purify it and test it on humans. Fleming was a bacteriologist working at St Mary's Hospital in London. In 1928, he was studying different strains of *Staphylococcus* bacteria. He had made some cultures on agar plates and left them on the laboratory bench during a 4-week holiday. When he returned he noticed that one of the plates had been contaminated by a mould fungus and that around the margins of the mould there was a clear zone with no bacteria growing (Figure 15.2).

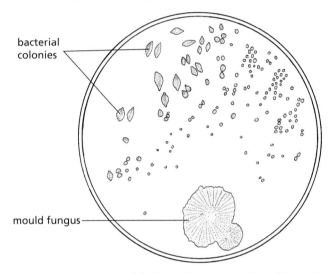

Figure 15.2 Appearance of the *Staphylococcus* colonies on Fleming's petri dish

Fleming reasoned that a substance had diffused out of the mould colony and killed the bacteria. The mould was identified as *Penicillium notatum* and the supposed anti-bacterial chemical was called penicillin. Fleming went on to culture the *Penicillium* on a liquid meat broth medium and showed that the broth contained penicillin, which suppressed the growth of a wide range of bacteria.

Two research assistants at St Mary's then tried to obtain a pure sample of penicillin, free from all the other substances in the broth. Although they succeeded, the procedure was cumbersome and the product was unstable. By this time, Fleming seemed to have lost interest and to assume that penicillin would be too difficult to extract and too unstable to be of medical value.

In 1939, **Howard Florey** (a pathologist) and **Ernst Chain** (a biochemist), working at Oxford University, succeeded in preparing reasonably pure penicillin and making it stable. Techniques of extraction had improved dramatically in 10 years and, in particular, freeze-drying enabled a stable water-soluble powder form of penicillin to be produced.

World War II was an urgent incentive for the production of penicillin in large quantities and this undoubtedly saved many lives that would otherwise have been lost as a result of infected wounds.

Once Ernst Chain had worked out the molecular structure of penicillin, it became possible to modify it chemically and produce other forms of penicillin that attacked a different range of bacteria or had different properties. For example, ampicillin is a modified penicillin that can be taken by mouth rather than by injection.

Because penicillin was the product of a mould, chemists searched for other moulds, particularly those present in the soil, which might produce antibiotics. A large number of these were discovered, including streptomycin (for tuberculosis), chloramphenicol (for typhoid), aureomycin and terramycin (broad spectrum antibiotics, which attack a wide range of bacteria). The ideal drug is one that kills or suppresses the growth of harmful cells, such as bacteria or cancer cells, without damaging the body cells. Scientists have been trying for years to find a 'magic bullet' that 'homes in' exclusively on its target cells. For bacterial diseases, antibiotics come pretty close to the ideal, though the bacteria do seem able to develop resistant forms after a few years.

Misused drugs

Narcotics

Heroin, morphine and codeine belong to a group of drugs called **narcotics**, made from opium. Heroin and morphine act as powerful depressants: they relieve severe pain and produce short-lived feelings of wellbeing and freedom from anxiety. They can both lead to tolerance and physical dependence within weeks, so they are prescribed with caution, to patients in severe pain.

The illegal use of heroin has terrible effects on the unfortunate addict. The overwhelming dependence on the drug leads many addicts into prostitution and crime in order to obtain the money to buy it.

There are severe withdrawal symptoms when an addict tries to give up the drug abruptly. These symptoms are called going 'cold turkey' and can include anxiety, muscle aches, sweating, abdominal cramping, diarrhoea, nausea and vomiting. A 'cure' is a long and often unsuccessful process.

Additional hazards are that blood poisoning, hepatitis and AIDS may result from the use of unsterilised needles when injecting the drug.

Codeine is a less effective analgesic than morphine, but does not lead so easily to dependence. It is still addictive if used in large enough doses.

Alcohol

The alcohol in wines, beer and spirits is a depressant of the central nervous system. Small amounts give a sense of wellbeing, with a release from anxiety. However, this is accompanied by a fall-off in performance in any activity requiring skill. It also gives a misleading sense of confidence in spite of the fact that one's judgement is clouded. A drunken driver usually thinks he or she is driving extremely well.

Even a small amount of alcohol in the blood increases our reaction time (the interval between receiving a stimulus and making a response). In some people, the reaction time is doubled even when the alcohol in the blood is well below the legal limit laid down for car drivers (Figure 15.3). This can make a big difference to the time needed for a driver to apply the brakes after seeing a hazard such as a child running into the road.

Alcohol causes vasodilation in the skin, giving a sensation of warmth but in fact leading to a greater loss of body heat (see 'Homeostasis' in Chapter 14). A concentration of 500 mg of alcohol in 100 cm³ of blood results in unconsciousness. More than this will cause death because it stops the breathing centre in the brain. The liver treats alcohol as a toxin: 90% of alcohol taken in is **detoxified** in the liver (along with other toxins). The process of detoxification involves the oxidation of alcohol to carbon dioxide and water. Only 10% is excreted by the kidneys. On average, the liver can oxidise about 75 mg alcohol per 1 kg body weight per hour. This rate varies considerably from one individual to the next but it indicates that it would take about 3 hours to oxidise the alcohol in a pint of beer or a glass of wine. If the alcohol intake exceeds this rate of oxidation, the level of alcohol in the blood builds up to toxic proportions; that is, it leads to **intoxication**.

Some people build up a tolerance to alcohol and this may lead to both emotional and physical dependence (alcoholism). High doses of alcohol can cause the liver cells to form too many fat droplets, leading to the disease called **cirrhosis**. A cirrhotic liver is less able to stop poisonous substances in the intestinal blood from reaching the general circulation.

Pregnancy

Drinking alcohol during pregnancy can present a major risk to the developing fetus. Further details are given in Chapter 16.

Behaviour

Alcohol reduces inhibitions because it depresses that part of the brain which causes shyness. This may be considered an advantage in 'breaking the ice' at parties. But it can also lead to irresponsible behaviour such as vandalism and aggression.

Moderate drinking

A moderate intake of alcoholic drink seems to do little physiological harm (except in pregnant women). But what is a 'moderate' intake?

A variety of drinks that all contain the same amount of alcohol is shown in Figure 15.4. Beer is a fairly dilute form of alcohol. Whisky, however, is about 40% alcohol. Even so, half a pint of beer contains the same amount of alcohol as a single whisky. This amount of alcohol can be called a 'unit'.

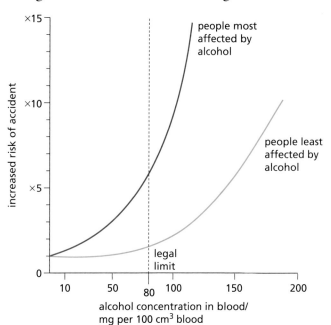

Figure 15.3 Increased risk of accidents after drinking alcohol. People vary in their reactions to alcohol. Body weight, for example, makes a difference.

It is the number of units of alcohol, not the type of drink, which has a physiological effect on the body. In Britain, the Health Development Agency recommends upper limits of 21–28 units for men and 14–21 units for women over a 1-week period at the time of publication of this book. Pregnant women should avoid alcohol altogether.

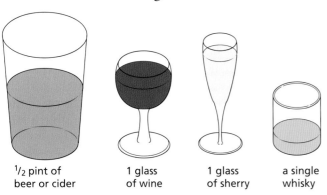

Figure 15.4 Alcohol content of drinks. All these drinks contain the same amount of alcohol (1 unit). Although the alcohol is more dilute in the beer than in the whisky, it has the same effect on the body.

Smoking

The short-term effects of smoking cause the bronchioles to constrict and the cilia lining the air passages to stop beating. The smoke also makes the lining produce more mucus. **Nicotine**, the addictive component of tobacco smoke, produces an increase in the rate of the heartbeat and a rise in blood pressure. It may, in some cases, cause an erratic and irregular heart beat. Tar in cigarette smoke is thought to be the main cause of lung cancer in smokers. **Carbon monoxide** permanently binds with haemoglobin in red blood cells, reducing the smoker's ability to provide oxygen to respiring cells. This results in a smoker getting out of breath more easily and it reduces physical fitness.

The long-term effects of smoking may take many years to develop but they are severe, disabling and often lethal.

Lung cancer

Cancer is a term used for diseases in which cells become abnormal and divide out-of-control. They can then move around the body and invade other tissues. A chemical that causes cancer is known as a **carcinogen**. Carcinogens present in cigarette smoke, such as tar, increase the risk of lung cells becoming cancerous. Tumours develop. These are balls of abnormal cells, which do not allow gaseous exchange like normal lung cells.

Many studies have now demonstrated how cigarette smoke damages lung cells, confirming that smoking does cause cancer. The higher the number of cigarettes smoked, the greater the risk of lung cancer.

Chronic obstructive pulmonary disease (COPD)

This term covers a number of lung diseases, which include chronic bronchitis, emphysema and chronic obstructive airways disease. A person suffering from COPD will experience difficulties with breathing, mainly because of narrowing of the airways (bronchi and bronchioles). Symptoms of COPD include breathlessness when active, frequent chest infections and a persistent cough with phlegm (sticky mucus).

Emphysema

Emphysema is a breakdown of the alveoli. The action of one or more of the substances in tobacco smoke weakens the walls of the alveoli. The irritant substances in the smoke cause a 'smokers' cough' and the coughing bursts some of the weakened alveoli. In time, the absorbing surface of the lungs is greatly reduced (Figure 15.5). Then the smoker cannot oxygenate his or her blood properly and the least exertion makes the person breathless and exhausted.

Chronic bronchitis

The smoke stops the cilia in the air passages from beating, so the irritant substances in the smoke and the excess mucus collect in the bronchi. This leads to inflammation known as **bronchitis**. Over 95% of people suffering from bronchitis are smokers and they have a 20 times greater chance of dying from bronchitis than non-smokers.

Heart disease

Coronary heart disease is the leading cause of death in most developed countries. It results from a blockage of coronary arteries by fatty deposits. This reduces the supply of oxygenated blood to the heart muscle and sooner or later leads to heart failure (see Chapter 9). High blood pressure, diets with too much animal fat and lack of exercise are also thought to be causes of heart attack, but about a quarter of all deaths due to coronary heart disease are thought to be caused by smoking (see Figure 9.12).

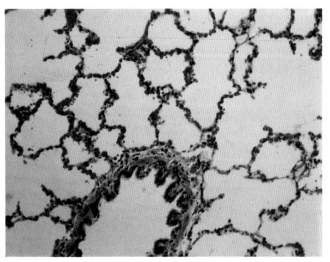

(a) Normal lung tissue showing a bronchiole and about 20 alveoli (×200)

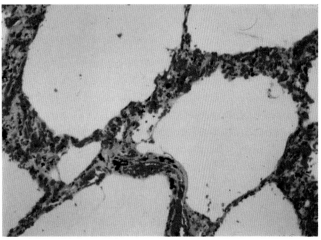

(b) Lung tissue from a person with emphysema. This is the same magnification as (a). The alveoli have broken down leaving only about five air sacs, which provide a much reduced absorbing surface.

Figure 15.5 Emphysema

The nicotine and carbon monoxide from cigarette smoke increase the tendency for the blood to clot and so block the coronary arteries, already partly blocked by fatty deposits. The carbon monoxide increases the rate at which the fatty material is deposited in the arteries.

Other risks

About 95% of patients with disease of the leg arteries are cigarette smokers; this condition is the most frequent cause of leg amputations.

Strokes due to arterial disease in the brain are more frequent in smokers.

Cancer of the bladder, ulcers in the stomach and duodenum, tooth decay, gum disease and tuberculosis all occur more frequently in smokers.

Babies born to women who smoke during pregnancy are smaller than average, probably as a result of reduced oxygen supply caused by the carbon monoxide in the blood. In smokers, there is twice the frequency of miscarriages, a 50% higher still-birth rate and a 26% higher death rate of babies.

A recent estimate is that one in every three smokers will die as a result of their smoking habits. Those who do not die at an early age will probably be seriously disabled by one of the conditions described above.

Passive smoking

It is not only the smokers themselves who are harmed by tobacco smoke. Non-smokers in the same room are also affected. One study has shown that children whose parents both smoke breathe in as much nicotine as if they were themselves smoking 80 cigarettes a year.

Statistical studies also suggest that the non-smoking wives of smokers have an increased chance of lung cancer.

Reducing the risks

By giving up smoking, a person who smokes up to 20 cigarettes a day will, after 10 years, be at no greater risk than a non-smoker of the same age. A pipe or cigar smoker, provided he or she does not inhale, is at less risk than a cigarette smoker but still at greater risk than a non-smoker.

How heroin affects the nervous system

As described in Chapter 14, heroin produces its effects by interacting with receptor molecules at synapses. Synapses are tiny gaps between neurones, across which electrical impulses cannot jump. To maintain the transmission of the impulse, a chemical messenger called a neurotransmitter is released into the gap. When it reaches the neurone on the other side, receptor molecules are stimulated to generate and release new electrical impulses. Heroin mimics the transmitter substances in synapses in the brain, causing the stimulation of receptor molecules. This causes the release of **dopamine** (a neurotransmitter), which gives a short-lived 'high'.

Evidence for a link between smoking and lung cancer

Although all forms of air pollution are likely to increase the chances of lung cancer, many scientific studies show, beyond all reasonable doubt, that the vast increase in lung cancer (4000% in the last century) is almost entirely due to cigarette smoking (Figure 15.6).

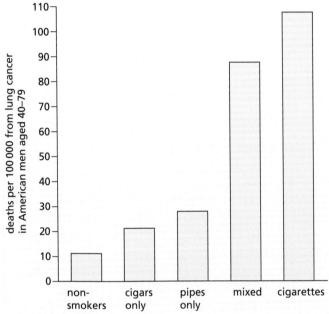

Figure 15.6 Smoking and lung cancer. Cigar and pipe smokers are probably at less risk because they often do not inhale. But notice that their death rate from lung cancer is still twice that of non-smokers. They are also at risk of other cancers such as mouth and throat cancer.

There are at least 17 substances in tobacco smoke known to cause cancer in experimental animals, and it is now thought that 90% of lung cancer is caused by smoking. Table 15.1 shows the relationship between smoking cigarettes and the risk of developing lung cancer.

Table 15.1 Cigarette smoking and lung cancer

Number of cigarettes per day	Increased risk of lung cancer
1–14	×8
15–24	×13
25+	×25

Correlations and causes

In Chapter 9 it was explained that a correlation between two variables does not prove that one of the variables causes the other. The fact that a higher risk of dying from lung cancer is correlated with heavy smoking does not actually prove that smoking is the cause of lung cancer. The alternative explanation is that people who become heavy smokers are, in some way, exposed to other potential causes of lung cancer, e.g. they live in areas of high air pollution or they have an inherited tendency to cancer of the lung. These alternatives are not very convincing, particularly when there is such an extensive list of ailments associated with smoking.

This is not to say that smoking is the only cause of lung cancer or that everyone who smokes will eventually develop lung cancer. There are likely to be complex interactions between life-styles, environments and genetic backgrounds which could lead, in some cases, to lung cancer. Smoking may be only a part, but a very important part, of these interactions.

Performance-enhancing hormones

In the last 30 years or so, some athletes and sports persons have made use of drugs to boost their performance. Some of these drugs are synthetic forms of hormones.

Testosterone is made in the testes of males and is responsible for promoting male primary and secondary sexual characteristics. Taking testosterone supplements (known as 'doping') leads to increased muscle and bone mass. The practice therefore has the potential to enhance a sportsperson's performance.

Anabolic steroids are synthetic derivatives of testosterone. They affect protein metabolism, increasing muscle development and reducing body fat. Athletic performance is thus enhanced. There are serious long-term effects of taking anabolic steroids. The list is a long one but the main effects are sterility, masculinisation in women, and liver and kidney malfunction.

An internationally famous athlete caught using performance enhancing drugs was Ben Johnson (Figure 15.7), who represented Canada as a sprinter. He gained medals in the 1987 World Championships and the 1988 Olympics, but these were withdrawn after a urine sample tested positive for anabolic steroids.

15 DRUGS

Figure 15.7 Ben Johnson (in red) beating his arch rival Carl Lewis (in blue). Johnson would later be banned from international athletics for life for using anabolic steroids.

Because these drugs enhance performance beyond what could be achieved by normal training, they are deemed unfair and banned by most sports organisations. Anabolic steroids are universally banned but different sports regulatory bodies have different rules for other substances.

The products of the steroid hormones can be detected in the urine and this is the basis of most tests for banned substances. Without these regulations, sport would become a competition between synthetic chemical substances rather than between individuals and teams.

Questions

Core
1 Why are doctors concerned about the over-use of antibiotics?
2 List at least four effects of the excessive consumption of alcohol.
3 Find out the cost of a packet of 20 cigarettes. If a person smokes 20 cigarettes a day, how much would this cost in a year?

Extended
4 What are:
   a the immediate effects and
   b the long-term effects
   of tobacco smoke on the trachea, bronchi and lungs?
5 Why does a regular smoker get out of breath sooner than a non-smoker of similar age and build?
6 If you smoke 20 cigarettes a day, by how much are your chances of getting lung cancer increased?
7 Apart from lung cancer, what other diseases are probably caused by smoking?

Checklist

After studying Chapter 15 you should know and understand the following:

- A drug is any substance taken into the body that modifies or affects chemical reactions in the body.
- Antibiotics are used in the treatment of bacterial infections.
- Some bacteria become resistant to antibiotics, which reduces their effectiveness.
- Antibiotics kill bacteria but not viruses.

- It is possible to minimise the development of resistant bacteria such as MRSA.
- Viruses have a different structure to bacteria, so they are not affected by antibiotics.

- Smoking and excessive drinking contribute to ill-health.
- Mood-influencing drugs may be useful for treating certain illnesses but are dangerous if used for other purposes.

- Tolerance means that the body needs more and more of a particular drug to produce the same effect.
- Dependence means that a person cannot do without a particular drug.
- Withdrawal symptoms are unpleasant physical effects experienced by an addict when the drug is not taken.
- Tobacco smoke affects the gaseous exchange system because it contains toxic components.
- Alcohol is a depressant drug, which slows down reaction time and reduces inhibitions.
- Alcohol in a pregnant woman's blood can damage her fetus.
- The liver is the site of breakdown of alcohol and other toxins.

- Heroin is a strongly addictive drug, which affects the nervous system.
- There is now strong enough evidence to provide a link between smoking and lung cancer.
- Some hormones are used to improve sporting performance.

16 Reproduction

Asexual reproduction
Define asexual reproduction
Examples of asexual reproduction
  Advantages and disadvantages of asexual reproduction
Sexual reproduction
Define sexual reproduction and fertilisation
  Haploid and diploid cells
  Advantages and disadvantages of sexual reproduction
Sexual reproduction in plants
Parts of insect-pollinated and wind-pollinated flowers and their functions
Define pollination
Fertilisation
Adaptations of insect-pollinated and wind-pollinated flowers
Investigate conditions needed for germination
  Define self-pollination and cross-pollination
  Implications of self-pollination to a species
  Growth of pollen tube and fertilisation
Sexual reproduction in humans
Parts of male and female reproductive systems
Describe fertilisation
Adaptive features of sperm and eggs
Development of embryo
Growth and development of fetus
	Antenatal care
	Labour and birth
	  Compare male and female gametes
	  Functions of the placenta and umbilical cord
	  Passage of toxins and viruses across placenta
	  Comparing breast feeding and bottle feeding
	Sex hormones in humans
	Puberty, hormones and secondary sexual characteristics
	Menstrual cycle
	  Sites of production and roles of hormones related to menstrual cycle and pregnancy
	Methods of birth control in humans
	Methods of birth control
	  Use of hormones in fertility treatment and contraception
	  Artificial insemination
	  In vitro fertilisation
	  Social implications of contraception and fertility treatments
	Sexually transmitted infections (STIs)
	Define sexually transmitted infection
	HIV
	Spread and control of STIs
	  How HIV affects the immune system

No organism can live for ever, but part of it lives on in its offspring. Offspring are produced by the process of reproduction. This process may be **sexual** or **asexual**, but in either case it results in the continuation of the species.

 ## Asexual reproduction

> **Key definition**
> **Asexual reproduction** is the process resulting in the production of genetically identical offspring from one parent.

Asexual means 'without sex' and this method of reproduction does not involve gametes (sex cells). In the single-celled protoctista or in bacteria, the cell simply divides into two and each new cell becomes an independent organism.

In more complex organisms, part of the body may grow and develop into a separate individual. For example, a small piece of stem planted in the soil may form roots and grow into a complete plant.

Bacteria reproduce by cell division or **fission**. Any bacterial cell can divide into two and each daughter cell becomes an independent bacterium (Figure 1.31). In some cases, this cell division can take place every 20 minutes so that, in a very short time, a large colony of bacteria can be produced. This is one reason why a small number of bacteria can seriously contaminate our food products (see Chapter 10). This kind of reproduction, without the formation of gametes (sex cells), is called **asexual reproduction**.

Asexual reproduction in fungi

Fungi have sexual and asexual methods of reproduction. In the asexual method they produce single-celled, haploid spores. These are dispersed, often by air currents and, if they reach a suitable situation, they grow new hyphae, which develop into a mycelium (see Figures 1.25 and 1.26).

Penicillium and *Mucor* are examples of mould fungi that grow on decaying food or vegetable matter. *Penicillium* is a genus of mould fungi that grows on decaying vegetable matter, damp leather and citrus fruits. The mycelium grows over the food, digesting it and absorbing nutrients. Vertical hyphae grow from the mycelium and, at their tips, produce chains of spores (Figures 16.1 and 16.2). These give the colony a blue-green colour and a powdery appearance (see Figure 19.17). The spores are dispersed by air currents and, if they reach a suitable substrate, grow into a new mycelium.

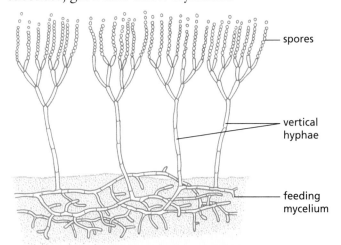

Figure 16.1 *Penicillium sp.*

Figure 16.3 Asexual reproduction in *Mucor*. The black spheres are sporangia that have not yet discharged their spores (×160).

Figure 16.4 Toadstools growing on a fallen tree. The toadstools are the reproductive structures that produce spores. The feeding hyphae are inside the tree, digesting the wood.

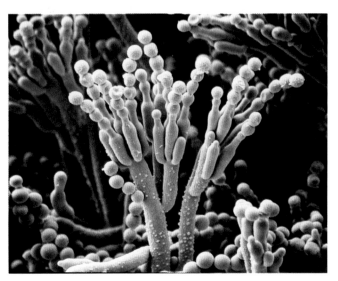

Figure 16.2 Scanning electron micrograph of *Penicillium* spores

Mucor feeds, grows and reproduces in a similar way to *Penicillium*, but *Mucor* produces spores in a slightly different way. Instead of chains of spores at the tips of the vertical hyphae, *Mucor* forms spherical sporangia, each containing hundreds of spores (Figure 16.3). These are dispersed on the feet of insects or by the splashes of rain drops.

The gills on the underside of a mushroom or toadstool (Figures 16.4 and 16.5) produce spores. Puffballs release clouds of spores (Figure 16.6).

Figure 16.5 A bracket fungus. The 'brackets' are the reproductive structures. The mycelium in the trunk feeds on living tissue and will eventually kill the tree.

Asexual reproduction

Figure 16.6 Puffball dispersing spores. When a raindrop hits the ripe puffball, a cloud of spores is ejected.

Figure 16.7 Bryophyllum. The plantlets are produced from the leaf margin. When they fall to the soil below, they grow into independent plants.

Asexual reproduction in flowering plants (vegetative propagation)

Although all flowering plants reproduce sexually (that is why they have flowers), many of them also have asexual methods.

Several of these asexual methods (also called '**vegetative propagation**') are described below. When vegetative propagation takes place naturally, it usually results from the growth of a lateral bud on a stem which is close to, or under, the soil. Instead of just making a branch, the bud produces a complete plant with roots, stem and leaves. When the old stem dies, the new plant is independent of the parent that produced it.

An unusual method of vegetative propagation is shown by Bryophyllum (Figure 16.7).

Stolons and rhizomes

The flowering shoots of plants such as the strawberry and the creeping buttercup are very short and, for the most part, below ground. The stems of shoots such as these are called **rootstocks**. The rootstocks bear leaves and flowers. After the main shoot has flowered, the lateral buds produce long shoots, which grow horizontally over the ground (Figure 16.8). These shoots are called **stolons** (or 'runners'), and have only small, scale-leaves at their nodes and very long internodes. At each node there is a bud that can produce not only a shoot, but roots as well. Thus a complete plant may develop and take root at the node, nourished for a time by food sent from the parent plant through the stolon. Eventually, the stolon dries up and withers, leaving an independent daughter plant growing a short distance away from the parent. In this way a strawberry plant can produce many daughter plants by vegetative propagation in addition to producing seeds.

In many plants, horizontal shoots arise from lateral buds near the stem base, and grow under the ground. Such underground horizontal stems are called **rhizomes**. At the nodes of the rhizome are buds, which may develop to produce shoots above the ground. The shoots become independent plants when the connecting rhizome dies.

Many grasses propagate by rhizomes; the couch grass (Figure 16.9) is a good example. Even a small piece of rhizome, provided it has a bud, can produce a new plant.

In the bracken, the entire stem is horizontal and below ground. The bracken fronds you see in summer are produced from lateral buds on a rhizome many centimetres below the soil.

16 REPRODUCTION

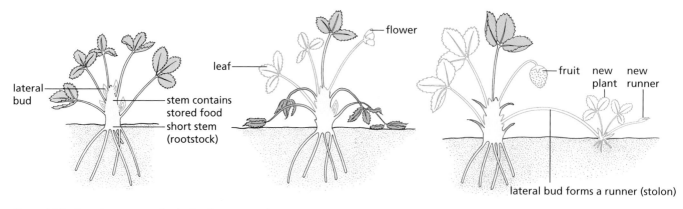

Figure 16.8 Strawberry runner developing from rootstock

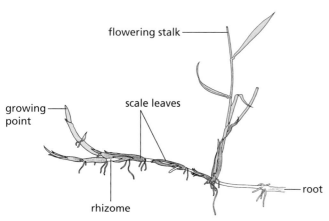

Figure 16.9 Couch grass rhizome

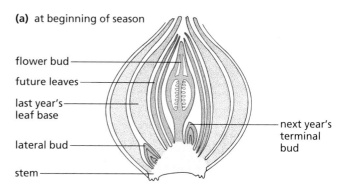

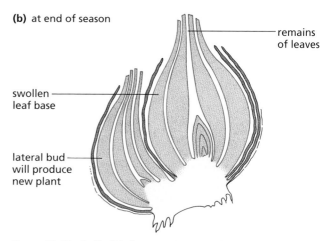

Figure 16.10 Daffodil bulb; vegetative reproduction

Bulbs and corms

Bulbs such as those of the daffodil and snowdrop are very short shoots. The stem is only a few millimetres long and the leaves which encircle the stem are thick and fleshy with stored food.

In spring, the stored food is used by a rapidly growing terminal bud, which produces a flowering stalk and a small number of leaves. During the growing season, food made in the leaves is sent to the leaf bases and stored. The leaf bases swell and form a new bulb ready for growth in the following year.

Vegetative reproduction occurs when some of the food is sent to a lateral bud as well as to the leaf bases. The lateral bud grows inside the parent bulb and, next year, will produce an independent plant (Figure 16.10).

The **corms** of crocuses and anemones have life cycles similar to those of bulbs but it is the stem, rather than the leaf bases, which swells with stored food. Vegetative reproduction takes place when a lateral bud on the short, fat stem grows into an independent plant.

In many cases the organs associated with asexual reproduction also serve as food stores. Food in the storage organs enables very rapid growth in the spring. A great many of the spring and early summer plants have bulbs, corms, rhizomes or tubers: daffodil, snowdrop and bluebell, crocus and cuckoo pint, iris and lily-of-the-valley and lesser celandine.

Potatoes are **stem tubers**. Lateral buds at the base of the potato shoot produce underground shoots

Asexual reproduction

(rhizomes). These rhizomes swell up with stored starch and form tubers (Figure 16.11(a)). Because the tubers are stems, they have buds. If the tubers are left in the ground or transplanted, the buds will produce shoots, using food stored in the tuber (Figure 16.11(b)). In this way, the potato plant can propagate vegetatively.

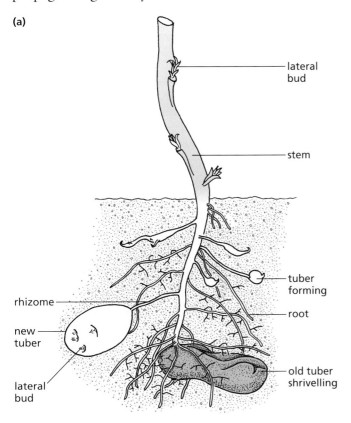

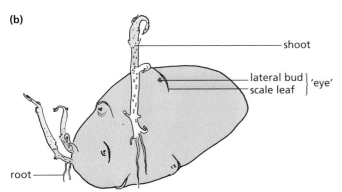

Figure 16.11 Stem tubers growing on a potato plant and a potato tuber sprouting

Artificial propagation

Agriculture and horticulture exploit vegetative reproduction in order to produce fresh stocks of plants. This can be done naturally, e.g. by planting potatoes, dividing up rootstocks or pegging down stolons at their nodes to make them take root. There are also methods that would not occur naturally in the plant's life cycle. Two methods of **artificial propagation** are by taking cuttings and by tissue culture.

Cuttings

It is possible to produce new individuals from certain plants by putting the cut end of a shoot into water or moist earth. Roots (Figure 16.12) grow from the base of the stem into the soil while the shoot continues to grow and produce leaves.

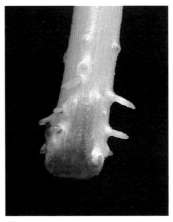

(a) **roots** developing from Busy Lizzie stem

(b) **roots** growing from Coleus cutting

Figure 16.12 Rooted cuttings

In practice, the cut end of the stem may be treated with a rooting 'hormone' (a type of auxin – see 'Tropic responses' in Chapter 14) to promote root growth, and evaporation from the shoot is reduced by covering it with polythene or a glass jar. Carnations, geraniums and chrysanthemums are commonly propagated from cuttings.

Tissue culture

Once a cell has become part of a tissue it usually loses the ability to reproduce. However, the nucleus of any cell in a plant still holds all the 'instructions' (Chapter 17) for making a complete plant and in certain circumstances they can be brought back into action.

In laboratory conditions, single plant cells can be induced to divide and grow into complete plants. One technique is to take small pieces of plant tissue

16 REPRODUCTION

from a root or stem and treat it with enzymes to separate it into individual cells. The cells are then provided with particular plant 'hormones', which induce cell division and, eventually, the formation of roots, stems and leaves.

An alternative method is to start with a small piece of tissue and place it on a nutrient jelly. Cells in the tissue start to divide and produce many cells, forming a shapeless mass called a **callus**. If the callus is then provided with the appropriate hormones it develops into a complete plant (Figure 16.13).

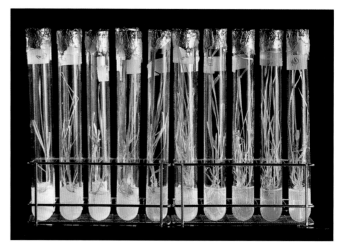

Figure 16.14 Tissue culture. Plants grown from small amounts of unspecialised tissue on an agar culture medium

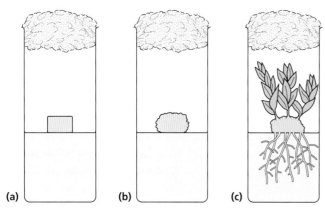

Figure 16.13 Propagation by tissue culture using nutrient jelly

Using the technique of tissue culture, large numbers of plants can be produced from small amounts of tissue (Figure 16.14) and they have the advantage of being free from fungal or bacterial infections. The plants produced in this way form **clones**, because they have been produced from a single parent plant.

Asexual reproduction in animals

Some species of invertebrate animals are able to reproduce asexually.

Hydra is a small animal, 5–10 mm long, which lives in ponds attached to pondweed. It traps small animals with its tentacles, swallows and digests them. *Hydra* reproduces sexually by releasing its male and female gametes into the water but it also has an asexual method, which is shown in Figure 16.15.

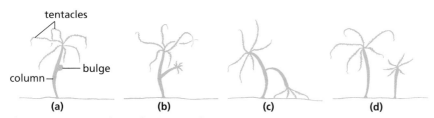

Figure 16.15 Asexual reproduction in *Hydra*

(a) a group of cells on the column start dividing rapidly and produce a bulge

(b) the bulge develops tentacles

(c) the daughter *Hydra* pulls itself off the parent

(d) the daughter becomes an independent animal

(e) *Hydra* with bud

The advantages and disadvantages of asexual reproduction

The advantages and disadvantages of asexual reproduction discussed below are in the context of flowering plants. However, the points made are equally applicable to most forms of asexual reproduction.

In asexual reproduction no gametes are involved and all the new plants are produced by cell division ('Mitosis', Chapter 17) from only one parent. Consequently they are genetically identical; there is no variation. A population of genetically identical individuals produced from a single parent is called a clone. This has the advantage of preserving the 'good' characteristics of a successful species from generation to generation. The disadvantage is that there is no variability for natural selection (Chapter 18) to act on in the process of evolution.

In **agriculture** and **horticulture**, asexual reproduction (vegetative propagation) is exploited to preserve desirable qualities in crops: the good characteristics of the parent are passed on to all the offspring. With a flower such as a daffodil, the bulbs produced can be guaranteed to produce the same shape and colour of flower from one generation to the next. In some cases, such as tissue culture, the young plants grown can be transported much more cheaply than, for example, potato tubers as the latter are much heavier and more bulky. Growth of new plants by asexual reproduction tends to be a quick process.

In natural conditions in the wild it might be a disadvantage to have no variation in a species. If the climate or other conditions change and a vegetatively produced plant has no resistance to a particular disease, the whole population could be wiped out.

Dispersal

A plant that reproduces vegetatively will already be growing in a favourable situation, so all the offspring will find themselves in a suitable environment. However, there is no vegetative dispersal mechanism and the plants will grow in dense colonies, competing with each other for water and minerals. The dense colonies, on the other hand, leave little room for competitors of other species.

As mentioned before, most plants that reproduce vegetatively also produce flowers and seeds. In this way they are able to colonise more distant habitats.

Food storage

The store of food in tubers, tap roots, bulbs, etc. enables the plants to grow rapidly as soon as conditions become favourable. Early growth enables the plant to flower and produce seeds before competition with other plants (for water, mineral salts and light) reaches its maximum. This must be particularly important in woods where, in summer, the leaf canopy prevents much light from reaching the ground and the tree roots tend to drain the soil of moisture over a wide area.

Table 16.1 Summary: advantages and disadvantages of asexual reproduction

Advantages	Disadvantages
No mate is needed. No gametes are needed. All the good characteristics of the parent are passed on to the offspring. Where there is no dispersal (e.g. with potato tubers), offspring will grow in the same favourable environment as the parent. Plants that reproduce asexually usually store large amounts of food that allow rapid growth when conditions are suitable.	There is little variation created, so adaptation to a changing environment (evolution) is unlikely. If the parent has no resistance to a particular disease, none of the offspring will have resistance. Lack of dispersal (e.g. with potato tubers) can lead to competition for nutrients, water and light.

● Sexual reproduction

> **Key definitions**
> **Sexual reproduction** is a process involving the fusion of two gametes (sex cells) to form a zygote and the production of offspring that are genetically different from each other.
> **Fertilisation** is the fusion of gamete nuclei.

The following statements apply equally to plants and animals. Sexual reproduction involves the production of sex cells. These sex cells are called **gametes** and they are made in reproductive organs. The process of cell division that produces the gametes is called **meiosis** (Chapter 17). In sexual reproduction, the male and female gametes come together and **fuse**, that is, their cytoplasm and nuclei join together to form a single cell called a **zygote**. The zygote then grows into a new individual (see Figure 16.30).

In flowering plants the male gametes are found in pollen grains and the female gametes, called **egg cells**, are present in **ovules**. In animals, male gametes are sperm and female gametes are eggs. Details of **fertilisation** are given later in this chapter.

In both plants and animals, the male gamete is microscopic and mobile (i.e. can move from one place to another). The sperm swim to the ovum; the pollen cell moves down the pollen tube (Figure 16.16). The female gametes are always larger than the male gametes and are not mobile. Pollination in seed-bearing plants and mating in most animals bring the male and female gametes close together.

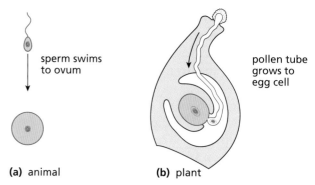

Figure 16.16 The male gamete is small and mobile; the female gamete is larger.

Chromosome numbers

In normal body cells (somatic cells) the chromosomes are present in the nucleus in pairs. Humans, for example, have 46 chromosomes: 23 pairs. Maize (sweetcorn) has 10 pairs. This is known as the **diploid** number. When gametes are formed, the number of chromosomes in the nucleus of each sex cell is halved. This is the **haploid** number. During fertilisation, when the nuclei of the sex cells fuse, a zygote is formed. It gains the chromosomes from both gametes, so it is a diploid cell (see Chapter 17).

The advantages and disadvantages of sexual reproduction

In plants, the gametes may come from the same plant or from different plants of the same species. In either case, the production and subsequent fusion of gametes produce a good deal of variation among the offspring (see Chapter 18). This may result from new combinations of characteristics, e.g. petal colour of one parent combined with fruit size of the other. It may also be the result of spontaneous changes in the gametes when they are produced.

Variation can have its disadvantages: some combinations will produce less successful individuals. On the other hand, there are likely to be some more successful combinations that have greater survival value or produce individuals which can thrive in new or changing environments.

In a population of plants that have been produced sexually, there is a chance that at least some of the offspring will have resistance to disease. These plants will survive and produce further offspring with disease resistance.

The seeds produced as a result of sexual reproduction will be scattered over a relatively wide range. Some will land in unsuitable environments, perhaps lacking light or water. These seeds will fail to germinate. Nevertheless, most methods of seed dispersal result in some of the seeds establishing populations in new habitats.

The seeds produced by sexual reproduction all contain some stored food but it is quickly used up during germination, which produces only a miniature plant. It takes a long time for a seedling to become established and eventually produce seeds of its own.

Sexual reproduction is exploited in agriculture and horticulture to produce new varieties of animals and plants by cross-breeding.

Cross-breeding

It is possible for biologists to use their knowledge of genetics (see 'Monohybrid inheritance' in Chapter 17) to produce new varieties of plants and animals. For example, suppose one variety of wheat produces a lot of grain but is not resistant to a fungus disease. Another variety is resistant to the disease but has only a poor yield of grain. If these two varieties are cross-pollinated (Figure 16.17), the F_1 (which means 'first filial generation') offspring should be disease-resistant and give a good yield of grain (assuming that the useful characteristics are controlled by dominant genes).

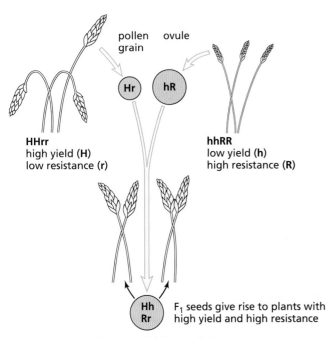

Figure 16.17 Combining useful characteristics

A long-term disadvantage of selective breeding is the loss of variability. By eliminating all the offspring who do not bear the desired characteristics, many genes are lost from the population. At some future date, when new combinations of genes are sought, some of the potentially useful ones may no longer be available.

You will find more information on cross-breeding in 'Selection', Chapter 18.

Table 16.2 Summary: advantages and disadvantages of sexual reproduction

Advantages	Disadvantages
There is variation in the offspring, so adaptation to a changing or new environment is likely, enabling survival of the species. New varieties can be created, which may have resistance to disease. In plants, seeds are produced, which allow dispersal away from the parent plant, reducing competition.	Two parents are usually needed (though not always – some plants can self-pollinate). Growth of a new plant to maturity from a seed is slow.

● Sexual reproduction in plants

Flowers are reproductive structures; they contain the reproductive organs of the plant. The male organs are the **stamens**, which produce pollen. The female organs are the **carpels**. After fertilisation, part of the carpel becomes the fruit of the plant and contains the seeds. In the flowers of most plants there are both stamens and carpels. These flowers are, therefore, both male and female, a condition known as **bisexual** or **hermaphrodite**.

Some species of plants have unisexual flowers, i.e. any one flower will contain either stamens or carpels but not both. Sometimes both male and female flowers are present on the same plant, e.g. the hazel, which has male and female catkins on the same tree. In the willow tree, on the other hand, the male and female catkins are on different trees.

The male gamete is a cell in the pollen grain. The female gamete is an egg cell in the ovule. The process that brings the male gamete within reach of the female gamete (i.e. from stamen to stigma) is called **pollination**. The pollen grain grows a microscopic tube, which carries the male gamete the last few millimetres to reach the female gamete for fertilisation. The zygote then grows to form the seed. These processes are all described in more detail later in this chapter.

Flower structure

The basic structure of a flower is shown in Figures 16.18 and 16.21.

Petals

Petals are usually brightly coloured and sometimes scented. They are arranged in a circle (Figure 16.18) or a cylinder. Most flowers have from four to ten petals. Sometimes they are joined together to form a tube (Figures 16.20 and 16.21) and the individual petals can no longer be distinguished. The colour and scent of the petals attract insects to the flower; the insects may bring about pollination.

The flowers of grasses and many trees do not have petals but small, leaf-like structures that enclose the reproductive organs (Figures 16.28 and 16.29).

16 REPRODUCTION

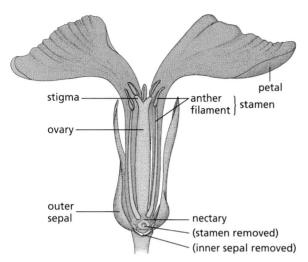

Figure 16.18 Wallflower; structure of flower (one sepal, two petals and stamen removed)

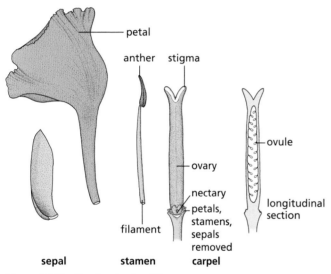

Figure 16.19 Floral parts of wallflower

Figure 16.20 Daffodil flower cut in half. The inner petals form a tube. Three stamens are visible round the long style and the ovary contains many ovules.

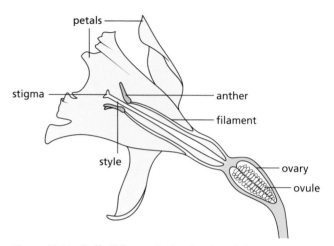

Figure 16.21 Daffodil flower. Outline drawing of Figure 16.20. In daffodils, lilies, tulips, etc. (monocots) there is no distinction between sepals and petals.

Sepals
Outside the petals is a ring of **sepals**. They are often green and much smaller than the petals. They may protect the flower when it is in the bud.

Stamens
The stamens are the male reproductive organs of a flower. Each stamen has a stalk called the **filament**, with an **anther** on the end. Flowers such as the buttercup and blackberry have many stamens; others such as the tulip have a small number, often the same as, or double, the number of petals or sepals. Each anther consists of four **pollen sacs** in which the pollen grains are produced by cell division. When the anthers are ripe, the pollen sacs split open and release their pollen (see Figure 16.26).

Pollen
Insect-pollinated flowers tend to produce smaller amounts of pollen grains (Figure 16.22(a)), which are often round and sticky, or covered in tiny spikes to attach to the furry bodies of insects.

Wind-pollinated flowers tend to produce larger amounts of smooth, light pollen grains (Figure 16.22(b)), which are easily carried by the wind. Large amounts are needed because much of the pollen is lost: there is a low chance of it reaching another flower of the same species.

Carpels
These are the female reproductive organs. Flowers such as the buttercup and blackberry have a large number of carpels while others, such as the lupin, have a single carpel. Each carpel consists of an **ovary**, bearing a **style** and a **stigma**.

Sexual reproduction in plants

(a) insect-borne pollen grains

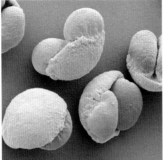
(b) wind-borne pollen grains

Figure 16.22 Pollen grains

Inside the wings are two more petals joined together to form a boat-shaped **keel**.

The single carpel is long, narrow and pod shaped, with about ten ovules in the ovary. The long style ends in a stigma just inside the pointed end of the keel. There are ten stamens: five long ones and five short ones. Their filaments are joined together at the base to form a sheath around the ovary.

The flowers of peas and beans are very similar to those of lupins.

Inside the ovary there are one or more ovules. Each blackberry ovary contains one ovule but the wallflower ovary contains several. The ovule will become a **seed**, and the whole ovary will become a **fruit**. (In biology, a fruit is the fertilised ovary of a flower, not necessarily something to eat.)

The style and stigma project from the top of the ovary. The stigma has a sticky surface and pollen grains will stick to it during pollination. The style may be quite short (e.g. wallflower, Figure 16.18) or very long (e.g. daffodil, Figures 16.20 and 16.21).

Receptacle
The flower structures just described are all attached to the expanded end of a flower stalk. This is called the **receptacle** and, in a few cases after fertilisation, it becomes fleshy and edible (e.g. apple and pear).

Lupin
The lupin flower is shown in Figures 16.23 to 16.25. There are five sepals but these are joined together forming a short tube. The five petals are of different shapes and sizes. The uppermost, called the **standard**, is held vertically. Two petals at the sides are called **wings** and are partly joined together.

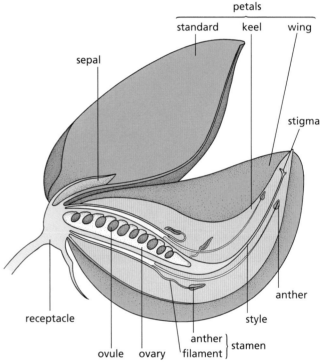
Figure 16.23 Half-flower of lupin

The shoots or branches of a plant carrying groups of flowers are called **inflorescences**. The flowering shoots of the lupin in Figure 16.25 are inflorescences, each one carrying about a hundred individual flowers.

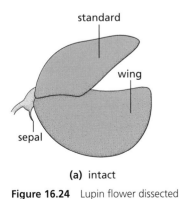
(a) intact

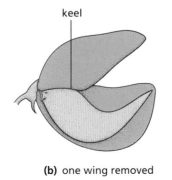
(b) one wing removed

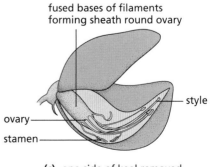
(c) one side of keel removed

Figure 16.24 Lupin flower dissected

16 REPRODUCTION

Figure 16.25 Lupin inflorescence. There are a hundred or more flowers in each inflorescence. The youngest flowers, at the top, have not yet opened. The oldest flowers are at the bottom and have already been pollinated.

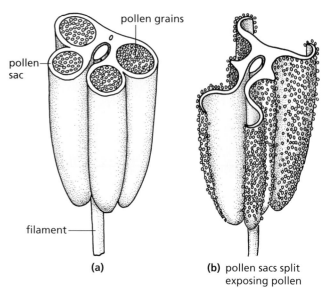

Figure 16.26 Structure of an anther (top cut off)

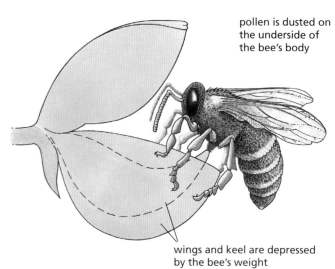

Figure 16.27 Pollination of the lupin

Pollination

> **Key definition**
> **Pollination** is the transfer of pollen grains from the anther to the stigma.

The transfer of pollen from the anthers to the stigma is called **pollination**. The anthers split open, exposing the microscopic pollen grains (Figure 16.26). The pollen grains are then carried away on the bodies of insects, or simply blown by the wind, and may land on the stigma of another flower.

Insect pollination

Lupin flowers have no nectar. The bees that visit them come to collect pollen, which they take back to the hive for food. Other members of the lupin family (Leguminosae, e.g. clover) do produce nectar.

The weight of the bee, when it lands on the flower's wings, pushes down these two petals and the petals of the keel. The pollen from the anthers has collected in the tip of the keel and, as the petals are pressed down, the stigma and long stamens push the pollen out from the keel on to the underside of the bee (Figure 16.27). The bee, with pollen grains sticking to its body, then flies to another flower. If this flower is older than the first one, it will already have lost its pollen. When the bee's weight pushes the keel down, only the stigma comes out and touches the insect's body, picking up pollen grains on its sticky surface.

Lupin and wallflower are examples of **insect-pollinated flowers**.

Wind pollination

Grasses, cereals and many trees are pollinated not by insects but by wind currents. The flowers are often quite small with inconspicuous, green, leaf-like bracts, rather than petals. They produce no nectar. The anthers and stigma are not enclosed by the bracts but are exposed to the air. The pollen grains, being light and smooth, may be carried long distances by the moving air and some of them will be trapped on the stigmas of other flowers.

In the grasses, at first, the feathery stigmas protrude from the flower, and pollen grains floating in the air are trapped by them. Later, the anthers hang outside the flower (Figures 16.28 and 16.29), the pollen sacs split and the wind blows the pollen away. This sequence varies between species.

If the branches of a birch or hazel tree with ripe male catkins, or the flowers of the ornamental pampas grass, are shaken, a shower of pollen can easily be seen.

Figure 16.28 Grass flowers. Note that the anthers hang freely outside the bracts.

Adaptation

Insect-pollinated flowers are considered to be adapted in various ways to their method of pollination. The term '**adaptation**' implies that, in the course of evolution, the structure and physiology of a flower have been modified in ways that improve the chances of successful pollination by insects.

Most insect-pollinated flowers have brightly coloured petals and scent, which attract a variety of insects. Some flowers produce nectar, which is also attractive to many insects. The dark lines ('honey guides') on petals are believed to help direct the insects to the nectar source and thus bring them into contact with the stamens and stigma.

These features are adaptations to insect pollination in general, but are not necessarily associated with any particular insect species. The various petal colours and the nectaries of the wallflower attract a variety of insects. Many flowers, however, have modifications that adapt them to pollination by only one type or species of insect. Flowers such as the honeysuckle, with narrow, deep petal tubes, are likely to be pollinated only by moths or butterflies, whose long 'tongues' can reach down the tube to the nectar.

Tube-like flowers such as foxgloves need to be visited by fairly large insects to effect pollination. The petal tube is often lined with dense hairs, which impede small insects that would take the nectar without pollinating the flower. A large bumble-bee, however, pushing into the petal tube, is forced to rub against the anthers and stigma.

Many tropical and sub-tropical flowers are adapted to pollination by birds, or even by mammals such as bats and mice.

Wind-pollinated flowers are adapted to their method of pollination by producing large quantities of light pollen, and having anthers and stigmas that project outside the flower (Figures 16.28 and 16.29). Many grasses have anthers that are not rigidly attached to the filaments and can be shaken by the wind. The stigmas of grasses are feathery, providing a large surface area, and act as a net that traps passing pollen grains.

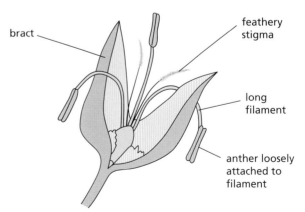

Figure 16.29 Wind-pollinated grass flower

Table 16.3 compares the features of wind- and insect-pollinated flowers.

16 REPRODUCTION

Table 16.3 Features of wind- and insect-pollinated flowers

Feature	Insect-pollinated	Wind-pollinated
petals	present – often large, coloured and scented, with guidelines to guide insects into the flower	absent, or small, green and inconspicuous
nectar	produced by nectaries, to attract insects	absent
stamen	present inside the flower	long filaments, allowing the anthers to hang freely outside the flower so the pollen is exposed to the wind
stigmas	small surface area; inside the flower	large and feathery; hanging outside the flower to catch pollen carried by the wind
pollen	smaller amounts; grains are often round and sticky or covered in spikes to attach to the furry bodies of insects	larger amounts of smooth and light pollen grains, which are easily carried by the wind
bracts (modified leaves)	absent	sometimes present

Practical work

The growth of pollen tubes

Method A

- Make a solution of 15 g sucrose and 0.1 g sodium borate in 100 cm³ water.
- Put a drop of this solution on a cavity slide and scatter some pollen grains on the drop. This can be done by scraping an anther (which must already have opened to expose the pollen) with a mounted needle, or simply by touching the anther on the liquid drop.
- Cover the drop with a coverslip and examine the slide under the microscope at intervals of about 15 minutes. In some cases, pollen tubes may be seen growing from the grains.
- Suitable plants include lily, narcissus, tulip, bluebell, lupin, wallflower, sweet pea or deadnettle, but a 15% sucrose solution may not be equally suitable for all of them. It may be necessary to experiment with solutions ranging from 5 to 20%.

Method B

- Cut the stigma from a mature flower, e.g. honeysuckle, crocus, evening primrose or chickweed, and place it on a slide in a drop of 0.5% methylene blue.
- Squash the stigma under a coverslip (if the stigma is large, it may be safer to squash it between two slides), and leave it for 5 minutes.
- Put a drop of water on one side of the slide, just touching the edge of the coverslip, and draw it under the coverslip by holding a piece of filter paper against the opposite edge. This will remove excess stain.
- If the squash preparation is now examined under the microscope, pollen tubes may be seen growing between the spread-out cells of the stigma.

Fertilisation

Pollination is complete when pollen from an anther has landed on a stigma. If the flower is to produce seeds, pollination has to be followed by a process called **fertilisation**. In all living organisms, fertilisation happens when a male sex cell and a female sex cell meet and join together (they are said to fuse together). The cell that is formed by this fusion is called a **zygote** and develops into an embryo of an animal or a plant (Figure 16.30). The sex cells of all living organisms are called **gametes**.

In flowering plants, the male gamete is in the pollen grain; the female gamete, called the egg cell, is in the ovule. For fertilisation to occur, the nucleus of the male cell from the pollen grain has to reach the female nucleus of the egg cell in the ovule, and fuse with it.

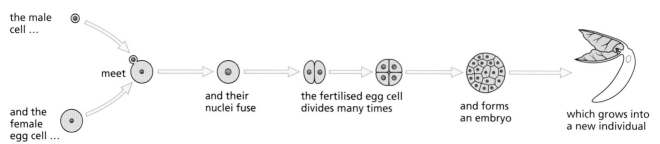

Figure 16.30 Fertilisation. The male and female gametes fuse to form a zygote, which grows into a new individual.

Sexual reproduction in plants

● Extension work

Germination

The stages of germination of a French bean are shown in Figure 16.31.

A seed just shed from its parent plant contains only 5–20% water, compared with 80–90% in mature plant tissues. Once in the soil, some seeds will absorb water and swell up, but will not necessarily start to germinate until other conditions are suitable.

The **radicle** grows first and bursts through the **testa** (Figure 16.31(a)). The radicle continues to grow down into the soil, pushing its way between soil particles and small stones. Its tip is protected by the root cap (see 'Water uptake' in Chapter 8). Branches, called lateral roots, grow out from the side of the main root and help to anchor it firmly in the soil. On the main root and the lateral roots, microscopic root hairs grow out. These are fine outgrowths from some of the outer cells. They make close contact with the soil particles and absorb water from the spaces between them.

In the French bean a region of the embryo's stem, the **hypocotyl**, just above the radicle (Figure 16.31(b)), now starts to elongate. The radicle is by now firmly anchored in the soil, so the rapidly growing hypocotyl arches upwards through the soil, pulling the **cotyledons** with it (Figure 16.31(c)). Sometimes the cotyledons are pulled out of the testa, leaving it below the soil, and sometimes the cotyledons remain enclosed in the testa for a time. In either case, the **plumule** is well protected from damage while it is being pulled through the soil, because it is enclosed between the cotyledons (Figure 16.31(d)).

Once the cotyledons are above the soil, the hypocotyl straightens up and the leaves of the plumule open out (Figure 16.31(e)). Up to this point, all the food needed for making new cells and producing energy has come from the cotyledons.

The main type of food stored in the cotyledons is starch. Before this can be used by the growing shoot and root, the starch has to be turned into soluble sugar. In this form, it can be transported by the phloem cells. The change from starch to sugar in the cotyledons is brought about by enzymes, which become active as soon as the seed starts to

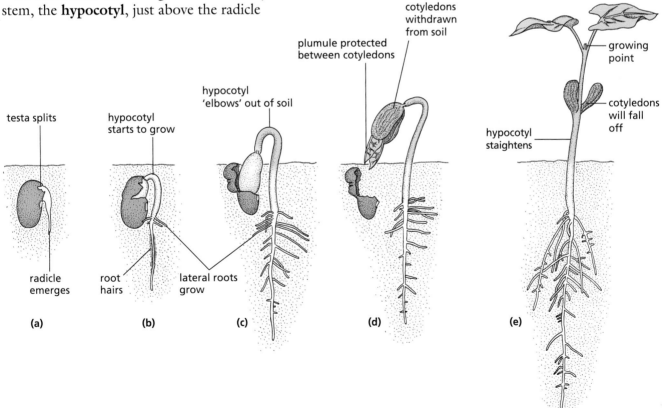

Figure 16.31 Germination of French bean

germinate. The cotyledons shrivel as their food reserve is used up, and they fall off altogether soon after they have been brought above the soil.

By now the plumule leaves have grown much larger, turned green and started to absorb sunlight and make their own food by photosynthesis (page 66). Between the plumule leaves is a growing point, which continues the upward growth of the stem and the production of new leaves. The embryo has now become an independent plant, absorbing water and mineral salts from the soil, carbon dioxide from the air and making food in its leaves.

The importance of water, oxygen and temperature in germination

Use of water in the seedling
Most seeds, when first dispersed, contain very little water. In this dehydrated state, their metabolism is very slow and their food reserves are not used up. The dry seeds can also resist extremes of temperature and desiccation. Before the metabolic changes needed for germination can take place, seeds must absorb water.

Water is absorbed firstly through the micropyle, in some species, and then through the testa as a whole. Once the radicle has emerged, it will absorb water from the soil, particularly through the root hairs. The water that reaches the embryo and cotyledons is used to:

- activate the enzymes in the seed
- help the conversion of stored starch to sugar, and proteins to amino acids
- transport the sugar in solution from the cotyledons to the growing regions
- expand the vacuoles of new cells, causing the root and shoot to grow and the leaves to expand
- maintain the turgor (Chapter 3) of the cells and thus keep the shoot upright and the leaves expanded
- provide the water needed for photosynthesis once the plumule and young leaves are above ground
- transport salts from the soil to the shoot.

Uses of oxygen
In some seeds the testa is not very permeable to oxygen, and the early stages of germination are probably anaerobic (Chapter 12). The testa when soaked or split open allows oxygen to enter. The oxygen is used in aerobic respiration, which provides the energy for the many chemical changes involved in mobilising the food reserves and making the new cytoplasm and cell walls of the growing seedling.

Importance of temperature
In Chapter 5 it was explained that a rise in temperature speeds up most chemical reactions, including those taking place in living organisms. Germination, therefore, occurs more rapidly at high temperatures, up to about 40°C. Above 45°C, the enzymes in the cells are denatured and the seedlings would be killed. Below certain temperatures (e.g. 0–4°C) germination may not start at all in some seeds. However, there is considerable variation in the range of temperatures at which seeds of different species will germinate.

● Extension work

Germination and light
Since a great many cultivated plants are grown from seeds which are planted just below soil level, it seems obvious that light is not necessary for germination. There are some species, however, in which the seeds need some exposure to light before they will germinate, e.g. foxgloves and some varieties of lettuce. In all seedlings, once the shoot is above ground, light is necessary for photosynthesis.

Dormancy
When plants shed their seeds in summer and autumn, there is usually no shortage of water, oxygen and warmth. Yet, in a great many species, the seeds do not germinate until the following spring. These seeds are said to be **dormant**, i.e. there is some internal control mechanism that prevents immediate germination even though the external conditions are suitable.

If the seeds did germinate in the autumn, the seedlings might be killed by exposure to frost, snow and freezing conditions. Dormancy delays the period of germination so that adverse conditions are avoided.

The controlling mechanisms are very varied and are still the subject of investigation and discussion. The factors known to influence dormancy are plant growth substances (see 'Tropic responses' in Chapter 14), the testa, low temperature and light, or a combination of these.

Sexual reproduction in plants

Practical work

Experiments on the conditions for germination

The environmental conditions that might be expected to affect germination are temperature, light intensity and the availability of water and air. The relative importance of some of these conditions can be tested by the experiments that follow.

1 The need for water

- Label three containers A, B and C and put dry cotton wool in the bottom of each.
- Place equal numbers of soaked seeds in all three.
- Leave A quite dry; add water to B to make the cotton wool moist; add water to C until all the seeds are completely covered (Figure 16.32).
- Put lids on the containers and leave them all at room temperature for a week.

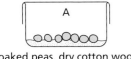
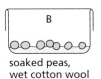 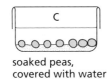

Figure 16.32 Experiment to show the need for water in germination

Result
The seeds in B will germinate normally. Those in A will not germinate. The seeds in C may have started to germinate but will probably not be as advanced as those in B and may have died and started to decay.

Interpretation
Although water is necessary for germination, too much of it may prevent germination by cutting down the oxygen supply to the seed.

2 The need for oxygen

- Set up the experiment as shown in Figure 16.33.

CARE: Pyrogallic acid and sodium hydroxide is a caustic mixture. Use eye shields, handle the liquids with care and report any spillage at once.

- If the moist cotton wool is rolled in some cress seeds, they will stick to it. The bungs must make an airtight seal in the flask and the cotton wool must not touch the solution. Pyrogallic acid and sodium hydroxide absorb oxygen from the air, so the cress seeds in flask A are deprived of oxygen. Flask B is the control (see 'Aerobic respiration' in Chapter 12). This is to show that germination can take place in these experimental conditions provided oxygen is present.

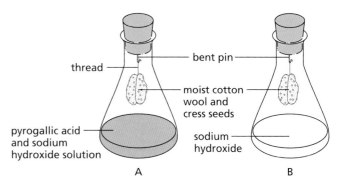

Figure 16.33 Experiment to show the need for oxygen

- Leave the flasks for a week at room temperature.

Result
The seeds in flask B will germinate but there will be little or no germination in flask A.

Interpretation
The main difference between flasks A and B is that A lacks oxygen. Since the seeds in this flask have not germinated, it looks as if oxygen is needed for germination.

To show that the chemicals in flask A had not killed the seeds, the cotton wool can be swapped from A to B. The seeds from A will now germinate.

Note: Sodium hydroxide absorbs carbon dioxide from the air. The mixture (sodium hydroxide + pyrogallic acid) in flask A, therefore, absorbs both carbon dioxide and oxygen from the air in this flask. In the control flask B, the sodium hydroxide absorbs carbon dioxide but not oxygen. If the seeds in B germinate, it shows that lack of carbon dioxide did not affect them, whereas lack of oxygen did.

3 Temperature and germination

- Soak some maize grains for a day and then roll them up in three strips of moist blotting paper as shown in Figure 16.34.
- Put the rolls into plastic bags. Place one in a refrigerator (about 4 °C), leave one upright in the room (about 20 °C) and put the third in a warm place such as over a radiator or, better, in an incubator set to 30 °C.
- Because the seeds in the refrigerator will be in darkness, the other seeds must also be enclosed in a box or a cupboard, to exclude light. Otherwise it could be objected that it was lack of light rather than low temperature that affected germination.
- After a week, examine the seedlings and measure the length of the roots and shoots.

Result
The seedlings kept at 30 °C will be more advanced than those at room temperature. The grains in the refrigerator may not have started to germinate at all.

Interpretation
Seeds will not germinate below a certain temperature. The higher the temperature, the faster the germination, at least up to 35–40 °C.

16 REPRODUCTION

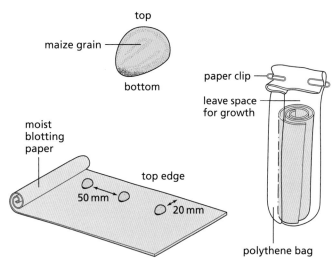

Figure 16.34 Experiment to show the influence of temperature on germination. Roll the seeds in moist blotting-paper and stand the rolls upright in plastic bags.

Controlling the variables

These experiments on germination illustrate one of the problems of designing biological experiments. You have to decide what conditions (the '**variables**') could influence the results and then try to change only one condition at a time. The dangers are that: (1) some of the variables might not be controllable, (2) controlling some of the variables might also affect the condition you want to investigate, and (3) there might be a number of important variables you have not thought of.

1 In your germination experiments, you were unable to control the quality of the seeds, but had to assume that the differences between them would be small. If some of the seeds were dead or diseased, they would not germinate in any conditions and this could distort the results. This is one reason for using as large a sample as possible in the experiments.
2 You had to ensure that, when temperature was the variable, the exclusion of light from the seeds in the refrigerator was not an additional variable. This was done by putting all the seeds in darkness.
3 A variable you might not have considered could be the way the seeds were handled. Some seeds can be induced to germinate more successfully by scratching or chipping the testa.

Self-pollination and cross-pollination

> **Key definitions**
> **Self-pollination** is the transfer of pollen grains from the anther of a flower to the stigma of the same flower, or a different flower on the same plant.
> **Cross-pollination** is the transfer of pollen grains from the anther of a flower to the stigma of a flower on a different plant of the same species.

In **self-pollinating** plants, the pollen that reaches the stigma comes from the same flower or another flower on the same plant. In **cross-pollination**, the pollen is carried from the anthers of one flower to the stigma in a flower of another plant of the same species.

If a bee carried pollen from one of the younger flowers near the middle of a lupin plant (Figure 16.25) to an older flower near the bottom, this would be self-pollination. If, however, the bee visited a separate lupin plant and pollinated its flowers, this would be cross-pollination.

The term 'cross-pollination', strictly speaking, should be applied only if there are genetic differences between the two plants involved. The flowers on a single plant all have the same genetic constitution. The flowers on plants growing from the same rhizome or rootstock (see 'Asexual reproduction' earlier in this chapter) will also have the same genetic constitution. Pollination between such flowers is little different from self-pollination in the same flower.

If a plant relies on self-pollination, the disadvantage will be that variation will not occur in subsequent generations. Those plants may not, therefore, be able to adapt to changing environmental conditions. However, self-pollination can happen even if there are no pollinators, since the flower's own pollen may drop onto its stigma. This means that even if pollinators are scarce (perhaps because of the reckless use of insecticides) the plant can produce seeds and prevent extinction.

Cross-pollination, on the other hand, will guarantee variation and give the plant species a better chance of adapting to changing conditions. Some plants maintain cross-pollination by producing stamens (male reproductive parts) at a different time to the carpels (female reproductive parts). However, cross-pollinated plants do have a reliance on pollinators to carry the pollen to other plants.

Fertilisation

The pollen grain absorbs liquid from the stigma and a microscopic **pollen tube** grows out of the grain. This tube grows down the style and into the ovary, where it enters a small hole, the **micropyle**, in an ovule (Figure 16.35). The nucleus of the pollen grain travels down the pollen tube and enters the ovule. Here it combines with the nucleus of the egg cell. Each ovule in an ovary needs to be fertilised by a separate pollen grain.

Although pollination must occur before the ovule can be fertilised, pollination does not necessarily result in fertilisation. A bee may visit many flowers on a Bramley apple tree, transferring pollen from

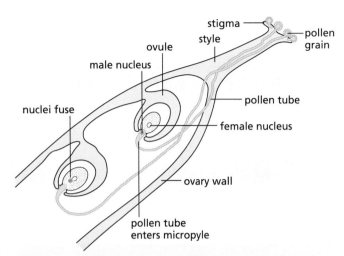

Figure 16.35 Diagram of fertilisation showing pollen tube

one flower to another. The Bramley, however, is 'self-sterile'; pollination with its own pollen will not result in fertilisation. Pollination with pollen from a different variety of apple tree, for example a Worcester, can result in successful fertilisation and fruit formation.

● Extension work

Fruit and seed formation

After the pollen and the egg nuclei have fused, the egg cell divides many times and produces a miniature plant called an **embryo**. This consists of a tiny root and shoot, with two special leaves called **cotyledons**. In dicot plants (see 'Features of organisms' in Chapter 1) food made in the leaves of the parent plant is carried in the phloem to the cotyledons.

The cotyledons eventually grow so large with this stored food that they completely enclose the embryo (see Figure 16.37). In monocot plants

(a) Tomato flowers – the petals of the older flowers are shrivelling

Figure 16.36 Tomato; fruit formation

(b) After fertilisation – the petals have dropped and the ovary is growing.

(c) Ripe fruit – the ovary has grown and ripened. The green sepals remain and the dried stigma is still attached.

16 REPRODUCTION

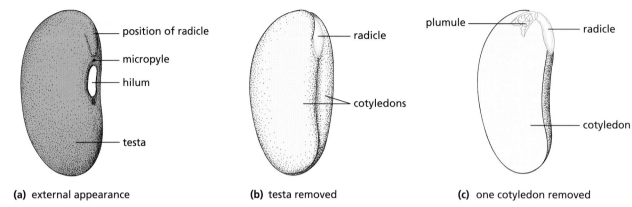

(a) external appearance (b) testa removed (c) one cotyledon removed

Figure 16.37 A French bean seed

Figure 16.38 Lupin flower after fertilisation. The ovary (still with the style and stigma attached) has grown much larger than the flower and the petals have shrivelled.

● Sexual reproduction in humans

Reproduction is the process of producing new individuals. In human reproduction the two sexes, male and female, each produce special types of reproductive cells, called gametes. The male gametes are the **sperm** (or **spermatozoa**) and the female gametes are the **ova** (singular = ovum) or eggs (Figure 16.39).

To produce a new individual, a sperm has to reach an ovum and join with it (fuse with it). The sperm nucleus then passes into the ovum and the two nuclei also fuse. This is fertilisation. The cell formed after the fertilisation of an ovum by a sperm is called a zygote. A zygote will grow by cell division

(see 'Features of organisms' in Chapter 1) the food store is laid down in a special tissue called endosperm, which is outside the cotyledons. In both cases the outer wall of the ovule becomes thicker and harder, and forms the seed coat or **testa**.

As the seeds grow, the ovary also becomes much larger and the petals and stamens shrivel and fall off (Figures 16.36(b) and 16.38). The ovary is now called a **fruit** (Figure 16.36). The biological definition of a fruit is a fertilised ovary. It is not necessarily edible – the lupin ovary forms a dry pod.

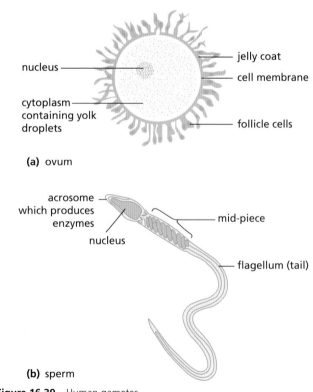

(a) ovum

(b) sperm

Figure 16.39 Human gametes

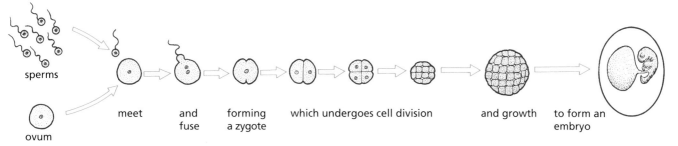

Figure 16.40 Fertilisation and development

to produce first an **embryo** and then a fully formed animal (Figure 16.40).

In humans, the male produces millions of sperm, while the female produces a smaller number of eggs (usually one a month for about 40 years). Usually only one egg is fertilised at a time; two eggs being fertilised at the same time produces (non-identical) twins.

To bring the sperm close enough to the ova for fertilisation to take place, there is an act of mating or **copulation**. In mammals this act results in sperm from the male animal being injected into the female. The sperm swim inside the female's reproductive system and fertilise any eggs that are present. The zygote then grows into an embryo inside the body of the female.

The human reproductive system
Female
Table 16.4 summarises the functions of parts of the female reproductive system. The eggs are produced from the female reproductive organs called **ovaries**. These are two whitish oval bodies, 3–4 cm long. They lie in the lower half of the abdomen, one on each side of the **uterus** (Figure 16.41 and Figure 16.42) Close to each ovary is the expanded, funnel-shaped opening of the **oviduct**, the tube down which the ova pass when released from the ovary. The oviduct is sometimes called the **Fallopian tube**.

The oviducts are narrow tubes that open into a wider tube, the uterus or womb, lower down in the abdomen. When there is no embryo developing in it, the uterus is only about 80 mm long. It leads to the outside through a muscular tube, the **vagina**. The **cervix** is a ring of muscle closing the lower end of the uterus where it joins the vagina. The urethra, from the bladder, opens into the **vulva** just in front of the vagina.

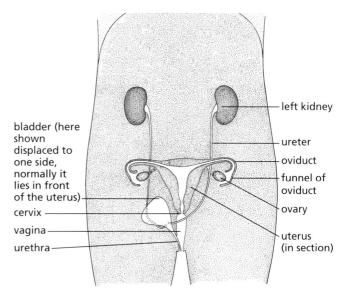

Figure 16.41 The female reproductive organs; front view

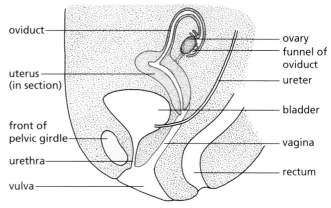

Figure 16.42 The female reproductive organs; side view

Male
Table 16.5 summarises the functions of parts of the male reproductive system. Sperm are produced in the male reproductive organs (Figures 16.43 and 16.44), called the **testes** (singular = testis). These lie outside the abdominal cavity in a special sac called the **scrotum**. In this position they are kept at a

16 REPRODUCTION

Table 16.4 Functions of parts of the female reproductive system

Part	Function
cervix	a ring of muscle, separating the vagina from the uterus
funnel of oviduct	directs an ovum (egg) from the ovary into the oviduct
ovary	contains follicles in which ova (eggs) are produced
oviduct	carries an ovum to the uterus, with propulsion provided by tiny cilia in the wall; also the site of fertilisation
urethra	carries urine from the bladder
uterus	where the fetus develops
vagina	receives the male penis during sexual intercourse; sperm are deposited here

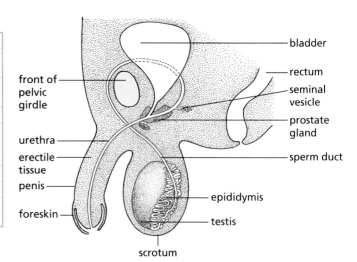

Figure 16.44 The male reproductive organs; side view

temperature slightly below the rest of the body. This is the best temperature for sperm production.

The testes consist of a mass of sperm-producing tubes (Figure 16.44). These tubes join to form ducts leading to the **epididymis**, a coiled tube about 6 metres long on the outside of each testis. The epididymis, in turn, leads into a muscular **sperm duct**.

Table 16.5 Functions of parts of the male reproductive system

Part	Function
epididymis	a mass of tubes in which sperm are stored
penis	can become firm, to insert into the vagina of the female during sexual intercourse in order to transfer sperm
prostate gland	adds fluid and nutrients to sperm to form semen
scrotum	a sac that holds the testes outside the body, keeping them cooler than body temperature
seminal vesicle	adds fluid and nutrients to sperm to form semen
sperm duct	muscular tube that links the testis to the urethra to allow the passage of semen containing sperm
testis	male gonad that produces sperm
urethra	passes semen containing sperm through the penis; also carries urine from the bladder

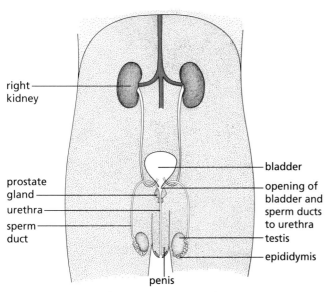

Figure 16.43 The male reproductive organs; front view

The two sperm ducts, one from each testis, open into the top of the urethra just after it leaves the bladder. A short, coiled tube called the **seminal vesicle** branches from each sperm duct just before it enters the **prostate gland**, which surrounds the urethra at this point.

The urethra passes through the **penis** and may conduct either urine or sperm at different times. The penis consists of connective tissue with many blood spaces in it. This is called **erectile tissue**.

Production of gametes

Sperm production

The lining of the sperm-producing tubules in the testis consists of rapidly dividing cells (Figure 16.45). After a series of cell divisions, the cells grow long tails called flagellae (singular: flagellum) and become sperm (Figure 16.46), which pass into the epididymis.

During copulation, the epididymis and sperm ducts contract and force sperm out through the urethra. The prostate gland and seminal vesicle add fluid to the sperm. This fluid plus the sperm it contains is called **semen**, and the ejection of sperm through the penis is called **ejaculation**.

Ovulation

The egg cells (ova) are present in the ovary from the time of birth. No more are formed during the female's lifetime, but between the ages of 10 and 14 some of

Sexual reproduction in humans

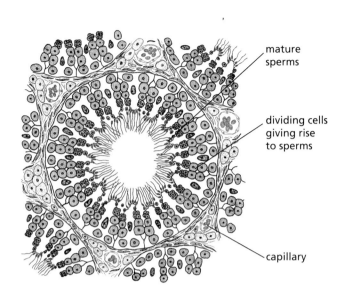

Figure 16.45 Section through sperm-producing tubules

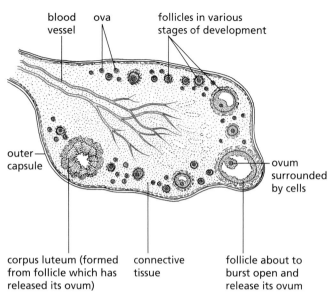

Figure 16.47 Section through an ovary

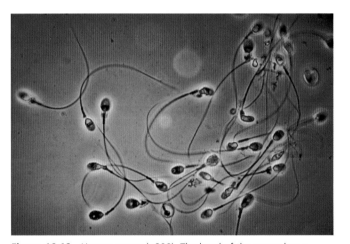

Figure 16.46 Human sperm (×800). The head of the sperm has a slightly different appearance when seen in 'side' view or in 'top' view.

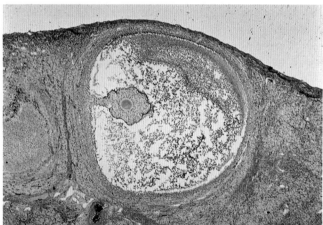

Figure 16.48 Mature follicle as seen in a section through part of an ovary (×30). The ovum is surrounded by follicle cells. These produce the fluid that occupies much of the space in the follicle.

the egg cells start to mature and are released, one at a time about every 4 weeks from alternate ovaries. As each ovum matures, the cells around it divide rapidly and produce a fluid-filled sac. This sac is called a **follicle** (Figure 16.47) and, when mature, it projects from the surface of the ovary like a small blister (Figure 16.48). Finally, the follicle bursts and releases the ovum with its coating of cells into the funnel of the oviduct. This is called **ovulation**. From here, the ovum is wafted down the oviduct by the action of cilia (see 'Levels of organisation' in Chapter 2) in the lining of the tube. If the ovum meets sperm cells in the oviduct, it may be fertilised by one of them.

The released ovum is enclosed in a jelly-like coat called the **zona pellucida** and is still surrounded by a layer of follicle cells. Before fertilisation can occur, sperm have to get through this layer of cells and the successful sperm has to penetrate the zona pellucida with the aid of enzymes secreted by the head of the sperm.

Mating and fertilisation
Mating
Sexual arousal in the male results in an erection. That is, the penis becomes firm and erect as a result of blood flowing into the erectile tissue. Arousal in the female stimulates the lining of the vagina to produce mucus. This lubricates the vagina and makes it easy for the erect penis to enter.

In the act of copulation, the male inserts the penis into the female's vagina. The sensory stimulus (sensation) that this produces causes a reflex (see 'Nervous control

16 REPRODUCTION

in humans' in Chapter 14) in the male, which results in the ejaculation of semen into the top of the vagina.

The previous paragraph is a very simple description of a biological event. In humans, however, the sex act has intense psychological and emotional importance. Most people feel a strong sexual drive, which has little to do with the need to reproduce. Sometimes the sex act is simply the meeting of an urgent physical need. Sometimes it is an experience that both man and woman enjoy together. At its 'highest' level it is both of these, and is also an expression of deeply felt affection within a lasting relationship.

Fertilisation

The sperm swim through the cervix and into the uterus by wriggling movements of their tails. They pass through the uterus and enter the oviduct, but the method by which they do this is not known for certain. If there is an ovum in the oviduct, one of the sperm may bump into it and stick to its surface. The acrosome at the head of the sperm secretes enzymes which digest part of the egg membrane. The sperm then enters the cytoplasm of the ovum and the male nucleus of the sperm fuses with the female nucleus. This is the moment of fertilisation and is shown in more detail in Figure 16.49. Although a single ejaculation may contain over three hundred million sperm, only a few hundred will reach the oviduct and only one will fertilise the ovum. The function of the others is not fully understood.

The released ovum is thought to survive for about 24 hours; the sperm might be able to fertilise an ovum for about 2 or 3 days. So there is only a short period of about 4 days each month when fertilisation might occur. If this fertile period can be estimated accurately, it can be used either to achieve or to avoid fertilisation (conception) (see 'Methods of birth control in humans').

The fertilised egg has 23 chromosomes from the mother and 23 from the father, bringing its chromosome number to 46 (the same as other human body cells). It is called a zygote.

Pregnancy and development

The fertilised ovum (zygote) first divides into two cells. Each of these divides again, so producing four cells. The cells continue to divide in this way to produce a solid ball of cells (Figure 16.50), an early stage in the development of the embryo. This early embryo travels down the oviduct to the uterus. Here it sinks into the lining of the uterus, a process called **implantation** (Figure 16.52(a)). The embryo continues to grow and produces new cells that form tissues and organs (Figure 16.51). After 8 weeks, when all the organs are formed, the embryo is called a **fetus**. One of the first organs to form is the heart, which pumps blood around the body of the embryo.

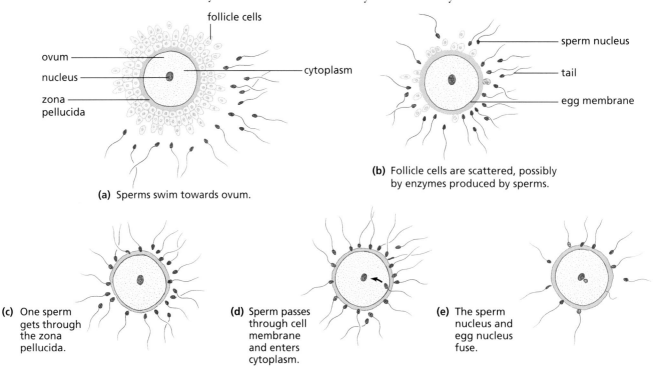

Figure 16.49 Fertilisation of an ovum

Sexual reproduction in humans

As the embryo grows, the uterus enlarges to contain it. Inside the uterus the embryo becomes enclosed in a fluid-filled sac called the **amnion** or water sac, which protects it from damage and prevents unequal pressures from acting on it (Figure 16.52(b) and (c)). The fluid is called **amniotic fluid**. The oxygen and food needed to keep the embryo alive and growing are obtained from the mother's blood by means of a structure called the **placenta**.

Placenta

Soon after the ball of cells reaches the uterus, some of the cells, instead of forming the organs of the embryo, grow into a disc-like structure, the placenta (Figure 16.52(c)). The placenta becomes closely attached to the lining of the uterus and is attached to the embryo by a tube called the **umbilical cord** (Figure 16.52(c)). The nervous system (brain, spinal cord and sense organs) start to develop very quickly. After a few weeks, the embryo's heart has developed and is circulating blood through the umbilical cord and placenta as well as through its own tissues (Figure 16.51(b)). Oxygen and nutrients such as glucose and amino acids pass across the placenta to the embryo's bloodstream. Carbon dioxide passes from the embryo's blood to that of the mother. Blood entering the placenta from the mother does not mix with the embryo's blood.

Figure 16.53 shows the human embryo at 7 weeks surrounded by the amnion and placenta.

Antenatal care

'Antenatal' or 'prenatal' refers to the period before birth. Antenatal care is the way a woman should look after herself during pregnancy, so that the birth will be safe and her baby healthy.

The mother-to-be should make sure that she eats properly, and perhaps takes more iron and folic acid (a vitamin), than she usually does to prevent anaemia. If her job is a light one, she may go on working for the first 6 months of pregnancy. She should not do heavy work, however, or repeated lifting or stooping.

Pregnant women who drink or smoke are more likely to have babies with low birth weights. These babies are more likely to be ill than babies of normal

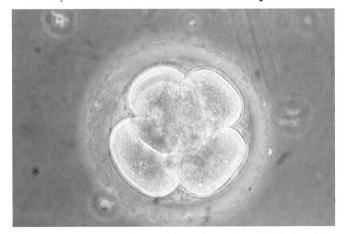

Figure 16.50 Human embryo at the 8-cell stage (×230) with five of the cells clearly visible. The embryo is surrounded by the zona pellucida.

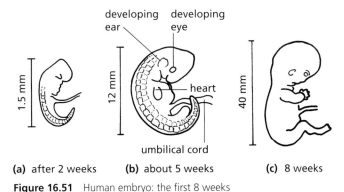

(a) after 2 weeks (b) about 5 weeks (c) 8 weeks

Figure 16.51 Human embryo: the first 8 weeks

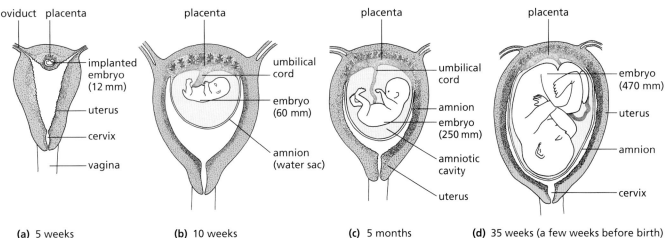

(a) 5 weeks (b) 10 weeks (c) 5 months (d) 35 weeks (a few weeks before birth)

Figure 16.52 Growth and development in the uterus (not to scale)

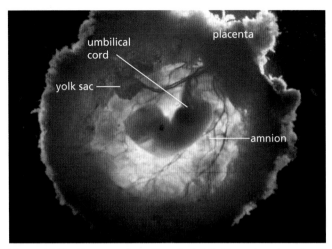

Figure 16.53 Human embryo, 7 weeks (×1.5). The embryo is enclosed in the amnion. Its limbs, eye and ear-hole are clearly visible. The amnion is surrounded by the placenta; the fluffy-looking structures are the placental villi, which penetrate into the lining of the uterus. The umbilical cord connects the embryo to the placenta.

Figure 16.54 Children suffering from the effects of thalidomide

weight. Smoking may also make a miscarriage more likely. So a woman who smokes should give up smoking during her pregnancy. Alcohol can cross the placenta and damage the fetus. Pregnant women who take as little as one alcoholic drink a day are at risk of having babies with lower than average birth weights. These underweight babies are more likely to become ill.

Heavy drinking during pregnancy, sometimes called 'binge drinking', can lead to deformed babies. This risk is particularly great in the early stages of pregnancy when the brain of the fetus is developing, and can result in a condition called fetal alcohol syndrome (FAS). At that stage the mother may not yet be aware of her pregnancy and continue to drink heavily. A child suffering from FAS can have a range of medical problems, many associated with permanent brain damage. All levels of drinking are thought to increase the risk of miscarriage.

During pregnancy, a woman should not take any drugs unless they are strictly necessary and prescribed by a doctor. In the 1950s, a drug called thalidomide was used to treat the bouts of early morning sickness that often occur in the first 3 months of pregnancy. Although tests had appeared to show the drug to be safe, it had not been tested on pregnant animals. About 20% of pregnant women who took thalidomide had babies with deformed or missing limbs (Figure 16.54).

If a woman catches **rubella** (German measles) during the first 4 months of pregnancy, there is a danger that the virus may affect the fetus and cause abortion or stillbirth. Even if the baby is born alive, the virus may have caused defects of the eyes (cataracts), ears (deafness) or nervous system. All girls should be vaccinated against rubella to make sure that their bodies contain antibodies to the disease (see Chapter 10).

Twins

Sometimes a woman releases two ova when she ovulates. If both ova are fertilised, they may form twin embryos, each with its own placenta and amnion. Because the twins come from two separate ova, each fertilised by a different sperm, it is possible to have a boy and a girl. Twins formed in this way are called **fraternal twins**. Although they are both born within a few minutes of each other, they are no more alike than other brothers or sisters.

Another cause of twinning is when a single fertilised egg, during an early stage of cell division, forms two separate embryos. Sometimes these may share a placenta and amnion. Twins formed from a single ovum and sperm must be the same sex, because only one sperm (X or Y) fertilised the ovum. These 'one-egg' twins are sometimes called **identical twins** because, unlike fraternal twins, they will closely resemble each other in every respect.

Birth

The period from fertilisation to birth takes about 38 weeks in humans. This is called the **gestation** period. A few weeks before the birth, the fetus has come to lie head downwards in the uterus, with its head just above the cervix (Figures 16.52(d) and

16.55). When birth starts, the uterus begins to contract rhythmically. This is the beginning of what is called 'labour'. These regular rhythmic contractions become stronger and more frequent. The opening of the cervix gradually widens (dilates) enough to let the baby's head pass through and the contractions of the muscles in the uterus wall are assisted by muscular contractions of the abdomen. The amniotic sac breaks at some stage in labour and the fluid escapes through the vagina. Finally, the muscular contractions of the uterus wall and abdomen push the baby head-first through the widened cervix and vagina (Figure 16.56). The umbilical cord, which still connects the child to the placenta, is tied and cut. Later, the placenta breaks away from the uterus and is pushed out separately as the 'afterbirth'.

The sudden fall in temperature felt by the newly born baby stimulates it to take its first breath and it usually cries. In a few days, the remains of the umbilical cord attached to the baby's abdomen shrivel and fall away, leaving a scar in the abdominal wall, called the navel.

Induced birth

Sometimes, when a pregnancy has lasted for more than 38 weeks or when examination shows that the placenta is not coping with the demands of the fetus, birth may be induced. This means that it is started artificially.

This is often done by carefully breaking the membrane of the amniotic sac. Another method is to inject a hormone, **oxytocin**, into the mother's veins. Either of these methods brings on the start of labour. Sometimes both are used together.

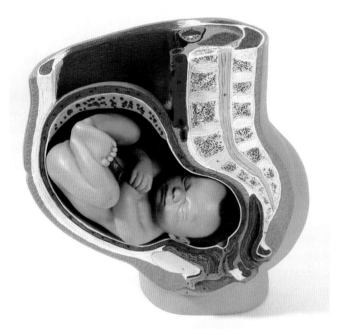

Figure 16.55 Model of human fetus just before birth. The cervix and vagina seem to provide narrow channels for the baby to pass through but they widen quite naturally during labour and delivery.

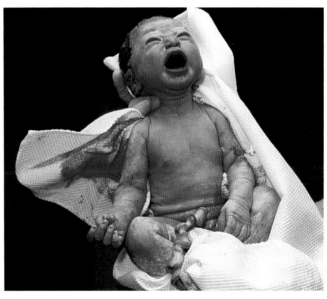

Figure 16.56 Delivery of a baby. The umbilical cord is still intact.

Comparing male and female gametes

Figure 2.13(g) shows a sperm cell in detail. Sperm are much smaller than eggs and are produced in much larger numbers (over 300 million in a single ejaculation). The tip of the cell carries an acrosome, which secretes enzymes capable of digesting a path into an egg cell, through the jelly coat, so the sperm nucleus can fuse with the egg nucleus. The cytoplasm of the mid-piece of the sperm contains many mitochondria. They carry out respiration, providing energy to make the tail (flagellum) move and propel the sperm forward.

The egg cell (see Figure 2.13(h)) is much larger than a sperm cell and only one egg is released each month while the woman is fertile. It is surrounded by a jelly coat, which protects the contents of the cell and prevents more than one sperm from entering and fertilising the egg. The egg cell contains a large amount of cytoplasm, which is rich in fats and proteins. The fats act as energy stores. Proteins are available for growth if the egg is fertilised.

16 REPRODUCTION

Functions of the placenta and umbilical cord

The blood vessels in the placenta are very close to the blood vessels in the uterus so that oxygen, glucose, amino acids and salts can pass from the mother's blood to the embryo's blood (Figure 16.57(a)). So the blood flowing in the umbilical vein from the placenta carries food and oxygen to be used by the living, growing tissues of the embryo. In a similar way, the carbon dioxide and urea in the embryo's blood escape from the vessels in the placenta and are carried away by the mother's blood in the uterus (Figure 16.57(b)). In this way the embryo gets rid of its excretory products.

There is no direct communication between the mother's blood system and that of the embryo. The exchange of substances takes place across the thin walls of the blood vessels. In this way, the mother's blood pressure cannot damage the delicate vessels of the embryo and it is possible for the placenta to select the substances allowed to pass into the embryo's blood. The placenta can prevent some harmful substances in the mother's blood from reaching the embryo. It cannot prevent all of them, however: alcohol and nicotine can pass to the developing fetus. If the mother is a heroin addict, the baby can be born addicted to the drug.

Some pathogens such as the rubella virus and HIV can pass across the placenta. Rubella (German measles), although a mild infection for the mother, can infect the fetus and results in major health problems, including deafness, congenital heart disease, diabetes and mental retardation. HIV is potentially fatal.

The placenta produces hormones, including oestrogens and progesterone. It is assumed that these hormones play an important part in maintaining the pregnancy and preparing for birth, but their precise function is not known. They may influence the development and activity of the muscle layers in the wall of the uterus and prepare the mammary glands in the breasts for milk production.

Feeding and parental care

Within the first 24 hours after birth, the baby starts to suck at the breast. During pregnancy the mammary glands (breasts) enlarge as a result of an increase in the number of milk-secreting cells. No milk is secreted during pregnancy, but the hormones that start the birth process also act on the milk-secreting cells of the breasts. The breasts are stimulated to release milk by the baby sucking the nipple. The continued production of milk is under the control of hormones, but the amount of milk produced is related to the quantity taken by the child during suckling.

Milk contains the proteins, fats, sugar, vitamins and salts that babies need for their energy requirements and tissue-building, but there is too little iron present for the manufacture of haemoglobin. All the iron needed for the first weeks or months is stored in the liver of the fetus during gestation.

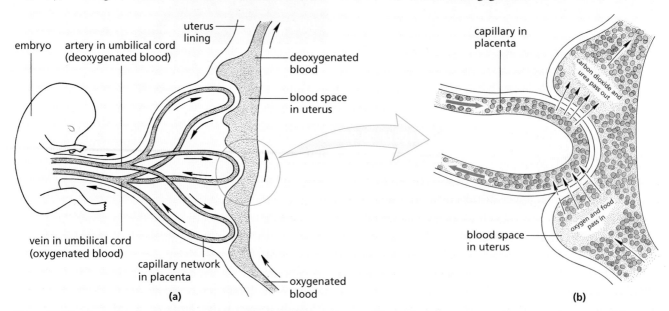

Figure 16.57 The exchange of substances between the blood of the embryo and the mother

The liquid produced in the first few days is called **colostrum**. It is sticky and yellow, and contains more protein than the milk produced later. It also contains some of the mother's antibodies. This provides passive immunity (see Chapter 10) to infection.

The mother's milk supply increases with the demands of the baby, up to 1 litre per day. It is gradually supplemented and eventually replaced entirely by solid food, a process known as **weaning**.

Cows' milk is not wholly suitable for human babies. It has more protein, sodium and phosphorus, and less sugar, vitamin A and vitamin C, than human milk. It is less easily digested than human milk. Manufacturers modify the components of dried cows' milk to resemble human milk more closely and this makes it more acceptable if the mother cannot breastfeed her baby.

Cows' milk and proprietary dried milk both lack human antibodies, whereas the mother's milk contains antibodies to any diseases from which she has recovered. It also carries white cells that produce antibodies or ingest bacteria. These antibodies are important in defending the baby against infection at a time when its own immune responses are not fully developed. Breastfeeding provides milk free from bacteria, whereas bottle-feeding carries the risk of introducing bacteria that cause intestinal diseases. Breastfeeding also offers emotional and psychological benefits to both mother and baby.

Other advantages of breastfeeding over bottle-feeding include the following:

- There is no risk of an allergic reaction to breast milk.
- Breast milk is produced at the correct temperature.
- There are no additives or preservatives in breast milk.
- Breast milk does not require sterilisation since there are no bacteria present that could cause intestinal disease.
- There is no cost involved in using breast milk.
- Breast milk does not need to be prepared.
- Breastfeeding triggers a reduction in the size of the mother's uterus.

● Sex hormones in humans

Puberty and the menstrual cycle

Puberty

Although the ovaries of a young girl contain all the ova she will ever produce, they do not start to be released until she reaches the age of about 10–14 years. This stage in her life is known as **puberty**.

At about the same time as the first ovulation, the ovary also releases female sex hormones into the bloodstream. These hormones are called **oestrogens** and when they circulate around the body, they bring about the development of **secondary sexual characteristics**. In a girl these are the increased growth of the breasts, a widening of the hips and the growth of hair in the pubic region and in the armpits. There is also an increase in the size of the uterus and vagina. Once all these changes are complete, the girl is capable of having a baby.

Puberty in boys occurs at about the same age as in girls. The testes start to produce sperm for the first time and also release a hormone, called **testosterone**, into the bloodstream. The male secondary sexual characteristics, which begin to appear at puberty, are enlargement of the testes and penis, deepening of the voice, growth of hair in the pubic region, armpits, chest and, later on, the face. In both sexes there is a rapid increase in the rate of growth during puberty.

In addition to the physical changes at puberty, there are emotional and psychological changes associated with the transition from being a child to becoming an adult, i.e. the period of **adolescence**. Most people adjust to these changes smoothly and without problems. Sometimes, however, a conflict arises between having the status of a child and the sexuality and feelings of an adult.

The menstrual cycle

The ovaries release an ovum about every 4 weeks. In preparation for this the lining of the uterus wall thickens, so that an embryo can embed itself if the released ovum is fertilised. If no implantation occurs, the uterus lining breaks down. The cells, along with blood are passed out of the vagina. This is called a **menstrual period**. The appearance of the first menstrual period is one of the signs of puberty in girls. After menstruation, the uterus lining starts to re-form and another ovum starts to mature.

Hormones and the menstrual cycle

At the start of the cycle, the lining of the uterus wall has broken down (menstruation). As each follicle in the ovaries develops, the amount of oestrogens produced by the ovary increases. The oestrogens act on the uterus and cause its lining to become thicker and develop more blood vessels. These are changes that help an early embryo to implant.

Two hormones, produced by the **pituitary gland** at the base of the brain, promote ovulation. The hormones are **follicle-stimulating hormone** (**FSH**) and **luteinising hormone**, or **lutropin** (**LH**). They act on a ripe follicle and stimulate maturation and release of the ovum.

Once the ovum has been released, the follicle that produced it develops into a solid body called the **corpus luteum**. This produces a hormone called **progesterone**, which affects the uterus lining in the same way as the oestrogens, making it grow thicker and produce more blood vessels.

If the ovum is fertilised, the corpus luteum continues to release progesterone and so keeps the uterus in a state suitable for implantation. If the ovum is not fertilised, the corpus luteum stops producing progesterone. As a result, the thickened lining of the uterus breaks down and loses blood, which escapes through the cervix and vagina. The events in the menstrual cycle are shown in Figure 16.58.

Menopause

Between the ages of 40 and 55, the ovaries cease to release ova or produce hormones. As a consequence, menstrual periods cease, the woman can no longer have children, and sexual desire is gradually reduced.

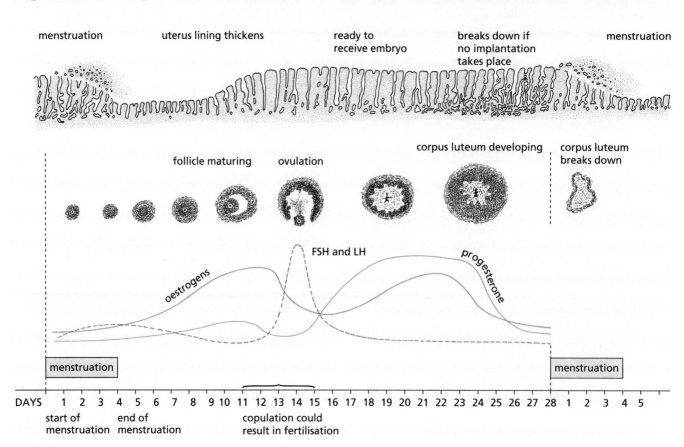

Figure 16.58 The menstrual cycle

Methods of birth control in humans

As little as 4 weeks after giving birth, it is possible, though unlikely, that a woman may conceive again. Frequent breastfeeding may reduce the chances of conception. Nevertheless, it would be possible to have children at about 1-year intervals. Most people do not want, or cannot afford, to have as many children as this. All human communities, therefore, practise some form of birth control to space out births and limit the size of the family.

Natural methods of family planning

Abstinence
This is the most obvious way of preventing a pregnancy. This involves a couple avoiding sexual intercourse. In this way, sperm cannot come into contact with an egg and fertilisation cannot happen.

Monitoring body temperature
If it were possible to know exactly when ovulation occurred, intercourse could be avoided for 3–4 days before and 1 day after ovulation. At the moment, however, there is no simple, reliable way to recognise ovulation, though it is usually 12–16 days before the onset of the next menstrual period. By keeping careful records of the intervals between menstrual periods, it is possible to calculate a potentially fertile period of about 10 days in mid-cycle, when sexual intercourse should be avoided if children are not wanted.

On its own, this method is not very reliable but there are some physiological clues that help to make it more accurate. During or soon after ovulation, a woman's temperature rises by about 0.5 °C. It is reasonable to assume that 1 day after the temperature returns to normal, a woman will be infertile.

Cervical mucus
Another clue comes from the type of mucus secreted by the cervix and lining of the vagina. As the time for ovulation approaches, the mucus becomes more fluid. Women can learn to detect these changes and so calculate their fertile period.

By combining the 'calendar', 'temperature' and 'mucus' methods, it is possible to achieve about 80% 'success', i.e. only 20% unplanned pregnancies. Highly motivated couples may achieve better rates of success and, of course, it is a very helpful way of finding the fertile period for couples who do want to conceive.

Artificial methods of family planning
Barrier methods
Sheath or condom
A thin rubber sheath is placed on the erect penis before sexual intercourse. The sheath traps the sperm and prevents them from reaching the uterus. It also prevents the transmission of sexually transmitted infections (STIs).

Diaphragm
A thin rubber disc, placed in the vagina before intercourse, covers the cervix and stops sperm entering the uterus. Condoms and diaphragms, used in conjunction with chemicals that immobilise sperm, are about 95% effective. However, a diaphragm does not prevent the risk of transmission of STIs.

Femidom
This is a female condom. It is a sheath or pouch, made of polyurethane or rubber, with a flexible ring at each end. The ring at the closed end of the sheath is inserted into the vagina to hold the femidom in place. The ring at the open end is placed outside the vagina. During sexual intercourse, semen is trapped inside the femidom. A femidom reduces the risk of infection by STIs.

Chemical methods
Spermicides
Spermicides are chemicals which, though harmless to the tissues, can kill or immobilise sperm. The spermicide, in the form of a cream, gel or foam, is placed in the vagina. On their own, spermicides are not very reliable but, in conjunction with condoms or diaphragms, they are effective.

Intra-uterine device (IUD)
A small T-shaped plastic and copper device, also known as a coil, can be inserted by a doctor or nurse into the wall of the uterus, where it probably prevents implantation of a fertilised ovum. It is about 98% effective but there is a small risk of developing uterine infections, and it does not protect against STIs.

Intra-uterine system (IUS)
This is similar to an IUD; is T-shaped and releases the hormone progesterone slowly over a long period of time (up to 5 years). The hormone prevents ovulation. An IUS does not protect against STIs.

Contraceptive pill
The pill contains chemicals, which have the same effect on the body as the hormones oestrogen and

progesterone. When mixed in suitable proportions these hormones suppress ovulation and so prevent conception. The pills need to be taken each day for the 21 days between menstrual periods.

There are many varieties of contraceptive pill in which the relative proportions of oestrogen- and progesterone-like chemicals vary. They are 99% effective, but long-term use of some types may increase the risk of cancer of the breast and cervix. The pill does not protect against STIs.

Contraceptive implant
This is a small plastic tube about 4 cm long, which is inserted under the skin of the upper arm of a woman by a doctor or nurse. Once in place it slowly releases the hormone progesterone, preventing pregnancy. It lasts for about 3 years. It does not protect against STIs, but has more than a 99% success rate in preventing pregnancy.

Contraceptive injection
This injection, given to women, contains progesterone and stays effective for between 8 and 12 weeks. It works by thickening the mucus in the cervix, stopping sperm reaching an egg. It also thins the lining of the uterus, making it unsuitable for implantation of an embryo. It does not protect against STIs.

Surgical methods
Male sterilisation – vasectomy
This is a simple and safe surgical operation in which the man's sperm ducts are cut and the ends sealed. This means that his semen contains the secretions of the prostate gland and seminal vesicle but no sperm, so cannot fertilise an ovum. Sexual desire, erection, copulation and ejaculation are quite unaffected.

The testis continues to produce sperm and testosterone. The sperm are removed by white cells as fast as they form. The testosterone ensures that there is no loss of masculinity.

The sperm ducts can be rejoined by surgery but this is not always successful.

Female sterilisation – laparotomy
A woman may be sterilised by an operation in which her oviducts are tied, blocked or cut. The ovaries are unaffected. Sexual desire and menstruation continue as before, but sperm can no longer reach the ova. Ova are released, but break down in the upper part of the oviduct.

The operation cannot usually be reversed.

The use of hormones in fertility and contraception treatments

Infertility
About 85–90% of couples trying for a baby achieve pregnancy within a year. Those that do not may be sub-fertile or infertile. Female infertility is usually caused by a failure to ovulate or a blockage or distortion of the oviducts. The latter can often be corrected by surgery.

Using hormones to improve fertility
Failure to produce ova can be treated with **fertility drugs**. These drugs are similar to hormones and act by increasing the levels of FSH and LH. Administration of the drug is timed to promote ovulation to coincide with copulation.

Artificial insemination (AI)
Male infertility is caused by an inadequate quantity of sperm in the semen or by sperm that are insufficiently mobile to reach the oviducts. There are few effective treatments for this condition, but pregnancy may be achieved by **artificial insemination** (AI). This involves injecting semen through a tube into the top of the uterus. In some cases, the husband's semen can be used but, more often, the semen is supplied by an anonymous donor.

With AI, the woman has the satisfaction of bearing her child rather than adopting, and 50% of the child's genes are from the mother. It also allows a couple to have a baby that is biologically theirs if the man is infertile.

Apart from religious or moral objections, the disadvantages are that the child can never know his or her father and there may be legal problems about the legitimacy of the child in some countries.

In vitro fertilisation
'*In vitro*' means literally 'in glass' or, in other words, the fertilisation is allowed to take place in laboratory glassware (hence the term 'test-tube babies'). This technique may be employed where surgery cannot be used to repair blocked oviducts.

In vitro fertilisation has received considerable publicity since the first 'test-tube' baby was born in 1978. The woman may be given fertility drugs, which cause her ovaries to release several mature ova simultaneously. These ova are then collected by laparoscopy, i.e. they are sucked up in a fine tube inserted through the abdominal wall. The ova are then mixed with the husband's seminal fluid and watched under the microscope to see if cell division takes place. (Figure 16.50 is a photograph of such an 'in vitro' fertilised ovum.)

One or more of the dividing zygotes are then introduced to the woman's uterus by means of a tube inserted through the cervix. Usually, only one (or none) of the zygotes develops, though occasionally there are multiple births.

The success rate for *in vitro* fertilisation is between 12 and 40% depending on how many embryos are transplanted. However, new research using time-lapse photography of the developing IVF embryos during the first few days of life could raise the success rate to up to 78%. It could also reduce the cost from between £5000 and £10 000 for each treatment cycle to £750 in Britain. The photographs are used to select the best embryos, based on their early development.

Using hormones for contraception

Oestrogen and progesterone control important events in the menstrual cycle.

Oestrogen encourages the re-growth of the lining of the uterus wall after a period and prevents the release of FSH. If FSH is blocked, no further ova are matured. The uterus lining needs to be thick to allow successful implantation of an embryo.

Progesterone maintains the thickness of the uterine lining. It also inhibits the secretion of luteinising hormone (LH), which is responsible for ovulation. If LH is suppressed, ovulation cannot happen, so there are no ova to be fertilised.

Because of the roles of oestrogen and progesterone, they are used, singly or in combination, in a range of contraceptive methods.

Social implications of contraception and fertility treatments

Some religions are against any artificial forms of contraception and actively discourage the use of contraceptives such as the sheath and femidom. However, these are important in the prevention of transmission of STDs in addition to their role as contraceptives.

Fertility treatments such as *in vitro* fertilisation are controversial because of the 'spare' embryos that are created and not returned to the uterus. Some people believe that since these embryos are potential human beings, they should not be destroyed or used for research. In some cases the 'spare' embryos have been frozen and used later if the first transplants did not work.

● Sexually transmitted infections (STIs)

> **Key definition**
> A **sexually transmitted infection** is an infection that is transmitted via body fluids through sexual contact.

AIDS and HIV

The initials of AIDS stand for **acquired immune deficiency syndrome**. (A 'syndrome' is a pattern of symptoms associated with a particular disease.) The virus that causes AIDS is the **human immunodeficiency virus** (**HIV**).

After a person has been infected, years may pass before symptoms develop. So people may carry the virus yet not show any symptoms. They can still infect other people, however. It is not known for certain what proportion of HIV carriers will eventually develop AIDS: perhaps 30–50%, or more.

HIV is transmitted by direct infection of the blood. Drug users who share needles contaminated with infected blood run a high risk of the disease. It can also be transmitted sexually, both between men and women and, especially, between homosexual men who practise anal intercourse. Prostitutes, who have many sexual partners, are at risk of being infected if they have sex without using condoms and are, therefore, a potential source of HIV to others.

Haemophiliacs have also fallen victim to AIDS. Haemophiliacs have to inject themselves with a blood product that contains a clotting factor. Before the risks were recognised, infected carriers sometimes donated blood, which was used to produce the clotting factor.

Babies born to HIV carriers may become infected with HIV, either in the uterus or during birth or from the mother's milk. The rate of infection varies from about 40% in parts of Africa to 14% in Europe. If the mother is given drug therapy during labour and the baby within 3 days, this method of transmission is reduced.

There is no evidence to suggest that the disease can be passed on by droplets (Chapter 10), by saliva or by normal everyday contact.

When AIDS first appeared, there were no effective drugs. Today, there is a range of drugs that can be given separately or as a 'cocktail', which slow the progress of the disease. Research to find a vaccine and more effective drugs is ongoing.

There is a range of blood tests designed to detect HIV infection. These tests do not detect the virus but do indicate whether antibodies to the virus are in the blood. If HIV antibodies are present, the person is said to be **HIV positive**. The tests vary in their reliability and some are too expensive for widespread use. The American Food and Drug Administration claims a 99.8% accuracy, but this figure is disputed.

Control of the spread of STIs

The best way to avoid sexually transmitted infections is to avoid having sexual intercourse with an infected person. However, the symptoms of the disease are often not obvious and it is difficult to recognise an infected individual. So the disease is avoided by not having sexual intercourse with a person who *might* have the disease. Such persons are:

- prostitutes who offer sexual intercourse for money
- people who are known to have had sexual relationships with many others
- casual acquaintances whose background and past sexual activities are not known.

These are good reasons, among many others, for being faithful to one partner.

The risk of catching a sexually transmitted disease can be greatly reduced if the man uses a condom or if a woman uses a femidom. These act as barriers to bacteria or viruses.

If a person suspects that he or she has caught a sexually transmitted disease, treatment must be sought at once. Information about treatment can be obtained by phoning one of the numbers listed under 'Venereal Disease' or 'Health Information Service' in the telephone directory. Treatment is always confidential. The patients must, however, ensure that anyone they have had sexual contact with also gets treatment. There is no point in one partner being cured if the other is still infected.

STIs that are caused by a bacterium, such as syphilis and gonorrhoea, can be treated with antibiotics if the symptoms are recognised early enough. However, HIV is viral so antibiotics are not effective.

The effects of HIV on the immune system

HIV attacks certain kinds of lymphocyte (see 'Blood' in Chapter 9), so the number of these cells in the body decreases. Lymphocytes produce antibodies against infections. If the body cannot respond to infections through the immune system, it becomes vulnerable to pathogens that might not otherwise be life-threatening. As a result, the patient has little or no resistance to a wide range of diseases such as influenza, pneumonia, blood disorders, skin cancer or damage to the nervous system, which the body cannot resist.

Questions

Core

1 Plants can often be propagated from stems but rarely from roots. What features of shoots account for this difference?
2 The plants that survive a heath fire are often those that have a rhizome (e.g. ferns). Suggest a reason why this is so.
3 Working from outside to inside, list the parts of a bisexual flower.
4 What features of flowers might attract insects?
5 Which part of a flower becomes:
   a the seed
   b the fruit?
6 Put the following events in the correct order for pollination in a lupin plant:
   A Bee gets dusted with pollen.
   B Pollen is deposited on stigma.
   C Bee visits older flower.
   D Bee visits young flower.
   E Anthers split open.
7 What are the functions in a seed of:
   a the radicle
   b the plumule
   c the cotyledons?

8 During germination of the broad bean, how are the following parts protected from damage as they are forced through the soil:
   a the plumule
   b the radicle?
9 List all the possible purposes for which a growing seedling might use the food stored in its cotyledons.
10 At what stage of development is a seedling able to stop depending on the cotyledons for its food?
11 What do you think are the advantages to a germinating seed of having its radicle growing some time before the shoot starts to grow?
12 a Describe the natural conditions in the soil that would be most favourable for germination.
   b How could a gardener try to create these conditions?
13 How do sperm differ from ova in their structure (see Figure 16.39)?
14 List the structures, in the correct order, through which the sperm must pass from the time they are produced in the testis, to the time they leave the urethra.
15 What structures are shown in Figure 16.44, but are not shown in Figure 16.43?
16 In what ways does a zygote differ from any other cell in the body?
17 If a woman starts ovulating at 13 years old and stops at 50:
   a how many ova are likely to be released from her ovaries
   b about how many of these are likely to be fertilised?
18 List, in the correct order, the parts of the female reproductive system through which sperm must pass before reaching and fertilising an ovum.
19 State exactly what happens at the moment of fertilisation.
20 Is fertilisation likely to occur if mating takes place:
   a 2 days before ovulation
   b 2 days after ovulation?
   Explain your answers.
21 Draw up a table with three columns as shown below. In the first column write:
   male reproductive organs
   female reproductive organs
   male gamete
   female gamete
   place where fertilisation occurs
   zygote grows into
   Now complete the other two columns.

	Flowering plants	Mammals
male reproductive organs		
female reproductive organs		
male gamete, etc.		

22 In what ways will the composition of the blood in the umbilical vein differ from that in the umbilical artery?
23 An embryo is surrounded with fluid, its lungs are filled with fluid and it cannot breathe. Why doesn't it suffocate?
24 If a mother gives birth to twin boys, does this mean that they are identical twins? Explain.
25 Study Figures 16.51 and 16.52. On each diagram the age and size of the developing embryo are stated.
   a Copy and complete the following table:

Age/weeks	Size/mm
0	0
2	
5	
8	
10	
20	
35	

   b Use the data in your table to plot a graph to show the growth of the embryo.

Extended
26 In what ways does asexual reproduction in *Mucor* differ from asexual reproduction in flowering plants?
27 A gardener finds a new and attractive plant produced as a result of a chance mutation. Should she attempt to produce more of the same plant by self-pollination or by vegetative propagation? Explain your reasoning.
28 Which of the following do not play a part in asexual reproduction?
   mitosis, gametes, meiosis, cell division, chromosomes, zygote
29 Revise asexual reproduction and then state how we exploit the process of asexual reproduction in plants.
30 Which structures in a flower produce:
   a the male gametes
   b the female gametes?
31 In not more than two sentences, distinguish between the terms *pollination* and *fertilisation*.
32 In flowering plants:
   a can pollination occur without fertilisation
   b can fertilisation occur without pollination?
33 Which parts of a tomato flower:
   a grow to form the fruit
   b fall off after fertilisation
   c remain attached to the fruit?
34 From the list of changes at puberty in girls, select those that are related to childbearing and say what part you think they play.
35 One of the first signs of pregnancy is that the menstrual periods stop. Explain why you would expect this.

16 REPRODUCTION

Checklist

After studying Chapter 16 you should know and understand the following:

Asexual reproduction

- Asexual reproduction is the process resulting in the production of genetically identical offspring from one parent.
- Asexual reproduction occurs without gametes or fertilisation.
- Fungi can reproduce asexually by single-celled spores.
- Many flowering plants reproduce asexually by vegetative propagation.
- Plants reproduce asexually when some of their buds grow into new plants.
- The stolon of the strawberry plant is a horizontal stem that grows above the ground, takes root at the nodes and produces new plants.
- The couch grass rhizome is a horizontal stem that grows below the ground and sends up shoots from its nodes.
- Bulbs are condensed shoots with circular fleshy leaves. Bulb-forming plants reproduce asexually from lateral buds.
- Rhizomes, corms, bulbs and tap roots may store food, which is used to accelerate early growth.
- A clone is a population of organisms produced asexually from a single parent.
- Whole plants can be produced from single cells or small pieces of tissue.

- Artificial propagation from cuttings or grafts preserves the desirable characteristics of a crop plant.
- Vegetative propagation produces (genetically) identical individuals.
- Asexual reproduction keeps the characteristics of the organism the same from one generation to the next, but does not result in variation to cope with environmental change.

Sexual reproduction

- Sexual reproduction is the process involving the fusion of the nuclei of two gametes (sex cells) to form a zygote and the production of offspring that are genetically different from each other.
- The male gamete is small and mobile. The female gamete is larger and not often mobile.
- The male gamete of an animal is a sperm. The male gamete of a flowering plant is the pollen nucleus.
- The female gamete of an animal is an ovum. The female gamete of a flowering plant is an egg cell in an ovule.
- Fertilisation is the fusion of gamete nuclei.

- The nuclei of gametes are haploid and the nucleus of the zygote is diploid.
- There are advantages and disadvantages of sexual reproduction to a species.
- There are advantages and disadvantages of sexual reproduction in crop production.

Sexual reproduction in plants

- Flowers contain the reproductive organs of plants.
- The stamens are the male organs. They produce pollen grains, which contain the male gamete.
- The carpels are the female organs. They produce ovules, which contain the female gamete and will form the seeds.
- The flowers of most plant species contain male and female organs. A few species have unisexual flowers.
- Brightly coloured petals attract insects, which pollinate the flower.
- Pollination is the transfer of pollen from the anthers of one flower to the stigma of a flower on the same or another plant.
- Pollination may be carried out by insects or by the wind.
- Flowers that are pollinated by insects are usually brightly coloured and have nectar.
- Flowers that are pollinated by the wind are usually small and green. Their stigmas and anthers hang outside the flower where they are exposed to air movements.
- Fertilisation occurs when a pollen tube grows from a pollen grain into the ovary and up to an ovule. The pollen nucleus passes down the tube and fuses with the ovule nucleus.
- After fertilisation, the ovary grows rapidly to become a fruit and the ovules become seeds.
- Germination is influenced by temperature and the amount of water and oxygen available.

- Self-pollination is the transfer of pollen grains from the anther of a flower to the stigma of the same flower.
- Cross-pollination is transfer of pollen grains from the anther of a flower to the stigma of a flower on a different plant of the same species.
- Self-pollination and cross-pollination have implications to a species.

Sexual reproduction in humans

- The male reproductive cells (gametes) are sperm. They are produced in the testes and expelled through the urethra and penis during mating.
- The female reproductive cells (gametes) are ova (eggs). They are produced in the ovaries. One is released each month. If sperm are present, the ovum may be fertilised as it passes down the oviduct to the uterus.
- Fertilisation happens when a sperm enters an ovum and the sperm and egg nuclei join up (fuse).
- The fertilised ovum (zygote) divides into many cells and becomes embedded in the lining of the uterus. Here it grows into an embryo.
- The embryo gets its food and oxygen from its mother.
- The embryo's blood is pumped through blood vessels in the umbilical cord to the placenta, which is attached to the uterus lining. The embryo's blood comes very close to the mother's blood so that food and oxygen can be picked up and carbon dioxide and nitrogenous waste can be got rid of.

- Good ante-natal care, in the form of special dietary needs and maintaining good health, is needed to support the mother and her fetus.
- When the embryo is fully grown, it is pushed out of the uterus through the vagina by contractions of the uterus and abdomen.
- Twins may result from two ova being fertilised at the same time or from a zygote forming two embryos.
- Eggs and sperm are different in size, structure, mobility and numbers produced.
- Sperm and eggs have special features to adapt them for their functions.
- The placenta and umbilical cord are involved in exchange of materials between the mother and fetus. Some toxins and viruses can also be passed across and affect the fetus.
- Human milk and breastfeeding are best for babies.

Sex hormones in humans

- At puberty, the testes and ovaries start to produce mature gametes and the secondary sexual characteristics develop.
- Each month, the uterus lining thickens up in readiness to receive a fertilised ovum. If an ovum is not fertilised, the lining and some blood are lost through the vagina. This is menstruation.
- Oestrogen and progesterone are secreted by endocrine glands.
- The release of ova and the development of an embryo are under the control of hormones like oestrogen, progesterone, follicle-stimulating hormone and luteinising hormone.

Methods of birth control in humans

- There are effective ways of spacing births and limiting the size of a family. These include natural, chemical, barrier and surgical methods.

- Hormones can be used to control fertility, including contraception and promoting egg-cell development.
- Female infertility may be relieved by surgery, fertility drugs or *in vitro* fertilisation.
- Male infertility can be by-passed by artificial insemination.
- There are social implications of using hormones in contraception and for increasing the chances of pregnancy.

Sexually transmitted infections (STIs)

- A sexually transmitted infection is an infection transmitted via bodily fluids through sexual contact.
- HIV is an example of an STI.
- HIV can be transmitted in a number of ways.
- The spread of HIV can be controlled.
- HIV infection may lead to AIDS.

- HIV affects the immune system by reducing the number of lymphocytes and decreasing the ability to produce antibodies.

17 Inheritance

Inheritance
Define inheritance

Chromosomes, genes and proteins
Define chromosome and gene
Inheritance of sex in humans

Genetic code for proteins
Role of DNA in cell function
How a protein is made
Gene expression
Define haploid nucleus, diploid nucleus
Diploid cells

Mitosis
Define mitosis
Role of mitosis
Duplication and separation of chromosomes

Meiosis
Define meiosis
Role of meiosis

The process of mitosis
The function of chromosomes
Stem cells
Gamete production and chromosomes
Meiosis

Monohybrid inheritance
Define allele, genotype, phenotype, homozygous, heterozygous, dominant, recessive
Use of genetic diagrams and Punnett squares

Use of test crosses
Co-dominance and incomplete dominance
Define sex-linked characteristic
Colour blindness
Genetic crosses involving co-dominance and sex linkage

● Inheritance

Key definition
Inheritance is the transmission of genetic information from generation to generation.

We often talk about people inheriting certain characteristics: 'Nathan has inherited his father's curly hair', or 'Fatima has inherited her mother's brown eyes'. We expect tall parents to have tall children. The inheritance of such characteristics is called **heredity** and the branch of biology that studies how heredity works is called **genetics**.

● Chromosomes, genes and proteins

Key definitions
A **chromosome** is a thread of DNA, made up of a string of genes.
A **gene** is a length of DNA that codes for a protein.

Inside a nucleus are thread-like structures called **chromosomes** which can be seen most clearly at the time when the cell is dividing. Each chromosome has certain characteristics when ready to divide: there are two **chromatids**, joined at one point called a **centromere** (Figure 17.1). Each chromatid is a string of **genes**, coding for the person's characteristics. The other chromatid carries the same genes in the same order.

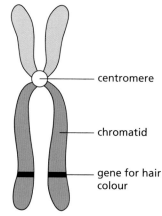

Figure 17.1 Structure of a chromosome

A human body (**somatic**) cell nucleus contains 46 chromosomes. These are difficult to distinguish when packed inside the nucleus, so scientists separate them and arrange them according to size and appearance. The outcome is called a **karyotype** (Figure 17.2). There are pairs of chromosomes. The only pair that do not necessarily match is chromosome pair 23: the 'sex chromosomes'. The Y chromosome is much smaller than the X chromosome.

The inheritance of sex

Whether you are a male or female depends on the pair of chromosomes called the 'sex chromosomes'. In females, the two sex chromosomes, called the X chromosomes, are the same size as each other. In males, the two sex chromosomes are of different sizes. One corresponds to the female sex

Chromosomes, genes and proteins

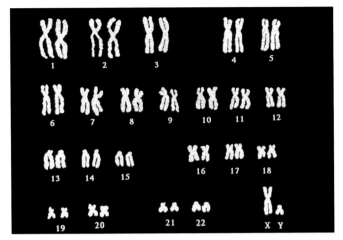

Figure 17.2 Human karyotype

chromosomes and is called the X chromosome. The other is smaller and is called the Y chromosome. So the female cells contain **XX** and male cells contain **XY**.

A process called **meiosis** takes place in the female's ovary. It makes gametes: sex cells, which have half the normal number of chromosomes. During the process, each ovum receives one of the X chromosomes, so all the ova are the same for this. Meiosis in the male's testes results in 50% of the sperms getting an X chromosome and 50% getting a Y chromosome (Figure 17.3). If an X sperm fertilises the ovum, the zygote will be XX and will grow into a girl. If a Y sperm fertilises the ovum, the zygote will be XY and will develop into a boy. There is an equal chance of an X or Y chromosome fertilising an ovum, so the numbers of girl and boy babies are more or less the same.

Figure 17.4 shows how sex is inherited.

parents	male		female
sex chromosomes of parents	XY	×	XX
gametes (sex cells)	X Y	×	X X
sex chromosomes of children	XX XX		XY XY
sex of children	female female		male male

The ratio is 1 female : 1 male

Figure 17.4 Determination of sex

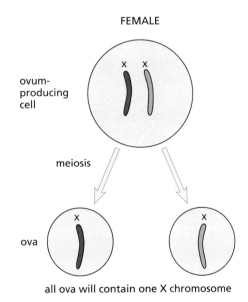

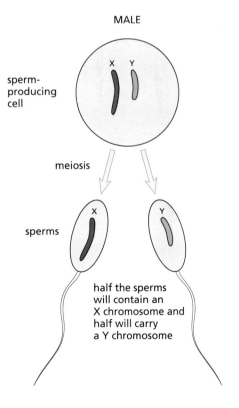

Figure 17.3 Determination of sex. Note that:
  (i) only the X and Y chromosomes are shown
  (ii) details of meiosis have been omitted
  (iii) in fact, four gametes are produced in each case, but two are sufficient to show the distribution of X and Y chromosomes

251

17 INHERITANCE

The genetic code

The structure of DNA has already been described in Chapter 4.

Each nucleotide carries one of four bases (A, T, C or G). A string of nucleotides therefore holds a sequence of bases. This sequence forms a code, which instructs the cell to make particular proteins. Proteins are made from amino acids linked together (Chapter 4). The type and sequence of the amino acids joined together will determine the kind of protein formed. For example, one protein molecule may start with the sequence *alanine–glycine–glycine* .... A different protein may start *glycine–serine–alanine* ....

It is the sequence of bases in the DNA molecule that decides which amino acids are used and in which order they are joined. Each group of three bases stands for one amino acid, e.g. the triplet of bases CGA specifies the amino acid *alanine*, the base triplet CAT specifies the amino acid *valine*, and the triplet CCA stands for *glycine*. The tri-peptide *valine–glycine–alanine* is specified by the DNA code CAT–CCA–CGA (Figure 17.5).

A gene, then, is a sequence of triplets of the four bases, which specifies an entire protein. Insulin is a small protein with only 51 amino acids. A sequence of 153 (i.e. 3 × 51) bases in the DNA molecule would constitute the gene that makes an islet cell in the pancreas produce insulin. Most proteins are much larger than this and most genes contain a thousand or more bases.

The chemical reactions that take place in a cell determine what sort of a cell it is and what its functions are. These chemical reactions are, in turn, controlled by enzymes. Enzymes are proteins. It follows, therefore, that the **genetic code** of DNA, in determining which proteins, particularly enzymes, are produced in a cell, also determines the cell's structure and function. In this way, the genes also determine the structure and function of the whole organism.

Other proteins coded for in DNA include antibodies and the receptors for neurotransmitters (see details of synapses in Chapter 14).

The manufacture of proteins in cells

DNA molecules remain in the nucleus, but the proteins they carry the codes for are needed elsewhere in the cell. A molecule called messenger RNA (**mRNA**) is used to transfer the information from the nucleus. It is much smaller than a DNA molecule and is made up of only one strand. Another difference is that mRNA molecules contain slightly different bases (A, C, G and U). Base U is **uracil**. It attaches to the DNA base A.

To pass on the protein code, the double helix of DNA (see Figure 4.12) unwinds to expose the chain of bases. One strand acts as template. A messenger RNA molecule is formed along part of this strand, made up of a chain of nucleotides with complementary bases to a section of the DNA strand (Figure 17.6). The mRNA molecule carrying the protein code then passes out of the nucleus, through a nuclear pore in the membrane. Once in the cytoplasm it attaches itself to a **ribosome**. Ribosomes make proteins. The mRNA molecule instructs the ribosome to put together a chain of amino acids in a specific sequence, thus making a protein. Other mRNA molecules will carry codes for different proteins.

Some proteins are made up of a relatively small number of amino acids. As stated, insulin is a chain of 51 amino acids. On the mRNA molecule each amino acid is coded by a sequence of three bases (a triplet), so the mRNA molecule coding for insulin will contain 153 bases. Other protein molecules are much bigger: haemoglobin in red blood cells is made of 574 amino acids.

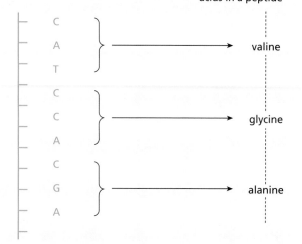

Figure 17.5 The genetic code (triplet code)

Chromosomes, genes and proteins

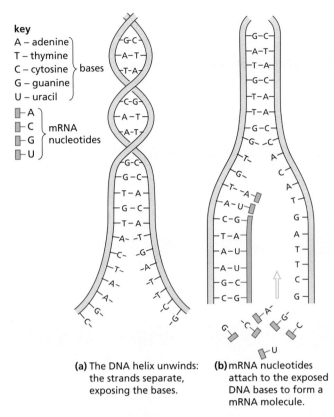

(a) The DNA helix unwinds: the strands separate, exposing the bases.

(b) mRNA nucleotides attach to the exposed DNA bases to form a mRNA molecule.

Figure 17.6 Formation of messenger RNA

Gene expression

Body cells do not all have the same requirements for proteins. For example, the function of some cells in the stomach is to make the protein pepsin (see 'Chemical digestion' in Chapter 7). Bone marrow cells make the protein haemoglobin, but do not need digestive enzymes. Specialised cells all contain the same genes in their nuclei, but only the genes needed to code for specific proteins are switched on (**expressed**). This enables the cell to make only the proteins it needs to fulfil its function.

> **Key definitions**
> A **haploid nucleus** is a nucleus containing a single set of unpaired chromosomes present, for example, in sperm and egg cells.
> A **diploid nucleus** is a nucleus containing two sets of chromosomes present, for example, in body cells.

Number of chromosomes

Figure 17.2 is a karyotype of a human body cell because there are 23 pairs of chromosomes present (they come from a diploid cell). Because the chromosomes are in pairs, the diploid number is always an even number. The karyotype of a sperm cell would show 23 single chromosomes (they come from a **haploid** cell). The sex chromosome would be either X or Y. The chromosomes have different shapes and sizes and can be recognised by a trained observer.

There is a fixed number of chromosomes in each species. Human body cells each contain 46 chromosomes, mouse cells contain 40 and garden pea cells 14 (see also Figure 17.7).

The number of chromosomes in a species is the same in all of its body cells. There are 46 chromosomes in each of your liver cells, in every nerve cell, skin cell and so on.

The chromosomes are always in pairs (Figure 17.7), e.g. two long ones, two short ones, two medium ones. This is because when the zygote is formed, one of each pair comes from the male gamete and one from the female gamete. Your 46 chromosomes consist of 23 from your mother and 23 from your father.

The chromosomes of each pair are called **homologous** chromosomes. In Figure 17.18(b), the two long chromosomes form one homologous pair and the two short chromosomes form another.

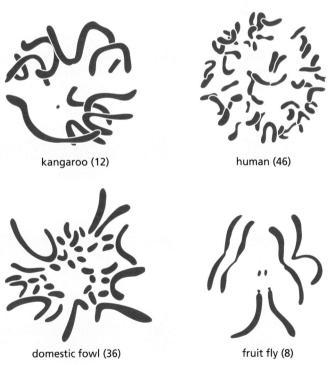

Figure 17.7 Chromosomes of different species. Note that the chromosomes are always in pairs.

253

17 INHERITANCE

● Mitosis

> **Key definitions**
> **Mitosis** is nuclear division giving rise to genetically identical cells.

Genetics is the study of inheritance. It can be used to forecast what sorts of offspring are likely to be produced when plants or animals reproduce sexually. What will be the eye colour of children whose mother has blue eyes and whose father has brown eyes? Will a mating between a black mouse and a white mouse produce grey mice, black-and-white mice or some black and some white mice?

To understand the method of inheritance, we need to look once again at the process of sexual reproduction and fertilisation. In sexual reproduction, a new organism starts life as a single cell called a zygote (Chapter 16). This means that you started from a single cell. Although you were supplied with oxygen and food in the uterus, all your tissues and organs were produced by cell division from this one cell. So, the 'instructions' that dictated which cells were to become liver or muscle or bone must all have been present in this first cell. The instructions that decided that you should be tall or short, dark or fair, male or female must also have been present in the zygote.

The process of **mitosis** is important in growth. We all started off as a single cell (a zygote). That cell divided into two cells, then four and so on, to create the organism we are now, made up of millions of cells. Cells have a finite life: they wear out or become damaged, so they need to be replaced constantly. The processes of **growth**, **repair** and **replacement** of cells all rely on mitosis. Organisms that reproduce asexually (see Chapter 16) also use mitosis to create more cells.

Cell division

When plants and animals grow, their cells increase in number by dividing. Typical growing regions are the ends of bones, layers of cells in the skin, root tips and buds (Figure 17.11). Each cell divides to produce two daughter cells. Both daughter cells may divide again, but usually one of the cells grows and changes its shape and structure and becomes adapted to do one particular job – in other words, it becomes **specialised** (Figure 17.8). At the same time it loses its ability to divide any more. The other cell is still able to divide and so continue the growth of the tissue. Growth is, therefore, the result of cell division, followed by cell enlargement and, in many cases, cell specialisation.

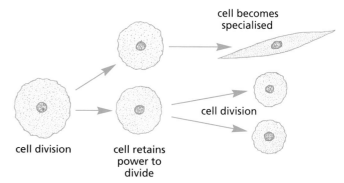

Figure 17.8 Cell division and specialisation. Cells that retain the ability to divide are sometimes called **stem cells.**

The process of cell division in an animal cell is shown in Figure 17.9. The events in a plant cell are shown in Figures 17.10 and 17.11. Because of the cell wall, the cytoplasm cannot simply pinch off in the middle, and a new wall has to be laid down between the two daughter cells. Also a new vacuole has to form.

Organelles such as mitochondria and chloroplasts are able to divide and are shared more or less equally between the daughter cells at cell division.

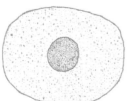 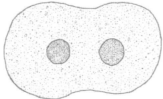 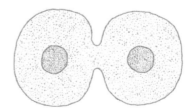 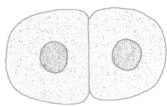

(a) Animal cell about to divide.

(b) The nucleus divides first.

(c) The daughter nuclei separate and the cytoplasm pinches off between the nuclei.

(d) Two cells are formed – one may keep the ability to divide, and the other may become specialised.

Figure 17.9 Cell division in an animal cell

Meiosis

 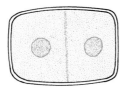 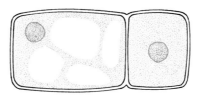 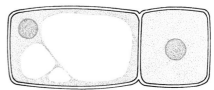

(a) A plant cell about to divide has a large nucleus and no vacuole.

(b) The nucleus divides first. A new cell wall develops and separates the two cells.

(c) The cytoplasm adds layers of cellulose on each side of the new cell wall. Vacuoles form in the cytoplasm of one cell.

(d) The vacuoles join up to form one vacuole. This takes in water and makes the cell bigger. The other cell will divide again.

Figure 17.10 Cell division in a plant cell

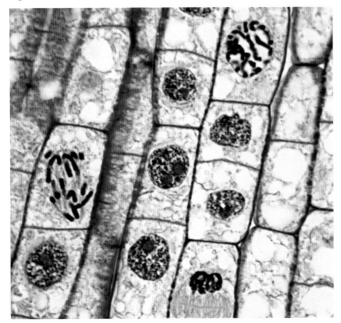

Figure 17.11 Cell division in an onion root tip (×250). The nuclei are stained blue. Most of the cells have just completed cell division.

Making the squash preparation

- Squash the softened, stained root tips by lightly tapping on the coverslip with a pencil: hold the pencil vertically and let it slip through the fingers to strike the coverslip (Figure 17.12).
- The root tip will spread out as a pink mass on the slide; the cells will separate and the nuclei, many of them with chromosomes in various stages of mitosis (because the root tip is a region of rapid cell division), can be seen under the high power of the microscope (×400).

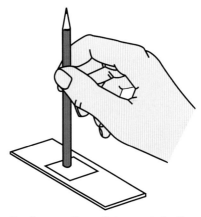

Figure 17.12 Tap the coverslip gently to squash the tissue

Practical work

Squash preparation of chromosomes using acetic orcein

Preparation of root tips

- Support *Allium cepa* (onion) root tips over beakers or jars of water.
- Keep the onions in darkness for several days until the roots growing into the water are 2–3 cm long.
- Cut off about 5 mm of the root tips and place them in a watch glass.
- Cover the root tips with nine drops acetic orcein and one drop molar hydrochloric acid.
- Heat the watch glass gently over a very small Bunsen flame till the steam rises from the stain, but do not boil.
- Leave the watch glass covered for at least 5 minutes.
- Place one of the root tips on a clean slide, cover with 45% ethanoic (acetic) acid and cut away all but the terminal 1 mm.
- Cover this root tip with a clean coverslip and make a squash preparation as described next.

● Meiosis

> **Key definitions**
> **Meiosis** is nuclear division, which gives rise to cells that are genetically different.

The process of meiosis takes place in the **gonads** of animals (e.g. the testes and ovaries of mammals, and the anthers and ovules of flowering plants). The cells formed are **gametes** (sperm and egg cells in mammals; egg cells and pollen grain nuclei in flowering plants). Gametes are different from other cells because they have half the normal number of chromosomes (they are **haploid**).

17 INHERITANCE

The process of mitosis

To understand how the 'instructions' are passed from cell to cell, we need to look in more detail at what happens when the zygote divides and produces an organism consisting of thousands of cells. This type of cell division is called mitosis. It takes place not only in a zygote but in all growing tissues.

When a cell is not dividing, there is very little detailed structure to be seen in the nucleus even if it is treated with special dyes called stains. Just before cell division, however, a number of long, thread-like structures appear in the nucleus and show up very clearly when the nucleus is stained (Figures 17.13 and 17.14). These thread-like structures are called chromosomes. Although they are present in the nucleus all the time, they show up clearly only at cell division because at this time they get shorter and thicker.

Each chromosome duplicates itself and is seen to be made up of two parallel strands, called chromatids (Figure 17.1). When the nucleus divides into two, one chromatid from each chromosome goes into each daughter nucleus. The chromatids in each nucleus now become chromosomes and later they will make copies of themselves ready for the next cell division. The process of copying is called **replication** because each chromosome makes a replica (an exact copy) of itself. As Figure 17.13 is a simplified diagram of mitosis, only two chromosomes are shown, but there are always more than this. Human cells contain 46 chromosomes.

Mitosis will be taking place in any part of a plant or animal that is producing new cells for growth or replacement. Bone marrow produces new blood cells by mitosis; the epidermal cells of the skin are replaced by mitotic divisions in the basal layer; new epithelial cells lining the alimentary canal are produced by mitosis; growth of muscle or bone in animals, and root, leaf, stem or fruit in plants, results from mitotic cell divisions.

An exception to this occurs in the final stages of gamete production in the reproductive organs of plants and animals. The cell divisions that give rise to gametes are not mitotic but meiotic.

Cells that are not involved in the production of gametes are called **somatic cells**. Mitosis takes place only in somatic cells.

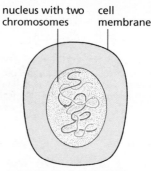

(a) Just before the cell divides, chromosomes appear in the nucleus.

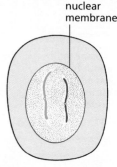

(b) The chromosomes get shorter and thicker.

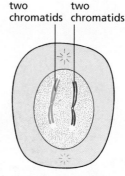

(c) Each chromosome is now seen to consist of two chromatids.

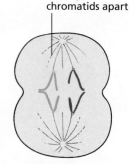

(d) The nuclear membrane disappears and the chromatids are pulled apart to opposite ends of the cell.

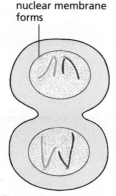

(e) A nuclear membrane forms round each set of chromatids, and the cell starts to divide.

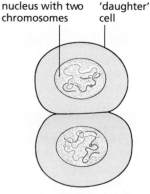

(f) Cell division completed, giving two 'daughter' cells, each containing the same number of chromosomes as the parent cell.

Figure 17.13 Mitosis. Only two chromosomes are shown. Three of the stages described here are shown in Figure 17.14.

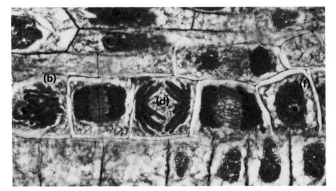

Figure 17.14 Mitosis in a root tip (×500). The letters refer to the stages described in Figure 17.13. (The tissue has been squashed to separate the cells.)

The function of chromosomes

When a cell is not dividing, its chromosomes become very long and thin. Along the length of the chromosome is a series of chemical structures called genes (Figure 17.15). The chemical that forms the genes is called DNA (which is short for deoxyribonucleic acid, Chapter 4). Each gene controls some part of the chemistry of the cell. It is these genes that provide the 'instructions' mentioned at the beginning of the chapter. For example, one gene may 'instruct' the cell to make the pigment that is formed in the iris of brown eyes. On one chromosome there will be a gene that causes the cells of the stomach to make the enzyme pepsin. When the chromosome replicates, it builds an exact replica of itself, gene by gene (Figure 17.16). When the chromatids separate at mitosis, each cell will receive a full set of genes. In this way, the chemical instructions in the zygote are passed on to all cells of the body. All the chromosomes, all the genes and, therefore, all the instructions are faithfully reproduced by mitosis and passed on complete to all the cells.

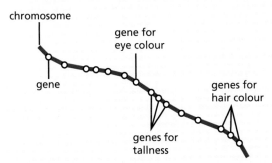

Figure 17.15 Relationship between chromosomes and genes. The drawing does not represent real genes or a real chromosome. There are probably thousands of genes on a chromosome.

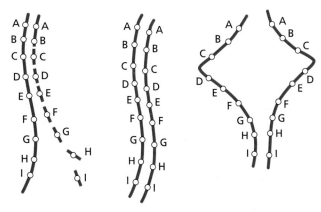

(a) A chromosome builds up a replica of itself.

(b) When the cell divides, the original and the replica are called chromatids.

(c) Mitosis separates the chromatids. Each new cell gets a full set of genes.

Figure 17.16 Replication. (A, B, C, etc. represent genes.)

Which of the instructions are used depends on where a cell finally ends up. The gene that causes brown eyes will have no effect in a stomach cell and the gene for making pepsin will not function in the cells of the eye. So a gene's chemical instructions are carried out only in the correct situation.

The genes that produce a specific effect in a cell (or whole organism) are said to be **expressed**. In the stomach lining, the gene for pepsin is expressed. The gene for melanin (the pigment in brown eyes) is not expressed.

Stem cells

Recent developments in tissue culture have involved **stem cells**. Stem cells are those cells in the body that have retained their power of division. Examples are the basal cells of the skin ('Homeostasis' in Chapter 14), which keep dividing to make new skin cells, and cells in the red bone marrow, which constantly divide to produce the whole range of blood cells ('Blood' in Chapter 9).

In normal circumstances this type of stem cell can produce only one type of tissue: epidermis, blood, muscle, nerves, etc. Even so, culture of these stem cells could lead to effective therapies by introducing healthy stem cells into the body to take over the function of diseased or defective cells.

Cells taken from early embryos (**embryonic stem cells**) can be induced to develop into almost any kind of cell, but there are ethical objections to using human embryos for this purpose. However, it has recently been shown that, given the right

17 INHERITANCE

conditions, brain stem cells can become muscle or blood cells, and liver cells have been cultured from blood stem cells. Scientists have also succeeded in reprogramming skin cells to develop into other types of cell, such as nerve cells. Bone marrow cells are used routinely to treat patients with leukaemia (cancer of white blood cells). The use of adult stem cells does not have the ethical problems of embryonic stem cells, since cells that could become whole organisms are not being destroyed.

Gamete production and chromosomes

The genes on the chromosomes carry the instructions that turn a single-cell zygote into a bird or a rabbit or an oak tree. The zygote is formed at fertilisation, when a male gamete fuses with a female gamete. Each gamete brings a set of chromosomes to the zygote. The gametes, therefore, must each contain only half the diploid number of chromosomes, otherwise the chromosome number would double each time an organism reproduced sexually. Each human sperm cell contains 23 chromosomes and each human ovum has 23 chromosomes. When the sperm and ovum fuse at fertilisation (Chapter 16), the diploid number of 46 (23 + 23) chromosomes is produced (Figure 17.17).

The process of cell division that gives rise to gametes is different from mitosis because it results in the cells containing only half the diploid number of chromosomes. This number is called the haploid number and the process of cell division that gives rise to gametes is called **meiosis**.

Meiosis takes place only in reproductive organs.

Meiosis

In a diploid cell that is going to divide and produce gametes, the chromosomes shorten and thicken as in mitosis. The pairs of homologous chromosomes, e.g. the two long ones and the two short ones in Figure 17.18(b), lie alongside each other and, when the nucleus divides for the first time, it is the chromosomes and not the chromatids that are separated. This results in only half the total number of chromosomes going to each daughter cell. In Figure 17.18(c), the diploid number of four chromosomes is being reduced to two chromosomes prior to the first cell division.

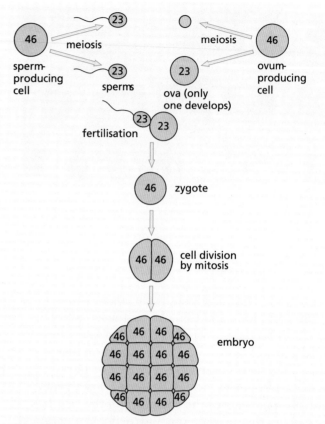

Figure 17.17 Chromosomes in gamete production and fertilisation

By now (Figure 17.18(d)), each chromosome is seen to consist of two chromatids and there is a second division of the nucleus (Figure 17.18(e)), which separates the chromatids into four distinct nuclei (Figure 17.18(f)).

This gives rise to four gametes, each with the haploid number of chromosomes. In the anther of a plant (Chapter 16), four haploid pollen grains are produced when a pollen mother cell divides by meiosis (Figure 17.19). In the testis of an animal, meiosis of each sperm-producing cell forms four sperm. In the cells of the ovule of a flowering plant or the ovary of a mammal, meiosis gives rise to only one mature female gamete. Four gametes may be produced initially, but only one of them turns into an egg cell that can be fertilised.

As a result of meiosis and fertilisation, the maternal and paternal chromosomes meet in different combinations in the zygotes. Consequently, the offspring will differ from their parents and from each other in a variety of ways.

Asexually produced organisms (Chapter 16) show no such variation because they are produced by mitosis and all their cells are identical to those of their single parent.

Monohybrid inheritance

Table 17.1 compares meiosis and mitosis.

Table 17.1 Mitosis and meiosis compared

Meiosis	Mitosis
occurs in the final stages of cell division leading to production of gametes	occurs during cell division of somatic cells
only half the chromosomes are passed on to the daughter cells, i.e. the haploid number of chromosomes	a full set of chromosomes is passed on to each daughter cell; this is the diploid number of chromosomes
homologous chromosomes and their genes are randomly assorted between the gametes	the chromosomes and genes in each daughter cell are identical
new organisms produced by meiosis in sexual reproduction will show variations from each other and from their parents	if new organisms are produced by mitosis in asexual reproduction (e.g. bulbs, Chapter 16) they will all resemble each other and their parents; they are said to be 'clones'

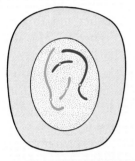

(a) The chromosomes appear. Those in red are from the organism's mother; the blue ones are from the father.

(b) Homologous chromosomes lie alongside each other.

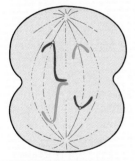

(c) The nuclear membrane disappears and corresponding chromosomes move apart to opposite ends of the cell.

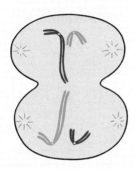

(d) By now each chromosome has become two chromatids.

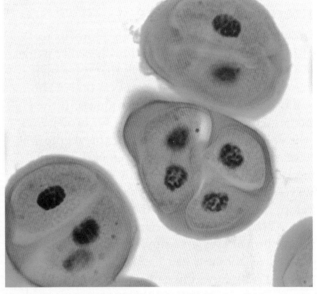

Figure 17.19 Meiosis in an anther (×1000). The last division of meiosis in the anther of a flower produces four pollen grains.

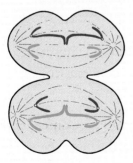

(e) A second division takes place to separate the chromatids.

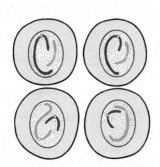

(f) Four gametes are formed. Each contains only half the original number of chromosomes.

Figure 17.18 Meiosis

● Monohybrid inheritance

Key definitions
An **allele** is a version of a gene.
Genotype is the genetic make-up of an organism in terms of the alleles present.
Phenotype is the features of an organism.
Homozygous means having two identical alleles of a particular gene e.g. **TT**, where **T** is tall. Note that two identical homozygous individuals that breed together will be pure-breeding.
Heterozygous means having two different alleles of a particular gene e.g. **Tt**. Note that a heterozygous individual will not be pure breeding.
An allele that is expressed if it is present is **dominant**.
An allele that is only expressed when there is no dominant allele of the gene present is **recessive**.

Alleles

The genes that occupy corresponding positions on homologous chromosomes and control the same characteristic are called **allelomorphic genes**, or **alleles**. The word 'allelomorph' means 'alternative form'. For example, there are two alternative forms of a gene for eye colour. One allele produces brown eyes and one allele produces blue eyes.

There are often more than two alleles of a gene. The human ABO blood groups are controlled by three alleles, though only two of these can be present in one genotype.

Patterns of inheritance

A knowledge of mitosis and meiosis allows us to explain, at least to some extent, how heredity works. The allele in a mother's body cells that causes her to have brown eyes may be present on one of the chromosomes in each ovum she produces. If the father's sperm cell contains an allele for brown eyes on the corresponding chromosome, the zygote will receive an allele for brown eyes from each parent. These alleles will be reproduced by mitosis in all the embryo's body cells and when the embryo's eyes develop, the alleles will make the cells of the iris produce brown pigment (melanin) and the child will have brown eyes. In a similar way, the child may receive alleles for curly hair.

Figure 17.20 shows this happening, but it does not, of course, show all the other chromosomes with thousands of genes for producing the enzymes, making different types of cell and all the other processes that control the development of the organism.

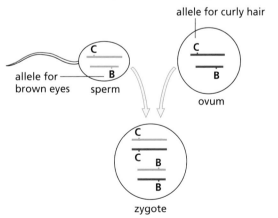

Figure 17.20 Fertilisation. Fertilisation restores the diploid number of chromosomes and combines the alleles from the mother and father.

Single-factor inheritance

Because it is impossible to follow the inheritance of the thousands of characteristics controlled by genes, it is usual to start with the study of a single gene that controls one characteristic. We have used eye colour as an example so far. Probably more than one allele pair is involved, but the simplified example will serve our purpose. It has already been explained how an allele for brown eyes from each parent results in the child having brown eyes. Suppose, however, that the mother has blue eyes and the father brown eyes. The child might receive an allele for blue eyes from its mother and an allele for brown eyes from its father (Figure 17.21). If this happens, the child will, in fact, have brown eyes. The allele for brown eyes is said to be **dominant** to the allele for blue eyes. Although the allele for blue eyes is present in all the child's cells, it is not expressed. It is said to be **recessive** to brown.

Eye colour is a useful 'model' for explaining inheritance but it is not wholly reliable because 'blue' eyes vary in colour and sometimes contain small amounts of brown pigment.

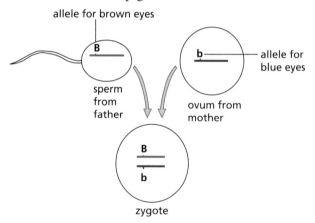

Figure 17.21 Combination of alleles in the zygote (only one chromosome is shown). The zygote has both alleles for eye colour; the child will have brown eyes.

This example illustrates the following important points:

- There is a pair of alleles for each characteristic, one allele from each parent.
- Although the allele pairs control the same characteristic, e.g. eye colour, they may have different effects. One tries to produce blue eyes, the other tries to produce brown eyes.
- Often one allele is dominant over the other.

- The alleles of each pair are on corresponding chromosomes and occupy corresponding positions. For example, in Figure 17.20 the alleles for eye colour are shown in the corresponding position on the two short chromosomes and the alleles for hair curliness are in corresponding positions on the two long chromosomes. In diagrams and explanations of heredity:
  - alleles are represented by letters
  - alleles controlling the same characteristic are given the same letter, and
  - the dominant allele is given the capital letter.

For example, in rabbits, the dominant allele for black fur is labelled **B**. The recessive allele for white fur is labelled **b** to show that it corresponds to **B** for black fur. If it were labelled **w**, we would not see any connection between **B** and **w**. **B** and **b** are obvious partners. In the same way **L** could represent the allele for long fur and **l** the allele for short fur.

Breeding true

A white rabbit must have both the recessive alleles **b** and **b**. If it had **B** and **b**, the dominant allele for black (**B**) would override the allele for white (**b**) and produce a black rabbit. A black rabbit, on the other hand, could be either **BB** or **Bb** and, by just looking at the rabbit, you could not tell the difference. When a male black rabbit **BB** produces sperm, each one of the pair of chromosomes carrying the **B** alleles will end up in different sperm cells. Since the alleles are the same, all the sperm will have the **B** allele for black fur (Figure 17.22(a)).

A black rabbit **BB** is called a true-breeding black and is said to be **homozygous** for black coat colour ('homo-' means 'the same'). If this rabbit mates with another black (**BB**) rabbit, all the babies will be black because all will receive a dominant allele for black fur. When all the offspring have the same characteristic as the parents, this is called '**breeding true**' for this characteristic.

When a **Bb** black rabbit produces gametes by meiosis, the chromosomes with the **B** allele and the chromosomes with the **b** allele will end up in different gametes. So 50% of the sperm cells will carry **B** alleles and 50% will carry **b** alleles (Figure 17.22(b)). Similarly, in the female, 50% of the ova will have a **B** allele and 50% will have a **b** allele. If a **b** sperm fertilises a **b** ovum, the offspring, with two **b** alleles (**bb**), will be white. The black **Bb** rabbits are not true-breeding because they may produce some white babies as well as black ones. The **Bb** rabbits are called **heterozygous** ('hetero-' means 'different').

The black **BB** rabbits are homozygous dominant. The white **bb** rabbits are homozygous recessive.

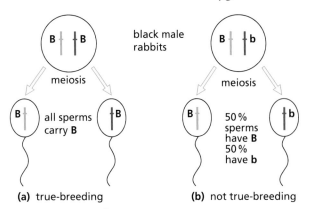

Figure 17.22 Breeding true

Genotype and phenotype

The two kinds of black rabbit **BB** and **Bb** are said to have the same **phenotype**. This is because their coat colours look exactly the same. However, because they have different allele pairs for coat colour they are said to have different **genotypes**, i.e. different combinations of alleles. One genotype is **BB** and the other is **Bb**.

You and your brother might both be brown-eyed phenotypes but your genotype could be **BB** and his could be **Bb**. You would be homozygous dominant for brown eyes; he would be heterozygous for eye colour.

The three to one ratio

The result of a mating between a true-breeding (homozygous) black mouse (**BB**) and a true-breeding (homozygous) brown mouse (**bb**) is shown in Figure 17.23(a). The illustration is greatly simplified because it shows only one pair of the 20 pairs of mouse chromosomes and only one pair of alleles on the chromosomes.

Because black is dominant to brown, all the offspring from this mating will be black phenotypes, because they all receive the dominant allele for black fur from the father. Their genotypes, however, will be **Bb** because they all receive the recessive **b** allele from the mother. They are heterozygous for coat colour. The offspring resulting from this first mating are called the F_1 **generation**.

Figure 17.23(b) shows what happens when these heterozygous, F_1 black mice are mated together to produce what is called the F_2 generation. Each sperm or ovum produced by meiosis can contain only one of the alleles for coat colour, either **B** or **b**. So there are two kinds of sperm cell, one kind with the **B** allele and one kind with the **b** allele. There are also two kinds of ovum, with either **B** or **b** alleles. When fertilisation occurs, there is no way of telling whether a **b** or a **B** sperm will fertilise a **B** or a **b** ovum, so we have to look at all the possible combinations as follows:

- A **b** sperm fertilises a **B** ovum. Result: **bB** zygote.
- A **b** sperm fertilises a **b** ovum. Result: **bb** zygote.
- A **B** sperm fertilises a **B** ovum. Result: **BB** zygote.
- A **B** sperm fertilises a **b** ovum. Result: **Bb** zygote.

There is no difference between **bB** and **Bb**, so there are three possible genotypes in the offspring – **BB**, **Bb** and **bb**. There are only two phenotypes – black (**BB** or **Bb**) and brown (**bb**). So, according to the laws of chance, we would expect three black baby mice and one brown. Mice usually have more than four offspring and what we really expect is that the **ratio** (proportion) of black to brown will be close to 3:1.

If the mouse had 13 babies, you might expect nine black and four brown, or eight black and five brown. Even if she had 16 babies you would not expect to find exactly 12 black and four brown because whether a **B** or **b** sperm fertilises a **B** or **b** ovum is a matter of chance. If you spun ten coins, you would not expect to get exactly five heads and five tails. You would not be surprised at six heads and four tails or even seven heads and three tails. In the same way, we would not be surprised at 14 black and two brown mice in a litter of 16.

To decide whether there really is a 3:1 ratio, we need a lot of results. These may come either from breeding the same pair of mice together for a year or so to produce many litters, or from mating 20 black and 20 brown mice, crossing the offspring and adding up the number of black and brown babies in the F_2 families (see also Figure 17.24).

When working out the results of a genetic cross, it is useful to display the outcomes in a '**Punnett square**' (Figure 17.25). This a box divided into four compartments. The two boxes along the top are labelled with the genotypes of the gametes of one parent. The genotypes are circled to show they are gametes. The parent's genotype is written above the gametes. The boxes down the left-hand side are labelled with the genotypes of the gametes of the other parent. The parent's genotype is written to the left. The genotypes of the offspring can then be predicted by completing the four boxes, as shown. In this example, two heterozygous tall organisms (**Tt**) are the parents. The genotypes of the offspring are **TT**, **Tt**, **Tt** and **tt**. We know that the allele **T** is dominant because the parents are tall, although they carry both tall and dwarf alleles. So, the phenotypes of the offspring will be three tall to one dwarf.

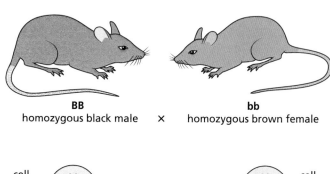

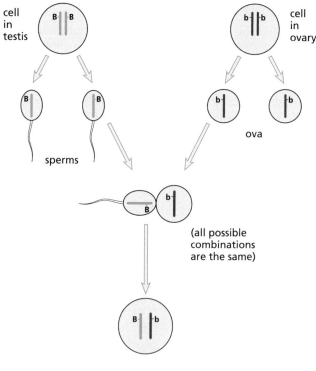

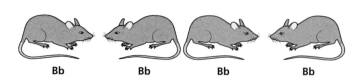

(a) all the F_1 generation are heterozygous black

Figure 17.23 Inheritance of coat colour in mice

Monohybrid inheritance

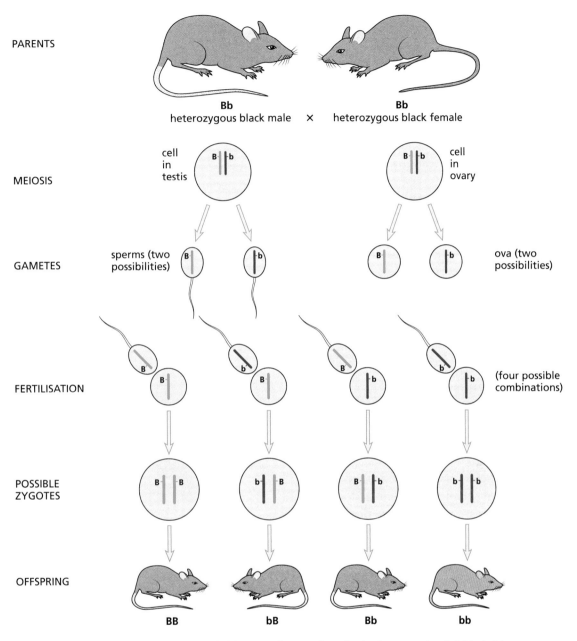

(b) the probable ratio of coat colours in the F₂ generation is 3 black : 1 brown

Figure 17.23 Inheritance of coat colour in mice *(continued)*

Figure 17.24 F$_2$ hybrids in maize. In the two left-hand cobs, the grain colour phenotypes appear in a 3:1 ratio (try counting single rows in the lighter cob). What was the colour of the parental grains for each of these cobs?

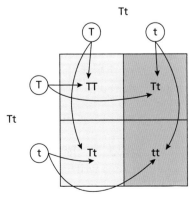

Figure 17.25 Using a Punnett square to predict the outcomes of a genetic cross

263

The recessive test-cross (back-cross)

A black mouse could have either the **BB** or the **Bb** genotype. One way to find out which is to cross the black mouse with a known homozygous recessive mouse, **bb**. The **bb** mouse will produce gametes with only the recessive **b** allele. A black homozygote, **BB**, will produce only **B** gametes. Thus, if the black mouse is **BB**, all the offspring from the cross will be black heterozygotes, **Bb**.

Half the gametes from a black **Bb** mouse would carry the **B** allele and half would have the **b** allele. So, if the black mouse is **Bb**, half of the offspring from the cross will, on average, be brown homozygotes, **bb**, and half will be black heterozygotes, **Bb**.

The term 'back-cross' refers to the fact that, in effect, the black, mystery mouse is being crossed with the same genotype as its brown grandparent, the **bb** mouse in Figure 17.23(a). Mouse ethics and speed of reproduction make the use of the actual grandparent quite feasible!

Co-dominance and incomplete dominance

Co-dominance

If both genes of an allelomorphic pair produce their effects in an individual (i.e. neither allele is dominant to the other) the alleles are said to be **co-dominant**.

The inheritance of the human ABO blood groups provides an example of co-dominance. In the ABO system, there are four phenotypic blood groups, A, B, AB and O. The alleles for groups A and B are co-dominant. If a person inherits alleles for group A and group B, his or her red cells will carry both antigen A and antigen B.

However, the alleles for groups A and B are both completely dominant to the allele for group O. (Group O people have neither A nor B antigens on their red cells.)

Table 17.2 shows the genotypes and phenotypes for the ABO blood groups. (Note that the allele for group O is sometimes represented as I^o and sometimes as i.)

Table 17.2 The ABO blood groups

Genotype	Blood group (phenotype)
I^AI^A or I^AI^o	A
I^BI^B or I^BI^o	B
I^AI^B	AB
I^oI^o	O

Since the alleles for groups A and B are dominant to that for group O, a group A person could have the genotype I^AI^A or I^AI^o. Similarly a group B person could be I^BI^B or I^BI^o. There are no alternative genotypes for groups AB and O.

Inheritance of blood group O

Blood group O can be inherited, even though neither parent shows this phenotype.

Two parents have the groups A and B. The father is I^AI^o and the mother is I^BI^o (Figure 17.26).

Phenotypes of parents	blood group A		blood group B
Genotypes of parents	I^AI^o	×	I^BI^o
Gametes	I^A I^o	×	I^B I^o

Punnett square:

	I^A	I^o
I^B	I^AI^B	I^BI^o
I^o	I^AI^o	I^oI^o

F_1 genotypes	I^AI^o	I^BI^o	I^AI^B	I^oI^o
F_1 phenotypes	A	B	AB	O
Ratio	1 :	1 :	1 :	1

Figure 17.26 Inheritance of blood group O

Some plants show co-dominance with regard to petal colour. For example, with the gene for flower colour in the geranium, the alleles are C^R (red) and C^W (white). The capital letter 'C' has been chosen to represent colour. Pure breeding (homozygous) flowers may be red (C^RC^R) or white (C^WC^W). If these are cross-pollinated, all the first filial (F_1) generation will be heterozygous (C^RC^W) and they are pink because both alleles have an effect on the phenotype.

Self-pollinating the pink (F_1) plants results in an unusual ratio in the next (F_2) generation of 1 red : 2 pink : 1 white.

Incomplete dominance

This term is sometimes taken to mean the same as 'co-dominance' but, strictly, it applies to a case where the effect of the recessive allele is not completely masked by the dominant allele.

An example occurs with sickle-cell anaemia (see 'Variation' in Chapter 18). If a person inherits both recessive alleles (Hb^SHb^S) for sickle-cell haemoglobin, then he or she will exhibit signs of the disease, i.e. distortion of the red cells leading to severe bouts of anaemia.

A heterozygote (Hb^AHb^S), however, will have a condition called 'sickle-cell trait'. Although there may be mild symptoms of anaemia the condition is not serious or life-threatening. In this case, the normal haemoglobin allele (Hb^A) is not completely dominant over the recessive (Hb^S) allele.

Sex linkage

> **Key definitions**
> A **sex-linked characteristic** is one in which the gene responsible is located on a sex chromosome, which makes it more common in one sex than the other.

The sex chromosomes, X and Y, carry genes that control sexual development. In addition they carry genes that control other characteristics. These tend to be on the X chromosome, which has longer arms to the chromatids. Even if the allele is recessive, because there is no corresponding allele on the Y chromosome, it is bound to be expressed in a male (XY). There is less chance of a recessive allele being expressed in a female (XX) because the other X chromosome may carry the dominant form of the allele.

One example of this is a form of colour blindness (Figure 17.27). In the following case, the mother is a carrier of colour blindness (X^CX^c). This means she shows no symptoms of colour blindness, but the recessive allele causing colour blindness is present on one of her X chromosomes. The father has normal colour vision (X^CY).

Phenotypes of parents	mother: normal vision	father: normal vision
Genotypes of parents	X^CX^c ×	X^CY
Gametes	X^C X^c ×	X^C Y

Punnett square:

	X^C	X^c
X^C	X^CX^C	X^CX^c
Y	X^CY	X^cY

F₁ genotypes	X^CX^C X^CX^c X^CY X^cY
F₁ phenotypes	2 females with normal vision; 2 males, one with normal vision, one with colour blindness

Figure 17.27 Inheritance of colour blindness

If the gene responsible for a particular condition is present only on the Y chromosome, only males can suffer from the condition because females do not possess the Y chromosome.

● Extension work

Ideas about heredity: Gregor Mendel (1822–84)

Mendel was an Augustinian monk from the town of Brünn (now Brno) in Czechoslovakia (now the Czech Republic). He studied maths and science at the University of Vienna in order to teach at a local school.

He was the first scientist to make a systematic study of patterns of inheritance involving single characteristics. This he did by using varieties of the pea plant, *Pisum sativum*, which he grew in the monastery garden. He chose pea plants because they were self-pollinating (Chapter 16). Pollen from the anthers reached the stigma of the same flower even before the flower bud opened.

Mendel selected varieties of pea plant that bore distinctive and contrasting characteristics, such as green seeds vs yellow seeds, dwarf vs tall, round seeds vs wrinkled (Figure 17.28). He used only plants that bred true.

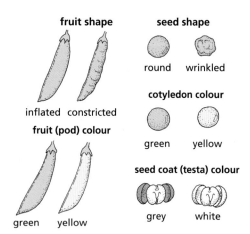

Figure 17.28 Some of the characteristics investigated by Mendel

He then crossed pairs of the contrasting varieties. To do this he had to open the flower buds, remove the stamens and use them to dust pollen on the stigmas of the contrasting variety. The offspring of this cross he called the 'first filial' generation, or F_1.

The first thing he noticed was that all the offspring of the F_1 cross showed the characteristic of only one of the parents. For example, tall plants crossed with dwarf plants produced only tall plants in the first generation.

Next he allowed the plants of the F_1 generation to self-pollinate and so produce a second filial generation, or F_2. Surprisingly, the dwarf characteristic that had, seemingly, disappeared in the F_1 reappeared in the F_2. This characteristic had not, in fact, been lost but merely concealed or suppressed in the F_1 to re-emerge in the F_2. Mendel called the repressed feature 'recessive' and the expressed feature 'dominant'.

Also, it must be noted, the plants were all either tall or dwarf; there were no intermediates, as might be expected if the characteristics blended.

Mendel noticed that pollen from tall plants, transferred to the stigmas of short plants, produced the same result as transferring pollen from short plants to the stigmas of tall plants. This meant that male and female gametes contributed equally to the observed characteristic.

When Mendel counted the number of contrasting offspring in the F_2, he found that they occurred in the ratio of three dominant to one recessive. For example, of 1064 F_2 plants from the tall × dwarf cross, 787 were tall and 277 dwarf, a ratio of 2.84:1. This F_2 ratio occurred in all Mendel's crosses, for example:

- round vs wrinkled seeds   5474:1850 = 2.96:1
- yellow vs green seeds   6022:2001 = 3.01:1
- green vs yellow pods   428:152 = 2.82:1

Two-thirds of the dominant tall F_2 plants did not breed true when self-pollinated but produced the 3:1 ratio of tall : dwarf. They were therefore similar to the plants of the F_1 generation.

It is not clear whether Mendel speculated on how the characteristics were represented in the gametes or how they achieved their effects. At one point he wrote of 'the differentiating elements of the egg and pollen cells', but it is questionable whether he envisaged actual structures being responsible.

Similarly, when Mendel wrote 'exactly similar factors must be at work', he meant that there must be similar processes taking place. He does not use the term 'factor' to imply particles or any entities that control heritable characteristics.

His symbols **A**, **Ab** and **b** seem to be shorthand for the types of plants he studied: **A** = true-breeding dominant, **b** = true-breeding recessive and **Ab** = the non-true-breeding 'hybrid'. The letters represented the visible characteristics, whereas today they represent the alleles responsible for producing the characteristic. For example, Mendel never refers to **AA** or **bb** so he probably did not appreciate that each characteristic is represented twice in the somatic cells but only once in the gametes.

When Mendel crossed plants, each carrying two contrasting characteristics, he found that the characteristics turned up in the offspring independently of each other. For example, in a cross between a tall plant with green seeds and a dwarf plant with yellow seeds, some of the offspring were tall with yellow seeds and some dwarf with green seeds.

So, Mendel's work was descriptive and mathematical rather than explanatory. He showed that certain characteristics were inherited in a predictable way, that the gametes were the vehicles, that these characteristics did not blend but retained their identity and could be inherited independently of each other. He also recognised dominant and recessive characteristics and, by 'hybridisation', that in the presence of the dominant characteristic the recessive characteristic, though not expressed, did not 'disappear'.

Mendel published his results in 1866 in '*Transactions of the Brünn Natural History Society*', which, understandably, did not have a wide circulation. Only when Mendel's work was rediscovered in 1900 was the importance and significance of his findings appreciated.

Mendel's observations are sometimes summarised in the form of 'Mendel's laws', but Mendel did not formulate any laws and these are the product of modern knowledge of genetics.

- The first 'law' (the law of segregation) is expressed as 'of a pair of contrasted characters only one can be represented in the gamete'.
- The second 'law' (the law of independent assortment) is given as 'each of a pair of contrasting characters may be combined with either of another pair'.

Questions

Core

1. A married couple has four girl children but no boys. This does not mean that the husband produces only X sperms. Explain why not.
2. Which sex chromosome determines the sex of a baby? Explain your answer.
3. Some plants occur in one of two sizes, tall or dwarf. This characteristic is controlled by one pair of genes. Tallness is dominant to shortness. Choose suitable letters for the gene pair.
4. Why are there two types of gene controlling one characteristic? Do the two types affect the characteristic in the same way as each other?
5. The allele for red hair is recessive to the allele for black hair. What colour hair will a person have if he inherits an allele for red hair from his mother and an allele for black hair from his father?
6. a Read Question 5 again. Choose letters for the alleles for red hair and black hair and write down the allele combination for having red hair.
   b Would you expect a red-haired couple to breed true?
   c Could a black-haired couple have a red-haired baby?
7. Use the words 'homozygous', 'heterozygous', 'dominant' and 'recessive' (where suitable) to describe the following allele combinations: **Aa**, **AA**, **aa**.
8. A plant has two varieties, one with red petals and one with white petals. When these two varieties are cross-pollinated, all the offspring have red petals. Which allele is dominant? Choose suitable letters to represent the two alleles.
9. Look at Figure 17.23(a). Why is there no possibility of getting a **BB** or a **bb** combination in the offspring?
10. In Figure 17.23(b) what proportion of the F_2 black mice are true-breeding?
11. Two black guinea-pigs are mated together on several occasions and their offspring are invariably black. However, when their black offspring are mated with white guinea-pigs, half of the matings result in all black litters and the other half produce litters containing equal numbers of black and white babies. From these results, deduce the genotypes of the parents and explain the results of the various matings, assuming that colour in this case is determined by a single pair of alleles.

Extended

12. How many bases will there be in an mRNA molecule coding for haemoglobin?

13. How many chromosomes would there be in the nucleus of:
    a a human muscle cell
    b a mouse kidney cell
    c a human skin cell that has just been produced by mitosis
    d a kangaroo sperm cell?
14. What is the diploid number in humans?
15. Suggest why sperm could be described as *male sperm* and *female sperm*.
16. a What are gametes?
    b What are the male and female gametes of
       i plants and
       ii animals called, and where are they produced?
    c What happens at fertilisation?
    d What is a zygote and what does it develop into?
17. How many chromatids will there be in the nucleus of a human cell just before cell division?
18. Why can chromosomes not be seen when a cell is not dividing?
19. In which human tissues would you expect mitosis to be going on, in:
    a a 5-year-old child
    b an adult?
20. What is the haploid number for:
    a a human
    b a fruit fly?
21. Which of the following cells would be haploid and which diploid: white blood cell, male cell in pollen grain, guard cell, root hair, ovum, sperm, skin cell, egg cell in ovule?
22. Where in the body of the following organisms would you expect meiosis to be taking place:
    a a human male
    b a human female
    c a flowering plant
23. How many chromosomes would be present in:
    a a mouse sperm cell
    b a mouse ovum?
24. Why are organisms that are produced by asexual reproduction identical to each other?
25. Two black rabbits thought to be homozygous for coat colour were mated and produced a litter that contained all black babies. The F_2, however, resulted in some white babies, which meant that one of the grandparents was heterozygous for coat colour. How would you find out which grandparent was heterozygous?
26. What combinations of blood groups can result in a child being born with blood group O? Use Punnett squares to show your reasoning.

27 A woman of blood group A claims that a man of blood group AB is the father of her child. A blood test reveals that the child's blood group is O.
   a Is it possible that the woman's claim is correct?
   b Could the father have been a group B man? Explain your reasoning.
28 A red cow has a pair of alleles for red hairs. A white bull has a pair of alleles for white hairs. If a red cow and a white bull are mated, the offspring are all 'roan', i.e. they have red and white hairs equally distributed over their body.
   a Is this an example of co-dominance or incomplete dominance?
   b What coat colours would you expect among the offspring of a mating between two roan cattle?
29 Predict the ratio of children with colour blindness resulting from a mother who is a carrier for colour blindness having children with a father who is colour blind.

Checklist

After studying Chapter 17 you should know and understand the following:

- Inheritance is the transmission of genetic information from generation to generation.

Chromosomes, genes and proteins

- A chromosome is a thread of DNA, made up of a string of genes.
- A gene is a length of DNA that codes for a protein.
- An allele is a version of a gene.
- Chromosomes are found as thread-like structures in the nuclei of all cells.
- Chromosomes are in pairs; one of each pair comes from the male and one from the female parent.
- Sex, in mammals, is determined by the X and Y chromosomes. Males are XY; females are XX.

- The DNA molecule is coiled along the length of the chromosome.
- A DNA molecule is made up of a double chain of nucleotides in the form of a helix.
- The nucleotide bases in the helix pair up A–T and C–G.
- Triplets of bases control production of the specific amino acids that make up a protein.
- Genes consist of specific lengths of DNA.
- Most genes control the type of enzyme that a cell will make.
- When proteins are made:
  - the DNA with the genetic code for the protein remains in the nucleus
  - mRNA molecules carry a copy of the genetic code to the cytoplasm
  - the mRNA passes through ribosomes in the cytoplasm and the ribosome puts together amino acids to form protein molecules.
- The specific order of amino acids is decided by the sequence of bases in the mRNA.
- All body cells in an organism contain the same genes, but many genes in a particular cell are not expressed because the cell only makes the specific proteins it needs.
- A haploid nucleus is a nucleus containing a single set of unpaired chromosomes (e.g. in sperm and egg cells).
- A diploid nucleus is a nucleus containing two sets of chromosomes (e.g. in body cells).
- In a diploid cell, there is a pair of each type of chromosome; in a human diploid cell there are 23 pairs.

Mitosis

- Mitosis is nuclear division giving rise to genetically identical cells.
- Mitosis is important in growth, repair of damaged tissues, replacement of cells and in asexual reproduction.

- Before mitosis, the exact duplication of chromosomes occurs.
- Each species of plant or animal has a fixed number of chromosomes in its cells.
- When cells divide by mitosis, the chromosomes and genes are copied exactly and each new cell gets a full set.
- Stem cells are unspecialised cells that divide by mitosis to produce daughter cells that can become specialised for specific purposes.

Meiosis

- Meiosis is reduction division in which the chromosome number is halved from diploid to haploid resulting in genetically different cells.
- Gametes are the result of meiosis.
- At meiosis, only one chromosome of each pair goes into the gamete.
- Meiosis produces variation by forming new combinations of maternal and paternal chromosomes.

Monohybrid inheritance

- The genotype of an organism is its genetic make-up.
- The phenotype of an organism is its features.
- Homozygous means having two identical alleles of a particular gene. Two identical homozygous individuals that breed together will be pure-breeding.
- Heterozygous means having two different alleles of a particular gene. A heterozygous individual will therefore not be pure-breeding.
- A dominant allele is one that is expressed if it is present.

- A recessive allele is one that is only expressed when there is no dominant allele of the gene present.
- Genetic diagrams are used to predict the results of monohybrid crosses and calculate phenotypic ratios.
- Punnett squares can be used in crosses to work out and show the possible different genotypes.
- A test-cross is used to identify an unknown genotype, for instance to find out if it is pure breeding or heterozygous.
- In some cases, neither one of a pair of alleles is fully dominant over the other. This is called co-dominance.
- The inheritance of ABO blood groups is an example of co-dominance.
- The phenotypes are A, B, AB and O blood groups.
- The genotypes are I^A, I^B and I^O.
- A sex-linked characteristic is a characteristic in which the gene responsible is located on a sex chromosome. This makes it more common in one sex than in the other.
- Colour blindness is an example of sex linkage.
- Genetic diagrams can be used to predict the results of monohybrid crosses involving co-dominance and sex linkage.

18 Variation and selection

Variation
Define variation
Discontinuous and continuous variation
Define mutation
Causes of mutations

Causes of discontinuous and continuous variation
Define gene mutation
Sickle-cell anaemia
Down's syndrome
Mutations in bacteria
Adaptive features
Define adaptive feature
Describe adaptive features of organisms

Define adaptive feature, fitness
Adaptive features of hydrophytes and xerophytes

Selection
Natural selection
Artificial selection
Selective breeding

Define the process of adaptation
Evolution
Development of strains of resistant bacteria
Use of selective breeding
Compare natural and artificial selection

● Variation

Key definition
Variation is the differences between individuals of the same species.

The term '**variation**' refers to observable differences within a species. All domestic cats belong to the same species, i.e. they can all interbreed, but there are many variations of size, coat colour, eye colour, fur length, etc. Those variations that can be inherited are determined by genes. They are **genetic variations**. **Phenotypic variations** may be brought about by genes, but can also be caused by the environment, or a combination of both genes and the environment.

So, there are variations that are not heritable, but determined by factors in the environment. A kitten that gets insufficient food will not grow to the same size as its litter mates. A cat with a skin disease may have bald patches in its coat. These conditions are not heritable. They are caused by environmental effects. Similarly, a fair-skinned person may be able to change the colour of his or her skin by exposing it to the Sun, so getting a tan. The tan is an **acquired characteristic**. You cannot inherit a suntan. Black skin, on the other hand, is an **inherited characteristic**.

Many features in plants and animals are a mixture of acquired and inherited characteristics (Figure 18.1). For example, some fair-skinned people never go brown in the Sun, they only become sunburned. They have not inherited the genes for producing the extra brown pigment in their skin. A fair-skinned person with the genes for producing pigment will only go brown if he or she exposes themselves to sunlight. So the tan is a result of both inherited and acquired characteristics.

Figure 18.1 Acquired characteristics. These apples have all been picked from different parts of the same tree. All the apples have similar genotypes, so the differences in size must have been caused by environmental effects.

Discontinuous variation

In **discontinuous variation**, the variations take the form of distinct, alternative phenotypes with no intermediates (Figures 18.2 and 18.4). The mice in Figure 17.23 are either black or brown; there are no intermediates. You are either male or female. Apart from a small number of abnormalities, sex is inherited in a discontinuous way. Some people can roll their tongue into a tube. Others are unable to do it. They are known as non-tongue rollers. Again, there are no intermediates (Figure 18.2).

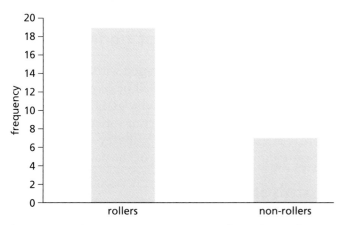

Figure 18.2 Discontinuous variation. Tongue rollers and non-rollers in a class

Discontinuous variation cannot usually be altered by the environment. You cannot change your eye colour by altering your diet. A genetic dwarf cannot grow taller by eating more food. You cannot learn how to roll your tongue.

Continuous variation

An example of **continuous variation** is height. There are no distinct categories of height; people are not either tall or short. There are all possible intermediates between very short and very tall (Figure 18.3).

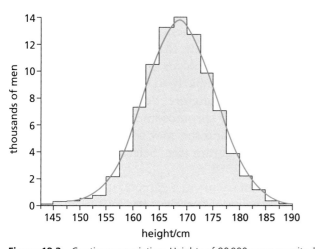

Figure 18.3 Continuous variation. Heights of 90 000 army recruits. The apparent 'steps' in the distribution are the result of arbitrarily chosen categories, differing in height by 1 cm. But heights do not differ by exactly 1 cm. If measurements could be made accurately to the nearest millimetre there would be a smooth curve like the one shown in colour.

There are many characteristics that are difficult to classify as either wholly continuous or discontinuous variations. Human eye colour has already been mentioned. People can be classified roughly as having blue eyes or brown eyes, but there are also categories described as grey, hazel or green. It is likely that there are a small number of genes for eye colour and a dominant gene for brown eyes, which overrides all the others when it is present. Similarly, red hair is a discontinuous variation but it is masked by genes for other colours and there is a continuous range of hair colour from blond to black.

Mutations

> **Key definition**
> A **mutation** is a spontaneous genetic change. Mutation is the way new alleles are formed.

Many of the cat coat variations mentioned overleaf may have arisen, in the first place, as mutations in a wild stock of cats. A recent variant produced by a mutation is the 'rex' variety, in which the coat has curly hairs.

Many of our high-yielding crop plants have arisen as a result of mutations in which the whole chromosome set has been doubled.

Exposure to **mutagens**, namely certain chemicals and radiation, is known to increase the rate of mutation. Some of the substances in tobacco smoke, such as tar, are mutagens, which can cause cancer.

Ionising radiation from X-rays and radioactive compounds, and ultraviolet radiation from sunlight, can both increase the mutation rate. It is uncertain whether there is a minimum dose of radiation below which there is negligible risk. It is possible that repeated exposure to low doses of radiation is as harmful as one exposure to a high dose. It has become clear in recent years that, in light-skinned people, unprotected exposure to ultraviolet radiation from the Sun can cause a form of skin cancer.

Generally speaking, however, exposure to natural and medical sources of radiation carries less risk than smoking cigarettes or driving a car, but it is sensible to keep exposure to a minimum.

18 VARIATION AND SELECTION

Genetic variation may be the result of new combinations of genes in the zygote, or mutations.

Discontinuous variation

Discontinuous variation is under the control of a single pair of alleles or a small number of genes. An example is human blood groups. These were discussed in Chapter 17.

A person is one of four blood groups: A, B, AB or O. There are no groups in between.

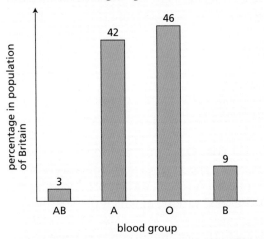

Figure 18.4 Discontinuous variation. Frequencies of ABO blood groups in Britain. The figures could not be adjusted to fit a smooth curve because there are no intermediates.

Continuous variation

Continuous variation is influenced by a combination of both genetic and environmental factors. Continuously variable characteristics are usually controlled by several pairs of alleles. There might be five pairs of alleles for height – (**Hh**), (**Tt**), (**Ll**), (**Ee**) and (**Gg**) – each dominant allele adding 4 cm to your height. If you inherited all ten dominant genes (**HH, TT**, etc.) you could be 40 cm taller than a person who inherited all ten recessive genes (**hh, tt**, etc.).

The actual number of genes that control height, intelligence, and even the colour of hair and skin, is not known.

Continuously variable characteristics are greatly influenced by the environment. A person may inherit genes for tallness and yet not get enough food to grow tall. A plant may have the genes for large fruits but not get enough water, minerals or sunlight to produce large fruits. Continuous variations in human populations, such as height, physique and intelligence, are always the result of interaction between the genotype and the environment.

New combinations of genes

If a grey cat with long fur is mated with a black cat with short fur, the kittens will all be black with short fur. If these offspring are mated together, in due course the litters may include four varieties: black–short, black–long, grey–short and grey–long. Two of these are different from either of the parents.

Mutation

> **Key definition**
> A **gene mutation** is a change in the base sequence in DNA.

A mutation may occur in a gene or a chromosome. In a gene mutation it may be that one or more genes are not replicated correctly. A chromosome mutation may result from damage to or loss of part of a chromosome during mitosis or meiosis, or even the gain of an extra chromosome, as in Down's syndrome (see page 273).

An abrupt change in a gene or chromosome is likely to result in a defective enzyme and will usually disrupt the complex reactions in the cells. Most mutations, therefore, are harmful to the organism.

Surprisingly, only about 3% of human DNA consists of genes. The rest consists of repeated sequences of nucleotides that do not code for proteins. This is sometimes called '**junk DNA**', but that term only means that we do not know its function. If mutations occur in these non-coding sequences they are unlikely to have any effect on the organism and are, therefore, described as 'neutral'.

Rarely, a gene or chromosome mutation produces a beneficial effect and this may contribute to the success of the organism (see 'Selection' later in this chapter).

If a mutation occurs in a gamete, it will affect all the cells of the individual that develops from the zygote. Thus the whole organism will be affected. If the mutation occurs in a somatic cell (body cell), it will affect only those cells produced, by mitosis, from the affected cell.

Thus, a mutation in a gamete may result in a genetic disorder, e.g. haemophilia or cystic fibrosis. Mutations in somatic cells may give rise to cancers by promoting uncontrolled cell division in the affected tissue. For example, skin cancer results from uncontrolled cell division in the basal layer of the skin.

A mutation may be as small as the substitution of one organic base for another in the DNA molecule, or as large as the breakage, loss or gain of a chromosome.

Sickle-cell anaemia

This condition has already been mentioned in Chapter 17. A person with sickle-cell disease has inherited both recessive alleles (**Hb^SHb^S**) for defective haemoglobin. The distortion and destruction of the red cells, which occurs in low oxygen concentrations, leads to bouts of severe anaemia (Figure 18.5). In many African countries, sufferers have a reduced chance of reaching reproductive age and having a family. There is thus a selection pressure, which tends to remove the homozygous recessives from the population. In such a case, you might expect the harmful **Hb^S** allele to be selected out of the population altogether. However, the heterozygotes (**Hb^AHb^S**) have virtually no symptoms of anaemia but do have the advantage that they are more resistant to malaria than the homozygotes **Hb^AHb^A**. It appears that the malaria parasite is unable to invade and reproduce in the sickle cells.

The selection pressure of malaria, therefore, favours the heterozygotes over the homozygotes and the potentially harmful **Hb^S** allele is kept in the population (Figure 18.6).

When Africans migrate to countries where malaria does not occur, the selective advantage of the **Hb^S** allele is lost and the frequency of this allele in the population diminishes.

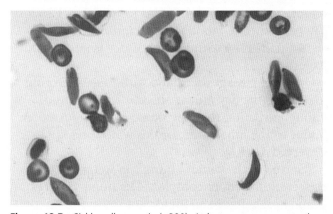

Figure 18.5 Sickle-cell anaemia (×800). At low oxygen concentration the red cells become distorted.

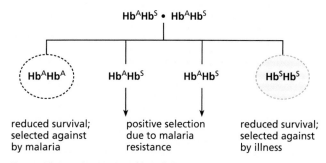

Figure 18.6 Selection in sickle-cell disease

With **sickle-cell anaemia**, the defective haemoglobin molecule differs from normal haemoglobin by only one amino acid (represented by a sequence of three bases), i.e. *valine* replaces *glutamic acid*. This could be the result of faulty replication at meiosis. When the relevant parental chromosome replicated at gamete formation, the DNA could have produced the triplet –CAT– (which specifies *valine*) instead of –CTT– (which specifies *glutamic acid*). In this case, a change of just one base (from A to T) makes a significant difference to the characteristics of the protein (haemoglobin).

Down's syndrome

Down's syndrome is a form of mental and physical disability, which results from a chromosome mutation. During the process of meiosis which produces an ovum, one of the chromosomes (chromosome 21) fails to separate from its homologous partner, a process known as **non-disjunction**. As a result, the ovum carries 24 chromosomes instead of 23, and the resulting zygote has 47 instead of the normal 46 chromosomes. The risk of having a baby with Down's syndrome increases as the mother gets older.

Mutations in bacteria

Mutations in bacteria often produce resistance to drugs. Bacterial cells reproduce very rapidly, perhaps as often as once every 20 minutes. Thus a mutation, even if it occurs only rarely, is likely to appear in a large population of bacteria. If a population of bacteria containing one or two drug-resistant mutants is subjected to that particular drug, the non-resistant bacteria will be killed but the drug-resistant mutants survive (see Figure 15.1). Mutant genes are inherited in the same way as normal genes, so when the surviving mutant bacteria reproduce, all their offspring will be resistant to the drug.

Mutations are comparatively rare events; perhaps only one in every 100 000 replications results in a mutation. Nevertheless they do occur naturally all the time.

18 VARIATION AND SELECTION

● Adaptive features

> **Key definition**
> An **adaptive feature** is an inherited feature that helps an organism to survive and reproduce in its environment.

Adaptation

When biologists say that a plant or animal is *adapted* to its habitat they usually mean that, in the course of evolution, changes have occurred in the organism, which make it more successful in exploiting its habitat, e.g. animals finding and digesting food, selecting nest sites or hiding places, or plants exploiting limited mineral resources or tolerating salinity or drought. It is tempting to assume that because we find a plant or animal in a particular habitat it must be adapted to its habitat. There is some logic in this; if an organism was not adapted to its habitat, presumably it would be eliminated by natural selection. However, it is best to look for positive evidence of **adaptation**.

Sometimes, just by looking at an organism and comparing it with related species, it is possible to make reasoned guesses about adaptation. For example, there seems little doubt that the long, hair-fringed hind legs of a water beetle are adaptations to locomotion in water when compared with the corresponding legs of a land-living relative (Figure 18.7).

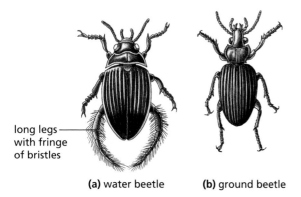

Figure 18.7 Adaptation to locomotion in water and on land

Similarly, in Figure 18.8 it seems reasonable to suppose that, compared with the generalised mammalian limb, the forelimbs of whales are adapted for locomotion in water.

By studying animals which live in extreme habitats, it is possible to suggest ways in which they might be adapted to these habitats especially if the observations are supported by physiological evidence.

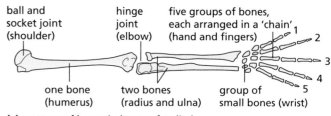

(a) pattern of bones in human forelimb

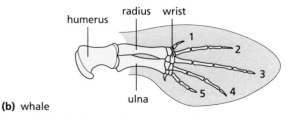

(b) whale

Figure 18.8 Skeletons of the forelimbs of human and whale

The camel

Camels are adapted to survive in a hot, dry and sandy environment. Adaptive physical features are closable nostrils and long eyelashes, which help keep out wind-blown sand (Figure 18.9). Their feet are broad and splay out under pressure, so reducing the tendency to sink into the sand. Thick fur insulates the body against heat gain in the intense sunlight.

Physiologically, a camel is able to survive without water for 6–8 days. Its stomach has a large water-holding capacity, though it drinks to replace water lost by evaporation rather than in anticipation of water deprivation.

The body temperature of a 'thirsty' camel rises to as much as 40 °C during the day and falls to about 35 °C at night. The elevated daytime temperature reduces the heat gradient between the body and the surroundings, so less heat is absorbed. A camel is able to tolerate water loss equivalent to 25% of its body weight, compared with humans for whom a 12% loss may be fatal. The blood volume and concentration are maintained by withdrawing water from the body tissues.

The nasal passages are lined with mucus. During exhalation, the dry mucus absorbs water vapour. During inhalation the now moist mucus adds water vapour to the inhaled air. In this way, water is conserved.

The role of the camel's humps in water conservation is more complex. The humps contain fat and are therefore an important reserve of energy-giving food. However, when the fat is metabolised during respiration, carbon dioxide and water

Adaptive features

Figure 18.9 Protection against wind-blown sand. The nostrils are slit-like and can be closed. The long eyelashes protect the eyes

Figure 18.11 The heavy coat and small ears also help the polar bear to reduce heat losses.

(metabolic water) are produced. The water enters the blood circulation and would normally be lost by evaporation from the lungs, but the water-conserving nasal mucus will trap at least a proportion of it.

The polar bear

Polar bears live in the Arctic, spending much of their time on snow and ice. Several physical features contribute to their adaptation to this cold environment.

It is a very large bear (Figure 18.10), which means that the ratio of its surface area to its volume is relatively small. The relatively small surface area means that the polar bear loses proportionately less heat than its more southerly relatives. Also its ears are small, another feature that reduces heat loss (Figure 18.11).

It has a thick coat with long, loosely packed coarse hairs (guard hairs) and a denser layer of shorter woolly hairs forming an insulating layer. The long hairs are oily and water-repellent and enable the bear to shake off water when it emerges from a spell of swimming.

Figure 18.10 The polar bear and the sun bear (from SE Asia). The smaller surface area/volume ratio in the polar bear helps conserve heat.

The principal thermal insulation comes from a 10 cm layer of fat (blubber) beneath the skin. The thermal conductivity of fat is little different from any other tissue but it has a limited blood supply. This means that very little warm blood circulates close to the skin surface.

The hollow hairs of the white fur are thought to transmit the Sun's heat to the black skin below. Black is an efficient colour for absorbing heat. The white colour is also probably an effective camouflage when hunting its prey, mainly seals.

A specific adaptation to walking on snow and ice is the heat-exchange arrangement in the limbs. The arteries supplying the feet run very close to the veins returning blood to the heart. Heat from the arteries is transferred to the veins before the blood reaches the feet (Figure 18.12). So, little heat is lost from the feet but their temperature is maintained above freezing point, preventing frost-bite.

Polar bears breed in winter when temperatures fall well below zero. However, the pregnant female excavates a den in the snow in which to give birth and rear her two cubs. In this way the cubs are protected from the extreme cold.

The female remains in the den for about 140 days, suckling her young on the rich milk, which is formed from her fat reserves.

Venus flytrap

Many plants show adaptions as well as animals. Insectivorous plants such as the Venus flytrap (Figure 18.13) live in habitats where there is often a shortage of nitrates for growth. They have developed pairs of leaves with tooth-like edges. The leaves have

18 VARIATION AND SELECTION

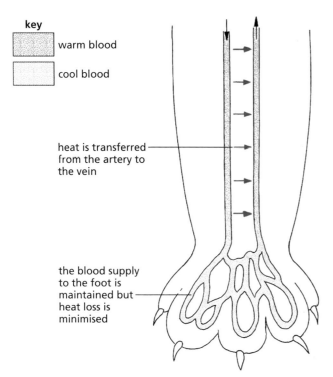

Figure 18.12 The heat-exchange mechanism in the polar bear's limb

sensitive hairs on their surface. When an insect walks inside the leaves, the hairs are triggered, causing the leaves to close very rapidly – trapping the animal. The leaves then secrete protease enzymes, which digest the insect's protein and produce soluble amino acids. These are absorbed by the leaf and used to build new proteins. It is unusual for a photosynthetic plant to show such rapid movement or to gain nourishment other than by photosynthesis.

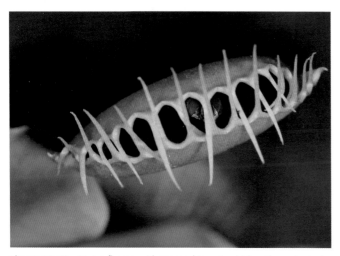

Figure 18.13 Venus flytrap with trapped insect, which will eventually be digested

Other adaptations
Adaptive features of the long-eared bat and the hare are illustrated in Figures 18.14 and 18.15.

Figure 18.14 Long-eared bat. The bat gives out high-pitched sounds, which are reflected back from its prey and from obstacles, to its ears and sensitive patches on its face. By timing these echoes the bat can judge its distance from the obstacle or prey. This allows it to fly and feed in the dark. Its body is covered in fur for insulation. Its forearms are covered by a membrane of skin to form a wing. The fingers are very long to stretch out the membrane to increase the surface area of the wing.

Figure 18.15 Hare. This animal is a herbivore and is hunted by predators such as foxes. Its fur is a good insulator and its colour provides excellent camouflage. The long ears help to pick up and locate sound vibrations. The eyes at the side of the head give the hare good all around vision. The hind legs are very long to enable the animal to run away from predators and its kick is a good defence mechanism. Some species of hare change the colour of their fur in winter from brown to white to provide better camouflage in snow.

Adaptive features

> **Key definitions**
> **Adaptive features** are the inherited functional features of an organism that increase its fitness.
> **Fitness** is the probability of that organism surviving and reproducing in the environment in which it is found.

Adaptations to arid conditions

In both hot and cold climates, plants may suffer from water shortage. High temperatures accelerate evaporation from leaves. At very low temperatures the soil water becomes frozen and therefore unavailable to the roots of plants. Plants modified to cope with lack of water are called **xerophytes**.

It is thought that the autumn leaf-fall of deciduous trees and shrubs is an essential adaptation to winter 'drought'. Loss of leaves removes virtually all evaporating surfaces at a time when water may become unavailable. Without leaves, however, the plants cannot make food by photosynthesis and so they enter a dormant condition in which metabolic activity is at a low level.

Pine tree

The pine tree (*Pinus*) (Figure 18.16) is an evergreen tree that survives in cold climates. It has small, compact, needle-like leaves. The small surface area of such leaves offers little resistance to high winds. This helps to resist wind damage and can reduce the amount of water lost in transpiration. However, photosynthesis can continue whenever water is available. Sunken stomata create high humidity and reduce transpiration. A thick waxy cuticle is present on the epidermis to prevent evaporation from the surface of the leaf.

Figure 18.16 Pine leaves, reduced to needles to lower the rate of transpiration

Some plants live in very sandy soil, which does not retain moisture well. Often this is combined with very low rainfall, making access to water difficult. Only plants with special adaptations, such as desert and sand dune species, can survive.

Cacti

Cacti are adapted to hot, dry conditions in several ways. Often they have no leaves, or the leaves are reduced to spines. This reduces the surface area for transpiration and also acts as a defence against herbivores. Photosynthesis is carried out by a thick green stem, which offers only a small surface area for evaporation. Cacti are succulent, i.e. they store water in their fleshy tissues and draw on this store for photosynthesis (Figure 18.17).

Figure 18.17 A cactus (succulent) growing in desert conditions in Arizona

The stomata of many cacti are closed during the day when temperatures are high, and open at night when evaporation is at a minimum. This strategy requires a slightly different form of photosynthesis. At night, carbon dioxide diffuses in through the open stomata and is 'fixed' (i.e. incorporated) into an organic acid. Little water vapour is lost at night. In the daytime the stomata are closed but the organic acid breaks down to yield carbon dioxide, which is then built into sugars by photosynthesis. Closure of the stomata in the daytime greatly reduces water loss.

18 VARIATION AND SELECTION

Marram grass

Marram grass (*Ammophila*) lives on sand dunes (Figure 18.18), where water drains away very quickly. It has very long roots to search for water deep down in the sand. Its leaves roll up into straw-like tubes in dry weather due to the presence of hinge cells, which become flaccid as they lose water (Figure 18.19). Leaf rolling, along with the fact that the stomata are sunken, helps to increase humidity around the stomata, reducing transpiration. The presence of fine hairs around the stomata reduces air movement so humidity builds up and transpiration is reduced.

Figure 18.18 Marram grass growing on a sand dune

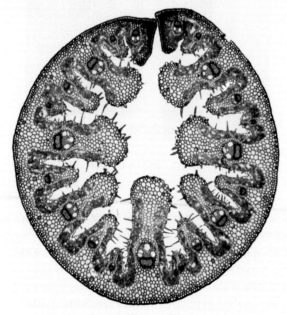

Figure 18.19 Transverse section of rolled up Marram grass leaf

Adaptations to living in water

Plants adapted to living in water are called **hydrophytes**. An example is the water lily (*Nymphaea*) (Figure 18.20). The leaves contain large air spaces to make them buoyant, so they float on or near the surface (Figure 18.21). This enables them to gain light for photosynthesis. The lower epidermis lacks stomata to prevent water entering the air spaces, while stomata are present on the upper epidermis for gas exchange. With land plants, most stomata are usually on the lower epidermis.

The roots of hydrophytes, which can be poorly developed, also contain air spaces. This is because the mud they grow in is poorly oxygenated and the root cells need oxygen for respiration. Stems lack much support as the water they are surrounded by provides buoyancy for the plant.

Figure 18.20 Water lily (*Nymphaea*)

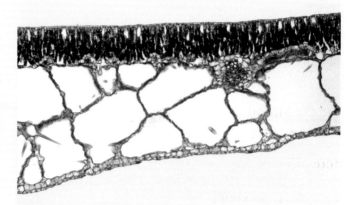

Figure 18.21 Section through water lily leaf

Selection

Natural selection

Theories of evolution have been put forward in various forms for hundreds of years. In 1858, Charles Darwin and Alfred Russel Wallace published a theory of evolution by natural selection, which is still an acceptable theory today.

The theory of evolution by natural selection is as follows:

- Individuals within a species are all slightly different from each other (Figure 18.22). These differences are called variations.
- If the climate or food supply changes, individuals possessing some of these variations may be better able to survive than others. For example, a variety of animal that could eat the leaves of shrubs as well as grass would be more likely to survive a drought than one that fed only on grass.
- If one variety lives longer than others, it is also likely to leave behind more offspring. A mouse that lives for 12 months may have ten litters of five babies (50 in all). A mouse that lives for 6 months may have only five litters of five babies (25 in all).
- If some of the offspring inherit alleles responsible for the variation that helped the parent survive better, they too will live longer and have more offspring.
- In time, this particular variety will outnumber and finally replace the original variety.

This is sometimes called 'the survival of the fittest'. However, 'fitness', in this case, does not mean good health but implies that the organism is well fitted to the conditions in which it lives.

Thomas Malthus, in 1798, suggested that the increase in the size of the human population would outstrip the rate of food production. He predicted that the number of people would eventually be regulated by famine, disease and war. When Darwin read the Malthus essay, he applied its principles to other populations of living organisms.

He observed that animals and plants produce vastly more offspring than can possibly survive to maturity and he reasoned that, therefore, there must be a 'struggle for survival'.

For example, if a pair of rabbits had eight offspring that grew up and formed four pairs, eventually having eight offspring per pair, in four generations the number of rabbits stemming from the original pair would be 512 (i.e. $2 \rightarrow 8 \rightarrow 32 \rightarrow 128 \rightarrow 512$). The population of rabbits, however, remains more or less constant. Many of the offspring in each generation must, therefore, have failed to survive to reproductive age.

Figure 18.22 Variation. The garden tiger moths in this picture are all from the same family. There is a lot of variation in the pattern on the wings.

Competition and selection

There will be **competition** between members of the rabbit population for food, burrows and mates. If food is scarce, space is short and the number of potential mates limited, then only the healthiest, most vigorous, most fertile and otherwise well-adapted rabbits will survive and breed.

The competition does not necessarily involve direct conflict. The best adapted rabbits may be able to run faster from predators, digest their food more efficiently, have larger litters or grow coats that camouflage them better or more effectively reduce heat losses. These rabbits will survive longer and leave more offspring. If the offspring inherit the advantageous characteristics of their parents, they may give rise to a new race of faster, different coloured,

18 VARIATION AND SELECTION

thicker furred and more fertile rabbits, which gradually replace the original, less well-adapted varieties. The new variations are said to have **survival value**.

This is natural selection; the better adapted varieties are 'selected' by the pressures of the environment (**selection pressures**).

For natural selection to be effective, the variations have to be heritable. Variations that are not heritable are of no value in natural selection. Training may give athletes more efficient muscles, but this characteristic will not be passed on to their children.

The peppered moth

A possible example of natural selection is provided by a species of moth called the peppered moth, found in Great Britain. The common form is speckled but there is also a variety that is black. The black variety was rare in 1850, but by 1895 in the Manchester area of England its numbers had risen to 98% of the population of peppered moths. Observation showed that the light variety was concealed better than the dark variety when they rested on tree-trunks covered with lichens (Figure 18.23). In the Manchester area of England, pollution had caused the death of the lichens and the darkening of the tree-trunks with soot. In this industrial area the dark variety was the better camouflaged (hidden) of the two and was not picked off so often by birds. So the dark variety survived better, left more offspring and nearly replaced the light form.

The selection pressure, in this case, was presumed to be mainly predation by birds. The adaptive variation that produced the selective advantage was the dark colour.

Although this is an attractive and plausible hypothesis of how natural selection could occur, some of the evidence does not support the hypothesis or has been called into question.

For example, the moths settle most frequently on the underside of branches rather than conspicuously on tree trunks, as in Figure 18.23. Also, in several unpolluted areas the dark form is quite abundant, for example 80% in East Anglia in England. Research is continuing in order to test the hypothesis.

Selective breeding

The process of selective breeding involves humans selecting individuals with desirable features. These individuals are then cross-bred to produce the next generation. Offspring with the most desirable features are chosen to continue the breeding programme and the process is repeated over a number of generations.

Human communities practise this form of selection when they breed plants and animals for specific characteristics. The many varieties of cat that you see today have been produced by selecting individuals with pointed ears, particular fur colour or length, or even no tail, etc. One of the kittens in a litter might vary from the others by having distinctly pointed ears. This individual, when mature, is allowed to breed. From the offspring, another very pointed-eared variant is selected for the next breeding stock, and so on, until the desired or 'fashionable' ear shape is established in a true-breeding population (Figure 18.24).

More important are the breeding programmes to improve agricultural livestock or crop plants. Animal-breeders will select cows for their high milk yield and

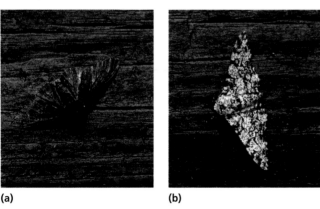

(a)   (b)   (c)   (d)

Figure 18.23 Selection for varieties of the peppered moth

sheep for their wool quality. Plant-breeders will select varieties for their high yield and resistance to fungus diseases (Figure 18.25).

Figure 18.24 Selective breeding. The Siamese cat, produced by artificial selection over many years

Figure 18.25 Selective breeding in tomatoes. Different breeding programmes have selected genes for fruit size, colour and shape. Similar processes have given rise to most of our cultivated plants and domesticated animals.

Evolution

> **Key definitions**
> **Adaptation** is the process, resulting from natural selection, by which populations become more suited to their environment over many generations.
> **Evolution** can be described as the change in adaptive features of a population over time as a result of natural selection.

Most biologists believe that natural selection, among other processes, contributes to the evolution of new species and that the great variety of living organisms on the Earth is the product of millions of years of evolution involving natural selection.

Antibiotic-resistant bacteria

Antibiotics are drugs used to treat infections caused by bacteria (see 'Medicinal drugs' in Chapter 15). Bacterial cells reproduce very rapidly, perhaps as often as once every 20 minutes. Thus a mutation, even if it occurs only rarely, is likely to appear in a large population of bacteria. If a population of bacteria containing one or two drug-resistant mutants is subjected to that particular drug, the non-resistant bacteria will be killed but the drug-resistant mutants survive (Figure 15.1). Mutant genes are inherited in the same way as normal genes, so when the surviving mutant bacteria reproduce, all their offspring will be resistant to the drug.

Selective breeding

An important part of any breeding programme is the selection of the desired varieties. The largest fruit on a tomato plant might be picked and its seeds planted next year. In the next generation, once again only seeds from the largest tomatoes are planted. Eventually it is possible to produce a true-breeding variety of tomato plant that forms large fruits. Figure 18.25 shows the result of such selective breeding. The same technique can be used for selecting other desirable qualities, such as flavour and disease resistance.

Similar principles can be applied to farm animals. Desirable characteristics, such as high milk yield and resistance to disease, may be combined. Stock-breeders will select calves from cows that give large quantities of milk. These calves will be used as breeding stock to build a herd of high yielders. A characteristic such as milk yield is probably under the control of many genes. At each stage of selective breeding the farmer, in effect, is keeping the beneficial genes and discarding the less useful genes from his or her animals.

Selective breeding in farm stock can be slow and expensive because the animals often have small numbers of offspring and breed only once a year.

By producing new combinations of genes, selective breeding achieves the same objectives as genetic engineering but it takes much longer and is less predictable.

In selective breeding, the transfer of genes takes place between individuals of the same or closely related species. Genetic engineering involves transfer between unrelated species.

Selective breeding and genetic engineering both endeavour to produce new and beneficial combinations of genes. Selective breeding, however, is much slower and less precise than genetic engineering. On the other hand, cross-breeding techniques have been around for a very long time and are widely accepted.

One of the drawbacks of selective breeding is that the whole set of genes is transferred. As well as the desirable genes, there may be genes that, in a homozygous condition, would be harmful. It is known that artificial selection repeated over a large number of generations tends to reduce the fitness of the new variety.

A long-term disadvantage of selective breeding is the loss of variability. By eliminating all the offspring that do not bear the desired characteristics, many genes are lost from the population. At some future date, when new combinations of genes are sought, some of the potentially useful ones may no longer be available.

In attempting to introduce, in plants, characteristics such as salt tolerance or resistance to disease or drought, the geneticist goes back to wild varieties, as shown in Figure 18.26. However, with the current rate of extinction, this source of genetic material is diminishing.

In the natural world, reduction of variability could lead to local extinction if the population was unable to adapt, by natural selection, to changing conditions.

Comparing natural and artificial selection

Natural selection occurs in groups of living organisms through the passing on of genes to the next generation by the best adapted organisms, without human interference. Those with genes that provide an advantage, to cope with changes in environmental conditions for example, are more likely to survive, while others die before they can breed and pass on their genes. However, variation within the population remains.

Artificial selection is used by humans to produce varieties of animals and plants that have an increased economic importance. It is considered a safe way of developing new strains of organisms, compared with genetic engineering, and is a much faster process than natural selection. However, artificial selection removes variation from a population, leaving it susceptible to disease and unable to cope with changes in environmental conditions. Potentially, therefore, artificial selection puts a species at risk of extinction.

Figure 18.26 The genetics of bread wheat. A primitive wheat (a) was crossed with a wild grass (b) to produce a better-yielding hybrid wheat (c). The hybrid wheat (c) was crossed with another wild grass (d) to produce one of the varieties of wheat (e) which is used for making flour and bread.

Questions

Core

1. Study the following photographs and captions, then make a list of the adaptations of each animal.
   a long-eared bat (Figure 18.14)
   b hare (Figure 18.15)
   b polar bear (Figure 18.11) (See also details in the text.)
2. What features of a bird's appearance and behaviour do you think might help it compete for a mate?
3. What selection pressures do you think might be operating on the plants in a lawn?

Extended

4. Suggest some good characteristics that an animal-breeder might try to combine in sheep by mating different varieties together.
5. A variety of barley has a good ear of seed but has a long stalk and is easily blown over. Another variety has a short, sturdy stalk but a poor ear of seed.
   Suggest a breeding programme to obtain and select a new variety that combines both of the useful characteristics. Choose letters to represent the genes and show the genotypes of the parent plants and their offspring.

Checklist

After studying Chapter 18 you should know and understand the following:

Variation

- Variation is the differences between individuals of the same species.
- Variations within a species may be inherited or acquired.
- Continuous variation results in a range of phenotypes between two extremes, e.g. height in humans.
- Discontinuous variation results in a limited number of phenotypes with no intermediates, e.g. tongue rolling.
- Mutation is the way in which new alleles are formed.
- Increases in the rate of mutation can be caused by ionising radiation and some chemicals.

- Discontinuous variation results, usually, from the effects of a single pair of alleles, and produces distinct and consistent differences between individuals.
- Blood groups are an example of discontinuous variation.
- Discontinuous variations cannot be changed by the environment.
- Phenotypic (continuous) variations are usually controlled by a number of genes affecting the same characteristic and can be influenced by the environment.
- A gene mutation is a change in the base sequence of DNA.
- Sickle-cell anaemia is caused by a change in the base sequence of the gene for haemoglobin. This results in abnormal haemoglobin, which changes shape when oxygen levels are low.
- The inheritance of sickle-cell anaemia can be predicted using genetic diagrams.
- People who are heterozygous for the sickle-cell allele have a resistance to malaria.

Adaptive features

- An adaptive feature is an inherited feature that helps an organism to survive and reproduce in its environment.
- Adaptive features of a species can be recognised from its image in a drawing or photograph.

- An adaptive feature is the inherited functional features of an organism that increase its fitness.
- Fitness is the probability of that organism surviving and reproducing in the environment in which it is found.
- Hydrophytes are plants that have adaptive features to live in a watery environment.
- Xerophytes are plants that have adaptive features to live in very dry environments.

Selection

- Some members of a species may have variations that enable them to compete more effectively.
- These variants will live longer and leave more offspring.
- If the beneficial variations are inherited, the offspring will also survive longer.
- The new varieties may gradually replace the older varieties.
- Natural selection involves the elimination of less well-adapted varieties by environmental pressures.
- Selective breeding is used to improve commercially useful plants and animals.

- Adaptation is the process, resulting from natural selection, by which populations become more suited to their environment over many generations.
- The development of strains of antibiotic-resistant bacteria is an example of natural selection.
- Selective breeding by artificial selection is carried out over many generations to improve crop plants and domesticated animals.
- Evolution is the change in adaptive features of a population over time as the result of natural selection.

19 Organisms and their environment

Energy flow
Sun as source of energy

Flow of energy through organisms

Food chains and food webs
Define food chain, food web, producer, consumer, herbivore, carnivore, decomposer
Interpret food chains, food webs and pyramids of number
Impact of over-harvesting and introduction of foreign species on food chains and webs

Transfer of energy between trophic levels
Define trophic level
Loss of energy between levels
Efficiency of supplying green plants as human food
Identify levels in food chains, webs, pyramids of number and biomass
Describe and interpret pyramids of biomass
Advantages of using pyramids of biomass
Recycling

Nutrient cycles
Carbon cycle
Water cycle

Nitrogen cycle
Roles of micro-organisms in nitrogen cycle

Population size
Define population
Factors affecting rate of population growth
Human population growth

Define community, ecosystem
Factors affecting the increase in size of the human population
Identify and explain phases on a sigmoid population growth curve

● Energy flow

Nearly all living things depend on the Sun to provide energy. This is harnessed by photosynthesising plants and the energy is then passed through food chains.

Dependence on sunlight

With the exception of atomic energy and tidal power, all the energy released on Earth is derived from sunlight. The energy released by animals comes, ultimately, from plants that they or their prey eat and the plants depend on sunlight for making their food. Photosynthesis is a process in which light energy is trapped by plants and converted into chemical energy (stored in molecules such as carbohydrates, fats and proteins). Since all animals depend, in the end, on plants for their food, they therefore depend indirectly on sunlight. A few examples of our own dependence on photosynthesis are given below.

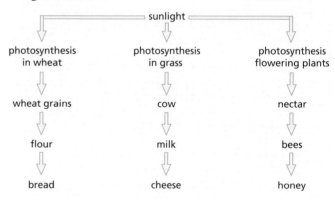

Nearly all the energy released on the Earth can be traced back to sunlight. Coal comes from tree-like plants, buried millions of years ago. These plants absorbed sunlight for their photosynthesis when they were alive. Petroleum was formed, also millions of years ago, probably from the partly decayed bodies of microscopic algae that lived in the sea. These, too, had absorbed sunlight for photosynthesis.

Today it is possible to use mirrors and solar panels to collect energy from the Sun directly, but the best way, so far, of trapping and storing energy from sunlight is to grow plants and make use of their products, such as starch, sugar, oil, alcohol and wood, for food or as energy sources. For example, sugar from sugar-cane can be fermented to alcohol, and used as a motor fuel instead of petrol.

Eventually, through one process or another, all the chemical energy in organisms is transferred to the environment. However, it is not a cyclical process like those described later in this chapter.

Food chains and food webs

> **Key definitions**
> A **food chain** shows the transfer of energy from one organism to the next, beginning with a producer.
> A **food web** is a network of interconnected food chains.
> A **producer** is an organism that makes its own organic nutrients, usually using energy from sunlight, through photosynthesis.
> A **consumer** is an organism that gets its energy from feeding on other organisms.
> A **herbivore** is an animal that gets its energy by eating plants.
> A **carnivore** is an animal that gets its energy by eating other animals.
> A **decomposer** is an organism that gets its energy from dead or waste organic material.

'Interdependence' means the way in which living organisms depend on each other in order to remain alive, grow and reproduce. For example, bees depend for their food on pollen and nectar from flowers. Flowers depend on bees for pollination (Chapter 16). Bees and flowers are, therefore, interdependent.

Food chains

One important way in which organisms depend on each other is for their food. Many animals, such as rabbits, feed on plants. Such animals are called **herbivores**. Animals that eat other animals are called **carnivores**. A **predator** is a carnivore that kills and eats other animals. A fox is a predator that preys on rabbits. **Scavengers** are carnivores that eat the dead remains of animals killed by predators. These are not hard and fast definitions. Predators will sometimes scavenge for their food and scavengers may occasionally kill living animals. Animals obtain their energy by ingestion.

Basically, all animals depend on plants for their food. Foxes may eat rabbits, but rabbits feed on grass. A hawk eats a lizard, the lizard has just eaten a grasshopper but the grasshopper was feeding on a grass blade. This relationship is called a food chain (Figure 19.1).

The organisms at the beginning of a food chain are usually very numerous while the animals at the end of

Figure 19.1 A food chain. The caterpillar eats the leaf; the blue tit eats the caterpillar but may fall prey to the kestrel.

the chain are often large and few in number. The **food pyramids** in Figure 19.2 show this relationship. There will be millions of microscopic, single-celled algae in a pond (Figure 19.3(a)). These will be eaten by the larger but less numerous water fleas and other crustacea (Figure 19.3(b)), which in turn will become the food of small fish such as minnow and stickleback. The hundreds of small fish may be able to provide enough food for only four or five large carnivores, like pike or perch.

The organisms at the base of the food pyramids in Figure 19.2 are plants. Plants produce food from carbon dioxide, water and salts (see 'Photosynthesis', Chapter 6), and are, therefore, called **producers**. The animals that eat the plants are called **primary consumers**, e.g. grasshoppers. Animals that prey on the plant-eaters are called **secondary consumers**, e.g. shrews, and these may be eaten by **tertiary consumers**, e.g. weasels or kestrels (Figure 19.4).

19 ORGANISMS AND THEIR ENVIRONMENT

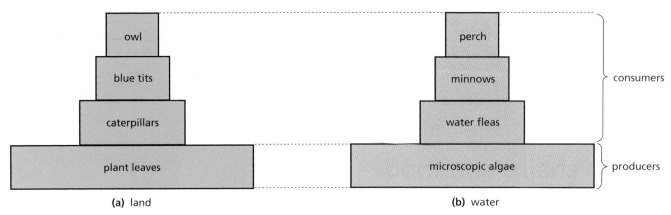

Figure 19.2 Examples of food pyramids (pyramids of numbers)

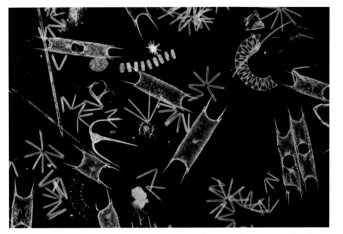

(a) phytoplankton (×100) These microscopic algae form the basis of a food pyramid in the water.

(b) zooplankton (×20) These crustacea will eat microscopic algae.

Figure 19.3 Plankton. The microscopic organisms that live in the surface waters of the sea or fresh water are called, collectively, plankton. The single-celled algae (see Chapter 1) are the phytoplankton. They are surrounded by water, salts and dissolved carbon dioxide. Their chloroplasts absorb sunlight and use its energy for making food by photosynthesis. Phytoplankton is eaten by small animals in the zooplankton, mainly crustacea (see Chapter 1). Small fish will eat the crustacea.

Figure 19.4 The kestrel, a secondary or tertiary consumer

Pyramids of numbers

The width of the bands in Figure 19.2 is meant to represent the relative number of organisms at each trophic level. So the diagrams are sometimes called **pyramids of numbers**.

However, you can probably think of situations where a pyramid of numbers would not show the same effect. For example, a single sycamore tree may provide food for thousands of greenfly. One oak tree may feed hundreds of caterpillars. In these cases the pyramid of numbers is upside-down, as shown in Figure 19.5.

Food webs

Food chains are not really as straightforward as described above, because most animals eat more than one type of food. A fox, for example, does not feed entirely on rabbits but takes beetles, rats and voles in its diet. To show these relationships more accurately, a **food web** can be drawn up (Figure 19.6).

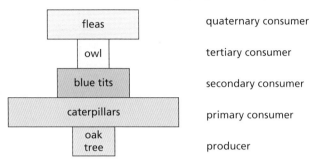

Figure 19.5 An inverted pyramid of numbers

The food webs for land, sea and fresh water, or for ponds, rivers and streams, will all be different. Food webs will also change with the seasons when the food supply changes.

If some event interferes with a food web, all the organisms in it are affected in some way. For example, if the rabbits in Figure 19.6 were to die out, the foxes, owls and stoats would eat more beetles and rats. Something like this happened in 1954 when the disease myxomatosis wiped out nearly all the rabbits in England. Foxes ate more voles, beetles and blackberries, and attacks on lambs and chickens increased. Even the vegetation was affected because the tree seedlings that the rabbits used to nibble on were able to grow. As a result, woody scrubland started to develop on what had been grassy downs. A similar effect is shown in Figure 19.7.

The effects of over-harvesting

Over-harvesting causes the reduction in numbers of a species to the point where it is endangered or made extinct. As a result biodiversity is affected. The species may be harvested for food, or for body parts such as tusks (elephants), horns (rhinos – Figure 19.8), bones and fur (tigers) or for selling as pets (reptiles, birds and fish, etc.). In parts of Africa, bush meat is used widely as a source of food. Bush meat is the flesh of primates, such as monkeys. However, hunting these animals is not always regulated or controlled and rare species can be threatened as a result of indiscriminate killing. (See also 'Habitat destruction' in Chapter 21.)

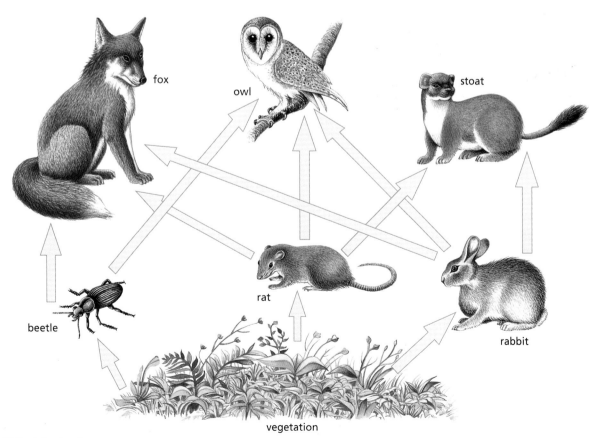

Figure 19.6 A food web

19 ORGANISMS AND THEIR ENVIRONMENT

(a) Sheep have eaten any seedlings that grew under the trees

(b) Ten years later, the fence has kept the sheep off and the tree seedlings have grown

Figure 19.7 Effect of grazing

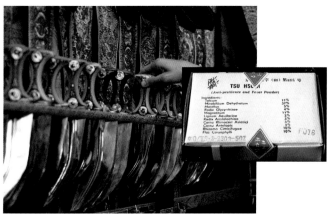

Figure 19.8 The rhinoceros is endangered because some people believe, mistakenly, that powdered rhino horn (Cornu Rhinoceri Asiatici) has medicinal properties, and others greatly prize rhino horn handles for their daggers.

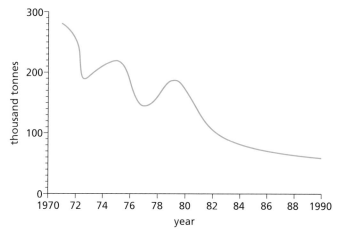

Figure 19.9 Landings of North Sea cod from 1970 to 1990

Overfishing

Small populations of humans, taking fish from lakes or oceans and using fairly basic methods of capture, had little effect on fish numbers. At present, however, commercial fishing has intensified to the point where some fish stocks are threatened or can no longer sustain fishing. In the past 100 years, fishing fleets have increased and the catching methods have become more sophisticated.

If the number of fish removed from a population exceeds the number of young fish reaching maturity, then the population will decline (Figure 19.9).

At first, the catch size remains the same but it takes longer to catch it. Then the catch starts to contain a greater number of small fish so that the return per day at sea goes down even more. Eventually the stocks are so depleted that it is no longer economical to exploit them. The costs of the boats, the fuel and the wages of the crew exceed the value of the catch. Men are laid off, boats lie rusting in the harbour and the economy of the fishing community and those who depend on it is destroyed. Overfishing has severely reduced stocks of many fish species: herring in the North Sea, halibut in the Pacific and anchovies off the Peruvian coast, for example. In 1965, 1.3 million tonnes of herring were caught in the North Sea. By 1977 the catch had diminished to 44 000 tonnes, i.e. about 3% of the 1965 catch.

Similarly, whaling has reduced the population of many whale species to levels that give cause for concern. Whales were the first marine organisms to face extinction through overfishing. This happened in the early 1800s when they were killed for their **blubber** (a thick fat layer around the body of the mammal) for use as lamp oil. The blue whale's numbers have been reduced from about 2 000 000 to 6000 as a result of intensive hunting.

Overfishing can reduce the populations of fish species and can also do great damage to the environment where they live. For example, the use of heavy nets dragged along the sea floor to catch the fish can wreck coral reefs, destroying the habitats of many other animal species. Even if the reef is not damaged, fishing for the top predators such as grouper fish has a direct effect on the food chain: fish lower down the chain increase in numbers, and overgraze on the reef. This process is happening on the Great Barrier Reef in Australia. Grouper fish are very slow growing and take a long time to become sexually mature, so the chances of them recovering from overfishing are low and they are becoming endangered.

Introducing foreign species to a habitat

One of the earliest examples of this process was the accidental introduction of rats to the Galapagos Islands by pirates or whalers in the 17th or 18th centuries. The rats had no natural predators and food was plentiful: they fed on the eggs of birds, reptiles and tortoises, along with young animals. The Galapagos Islands provide a habitat for many rare species, which became endangered as a result of the presence of the rats. A programme of rat extermination is now being carried out on the islands to protect their unique biodiversity.

The prickly pear cactus, *Opuntia*, was introduced to Australia in 1839 for use as a living fence to control the movement of cattle, but its growth got out of control because of the lack of herbivores that eat it. Millions of acres of land became unusable. A moth, *Cactoblastis cactorum*, whose young feed on the cactus, was successfully introduced from Argentina and helped to control the spread of the cactus. Other places with similar problems, for example the island of Nevis in the West Indies, followed Australia's example, but with less successful results. The moth had no natural predators and ate other native cactus species as well as the prickly pear, bringing them to the brink of extinction. The moth is now spreading to parts of the United States of America and poses a threat to other cactus species.

Food chains and webs can also be disrupted by the use of pesticides and other poisons, sometimes released accidentally during human activities. More details can be found in Chapter 21.

Energy transfer

Study Figure 19.1. When an herbivorous animal eats a plant (the caterpillar feeding on a leaf), the chemical energy stored in that plant leaf is transferred to the herbivore. Similarly, when a carnivore (the blue tit) eats the herbivore, the carnivore gains the energy stored in the herbivore. If the carnivore is eaten by another carnivore (the kestrel), the energy is transferred again.

Use of sunlight

To try and estimate just how much life the Earth can support it is necessary to examine how efficiently the Sun's energy is used. The amount of energy from the Sun reaching the Earth's surface in 1 year ranges from 2 million to 8 million kilojoules per m² ($2–8 \times 10^9 \, J \, m^{-2} \, yr^{-1}$) depending on the latitude. When this energy falls onto grassland, about 20% is reflected by the vegetation, 39% is used in evaporating water from the leaves (transpiration), 40% warms up the plants, the soil and the air, leaving only about 1% to be used in photosynthesis for making new organic matter in the leaves of the plants (Figure 19.10).

This figure of 1% will vary with the type of vegetation being considered and with climatic factors, such as availability of water and the soil temperature. Sugar-cane grown in ideal conditions can convert 3% of the Sun's energy into photosynthetic products; sugar-beet at the height of its growth has nearly a 9% efficiency. Tropical forests and swamps are far more productive than grassland but it is difficult, and, in some cases undesirable, to harvest and utilise their products.

In order to allow crop plants to approach their maximum efficiency they must be provided with sufficient water and mineral salts. This can be achieved by irrigation and the application of fertiliser.

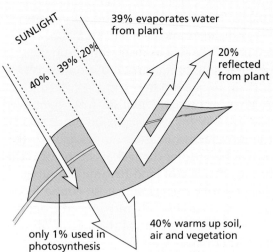

Figure 19.10 Absorption of Sun's energy by plants

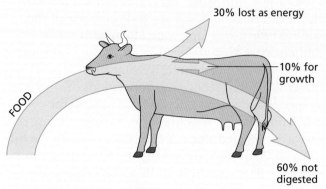

Figure 19.11 Energy transfer from plants to animals

Energy transfer between organisms

Having considered the energy conversion from sunlight to plant products, the next step is to study the efficiency of transmission of energy from plant products to primary consumers. On land, primary consumers eat only a small proportion of the available vegetation. In a deciduous forest only about 2% is eaten; in grazing land, 40% of the grass may be eaten by cows. In open water, however, where the producers are microscopic plants (phytoplankton, see Figure 19.3(a)) and are swallowed whole by the primary consumers in the zooplankton (see Figure 19.3(b)), 90% or more may be eaten. In the land communities, the parts of the vegetation not eaten by the primary consumers will eventually die and be used as a source of energy by the decomposers.

A cow is a primary consumer; over 60% of the grass it eats passes through its alimentary canal (Chapter 7) without being digested. Another 30% is used in the cow's respiration to provide energy for its movement and other life processes. Less than 10% of the plant material is converted into new animal tissue to contribute to growth (Figure 19.11). This figure will vary with the diet and the age of the animal. In a fully grown animal all the digested food will be used for energy and replacement and none will contribute to growth. Economically it is desirable to harvest the primary consumers before their rate of growth starts to fall off.

The transfer of energy from primary to secondary consumers is probably more efficient, since a greater proportion of the animal food is digested and absorbed than is the case with plant material. The transfer of energy at each stage in a food chain may be represented by classifying the organisms in a community as producers, or primary, secondary or tertiary consumers, and showing their relative masses in a pyramid such as the one shown in Figure 19.2 but on a more accurate scale. In Figure 19.12 the width of the horizontal bands is proportional to the masses (dry weight) of the organisms in a shallow pond.

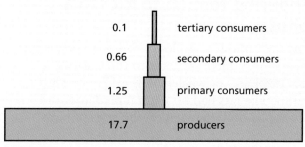

Figure 19.12 Biomass (dry weight) of living organisms in a shallow pond (grams per square metre)

> **Key definitions**
> The **trophic level** of an organism is its position in a food chain, food web or pyramid of numbers or biomass.

It is very unusual for food chains to have more than five trophic levels because, on average, about 90% of the energy is lost at each level. Consequently, very little of the energy entering the chain through the producer is available to the top consumer. The food chain below shows how the energy reduces through the chain. It is based on grass obtaining 100 units of energy.

grass → locust → lizard → snake → mongoose
100 units   10 units   1 unit   0.1 unit   0.01 unit

Energy transfer in agriculture

In human communities, the use of plant products to feed animals that provide meat, eggs and dairy products is wasteful, because only 10% of the plant material is converted to animal products. It is more economical to eat bread made from the wheat than to feed the wheat to hens and then eat the eggs and chicken meat. This is because eating the wheat as bread avoids using any part of its energy to keep the chickens alive and active. Energy losses can be reduced by keeping hens indoors in small cages, where they lose little heat to the atmosphere and cannot use much energy in movement (Figure 19.13). The same principles can be applied in 'intensive' methods of rearing calves. However, many people feel that these methods are less than humane, and the saving of energy is far less than if the plant products were eaten directly by humans, as is the case in vegetarians.

Figure 19.13 Battery chickens. The hens are well fed but kept in crowded and cramped conditions with no opportunity to move about or scratch in the soil as they would normally do.

Consideration of the energy flow of a modern agricultural system reveals other sources of inefficiency. To produce 1 tonne of nitrogenous fertiliser takes energy equivalent to burning 5 tonnes of coal. Calculations show that if the energy needed to produce the fertiliser is added to the energy used to produce a tractor and to power it, the energy derived from the food so produced is less than that expended in producing it.

Pyramids of biomass

As stated earlier, displaying food chains using pyramids of number, such as those shown in Figure 19.5, can produce inverted pyramids. This is because the top consumers may be represented by large numbers of very small organisms, for example, fleas feeding on an owl. The way around this problem is to consider not the single tree, but the mass of the leaves that it produces in the growing season, and the mass of the insects that can live on them. **Biomass** is the term used when the mass of living organisms is being considered, and pyramids of biomass can be constructed as in Figure 19.12. A pyramid of biomass is nearly always the correct pyramid shape.

An alternative is to calculate the energy available in a year's supply of leaves and compare this with the energy needed to maintain the population of insects that feed on the leaves. This would produce a **pyramid of energy**, with the producers at the bottom having the greatest amount of energy. Each successive trophic level would show a reduced amount of energy.

The elements that make up living organisms are recycled, i.e. they are used over and over again (see next section). This is not the case with energy, which flows from producers to consumers and is eventually lost to the atmosphere as heat.

Recycling

There are a number of organisms that have not been fitted into the food webs or food chains described so far. Among these are the **decomposers**. Decomposers do not obtain their food by photosynthesis, nor do they kill and eat living animals or plants. Instead they feed on dead and decaying matter such as dead leaves in the soil or rotting tree-trunks (Figure 19.14). The most numerous examples are the fungi, such as mushrooms, toadstools or moulds, and the bacteria, particularly those that live in the soil. They produce extracellular enzymes that digest the decaying matter and then they absorb the soluble products back into their cells. In so doing, they remove the dead remains of plants and animals, which would otherwise collect on the Earth's surface. They also break these remains down into substances that can be used by other organisms. Some bacteria, for example, break down the protein of dead plants and animals and release nitrates, which are taken up by

19 ORGANISMS AND THEIR ENVIRONMENT

Figure 19.14 Decomposers. These toadstools are getting their food from the rotting log.

and the animals that eat the plants and each other are the consumers. The bacteria and fungi, especially those in the soil, are called the decomposers because they break down the dead remains and release the chemicals for the plants to use again. Three examples of recycling, for water, carbon and nitrogen, are described in the next section.

plant roots and are built into new amino acids and proteins. This use and reuse of materials in the living world is called **recycling**.

The general idea of recycling is illustrated in Figure 19.15. The green plants are the producers,

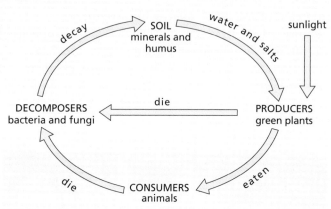

Figure 19.15 Recycling in an ecosystem

● Nutrient cycles

The carbon cycle

Carbon is an element that occurs in all the compounds which make up living organisms. Plants get their carbon from carbon dioxide in the atmosphere and animals get their carbon from plants. The carbon cycle, therefore, is mainly concerned with what happens to carbon dioxide (Figure 19.16).

Removal of carbon dioxide from the atmosphere
Photosynthesis
Green plants remove carbon dioxide from the atmosphere as a result of their photosynthesis. The carbon from the carbon dioxide is built first into a carbohydrate such as sugar. Some of this is changed into starch or the cellulose of cell walls, and the proteins, pigments and other compounds of a plant. When the plants are eaten by animals, the organic plant material is digested, absorbed and built into the compounds making up the animals' tissues. Thus the carbon atoms from the plant become part of the animal.

Fossilisation
Any environment that prevents rapid decay may produce **fossils**. The carbon in the dead

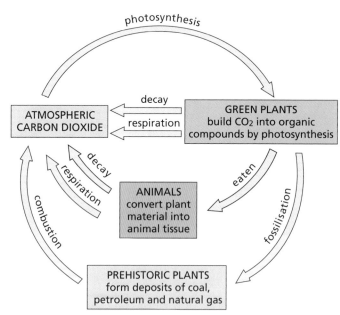

Figure 19.16 The carbon cycle

organisms becomes trapped and compressed and can remain there for millions of years. The carbon may form **fossil fuels** such as coal, oil and natural gas. Some animals make shells or exoskeletons containing carbon and these can become fossils.

Addition of carbon dioxide to the atmosphere

Respiration
Plants and animals obtain energy by oxidising carbohydrates in their cells to carbon dioxide and water (Chapter 12). The carbon dioxide and water are excreted so the carbon dioxide returns once again to the atmosphere.

Decomposition
A crucial factor in carbon recycling is the process of decomposition, or decay. If it were not for decay, essential materials would not be released from dead organisms. When an organism dies, the enzymes in its cells, freed from normal controls, start to digest its own tissues (auto-digestion). Soon, scavengers appear on the scene and eat much of the remains; blowfly larvae devour carcases, earthworms consume dead leaves.

Finally the decomposers, fungi and bacteria (collectively called **micro-organisms**), arrive and invade the remaining tissues (Figure 19.17). These saprophytes secrete extracellular enzymes (Chapter 5) into the tissues and reabsorb the liquid products of digestion. When the micro-organisms themselves die, auto-digestion takes place, releasing the products such as nitrates, sulfates, phosphates, etc. into the soil or the surrounding water to be taken up again by the producers in the ecosystem.

Figure 19.17 Mould fungus growing on over-ripe oranges

The speed of decay depends on the abundance of micro-organisms, temperature, the presence of water and, in many cases, oxygen. High temperatures speed up decay because they speed up respiration of the micro-organisms. Water is necessary for all living processes and oxygen is needed for aerobic respiration of the bacteria and fungi. Decay can take place in anaerobic conditions but it is slow and incomplete, as in the waterlogged conditions of peat bogs.

Combustion (burning)
When carbon-containing fuels such as wood, coal, petroleum and natural gas are burned, the carbon is oxidised to carbon dioxide ($C + O_2 \rightarrow CO_2$). The hydrocarbon fuels, such as coal and petroleum, come from ancient plants, which have only partly decomposed over the millions of years since they were buried.

So, an atom of carbon which today is in a molecule of carbon dioxide in the air may tomorrow be in a molecule of cellulose in the cell wall of a blade of grass. When the grass is eaten by a cow, the carbon atom may become part of a glucose molecule in the cow's bloodstream. When the glucose molecule is used for respiration, the carbon atom will be breathed out into the air once again as carbon dioxide.

The same kind of cycling applies to nearly all the elements of the Earth. No new matter is created, but it is repeatedly rearranged. A great proportion of the atoms of which you are composed will, at one time, have been part of other organisms.

The effects of the combustion of fossil fuels
If you look back at the carbon cycle, you will see that the natural processes of photosynthesis, respiration and decomposition would be expected to keep the CO_2 concentration at a steady level. However, since the Industrial Revolution, we have been burning the fossil fuels such as coal and petroleum and releasing extra CO_2 into the atmosphere. As a result, the concentration of CO_2 has increased from 0.029% to 0.035% since 1860. It is likely to go on increasing as we burn more and more fossil fuel.

Although it is not possible to prove beyond all reasonable doubt that production of CO_2 and other 'greenhouse gases' is causing a rise in the Earth's temperature, i.e. global warming, the majority of scientists and climatologists agree that it is happening now and will get worse unless we take drastic action to reduce the output of these gases (see 'Pollution' in Chapter 21 for further details of the greenhouse effect and global warming).

Another factor contributing to the increase in atmospheric CO_2 is **deforestation**. Trees are responsible for removing gaseous CO_2 and trapping the carbon in organic molecules (carbohydrates, proteins and fats – see Chapter 4). When they are cut down the amount of photosynthesis globally is reduced. Often deforestation is achieved by a process called 'slash and burn', where the felled trees are burned to provide land for agriculture (see 'Habitat destruction' in Chapter 21) and this releases even more atmospheric CO_2.

19 ORGANISMS AND THEIR ENVIRONMENT

The water cycle

The **water cycle** (Figure 19.18) is somewhat different from other cycles because only a tiny proportion of the water that is recycled passes through living organisms.

Animals lose water by **evaporation** (Chapter 14), defecation (Chapter 7), urination (Chapter 13) and exhalation (Chapter 11). They gain water from their food and drink. Plants take up water from the soil and lose it by **transpiration** (Chapter 8). Millions of tonnes of water are transpired, but only a tiny fraction of this has taken part in the reactions of respiration (Chapter 12) or photosynthesis (Chapter 6).

The great proportion of water is recycled without the intervention of animals or plants. The Sun shining and the wind blowing over the oceans **evaporate** water from their vast, exposed surfaces. The water vapour produced in this way enters the atmosphere and eventually **condenses** to form clouds. The clouds release their water in the form of rain or snow (**precipitation**). The rain collects in streams, rivers and lakes and ultimately finds its way back to the oceans. The human population diverts some of this water for drinking, washing, cooking, irrigation, hydroelectric schemes and other industrial purposes, before allowing it to return to the sea.

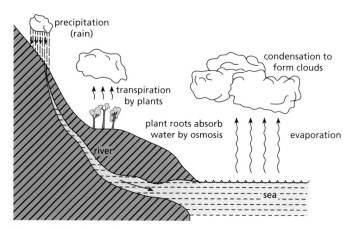

Figure 19.18 The water cycle

The nitrogen cycle

When a plant or animal dies, its tissues **decompose**, partly as a result of the action of saprotrophic bacteria. One of the important products of the decay of animal and plant protein is ammonia (NH_3, a compound of nitrogen), which is washed into the soil (Figure 19.20). It dissolves readily in water to form ammonium ions (NH_4^-).

The excretory products of animals contain nitrogenous waste products such as ammonia, urea and uric acid (Chapter 13). Urea is formed in the liver of humans as a result of **deamination**. The organic matter in animal droppings is also decomposed by soil bacteria.

Processes that add nitrates to soil
Nitrifying bacteria
These are bacteria living in the soil, which use the ammonia from excretory products and decaying organisms as a source of energy (as we use glucose in respiration). In the process of getting energy from ammonia, called **nitrification**, the bacteria produce **nitrates**.

- The 'nitrite' bacteria oxidise ammonium compounds to nitrites ($NH_4^- \rightarrow NO_2^-$).
- 'Nitrate' bacteria oxidise nitrites to nitrates ($NO_2^- \rightarrow NO_3^-$).

Although plant roots can take up ammonia in the form of its compounds, they take up nitrates more readily, so the nitrifying bacteria increase the fertility of the soil by making nitrates available to the plants.

Nitrogen-fixing bacteria

This is a special group of nitrifying bacteria that can absorb nitrogen as a gas from the air spaces in the soil, and build it into compounds of ammonia. Nitrogen gas cannot itself be used by plants. When it has been made into a compound of ammonia, however, it can easily be changed to nitrates by other nitrifying bacteria. The process of building the gas, nitrogen, into compounds of ammonia is called **nitrogen fixation**. Some of the nitrogen-fixing bacteria live freely in the soil. Others live in the roots of **leguminous plants** (peas, beans, clover), where they cause swellings called **root nodules** (Figure 19.19). These leguminous plants are able to thrive in soils where nitrates are scarce, because the nitrogen-fixing bacteria in their nodules make compounds of nitrogen available for them. Leguminous plants are also included in crop rotations to increase the nitrate content of the soil.

Lightning

The high temperature of lightning discharge causes some of the nitrogen and oxygen in the air to combine and form oxides of nitrogen. These dissolve in the rain and are washed into the soil as weak acids, where they form nitrates. Although several million tonnes of nitrate may reach the Earth's surface in this way each year, this forms only a small fraction of the total nitrogen being recycled.

Processes that remove nitrates from the soil

Uptake by plants

Plant roots absorb nitrates from the soil and combine them with carbohydrates to make amino acids, which are built up into proteins (Chapter 6). These proteins are then available to animals, which feed on the plants and digest the proteins in them.

Leaching

Nitrates are very soluble (i.e. dissolve easily in water), and as rainwater passes through the soil it dissolves the nitrates and carries them away in the run-off or to deeper layers of the soil. This is called **leaching**. (See Chapter 21 for some of the implications of leaching.)

Denitrifying bacteria

These are bacteria that obtain their energy by breaking down nitrates to nitrogen gas, which then escapes from the soil into the atmosphere.

All of these processes are summed up in Figure 19.20.

Figure 19.19 Root nodules of white clover – a leguminous plant

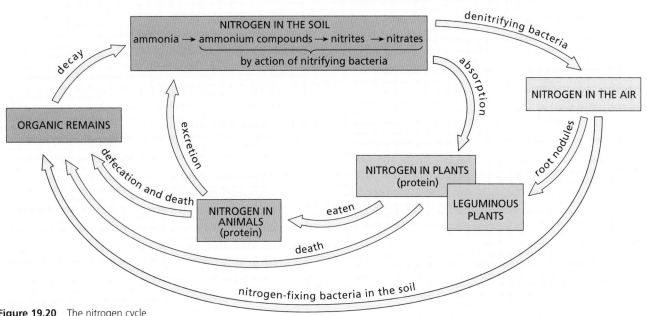

Figure 19.20 The nitrogen cycle

19 ORGANISMS AND THEIR ENVIRONMENT

Population size

> **Key definition**
> A **population** is a group of organisms of one species, living and interacting in the same area at the same time.

In biology, the term population always refers to a single species. A biologist might refer to the population of sparrows in a farmyard or the population of carp in a lake. In each case this would mean the total numbers of sparrows or the total numbers of carp in the stated area.

Population changes

If conditions are ideal, a population can increase in size. For this to happen there needs to be a good **food supply**. This will enable organisms to breed more successfully to produce more offspring; shortage of food can result in starvation, leading to death, or force emigration, reducing the population. The food shortage may be because the food source has all been eaten, or died out, or completed its growing season, or there is competition for it with other species in the same habitat.

In a habitat there are likely to be predators. If heavy **predation** of a population happens, the rate of breeding may be unable to produce enough organisms to replace those eaten, so the population will drop in numbers. There tends to be a time lag in population size change for predators and their prey: as predator numbers increase, prey numbers drop and as predator numbers drop, prey numbers rise again (unless there are other factors that prevent this happening) (see 'Predator–prey relationships' later in this chapter).

Disease can be a particular problem in large populations because it can spread easily from one individual to another. Epidemics can reduce population sizes very rapidly. An example was given in the section on food webs: the disease myxomatosis is caused by a virus. It wiped out nearly all the rabbits in England in 1954 and then spread to other parts of Europe, carried by fleas. It was first discovered in 1896 in Uruguay and was deliberately introduced to Australia in 1951 in an attempt to control its large rabbit populations.

When a disease spreads globally it is called a **pandemic**. One of the worst cases experienced by humans was known as Spanish flu. This virus killed between 40 and 50 million people in 1918.

The World Health Organization (WHO) estimates that there were 660 000 malaria deaths in 2010 and there were about 219 million cases of the disease. Malaria (Chapter 10) is caused by a single-celled parasite, spread by mosquitos. It is a treatable disease and drugs are gradually becoming more widely available to prevent it being fatal.

Human population

In AD 1000, the world population was probably about 300 million. In the early 19th century it rose to 1000 million (1 billion), and by 1984 it had reached 4.7 billion. In 2000 it reached about 6 billion and rose to 7.2 billion in 2014. The United Nations predicts that the global population will decline steadily by 2050, quoting predictions of between 8.3 and 10.9 billion people by that date. The graph in Figure 19.21 shows that the greatest population surge has taken place in the last 300 years.

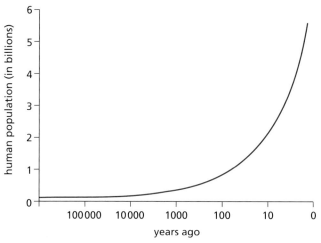

Figure 19.21 World population growth. The time scale (horizontal axis) is logarithmic. The right-hand space (0–10) represents only 10 years, but the left-hand space (100 000–1 million) represents 900 000 years. The greatest population growth has taken place in the last 300 years.

Population growth

About 20 years ago, the human population was increasing at the rate of 2% a year. This may not sound very much, but it means that the world population was doubling every 35 years. This doubles the demand for food, water, space and other resources. Recently, the growth rate has slowed to 1%. However, it is not the same everywhere. Nigeria's population is growing by 2.9% each year, but Western Europe's grows at only 0.1%.

Traditionally, it is assumed that population growth is limited by famine, disease or war. These factors are affecting local populations in some parts of the world today but they are unlikely to have a limiting effect on the rate of overall population growth.

Diseases such as malaria (see Chapter 10) and sleeping sickness (spread by tsetse flies) have for many years limited the spread of people into areas where these insects carry the infections.

Diseases such as bubonic plague and influenza have checked population growth from time to time, and the current AIDS epidemic in sub-Saharan Africa is having significant effects on population growth and life expectancy.

Factors affecting population growth

If a population is to grow, the birth rate must be higher than the death rate. Suppose a population of 1000 people produces 100 babies each year but only 50 people die each year. This means that 50 new individuals are added to the population each year and the population will double in 20 years (or less if the new individuals start reproducing at 16) (Figure 19.22).

One of the factors affecting population growth is **infant mortality**, i.e. the death rate for children less than 1 year old. Populations in the developing

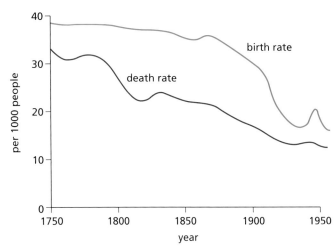

Figure 19.22 Birth and death rates in England and Wales from 1750 to 1950. Although the birth rate fell during this period, so did the death rate. As a result, the population continued to grow. Note the 'baby boom' after the Second World War. (Used by permission of Carolina Biological Supply Company.)

world are growing, not because of an increase in the number of babies born per family, but because more babies are surviving to reach reproductive age. Infant mortality is falling and more people are living longer. That is, **life expectancy** is increasing.

> **Key definition**
> A **community** is all of the populations of different species in an ecosystem.
> An **ecosystem** is a unit containing the community of organisms and their environment, interacting together. Examples include a decomposing log or a lake.

Communities

A **community** is made up of all the plants and animals living in an ecosystem. In the soil there is a community of organisms, which includes earthworms, springtails and other insects, mites, fungi and bacteria. In a lake, the animal community will include fish, insects, crustacea, molluscs and protoctista.

The plant community will consist of rooted plants with submerged leaves, rooted plants with floating leaves, reed-like plants growing at the lake margin, plants floating freely on the surface, filamentous algae and single-celled algae in the surface waters.

Ecosystems

The community of organisms in a habitat, plus the non-living part of the environment (air, water, soil, light, etc.) make up an **ecosystem**. A lake is an ecosystem, which consists of the plant and animal communities mentioned above, and the water, minerals, dissolved oxygen, soil and sunlight on which they depend. An ecosystem is self-supporting (Figure 19.23).

$$\left.\begin{array}{l}\text{individuals of the same species}\end{array}\right\} = \text{POPULATION} \atop + \atop \left.\begin{array}{l}\text{populations of other species}\end{array}\right\} \right\} = \text{COMMUNITY} \atop + \atop \left.\begin{array}{l}\text{non-living part of environment}\end{array}\right\} \right\} = \text{ECOSYSTEM}$$

In a woodland ecosystem, the plants absorb light and rainwater for photosynthesis, the animals feed on the plants and on each other. The dead remains of animals and plants, acted upon by fungi and bacteria, return nutrients to the soil.

Lakes and ponds are clear examples of ecosystems. Sunlight, water and minerals allow the plants to grow and support animal life. The recycling of materials from the dead organisms maintains the supply of nutrients.

So, a *population* of carp forms part of the animal *community* living in a *habitat* called a lake. The communities in this habitat, together with their watery *environment*, make up a self-supporting *ecosystem*.

Figure 19.23 An 'ecosphere'. The 5-inch globe contains seawater, bacteria, algae, snails and a few Pacific shrimps. Given a source of light it is a self-supporting system and survives for several years (at least). The shrimps live for up to 7 years but few reproduce.

A carp is a *secondary consumer* at the top of a *food chain*, where it is in *competition* with other species of fish for food and with other carp for food and mates.

The whole of that part of the Earth's surface which contains living organisms (called the **biosphere**) may be regarded as one vast ecosystem.

No new material (in significant amounts) enters the Earth's ecosystem from space and there is no significant loss of materials. The whole system depends on a constant input of energy from the Sun and recycling of the chemical elements.

Distribution in an ecosystem

All ecosystems contain producers, consumers and decomposers. The organisms are not distributed uniformly throughout the ecosystem but occupy habitats that suit their way of life.

For example, fish may range freely within an aquatic ecosystem but most of them will have preferred habitats in which they feed and spend most of their time. Plaice, sole and flounders feed on molluscs and worms on the sea floor, whereas herring and mackerel feed on plankton in the surface waters. In a pond, the snails do not range much beyond the plants where they feed. On a rocky coast, limpets and barnacles can withstand exposure between the tides and colonise the rocks. Sea anemones, on the other hand, are restricted mainly to the rocky pools left at low tide.

Factors affecting the increase in size of the human population

Increase in life expectancy

The life expectancy is the average age to which a newborn baby can be expected to live. In Europe between 1830 and 1900 the life expectancy was 40–50 years. Between 1900 and 1950 it rose to 65 and now stands at 73–74 years. In sub-Saharan Africa, life expectancy was rising to 58 years until the AIDS epidemic reduced it to about 45 years.

These figures are averages. They do not mean, for example, that everyone in the developing world will live to the age of 58. In the developing world, 40% of the deaths are of children younger than 5 years and only 25–30% are deaths of people over 60. In Europe, only 5–20% of deaths are those of children below the age of 5, but 70–80% are of people over 60.

An increase in the number of people over the age of 60 does not change the rate of population growth much, because these people are past child-bearing age. On the other hand, if the death rate among children falls and the extra children survive to reproduce, the population will continue to grow. This is the main reason for the rapid population growth in the developing world since 1950.

Causes of the reduction in death rate

The causes are not always easy to identify and vary from one community to the next. In 19th century Europe, agricultural development and economic expansion led to improvements in nutrition, housing and sanitation, and to clean water supplies. These improvements reduced the incidence of infectious diseases in the general population, and better-fed children could resist these infections when they did meet them. The drop in deaths from infectious diseases probably accounted for three-quarters of the total fall in deaths.

The social changes probably affected the population growth more than did the discovery of new drugs or improved medical techniques. Because of these techniques – particularly immunisation – diphtheria, tuberculosis and polio are now rare (Figure 19.24), and by 1977 smallpox had been wiped out by the World Health Organization's vaccination campaign.

In the developing world, sanitation, clean water supplies and nutrition are improving slowly. The surge in the population since 1950 is likely to be at least 50% due to modern drugs, vaccines and insecticides.

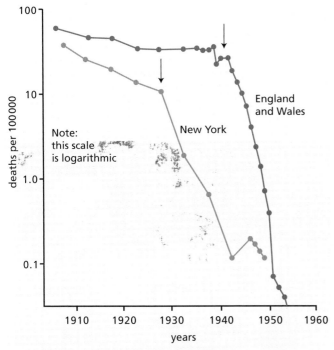

Figure 19.24 Fall in death rate from diphtheria as a result of immunisation. The arrows show when 50% or more of children were vaccinated. Note that the rate was already falling but was greatly increased by immunisation.

Stability and growth

Up to 300 years ago, the world population was relatively stable. Fertility (the birth rate) was high and so was the mortality rate (death rate). Probably less than half the children born lived to have children of their own. Many died in their first year (infant mortality), and many mothers died during childbirth.

No one saw any point in reducing the birth rate. If you had a lot of children, you had more help on your land and a better chance that some of them would live long enough to care for you in your old age.

In the past 300 years, the mortality rate has fallen but the birth rate has not gone down to the same extent. As a result the population has expanded rapidly.

In 18th century Europe, the **fertility rate** was about 5. This means that, on average, each woman would have five children. When the death rate fell, the fertility rate lagged behind so that the population increased. However, the fertility rate has now fallen to somewhere between 1.4 and 2.6 and the European population is more or less stable.

A fall in the fertility rate means that young people will form a smaller proportion of the population. There will also be an increasing proportion of old people for the younger generation to look after. In Britain it is estimated that, between 1981 and 1991, the number of people aged 75–84 increased by 16%. The number of those over 85 increased by about 46% (Figure 19.25).

In the developing world, the fertility rate has dropped from about 6.2 to 3.0. This is still higher than the mortality rate. An average fertility rate of 2.1 is necessary to keep the population stable.

As a community grows wealthier, the birth rate goes down. There are believed to be four reasons:

- **Longer and better education:** Marriage is postponed and a better-educated couple will have learned about methods of family limitation.
- **Better living conditions:** Once people realise that half their offspring are not going to die from disease or malnutrition, family sizes fall.
- **Agriculture and cities:** Modern agriculture is no longer labour intensive. Farmers do not need large families to help out on the land. City dwellers do not depend on their offspring to help raise crops or herd animals.
- **Application of family planning methods:** Either natural methods of birth control or the use of contraceptives is much more common.

It takes many years for social improvements to produce a fall in the birth rate. Some countries are trying to speed up the process by encouraging couples to limit their family size (Figure 19.26), or by penalising families who have too many children.

Meanwhile the population goes on growing. The United Nations expect that the birth rate and death rate will not be in balance until the year 2100.

19 ORGANISMS AND THEIR ENVIRONMENT

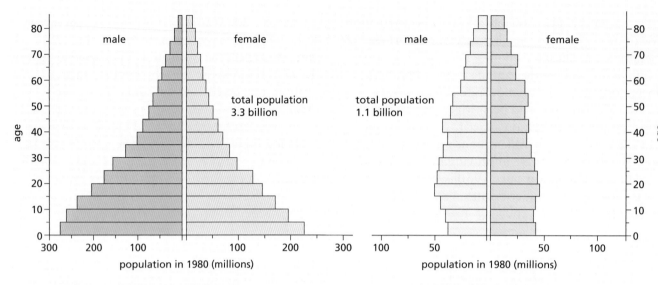

(a) The developing regions. The tapering pattern is characteristic of a population with a high birth rate and low average life expectancy. The bulk of the population is under 25.

(b) The developed regions. The almost rectangular pattern is characteristic of an industrialised society, with a steady birth rate and a life expectancy of about 70. (The horizontal scale is not the same as in a.)

Figure 19.25 Age distribution of population in 1980

By that time the world population may have reached 10 billion, assuming that the world supply of food will be able to feed this population.

In the past few decades, the world has produced enough food to feed, in theory, all the extra people. But the extra food and the extra people are not always in the same place. As a result, 72% of the world's population has a diet that lacks energy, as well as other nutrients.

Every year between 1965 and 1975, food production in the developed nations rose by 2.8%, while the population rose by 0.7%. In the developing nations during the same period, food production rose by only 1.5% each year, while the annual population rise was 2.4%.

The Western world can produce more food than its people can consume. Meanwhile people in the drier regions of Africa face famine due to drought and population pressure on the environment. Even if the food could be taken to the developing world, people there are often too poor to buy it. Ideally, each region needs to grow more food or reduce its population until the community is self-supporting. Some countries grow tobacco, cotton, tea and coffee (cash crops) in order to obtain foreign currency for imports from the Western world. This is fine, so long as they can also feed their people. But when food is scarce, people cannot live on the cash crops.

Figure 19.26 Family planning. A health worker in Bangladesh explains the use of a condom.

Population pressures

More people, more agriculture and more industrialisation will put still more pressure on the environment unless we are very watchful. If we damage the ozone layer, increase atmospheric carbon dioxide, release radioactive products or allow farmland to erode, we may meet with additional limits to population growth.

Sigmoid population growth curves

Population growth

A population will not necessarily be evenly spread throughout its habitat, nor will its numbers remain steady. The population will also be made up of a wide variety of individuals: adults (male and female), juveniles, larvae, eggs or seeds, for example. In studying populations, these variables often have to be simplified.

In the simplest case, where a single species is allowed to grow in laboratory conditions, the population develops more or less as shown in Figure 19.27.

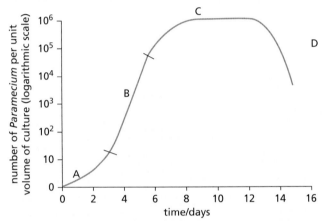

Figure 19.27 The sigmoid curve (*Paramecium caudatum*). This is the characteristic growth pattern of a population when food is abundant at first.

The population might be of yeast cells growing in a sugar solution, flour beetles in wholemeal flour or weevils in a grain store. The curve shown in Figure 19.27 was obtained using a single-celled organism called *Paramecium* (see Chapter 1), which reproduces by dividing into two (binary fission). The **sigmoid** (S-shaped) form of the graph can be explained as follows:

- **A: Lag phase**. The population is small. Although the numbers double at each generation, this does not result in a large increase.
- **B: Exponential phase (log phase)**. Continued doubling of the population at each generation produces a logarithmic growth rate (e.g. 64 – 128 – 256 – 512 – 1024). When a population of four doubles, it is not likely to strain the resources of the habitat, but when a population of 1024 doubles there is likely to be considerable competition for food and space and the growth rate starts to slow down.
- **C: Stationary phase**. The resources will no longer support an increasing population. At this stage, limiting factors come into play. The food supply may limit further expansion of the population, diseases may start to spread through the dense population and overcrowding may lead to a fall in reproduction rate. Now the mortality rate (death rate) equals the reproduction rate, so the population numbers stay the same.
- **D: Death phase**. The mortality rate (death rate) is now greater than the reproduction rate, so the population numbers begin to drop. Fewer offspring will live long enough to reproduce. The decline in population numbers can happen because the food supply is insufficient, waste products contaminate the habitat or disease spreads through the population.

Limits to population growth

The sigmoid curve is a very simplified model of population growth. Few organisms occupy a habitat on their own, and the conditions in a natural habitat will be changing all the time. The steady state of the population in part C of the sigmoid curve is rarely reached in nature. In fact, the population is unlikely to reach its maximum theoretical level because of the many factors limiting its growth. These are called **limiting factors**.

Competition

If, in the laboratory, two species of *Paramecium* (*P. aurelia* and *P. caudatum*) are placed in an aquarium tank, the population growth of *P. aurelia* follows the sigmoid curve but the population of *P. caudatum* soon declines to zero because *P. aurelia* takes up food more rapidly than *P. caudatum* (Figure 19.28).

This example of competition for food is only one of many factors in a natural environment that will limit a population or cause it to change.

Abiotic and biotic limiting factors

Plant populations will be affected by **abiotic** (non-biological) factors such as rainfall, temperature and light intensity. The population of small annual plants may be greatly reduced by a period of drought; a severe winter can affect the numbers of more hardy perennial plants. **Biotic** (biological) factors affecting plants include their leaves being eaten by browsing and grazing animals or by caterpillars and other insects, and the spread of fungus diseases.

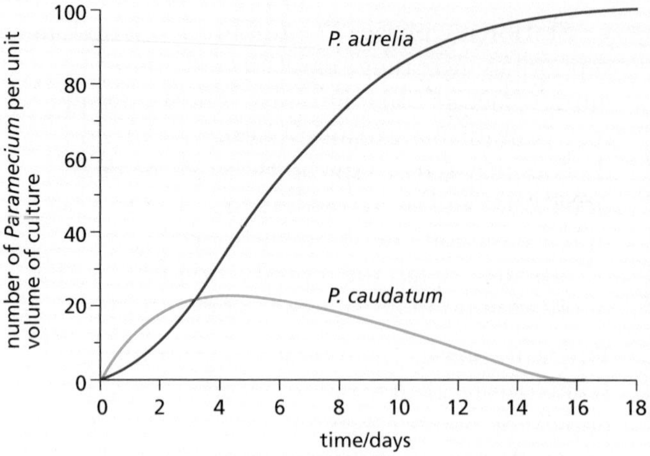

19 ORGANISMS AND THEIR ENVIRONMENT

Figure 19.28 The effect of competition. *Paramecium aurelia* and *P. caudatum* eat the same food but *P. aurelia* can capture and ingest it faster than *P. caudatum*.

Figure 19.29 A colony of nesting gannets. Availability of suitable nest sites is one of the factors that limits the population.

Animal populations, too, will be limited by abiotic factors such as seasonal changes. A cold winter can severely reduce the populations of small birds. However, animal populations are also greatly affected by biotic factors such as the availability of food, competition for nest sites (Figure 19.29), predation (i.e. being eaten by other animals), parasitism and diseases.

The size of an animal population will also be affected by the numbers of animals entering from other localities (immigration) or leaving the population (emigration).

In a natural environment, it is rarely possible to say whether the fluctuations observed in a population are mainly due to one particular factor because there are so many factors at work. In some cases, however, the key factors can be identified as mainly responsible for limiting the population.

Predator–prey relationships

A classic example of predator–prey relationships comes from an analysis of the fluctuating populations of lynxes and snowshoe hares in Canada. The figures are derived from the numbers of skins sold by trappers to the Hudson's Bay Company between 1845 and 1945.

The lynx preys on the snowshoe hare, and the most likely explanation of the graph in Figure 19.30 is that an increase in the hare population allowed the predators to increase. Eventually the increasing numbers of lynxes caused a reduction in the hare population.

However, seasonal or other changes affecting one or both of the animals could not be ruled out.

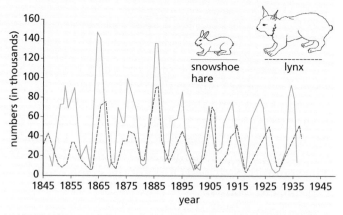

Figure 19.30 Prey–predator relationships: fluctuations in the numbers of pelts received by the Hudson's Bay Company for lynx (predator) and snowshoe hare (prey) over a 100-year period.

Questions

Core

1. Construct a simple food web using the following:
   sparrow, fox, wheat seeds, cat, kestrel, mouse.
2. Describe briefly all the possible ways in which the following might depend on each other:
   grass, earthworm, blackbird, oak tree, soil.
3. Explain how the following foodstuffs are produced as a result of photosynthesis:
   wine, butter, eggs, beans.
4. An electric motor, a car engine and a race horse can all produce energy.
   a Show how this energy could come, originally, from sunlight.
   b What forms of energy on the Earth are *not* derived from sunlight?
5. How do you think evidence is obtained in order to place animals such as a fox and a pigeon in a food web?
6. When humans colonised islands they often introduced their domestic animals, such as goats or cats. This usually had a devastating effect on the natural food webs. Suggest reasons for this.
7. a Why do living organisms need a supply of carbon?
   b Give three examples of carbon-containing compounds that occur in living organisms (see Chapter 4).
   c Where do these organisms get their carbon from?
      i animals
      ii plants
8. Write three chemical equations:
   a to illustrate that respiration produces carbon dioxide (see Chapter 12)
   b to show that burning produces carbon dioxide
   c to show that photosynthesis uses up carbon dioxide (see Chapter 6).
9. Outline the events that might happen to a carbon atom in a molecule of carbon dioxide, which entered the stoma in the leaf of a potato plant and became part of a starch molecule in a potato tuber, which was then eaten by a man. Finally the carbon atom is breathed out again in a molecule of carbon dioxide.
10. Look at the graph in Figure 19.22.
    a When did the post-war 'baby-boom' occur?
    b What was the growth rate of the population in 1800?
11. Which of the following causes of death are likely to have most effect on the growth rate of a population: smallpox, tuberculosis, heart disease, polio, strokes, measles?
    Give reasons for your answer.
12. Suggest some reasons why the birth rate tends to fall as a country becomes wealthier.
13. a Give examples of the kind of demands that an increasing population makes on the environment.
    b In what ways can these demands lead to environmental damage?
14. If there are 12 000 live births in a population of 400 000 in 1 year, what is the birth rate?
15. Try to explain why, on average, couples need to have just over two children if the population is to remain stable.
16. Study Figure 19.25 and then comment on:
    a the relative number of boy and girl babies
    b the relative number of men and women of reproductive age (20–40)
    c the relative numbers of the over-70s.
17. In Figure 19.24, what might be the reasons for the fall in death rate from diphtheria even before 50% immunisation was achieved?

Extended

18. It can be claimed that the Sun's energy is used indirectly to produce a muscle contraction in your arm. Trace the steps in the transfer of energy that would justify this claim.
19. Discuss the advantages and disadvantages of human attempts to exploit a food chain nearer to its source, e.g. the plankton in Figure 19.3.
20. On a lawn growing on nitrate-deficient soil, the patches of clover often stand out as dark green and healthy against a background of pale green grass. Suggest a reason for this contrast.
21. Very briefly explain the difference between nitrifying, nitrogen-fixing and denitrifying bacteria.
22. Study Figure 19.27.
    a How many days does it take for the mortality rate to equal the replacement rate?
    b What is the approximate increase in the population of *Paramecium*:
       i between day 0 and day 2
       ii between day 2 and day 4
       iii between day 8 and day 10?
    c In section B of the graph, what is the approximate reproduction rate of *Paramecium* (i.e. the number of new individuals per day)?
23. In 1937, two male and six female pheasants were introduced to an island off the NW coast of America. There were no other pheasants and no natural predators. The population for the next 6 years increased as follows:

Year	Population
1937	24
1938	65
1939	253
1940	563
1941	1122
1942	1611

Plot a graph of these figures and say whether it corresponds to any part of the sigmoid curve.

24. In Figure 19.28, which part of the curve approximately represents the exponential growth of the *P. aurelia* population? Give the answer in days.
25. What forms of competition might limit the population of sticklebacks in a pond?
26. Suggest some factors that might prevent an increase in the population of sparrows in a farmyard:
    a abiotic factors
    b biotic factors

19 ORGANISMS AND THEIR ENVIRONMENT

Checklist

After studying Chapter 19 you should know and understand the following:

Energy flow

- The Sun is the principal source of energy input to biological systems.
- Energy from the Sun flows through living organisms.
- First, light energy is converted into chemical energy in photosynthetic organisms. Then they are eaten by herbivores. Carnivores eat herbivores.
- As organisms die, the energy is transferred to the environment.

Food chains and food webs

- A food chain shows the transfer of energy from one organism to the next, beginning with a producer.
- A food web is a network of interconnected food chains.
- Producers are organisms that make their own organic nutrients, usually using energy from sunlight, through photosynthesis.
- Consumers are organisms that get their energy from feeding on other organisms.
- A herbivore is an animal that gets its energy by eating plants.
- A carnivore is an animal that gets its energy by eating other animals.
- All animals depend, ultimately, on plants for their source of food.
- Plants are the producers in a food web; animals may be primary, secondary or tertiary consumers.
- A pyramid of numbers has levels which represent the number of each species in a food chain. There are usually fewer consumers than producers, forming a pyramid shape.
- Over-harvesting unbalances food chains and webs, as does the introduction of foreign species to a habitat.
- Energy is transferred between trophic levels through feeding.
- The trophic level of an organism is its position in a food chain.
- The transfer of energy from one trophic level to another is inefficient.
- Only about 1% of the Sun's energy that reaches the Earth's surface is trapped by plants during photosynthesis.
- At each step in a food chain, only a small proportion of the food is used for growth. The rest is used for energy to keep the organism alive.
- Food chains usually have fewer than five trophic levels.
- Feeding crop plants to animals uses up a lot of energy and makes the process inefficient.
- There is an increased efficiency in supplying green plants as human food.
- A decomposer is an organism that gets its energy from dead or waste organic material.
- A pyramid of biomass is more useful than a pyramid of numbers in representing a food chain.

Nutrient cycles

- The materials that make up living organisms are constantly recycled.
- Plants take up carbon dioxide during photosynthesis; all living organisms give out carbon dioxide during respiration; the burning of carbon-containing fuels produces carbon dioxide.
- The uptake of carbon dioxide by plants balances the production of carbon dioxide from respiration and combustion.
- The water cycle involves evaporation, transpiration, condensation and precipitation (rain).
- The combustion of fossil fuels and the cutting down of forests increases the carbon dioxide concentrations in the atmosphere.
- Soil nitrates are derived naturally from the excretory products of animals and the dead remains of living organisms.
- Nitrifying bacteria turn these products into nitrates, which are taken up by plants.
- Nitrogen-fixing bacteria can make nitrogenous compounds from gaseous nitrogen.
- Plants make amino acids and proteins.
- Animals eat the proteins.
- Proteins are broken down to remove the nitrogen by the process of deamination.
- Micro-organisms play an important part in the nitrogen cycle. They are involved in decomposition, nitrification, nitrogen fixation and denitrification.

Population size

- A population is a group of organisms of one species, living and interacting in the same area at the same time.
- The factors affecting the rate of population growth for a population of an organism include food supply, predation and disease.
- The human population has increased in size rapidly over the past 250 years.
- The world population is growing at the rate of 1.7% each year. At this rate, the population more than doubles every 50 years.
- The rate of increase is slowing down and the population may stabilise at 10 billion by the year 2100.
- A population grows when the birth rate exceeds the death rate, provided the offspring live to reproduce.
- A community is all of the populations of different species in an ecosystem.
- An ecosystem is a unit containing the community of organisms and their environment, interacting together.
- A sigmoid population growth curve for a population growing in an environment with limited resources has lag, exponential (log), stationary and death phases.
- In the developed countries, the birth rate and the death rate are now about the same.
- In the developing countries, the birth rate exceeds the death rate and their populations are growing. This is not because more babies are born, but because more of them survive.
- The increased survival rate may be due to improved social conditions, such as clean water, efficient sewage disposal, better nutrition and better housing.
- It is also the result of vaccination, new drugs and improved medical services.
- As a population becomes wealthier, its birth rate tends to fall.

20 Biotechnology and genetic engineering

Biotechnology and genetic engineering
Use of bacteria in biotechnology and genetic engineering

Reasons why bacteria are useful in biotechnology and genetic engineering

Biotechnology
Role of anaerobic respiration in yeast in production of ethanol for biofuels and bread-making
Investigate use of pectinase in fruit juice production
Investigate use of biological washing powders containing enzymes

Investigate use of lactase to produce lactose-free milk
Production of antibiotics
Use of fermenters in penicillin production

Genetic engineering
Define genetic engineering
Examples of genetic engineering

Outline genetic engineering
Advantages and disadvantages of genetically modifying crops

● Biotechnology and genetic engineering

Biotechnology is the application of biological organisms, systems or processes to manufacturing and service industries. **Genetic engineering** involves the transfer of genes from one organism to (usually) an unrelated species.

Both processes often make use of bacteria because of their ability to make complex molecules (proteins for example) and their rapid reproduction rate.

Use of bacteria in biotechnology and genetic engineering

Bacteria are useful in biotechnology and genetic engineering because they can be grown and manipulated without raising ethical concerns. They have a genetic code that is the same as all other organisms, so genes from other animals or plants can be successfully transferred into bacterial DNA.

Bacterial DNA is in the form of a circular strand and also small circular pieces called **plasmids**. Scientists have developed techniques to cut open these plasmids and insert sections of DNA from other organisms into them. When the bacterium divides, the DNA in the modified plasmid is copied, including the 'foreign' DNA. This may contain a gene to make a particular protein such as insulin, which can be extracted and used as a medicine to treat diabetes.

● Biotechnology

Although biotechnology is 'hot news', we have been making use of it for hundreds of years. Wine-making, the brewing of beer, the baking of bread and the production of cheese all depend on fermentation processes brought about by yeasts, other fungi and bacteria, or enzymes from these organisms.

Antibiotics, such as penicillin, are produced by mould fungi or bacteria. The production of industrial chemicals such as citric acid or lactic acid needs bacteria or fungi to bring about essential chemical changes.

Sewage disposal (Chapter 21) depends on bacteria in the filter beds to form the basis of the food chain that purifies the effluent.

Biotechnology is not concerned solely with the use of micro-organisms. Cell cultures and enzymes also feature in modern developments. In this chapter, however, there is space to consider only a representative sample of biotechnological processes that use micro-organisms.

Biofuels

The term 'fermentation' does not apply only to alcoholic fermentation but to a wide range of reactions, brought about by enzymes or micro-organisms. In Chapter 12, the anaerobic respiration of glucose to alcohol or lactic acid was described as a form of fermentation.

Micro-organisms that bring about fermentation are using the chemical reaction to produce energy, which they need for their living processes. The reactions that are useful in fermentation biotechnology are mostly those that produce incompletely oxidised compounds. A reaction that goes all the way to carbon dioxide and water is not much use in this context.

The micro-organisms are encouraged to grow and multiply by providing nutrients such as glucose, with added salts and, possibly, vitamins. Oxygen or air is bubbled through the culture if the reaction is aerobic, or excluded if the process is anaerobic. An optimum pH and temperature are maintained for the species of microbe being cultured.

In 'Conservation' in Chapter 21, it is pointed out that ethanol (alcohol), produced from fermented sugar or surplus grain, could replace, or at least supplement, petrol.

Brazil, Zimbabwe and the USA produce ethanol as a renewable source of energy for the motor car. Since 1990, 30% of new cars in Brazil can use ethanol and many more use a mixture of petrol and ethanol. As well as being a renewable resource, ethanol produces less pollution than petrol.

However, biofuels are not yet economical to produce. For example, the energy used to grow, fertilise and harvest sugar-cane, plus the cost of extracting the sugar and converting it to ethanol, uses more energy than the ethanol releases when burned.

In addition, there are also environmental costs, some of which will be outlined in Chapter 21. Forests are being destroyed to plant soy beans or oil palms, removing the habitats of thousands of organisms, some of which, such as the orang-utan, are on the verge of extinction.

Another biofuel, oil from rapeseed or sunflower seed, can with suitable treatment replace diesel fuel. It is less polluting than diesel but more expensive to produce.

Bread

Yeast is the micro-organism used in bread-making but the only fermentation product needed is carbon dioxide. The carbon dioxide makes bubbles in the bread dough. These bubbles make the bread 'light' in texture.

Flour, water, salt, oil and yeast are mixed to make a dough. Yeast has no enzymes for digesting the starch in flour but the addition of water activates the amylases already present in flour and these digest some of the starch to sugar. With highly refined white flour, it may be necessary to add sugar to the dough. The yeast then ferments the sugar to alcohol and carbon dioxide.

A protein called **gluten** gives the dough a sticky, plastic texture, which holds the bubbles of gas. The dough is repeatedly folded and stretched ('kneaded') either by hand, in the home, or mechanically in the bakery. The dough is then left for an hour or two at a temperature of about 27 °C while the yeast does its work. The accumulating carbon dioxide bubbles make the dough rise to about double its volume (Figure 20.1). The dough may then be kneaded again or put straight into baking tins and into an oven at about 200 °C. This temperature makes the bubbles expand more, kills the yeast and evaporates the small quantities of alcohol before the dough turns into bread.

Figure 20.1 Carbon dioxide produced by the yeast has caused the dough to rise.

Enzymes

Enzymes can be produced by commercial fermentation using readily available feedstocks such as corn-steep liquor or molasses. Fungi (e.g. *Aspergillus*) or bacteria (e.g. *Bacillus*) are two of the commonest organisms used to produce the enzymes.

These organisms are selected because they are non-pathogenic and do not produce antibiotics. The fermentation process is similar to that described for penicillin. If the enzymes are extracellular (Chapter 5) then the liquid feedstock is filtered from the organism and the enzyme is extracted (Figure 20.2). If the enzymes are intracellular, the micro-organisms have to be filtered from the feedstock. They are then crushed and the enzymes extracted with water or other solvents.

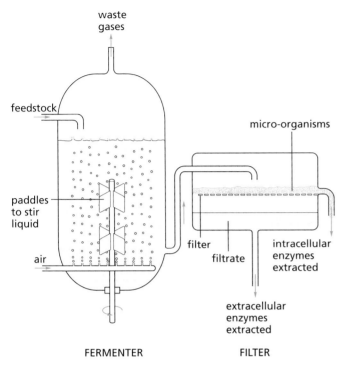

Figure 20.2 Principles of enzyme production from micro-organisms

Using the techniques of genetic engineering, new genes can be introduced into the microbes to 'improve' the action of the enzymes coded for by the genes (e.g. making the enzymes more heat stable).

One effective way of using enzymes is by 'immobilising' them. The enzymes or the micro-organisms that produce them are held in or on beads or membranes of an insoluble and inert substance, e.g. plastic. The beads or membranes are packed into columns and the substrate is poured over them at the optimum rate. This method has the advantage that the enzyme is not lost every time the product is extracted. Immobilised enzymes also allow the process to take place in a continuous way rather than a batch at a time.

Some commercial uses of enzymes are listed below.

- **Proteases**: In washing powders for dissolving stains from, e.g. egg, milk and blood; removing hair from animal hides; cheese manufacture; tenderising meat.
- **Lipases**: Flavour enhancer in cheese; in washing powders for removal of fatty stains.
- **Pectinases**: Clarification of fruit juices; maximising juice extraction.
- **Amylases**: Production of glucose from starch.

Pectinases are used to separate the juices from fruit such as apples. The enzymes can be extracted from fungi such as *Aspergillus niger*. They work by breaking down pectin, the jelly-like substance that sticks plant cell walls to each other. The enzymes can also be used to clarify fruit juice and wine (make it more transparent). During the breakdown process, a number of different polysaccharides are released, which make the juice cloudy, but pectinases break these down to make the juice clearer. The sugars produced also make the juice sweeter.

Biological washing powders

The majority of commercial enzyme production involves protein-digesting enzymes (proteases) and fat-digesting enzymes (lipases) for use in the food and textile industries. When combined in washing powders they are effective in removing stains in clothes caused by proteins, e.g. blood, egg and gravy, and fats, e.g. grease. Protein and fat molecules tend to be large and insoluble. When they have been digested the products are small, soluble molecules, which can pass out of the cloth.

Biological washing powders save energy because they can be used to wash clothes at lower temperatures, so there is no need to boil water. However, if they are put in water at higher temperatures the enzymes become denatured (see Chapter 5) and they lose their effectiveness.

Practical work

Investigating the use of pectinase in fruit juice production

- Make 100 cm³ of apple purée using a liquidiser, or use a tin of apple purée.
- Transfer the purée to a 250 cm³ beaker.
- Add one level teaspoon of powdered pectinase enzyme (care needed – see safety note), stir the mixture and leave it for about 5 minutes.
- Place a funnel in the top of a 100 cm³ measuring cylinder and line the funnel with a folded filter paper.
- Transfer the purée into the filter funnel and leave it in a warm place for up to 24 hours.
- Other measuring cylinders could be set up in the same way, with purée left to stand at different temperatures to compare the success of juice extraction.

Safety note: Take care to avoid skin or eye contact with the enzyme powder. Enzyme powder can cause allergies. Wipe up any spillages immediately and rinse the cloth thoroughly with water. Do not allow spillages to dry up.

20 BIOTECHNOLOGY AND GENETIC ENGINEERING

Result
Juice is extracted from the purée. It collects in the measuring cylinder and is transparent (it has been clarified).

Interpretation
Pectinase breaks down the apple tissue, releasing sugars in solution. More juice collects in the measuring cylinder when the purée has been kept in warm conditions; colder temperatures slow down the process.

Further investigation
If other enzymes are available, try comparing cellulase and amylase with pectinase. Combinations of these could be used to find out which is the most effective in extracting the juice. Remember to control variables to make a fair comparison.

Investigating the use of biological washing powder

- Break an egg into a plastic beaker and whisk it with a fork, spatula or stirring rod until thoroughly mixed.
- Cut up four pieces of white cloth to make squares 10 cm × 10 cm, smear egg evenly onto each of them and leave to dry.
- Set up four 250 cm^3 beakers as follows:
  A 100 cm^3 warm water, with no washing powder.
  B 5 cm^3 (1 level teaspoon) of non-biological washing powder dissolved in 100 cm^3 warm water.
  C 5 cm^3 (1 level teaspoon) of biological washing powder dissolved in 100 cm^3 warm water.
  D 5 cm^3 (1 level teaspoon) of biological washing powder dissolved in 100 cm^3 water and boiled for 5 minutes, then left to cool until warm.
- Place a piece of egg-stained cloth in each beaker and leave for 30 minutes.
- Remove the pieces of cloth and compare the effectiveness of each washing process.

Results
The piece of cloth in beaker **C** is most effectively cleaned, followed by **B** and then **D**. The cloth in **A** is largely unchanged.

Interpretation
The enzymes in the biological washing powder break down the proteins and fats in the egg stain to amino acids and fatty acids and glycerol. These are smaller, soluble molecules, which can escape from the cloth and dissolve in the water. Non-biological washing powder is less effective because it does not contain enzymes. Boiled biological washing powder is not very effective because the enzymes in it have been denatured. Beaker **A** was a control, with no active detergent or enzymes. Soaking the cloth in warm water alone does not remove the stain.

Lactose-free milk

Lactose is a type of disaccharide sugar found in milk and dairy products. Some people suffer from **lactose intolerance**, a digestive problem where the body does not produce enough of the enzyme lactase. As a result, the lactose remains in the gut, where it is fermented by bacteria, causing symptoms such as flatulence (wind), diarrhoea and stomach pains. Many foods contain dairy products, so people with lactose intolerance cannot eat them, or suffer the symptoms described above. However, lactose-free milk is now produced using the enzyme lactase.

The lactase can be produced on a large scale by fermenting yeasts such as *Kluyveromyces fragilis* or fungi such as *Aspergillus niger*. The fermentation process is shown in Figure 20.2.

A simple way to make lactose-free milk is to add lactase to milk. The enzyme breaks down lactose sugar into two monosaccharide sugars: glucose and galactose. Both can be absorbed by the intestine.

An alternative, large-scale method is to immobilise lactase on the surface of beads. The milk is then passed over the beads and the lactose sugar is effectively removed. This method avoids having the enzyme molecules in the milk because they remain on the beads.

The food industry uses lactase in the production of milk products such as yoghurt: it speeds up the process and makes the yoghurt taste sweeter.

Practical work

Action of lactase

This investigation uses glucose test strips (diastix). They are used by people with diabetes to test for glucose in their urine (see 'Homeostasis' in Chapter 14 for details of diabetes). The strips do not react to the presence of other sugars (lactose, sucrose, etc.)

- Pour 25 cm^3 warm, fresh milk into a 100 cm^3 beaker.
- Test the milk for glucose with a glucose test strip.
- Measure out 2 cm^3 of 2% lactase using a syringe or pipette and add this to the milk.

- Stir the mixture and leave for a few minutes.
- Test the milk again with a new glucose test strip.

Result

Milk gives a negative result for glucose, but milk exposed to lactase gives a positive result.

Interpretation

Lactase breaks down the lactose in milk, as shown in the equation below.

$$\text{lactose} \xrightarrow{\text{lactase}} \text{glucose} + \text{galactose}$$

Note: milk sometimes contains traces of glucose. If the milk gives a positive result with the glucose test strip, an alternative method would be to use a solution of lactose instead of milk. However, the amount of glucose in the milk, as indicated by the colour change on the test strip, should increase after treatment with lactase.

Antibiotics

When micro-organisms are used for the production of antibiotics, it is not their fermentation products that are wanted, but complex organic compounds, called **antibiotics**, that they synthesise.

Most of the antibiotics we use come from bacteria or fungi that live in the soil. The function of the antibiotics in this situation is not clear. One theory suggests that the chemicals help to suppress competition for limited food resources, but the evidence does not support this theory.

One of the most prolific sources of antibiotics is *Actinomycetes*. These are filamentous bacteria that resemble microscopic mould fungi. The actinomycete *Streptomyces* produces the antibiotic **streptomycin**.

Perhaps the best known antibiotic is **penicillin**, which is produced by the mould fungus *Penicillium* and was discovered by Sir Alexander Fleming in 1928. Penicillin is still an important antibiotic but it is produced by mutant forms of a different species of *Penicillium* from that studied by Fleming (Figure 20.3). The different mutant forms of the fungus produce different types of penicillin.

The penicillin types are chemically altered in the laboratory to make them more effective and to 'tailor' them for use with different diseases. 'Ampicillin', 'methicillin' and 'oxacillin' are examples.

Antibiotics attack bacteria in a variety of ways. Some of them disrupt the production of the cell wall and so prevent the bacteria from reproducing, or

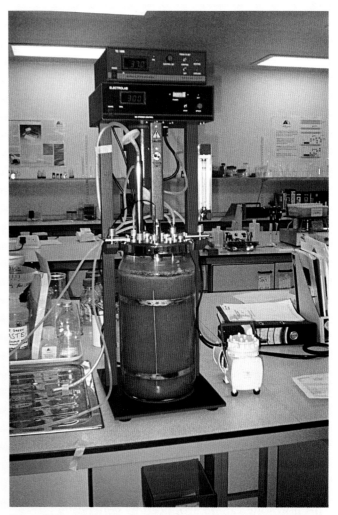

Figure 20.3 A laboratory fermenter for antibiotic production, which will eventually be scaled up to 10 000-litre fermentation vessels.

even cause them to burst open; some interfere with protein synthesis and thus arrest bacterial growth. Those that stop bacteria from reproducing are said to be **bacteriostatic**; those that kill the bacteria are **bacteriocidal**.

Animal cells do not have cell walls, and the cell structures involved in protein production are different. Consequently, antibiotics do not damage human cells although they may produce some side-effects such as allergic reactions.

Commercial production of penicillin

Antibiotics are produced in giant fermenting tanks, up to 100 000 litres in capacity. The tanks are filled with a nutrient solution. For penicillin production, the carbohydrate source is sugar, mainly lactose or 'corn-steep liquor' – a by-product of the manufacture of cornflour and maize starch; it contains amino acids as well as sugars. Mineral salts are added, the pH is adjusted to between 5 and 6, the temperature is maintained at about 26 °C, air is blown through the liquid and it is stirred. The principles of industrial fermentation are shown in Figure 20.2. The nutrient liquid is seeded with a culture of the appropriate micro-organism, which is allowed to grow for a day or two. Sterile conditions are essential. If 'foreign' bacteria or fungi get into the system they can completely disrupt the process. As the nutrient supply diminishes, the micro-organisms begin to secrete their antibiotics into the medium.

The nutrient fluid containing the antibiotic is filtered off and the antibiotic extracted by crystallisation or other methods.

● Genetic engineering

Key definition
Genetic engineering is changing the genetic material of an organism by removing, changing or inserting individual genes.

Applications of genetic engineering

The following section gives only a few examples of genetic engineering, a rapidly advancing process. Some products, such as insulin, are in full-scale production. A few **genetically modified** (**GM**) crops, e.g. maize and soya bean, are being grown on a large scale in the USA. Many other projects are still at the experimental stage, undergoing trials, awaiting approval by regulatory bodies or simply on a 'wish list'.

Production of human insulin

This hormone can be produced by genetically modified bacteria and has been in use since 1982. The human insulin gene is inserted into bacteria, which then secrete human insulin. The human insulin produced in this way (Figure 20.4) is purer than insulin prepared from pigs or cattle, which sometimes provokes allergic reactions owing to traces of 'foreign' protein. The GM insulin is acceptable to people with a range of religious beliefs who may not be allowed to use insulin from cows or pigs.

GM crops

Genetic engineering has huge potential benefits in agriculture but, apart from a relatively small range of crop plants, most developments are in the experimental or trial stages. In the USA, 50% of the soya bean crop and 30% of the maize harvest consist of genetically modified plants, which are resistant to herbicides and insect pests.

In the UK at the moment, GM crops are grown only on a trial basis and there is resistance to their growth and the presence of GM products in food.

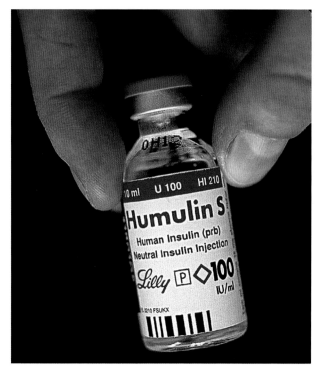

Figure 20.4 Human insulin prepared from genetically engineered bacteria. Though free from foreign proteins, it does not suit all patients.

Pest resistance

The bacterium, *Bacillus thuringiensis*, produces a toxin that kills caterpillars and other insect larvae. The toxin has been in use for some years as an insecticide. The gene for the toxin has been successfully introduced into some plant species using a bacterial vector. The plants produce the toxin and show increased resistance to attack by insect larvae. The gene is also passed on to the plant's offspring. Unfortunately there are signs that insects are developing immunity to the toxin.

Most American GM maize, apart from its herbicide-resistant gene, also carries a pesticide gene, which reduces the damage caused by a stem-boring larva of a moth (Figure 20.5).

Genetic engineering

Figure 20.5 The maize stem borer can cause considerable losses by killing young plants.

Figure 20.6 Genetically engineered tomatoes. In the three engineered tomatoes on the right, biologists have deleted the gene that produces the enzyme which makes fruit go soft.

Herbicide resistance

Some of the safest and most effective herbicides are those, such as glyphosate, which kill any green plant but become harmless as soon as they reach the soil. These herbicides cannot be used on crops because they kill the crop plants as well as the weeds. A gene for an enzyme that breaks down glyphosate can be introduced into a plant cell culture (Chapter 16). This should lead to a reduced use of herbicides.

Modifying plant products

A gene introduced to oilseed rape and other oil-producing plants can change the nature of the oils they produce to make them more suitable for commercial processes, e.g. detergent production. This might be very important when stocks of petroleum run out. It could be a renewable source of oil, which would not contribute to global warming (see 'Pollution' in Chapter 21).

The tomatoes in Figure 20.6 have been modified to improve their keeping qualities.

● Extension work

Other applications of genetic engineering

One of the objections to GM crops is that, although they show increased yields, this has benefited only the farmers and the chemical companies in the developed world. So far, genetic engineering has done little to improve yields or quality of crops in the developing world, except perhaps in China. In fact, there are a great many trials in progress, which hold out hopes of doing just that. Here are just a few.

Inadequate intake of iron is one of the major dietary deficiencies (Chapter 7) worldwide. An enzyme in some plant roots enables them to extract more iron from the soil. The gene for this enzyme can be transferred to plants, such as rice, enabling them to extract iron from iron-deficient soils.

Over 100 million children in the world are deficient in vitamin A. This deficiency often leads to blindness. A gene for beta-carotene, a precursor of vitamin A, can be inserted into plants to alleviate this widespread deficiency. This is not, of course, the only way to increase vitamin A availability but it could make a significant contribution.

Some acid soils contain levels of aluminium that reduce yields of maize by up to 8%. About 40% of soils in tropical and subtropical regions have this problem. A gene introduced into maize produces citrate, which binds the aluminium in the soil and releases phosphate ions. After 15 years of trials, the GM maize was made available to farmers, but pressure from environmental groups has blocked its adoption.

As a result of irrigation, much agricultural land has become salty and unproductive. Transferring a gene for salt tolerance from, say, mangrove plants to crop plants could bring these regions back into production.

If the gene, or genes, for nitrogen fixation (Chapter 19) from bacteria or leguminous plants could be introduced to cereal crops, yields could be increased without the need to add fertilisers.

Similarly, genes for drought resistance would make arid areas available for growing crops.

Genes coding for human vaccines have been introduced into plants.

Hepatitis B vaccine

The gene for the protein coat of the hepatitis virus is inserted into yeast cells. When these are cultured, they produce a protein that acts as an antigen (a vaccine, Chapter 10) and promotes the production of antibodies to the disease.

Transgenic plants have been engineered to produce vaccines that can be taken effectively by mouth. These include vaccines against rabies and cholera. Several species of plant have been used, including the banana, which is cheap and widespread in the tropics, can be eaten without cooking and does not produce seeds (Figure 20.7).

Figure 20.7 It is important to ensure that plants engineered to produce drugs and vaccines cannot find their way, by chance, into the human food chain. Strict control measures have to be applied.

Possible hazards of GM crops

One of the possible harmful effects of planting GM crops is that their modified genes might get into wild plants. If a gene for herbicide resistance found its way, via pollination, into a 'weed' plant, this plant might become resistant to herbicides and so become a 'super weed'. The purpose of field trials is to assess the likelihood of this happening. Until it is established that this is a negligible risk, licences to grow GM crops will not be issued.

To prevent the transfer of pollen from GM plants, other genes can be introduced, which stop the plant from producing pollen and induce the seeds and fruits to develop without fertilisation. This is a process that occurs naturally in many cultivated and wild plants.

Apart from specific hazards, there is also a sense of unease about introducing genes from one species into a totally different species. This is something that does not happen 'in nature' and therefore long-term effects are not known. In conventional cross-breeding, the genes transferred come from the same, or a closely related, species. However, in cross-breeding the whole raft of genes is transferred and this has sometimes had bad results when genes other than the target genes have combined to produce harmful products. Genetic engineering offers the advantage of transferring only those genes that are required.

The differences between the genetic make-up of different organisms is not as great as we tend to think. Plants and animals share 60% of their genes and humans have 50% of their genes in common with fruit flies. Not all genetic engineering involves transfer of 'alien' genes. In some cases it is the plant's own genes that are modified to improve its success in the field.

At least some of the protests against GM crops may be ill-judged (Figure 20.8).

Figure 20.8 Ill-judged protest. These vandalised poplars carried a gene that softened the cell walls, reducing the need for environmentally damaging chemicals used in paper making. They were also all female plants so no pollen could have been produced.

Use of bacteria and restriction enzymes in genetic engineering

To understand the principles of genetic engineering you need to know something about bacteria (Figure 1.29) and **restriction enzymes**.

Bacteria are microscopic single-celled organisms with cytoplasm, cell membranes and cell walls, but without a proper nucleus. Genetic control in a bacterium is exercised by a double strand of deoxyribonucleic acid (DNA) in the form of a circle, but not enclosed in a nuclear membrane. This circular DNA strand carries the genes that control bacterial metabolism.

In addition, there are present in the cytoplasm a number of small, circular pieces of DNA called **plasmids**. The plasmids often carry genes that give the bacterium resistance to particular antibiotics such as tetracycline and ampicillin.

Restriction enzymes are produced by bacteria. They 'cut' DNA molecules at specific sites, e.g. between the A and the T in the sequence GAA–TTC. Restriction enzymes can be extracted from bacteria and purified. By using a selected restriction enzyme, DNA molecules extracted from different organisms can be cut at predictable sites and made to produce lengths of DNA that contain specific genes.

DNA from human cells can be extracted and restriction enzymes used to 'cut' out a sequence of DNA that includes a gene, e.g. the gene for production of insulin (Figure 20.9). These lengths have sticky ends. Plasmids are extracted from bacteria and 'cut open' with the same restriction enzyme. If the human DNA is then added to a suspension of the plasmids, some of the human DNA will attach to some of the plasmids by their sticky ends, and the plasmids will then close up again, given suitable enzymes such as **ligase**. The DNA in these plasmids is called **recombinant DNA**.

The bacteria can be induced to take up the plasmids and, by ingenious culture methods using antibiotics, it is possible to select the bacteria that contain the recombinant DNA. The human DNA in the plasmids continues to produce the same protein as it did in the human cells. In the example mentioned, this would be the protein, insulin (Chapter 14). The plasmids are said to be the **vectors** that carry the human DNA into the bacteria and the technique is sometimes called **gene-splicing**.

Given suitable nutrient solutions, bacteria multiply rapidly and produce vast numbers of offspring. The bacteria reproduce by mitosis (Chapter 17) and so each daughter bacterium will contain the same DNA and the same plasmids as the parent. The offspring form a clone and the insulin gene is said to be cloned by this method.

The bacteria are cultured in special vessels called **fermenters** (Figure 20.2) and the insulin that they produce can be extracted from the culture medium and purified for use in treating diabetes (Chapter 14).

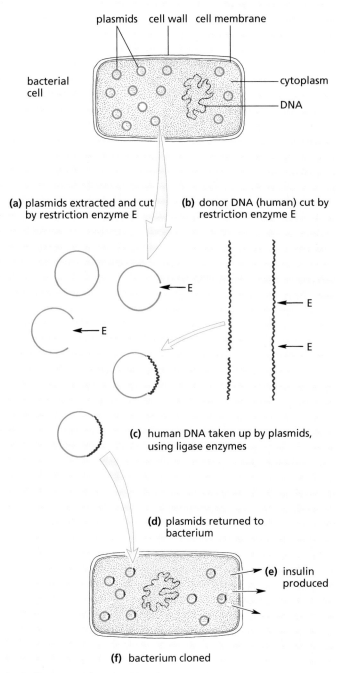

Figure 20.9 The principles of genetic engineering

This is only one type of genetic engineering. The vector may be a virus rather than a plasmid; the DNA may be inserted directly, without a vector; the donor DNA may be synthesised from nucleotides rather than extracted from cells; yeast may be used instead of bacteria. The outcome, however, is the same. DNA from one species is inserted into a different species and made to produce its normal proteins (Figure 20.9).

In the example shown in Figure 20.9, the gene product, insulin, is harvested and used to treat diabetes. In other cases, genes are inserted into organisms to promote changes that may be beneficial. Bacteria or viruses are used as vectors to deliver the genes. For example, a bacterium is used to deliver a gene for herbicide resistance in crop plants.

GM food

This is food prepared from GM crops. Most genetic modifications are aimed at increasing yields rather than changing the quality of food. However, it is possible to improve the protein, mineral or vitamin content of food and the keeping qualities of some products (Figure 20.6).

Possible hazards of GM food

One of the worries is that the vectors for delivering recombinant DNA contain genes for antibiotic resistance. The antibiotic-resistant properties are used to select only those vectors that have taken up the new DNA. If, in the intestine, the DNA managed to get into potentially harmful bacteria, it might make them resistant to antibiotic drugs.

Although there is no evidence to suggest this happens in experimental animals, the main biotech companies are trying to find methods of selecting vectors without using antibiotics.

Another concern is that GM food could contain pesticide residues or substances that cause allergies (allergens). However, it has to be said that all GM products are rigorously tested for toxins and allergens over many years, far more so than any products from conventional cross-breeding. The GM products have to be passed by a series of regulatory and advisory bodies before they are released on to the market. In fact only a handful of GM foods are available. One of these is soya, which is included, in one form or another, in 60% of processed foods.

Golden rice was a variety of rice developed through genetic engineering to carry a gene that is responsible for making beta-carotene, a precursor of vitamin A. In countries where rice is a staple food, the use of golden rice could reduce the incidence of a condition called night blindness – a serious problem which is estimated to kill 670 000 children under the age of 5 each year.

However, some argue that there is a danger of the precursor changing into other, toxic chemicals once eaten. There were also concerns about a reduction in biodiversity as a result of the introduction of GM species. Subsistence farmers could also be tied to large agricultural suppliers who may then manipulate seed prices.

Questions

Core

1. Outline the biology involved in making bread.
2. How is DNA in a bacterium different from DNA in an animal cell?
3. Outline three commercial uses of enzymes.

Extension

4. Give two reasons why bacteria are more suitable for use in genetic engineering than, for example, mammals.
5. a With reference to their sources, explain why ethanol is described as a renewable energy source while petrol is described as a non-renewable source.
   b Use of a renewable source of energy such as ethanol for fuel in motor cars seems like a good solution to fuel shortages. What are the disadvantages of using ethanol?
6. Some people are lactose-intolerant. Explain how biotechnology can be used to allow people with this condition to eat milk products.
7. Make a table to outline the advantages and disadvantages of GM crops.
8. How can genetic engineering be used to solve major worldwide dietary deficiencies such as vitamin and mineral deficiencies?

Checklist

After studying Chapter 20 you should know and understand the following:

Biotechnology and genetic engineering

- Bacteria are useful in biotechnology and genetic engineering because of their ability to make complex molecules and their rapid reproduction.

- Bacteria are useful in biotechnology and genetic engineering because of lack of ethical concerns over their manipulation and growth.
- The genetic code in bacteria is shared with all other organisms.
- Bacteria contain DNA in the form of plasmids, which can be cut open to insert genes.

Biotechnology

- Biotechnology is the application of living organisms, systems or processes in industry.
- Many biotechnological processes use micro-organisms (fungi and bacteria) to bring about the reactions.
- Most biotechnological processes are classed as 'fermentations'.
- Fermentation may be aerobic or anaerobic.
- The required product of biotechnology may be the organism itself (e.g. mycoprotein) or one of its products (e.g. alcohol).
- Yeasts produce ethanol by anaerobic respiration. The ethanol can be produced commercially for biofuel.
- Anaerobic respiration by yeast is also involved in bread-making.
- Pectinase can be used to extract fruit juices.
- Lipase and protease enzymes are used in biological washing powders to remove fat and protein stains.

- Lactase is used to produce lactose-free milk.
- Antibiotics are produced from bacteria and fungi.
- The fungus *Penicillium* is used in the production of the antibiotic penicillin.

- Fermenters are used in the production of penicillin.
- Enzymes from micro-organisms can be produced on an industrial scale and used in other biotechnology processes.
- Sterile conditions are essential in biotechnology to avoid contamination by unwanted microbes.

Genetic engineering

- Genetic engineering is changing the genetic material of an organism by removing, changing or inserting individual genes.
- Examples of genetic engineering include:
  - the insertion of humans genes into bacteria to produce human insulin
  - the insertion of genes into crop plants to confer resistance to herbicides or insect pests
  - the insertion of genes into crop plants to provide additional vitamins.
- Plasmids and viruses are vectors used to deliver the genes.
- Genetic engineering is used in the production of enzymes, hormones and drugs.
- Crop plants can be genetically modified to resist insect pests and herbicides.
- There is concern that the genes introduced into crop plants might spread to wild plants.

- Genetic engineering can use bacteria to produce human protein, such as insulin.
- Human gene DNA is isolated using restriction enzymes, forming sticky ends.
- Bacterial plasmid DNA is cut with same restriction enzymes, forming matching sticky ends.
- Human gene DNA is inserted into the bacterial plasmid DNA using DNA ligase to form a recombinant plasmid.
- The plasmid is inserted into bacteria.
- The bacteria containing the recombinant plasmid are replicated.
- They make a human protein as they express the gene.
- There are advantages and disadvantages of genetically modifying crops, such as soya, maize and rice.

21 Human influences on ecosystems

- **Food supply**
- Use of modern technology in increased food production
- Negative impacts of monocultures and intensive livestock production to an ecosystem
- Social, environmental and economic implications of providing sufficient food for an increasing human global population

- **Habitat destruction**
- Reasons for habitat destruction
- Effects of altering food chains and webs on habitats
- Effects of deforestation on habitats
- Explain undesirable effects of deforestation on the environment

- **Pollution**
- Sources and effects of land and water pollution
- Sources and effects of air pollution

- Eutrophication
- Effects of non-biodegradable plastics on the environment
- Acid rain
- Greenhouse effect and climate change
- Negative impacts of female hormones in water courses

- **Conservation**
- Define sustainable resource
- The need to conserve non-renewable resources
- Maintenance of forest and fish stocks
- Reuse and recycling of products
- Treatment of sewage
- Reasons why species are becoming endangered or extinct
- Conservation of endangered species

- Define sustainable development
- Methods for sustaining forest and fish stocks
- Strategies for sustainable development
- Reasons for conservation programmes

● Food supply

A few thousand years ago, most of the humans on the Earth obtained their food by gathering leaves, fruits or roots and by hunting animals. The population was probably limited by the amount of food that could be collected in this way.

Human faeces, urine and dead bodies were left on or in the soil and so played a part in the nitrogen cycle (Chapter 19). Life may have been short, and many babies may have died from starvation or illness, but humans fitted into the food web and nitrogen cycle like any other animal.

Once agriculture had been developed, it was possible to support much larger populations and the balance between humans and their environment was upset.

Intensification of agriculture

Forests and woodland are cut down in order to grow more food. This destroys important wildlife habitats and may affect the climate. Tropical rainforest is being cut down at the rate of 111 400 square kilometres per year. Since 1950, between 30 and 50% of British deciduous woodlands have been felled to make way for farmland or conifer plantations.

Modern **agricultural machinery** is used to clear the land, prepare the soil and plant, maintain and harvest crops to improve efficiency. To make the process even more efficient, fields are made larger by taking out hedges (Figure 21.1).

Figure 21.1 Destruction of a hedgerow. Permission now has to be sought from the local authority before this can happen. Grants are available in some countries to replant hedges.

Larger vehicles such as tractors (see Figure 21.6) and combine harvesters (see Figure 21.5) can then be used in the fields to speed up the farming processes. However, studies have shown that repeated ploughing of a pasture reduces the number of species in the soil.

Food supply

The use of chemical fertilisers to improve yield

In a natural community of plants and animals, the processes that remove and replace mineral elements in the soil are in balance. In agriculture, most of the crop is usually removed so that there is little or no organic matter for nitrifying bacteria to act on. In a farm with animals, the animal manure, mixed with straw, is ploughed back into the soil or spread on the pasture. The manure thus replaces the nitrates and other minerals removed by the crop. It also gives the soil a good structure and improves its water-holding properties.

When animal manure is not available in large enough quantities, **chemical fertilisers** are used. These are mineral salts made on an industrial scale. Examples are ammonium sulfate (for nitrogen and sulfur), ammonium nitrate (for nitrogen) and compound NPK fertiliser for nitrogen, phosphorus and potassium. These are spread on the soil in carefully calculated amounts to provide the minerals, particularly nitrogen, phosphorus and potassium, which the plants need. These artificial fertilisers increase the yield of crops from agricultural land, but they do little to maintain a good soil structure because they contain no organic matter (Figures 21.2 and 21.3). In some cases, their use results in the pollution of rivers and streams (see 'Pollution' later in this chapter).

Monoculture

The whole point of crop farming is to remove a mixed population of trees, shrubs, wild flowers and grasses (Figure 21.4) and replace it with a dense population of only one species such as wheat or beans (Figure 21.5). When a crop of a single species is grown on the same land, year after year, it is called a **monoculture**.

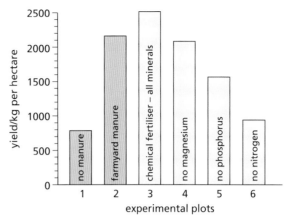

Figure 21.3 Average yearly wheat yields from 1852 to 1925, Broadbalk field, Rothamsted Experimental Station. Plot 1 received no manure or chemical fertiliser for 73 years. Plot 2 received an annual application of farmyard manure. Plot 3 received chemical fertiliser with all necessary minerals. Plots 4 to 6 received chemical fertiliser lacking one element.

Figure 21.4 Natural vegetation. Uncultivated land carries a wide variety of species.

Figure 21.2 Experimental plots of wheat. The rectangular plots have been treated with different fertilisers.

Figure 21.5 A monoculture. Only wheat is allowed to grow. All competing plants are destroyed.

21 HUMAN INFLUENCES ON ECOSYSTEMS

Figure 21.6 Weed control by herbicide spraying. A young wheat crop is sprayed with herbicide to suppress weeds.

Figure 21.7 Effect of a herbicide spray. The crop has been sprayed except for a strip which the tractor driver missed.

The negative impact of monocultures

In a monoculture, every attempt is made to destroy organisms that feed on, compete with or infect the crop plant. So, the balanced life of a natural plant and animal community is displaced from farmland and left to survive only in small areas of woodland, heath or hedgerow. We have to decide on a balance between the amount of land to be used for agriculture and the amount of land left alone in order to keep a rich variety of wildlife on the Earth's surface.

Pesticides: insecticides and herbicides

Monocultures, with their dense populations of single species and repeated planting, are very susceptible to attack by insects or the spread of fungus diseases. To combat these threats, **pesticides** are used. A pesticide is a chemical that destroys agricultural pests or competitors.

For a monoculture to be maintained, plants that compete with the crop plant for root space, soil minerals and sunlight are killed by chemicals called **herbicides** (Figures 21.6 and 21.7). To destroy insects that eat and damage the plants, the crops are sprayed with **insecticides**.

The trouble with most pesticides is that they kill indiscriminately. Insecticides, for example, kill not only harmful insects but the harmless and beneficial ones, such as bees, which pollinate flowering plants, and ladybirds, which eat aphids.

In about 1960, a group of chemicals, including **aldrin** and **dieldrin**, were used as insecticides to kill wireworms and other insect pests in the soil. However, aldrin was found to reduce the number of species of soil animals in a pasture to half the original number (Figure 21.8). Dieldrin was also used as a seed dressing. If seeds were dipped in the chemical before planting, it prevented certain insects from attacking the seedlings. This was thought to be better than spraying the soil with dieldrin, which would have killed all the insects in the soil. Unfortunately pigeons, rooks, pheasants and partridges dug up and ate so much of the seed that the dieldrin poisoned them. Thousands of these birds were poisoned and, because they were part of a food web, birds of prey and foxes, which fed on them, were also killed. The use of dieldrin and aldrin was restricted in 1981 and banned in 1992.

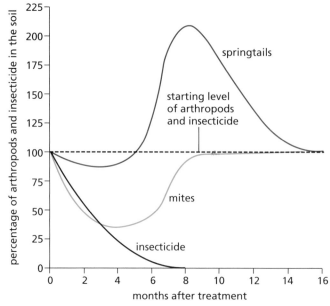

Figure 21.8 The effect of insecticide on some soil organisms

One alternative to pesticides is the use of biological control, though this also is not without its drawbacks unless it is thoroughly researched and tested. It may involve the introduction of foreign species, which could interfere with food chains and webs (see Chapter 19).

Selective breeding

An important part of any breeding programme is the selection of the desired varieties. The largest fruit on a tomato plant might be picked and its seeds planted next year. In the next generation, once again only seeds from the largest tomatoes are planted. Eventually it is possible to produce a true-breeding variety of tomato plant that forms large fruits. Figure 18.25 shows the result of such selective breeding for different characteristics. The same technique can be used for selecting other desirable qualities, such as flavour and disease resistance.

Similar principles can be applied to farm animals. Desirable characteristics, such as high milk yield and resistance to disease, may be combined. Stock-breeders will select calves from cows that give large quantities of milk. These calves will be used as breeding stock to build a herd of high yielders. A characteristic such as milk yield is probably under the control of many genes. At each stage of selective breeding the farmer, in effect, is keeping the beneficial genes and discarding the less useful genes from his or her animals.

Selective breeding in farm stock can be slow and expensive because the animals often have small numbers of offspring and breed only once a year.

One of the drawbacks of selective breeding is that the whole set of genes is transferred. As well as the desirable genes, there may be genes which, in a homozygous condition, would be harmful. It is known that artificial selection repeated over a large number of generations tends to reduce the fitness of the new variety (Chapter 18).

A long-term disadvantage of selective breeding is the loss of variability. By eliminating all the offspring who do not bear the desired characteristics, many genes are lost from the population. At some future date, when new combinations of genes are sought, some of the potentially useful ones may no longer be available.

In attempting to introduce, in plants, characteristics such as salt tolerance or resistance to disease or drought, the plant breeder goes back to wild varieties, as shown in Figure 18.26. However, with the current rate of extinction, this source of genetic material is diminishing.

In the natural world, reduction of variability could lead to local extinction if the population was unable to adapt, by natural selection, to changing conditions.

The negative impacts of intensive livestock production

Intensive livestock production is also known as 'factory farming'. Chickens (Figure 19.13) and calves are often reared in large sheds instead of in open fields. Their urine and faeces are washed out of the sheds with water forming 'slurry'. If this slurry gets into streams and rivers it supplies an excess of nitrates and phosphates for the microscopic algae. This starts a chain of events, which can lead to **eutrophication** of the water system (see later in this chapter).

Overgrazing can result if too many animals are kept on a pasture. They eat the grass down almost to the roots, and their hooves trample the surface soil into a hard layer. As a result, the rainwater will not penetrate the soil so it runs off the surface, carrying the soil with it. The soil becomes eroded.

The problems of world food supplies

There is not always enough food available in a country to feed the people living there. A severe food shortage can lead to famine. Food may have to be brought in (imported). Fresh food can have a limited storage life, so it needs to be transported quickly or treated to prevent it going rotten. Methods to increase the life of food include transport in chilled containers, or picking the produce before it is ripe. When it has reached its destination, it is exposed to chemicals such as plant auxins to bring on the ripening process. The use of aeroplanes to transport food is very expensive. The redistribution of food from first world countries to a poorer one can have a detrimental effect on that country's local economy by reducing the value of food grown by local farmers. Some food grown by countries with large debts may be exported as cash crops, even though the local people desperately need the food.

21 HUMAN INFLUENCES ON ECOSYSTEMS

Other problems that can result in famine include:

- climate change and natural disasters such as flooding (caused by excessive rainfall or tsunamis) or drought; waterlogged soil can become infertile due to the activities of denitrifying bacteria, which break down nitrates
- pollution
- shortage of water through its use for other purposes, the diversion of rivers, building dams to provide hydroelectricity
- eating next year's seeds through desperation for food
- poor soil, lack of inorganic ions or fertiliser
- desertification due to soil erosion as a result of deforestation
- lack of money to buy seeds, fertiliser, pesticides or machinery
- war, which can make it too dangerous to farm, or which removes labour
- urbanisation (building on farmland); the development of towns and cities makes less and less land available for farmland
- an increasing population
- pest damage or disease
- poor education of farmers and outmoded farming practices
- the destruction of forests, so there is nothing to hunt and no food to collect
- use of farmland to grow cash crops, or plants for biofuel.

● Habitat destruction

Removal of habitats

Farmland is not a natural habitat but, at one time, hedgerows, hay meadows and stubble fields were important habitats for plants and animals. Hay meadows and hedgerows supported a wide range of wild plants as well as providing feeding and nesting sites for birds and animals.

Intensive agriculture has destroyed many of these habitats; hedges have been grubbed out (see Figure 21.1) to make fields larger, a monoculture of silage grasses (Figure 21.9) has replaced the mixed population of a hay meadow (Figure 21.10) and planting of winter wheat has denied animals access to stubble fields in autumn. As a result, populations of butterflies, flowers and birds such as skylarks, grey partridges, corn buntings and tree sparrows have crashed.

Recent legislation now prohibits the removal of hedgerows without approval from the local authority but the only hedges protected in this way are those deemed to be 'important' because of species diversity or historical significance.

In Britain, the **Farming and Wildlife Advisory Group** (**FWAG**) can advise farmers how to manage their land in ways that encourage wildlife. This includes, for example, leaving strips of uncultivated land around the margins of fields or planting new hedgerows. Even strips of wild grasses and flowers between fields significantly increase the population of beneficial insects.

The development of towns and cities (**urbanisation**) makes a great demand on land, destroying natural habitats. In addition, the crowding of growing populations into towns leads to problems of waste disposal. The sewage and domestic waste from a town of several thousand people can cause disease and pollution in the absence of effective means of disposal, damaging surrounding habitats.

Extraction of natural resources

An increasing population and greater demands on modern technology means we need more raw materials for the manufacturing industry and greater energy supplies.

Fossil fuels such as coal can be mined, but this can permanently damage habitats, partly due to the process of extraction, but also due to dumping of the rock extracted in spoil heaps. Some methods of coal extraction involve scraping off existing soil from the surface of the land. Spoil heaps created from waste rock can contain toxic metals, which prevent re-colonisation of the land. Open-pit mining puts demands on local water sources, affecting habitats in lakes and rivers. Water can become contaminated with toxic metals from the mining site, damaging aquatic habitats.

Oil spillages around oil wells are extremely toxic. Once the oil seeps into the soil and water systems, habitats are destroyed (Figure 21.11)

Habitat destruction

Figure 21.9 Grass for silage. There is no variety of plant life and, therefore, an impoverished population of insects and other animals.

Figure 21.10 The variety of wild flowers in a traditional hay meadow will attract butterflies and other insects.

Figure 21.11 Habitat destruction caused by an oil spillage in Nigeria

Mining for raw materials such as gold, iron aluminium and silicon leaves huge scars in the landscape and destroys large areas of natural habitat (Figure 21.12). The extraction of sand and gravel also leaves large pits that prevent previous habitats redeveloping.

Figure 21.12 Open-pit gold mine in New Zealand

In response to this increased human activity, in 1982 the United Nations developed the **World Charter for Nature**. This was followed in 1990 by **The World Ethic of Sustainability**, created by the World Wide Fund for Nature (WWF), the International Union for Conservation of Nature (IUCN) and the United Nations Environment Programme (UNEP). Included in this charter were habitat conservation and the need to protect natural resources from depletion.

Marine pollution

Marine habitats around the world are becoming contaminated with human debris. This includes untreated sewage, agricultural fertilisers and pesticides. Oil spills still cause problems, but this source of marine pollution is gradually reducing. Plastics are a huge problem: many are non-biodegradable so they persist in the environment. Others form micro-particles as they break down and these are mistaken by marine organisms for food and are indigestible. They stay in the stomach, causing sickness, or prevent the gills from working efficiently. Where fertilisers and sewage enter the marine environment, 'dead zones' develop where there is insufficient oxygen to sustain life. This destroys habitats (see next section).

Oil spills wash up on the intertidal zone, killing the seaweeds that provide nutrients for food chains. Filter-feeding animals such as barnacles and some species of mollusc die from taking in the oil (see Figure 1.8).

Any form of habitat destruction by humans, even where a single species is wiped out, can have an impact on food chains and food webs because other organisms will use that species as a food source, or their numbers will be controlled through its predation.

Deforestation

The removal of large numbers of trees results in habitat destruction on a massive scale.

- Animals living in the forest lose their homes and sources of food; species of plant become extinct as the land is used for other purposes such as agriculture, mining, housing and roads.
- Soil erosion is more likely to happen as there are no roots to hold the soil in place. The soil can end up in rivers and lakes, destroying habitats there.
- Flooding becomes more frequent as there is no soil to absorb and hold rainwater. Plant roots rot and animals drown, destroying food chains and webs.
- Carbon dioxide builds up in the atmosphere as there are fewer trees to photosynthesise, increasing global warming. Climate change affects habitats.

The undesirable effects of deforestation on the environment

Forests have a profound effect on climate, water supply and soil maintenance. They have been described as environmental buffers. For example, they intercept heavy rainfall and release the water steadily and slowly to the soil beneath and to the streams and rivers that start in or flow through them. The tree roots hold the soil in place.

At present, we are destroying forests, particularly tropical forests, at a rapid rate (1) for their timber, (2) to make way for agriculture, roads (Figure 21.13) and settlements, and (3) for firewood. The Food and Agriculture Organisation, run by the United Nations, reported that the overall tropical deforestation rates in the decade up to 2010 were 8.5% higher than during the 1990s. At the current rate of destruction, it is estimated that all tropical rainforests will have disappeared in the next 75 years.

Removal of forests allows soil erosion, silting up of lakes and rivers, floods and the loss for ever of thousands of species of animals and plants.

Trees can grow on hillsides even when the soil layer is quite thin. When the trees are cut down and the soil is ploughed, there is less protection from the wind and rain. Heavy rainfall washes the soil off the hillsides into the rivers. The hillsides are left bare and useless and the rivers become choked up with mud and silt, which can cause floods (Figures 21.14 and 21.15). For example, Argentina spends 10 million dollars a year on dredging silt from the River Plate estuary to keep the port of Buenos Aires open to shipping. It has been found that 80% of this sediment comes from a deforested and overgrazed region 1800 km upstream, which represents only 4% of the river's total catchment area. Similar sedimentation has halved the lives of reservoirs, hydroelectric schemes and irrigation programmes. The disastrous floods in India and Bangladesh in recent years may be attributed largely to deforestation.

Figure 21.13 Cutting a road through a tropical rainforest. The road not only destroys the natural vegetation, it also opens up the forest to further exploitation.

Habitat destruction

Forests and climate

About half the rain that falls in tropical forests comes from the transpiration of the trees themselves. The clouds that form from this transpired water help to reflect sunlight and so keep the region relatively cool and humid. When areas of forest are cleared, this source of rain is removed, cloud cover is reduced and the local climate changes quite dramatically. The temperature range from day to night is more extreme and the rainfall diminishes.

In North Eastern Brazil, for example, an area which was once rainforest is now an arid wasteland. If more than 60% of a forest is cleared, it may cause irreversible changes in the climate of the whole region. This could turn the region into an unproductive desert.

Removal of trees on such a large scale also reduces the amount of carbon dioxide removed from the atmosphere in the process of photosynthesis (see 'Nutrient cycles', Chapter 19, and 'Photosynthesis', Chapter 6). Most scientists agree that the build-up of CO_2 in the atmosphere contributes to global warming.

Figure 21.14 Soil erosion. Removal of forest trees from steeply sloping ground has allowed the rain to wash away the topsoil.

The soil of tropical forests is usually very poor in nutrients. Most of the organic matter is in the leafy canopy of the tree tops. For a year or two after felling and burning, the forest soil yields good crops but the nutrients are soon depleted and the soil eroded. The agricultural benefit from cutting down forests is very short-lived, and the forest does not recover even if the impoverished land is abandoned.

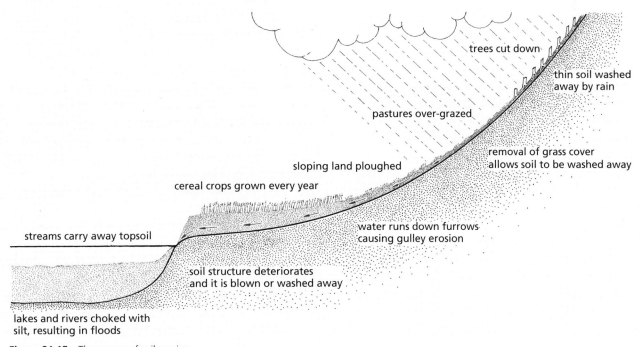

Figure 21.15 The causes of soil erosion

Forests and biodiversity

One of the most characteristic features of tropical rainforests is the enormous diversity of species they contain. In Britain, a forest or wood may consist of only one or two species of tree such as oak, ash, beech or pine. In tropical forests there are many more species and they are widely dispersed throughout the habitat. It follows that there is also a wide diversity of animals that live in such habitats. In fact, it has been estimated that half of the world's 10 million species live in tropical forests.

Destruction of tropical forest, therefore, destroys a large number of different species, driving many of them to the verge of extinction, and also drives out the indigenous populations of humans. In addition, we may be depriving ourselves of many valuable sources of chemical compounds that the plants and animals produce. The US National Cancer Institute has identified 3000 plants having products active against cancer cells and 70% of them come from rainforests (Figure 21.16).

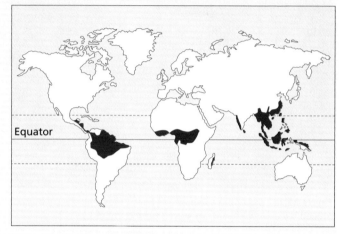

Figure 21.16 The world's rainforests

● Pollution

Insecticides

The effects of the insecticides aldrin and dieldrin were discussed earlier in this chapter. Most insecticide pollution is as a result of their use in agriculture. However, one pesticide, called DDT, was used to control the spread of malaria by killing mosquitos, which carry the protoctist parasites that cause the disease. Unfortunately, DDT remains in the environment after it has been sprayed and can be absorbed in sub-lethal doses by microscopic organisms. Hence, it can enter food chains and accumulate as it moves up them.

The concentration of insecticide often increases as it passes along a food chain (Figure 21.17). Clear Lake in California was sprayed with DDT to kill gnat larvae. The insecticide made only a weak solution of 0.015 parts per million (ppm) in the lake water. The microscopic plants and animals that fed in the lake water built up concentrations of about 5 ppm in their bodies. The small fish that fed on the microscopic animals had 10 ppm. The small fish were eaten by larger fish, which in turn were eaten by birds called grebes. The grebes were found to have 1600 ppm of DDT in their body fat and this high concentration killed large numbers of them.

Larger scale pollution of water by insecticides, for instance by leakage from storage containers, may kill aquatic insects, destroying one or more levels in a food chain or food web, with serious consequences to the ecosystem.

A build-up of pesticides can also occur in food chains on land. In the 1950s in the USA, DDT was sprayed on to elm trees to try and control the beetle that spread Dutch elm disease. The fallen leaves, contaminated with DDT, were eaten by earthworms. Because each worm ate many leaves, the DDT concentration in their bodies was increased ten times. When birds ate a large number of worms, the concentration of DDT in the birds' bodies reached lethal proportions and there was a 30–90% mortality among robins and other song birds in the cities.

Even if DDT did not kill the birds, it caused them to lay eggs with thin shells. The eggs broke easily and fewer chicks were raised. In Britain, the numbers of peregrine falcons and sparrow hawks declined drastically between 1955 and 1965. These birds are at the top of a food web and so accumulate very high doses of the pesticides that are present in their prey, such as pigeons. After the use of DDT was restricted, the population of peregrines and sparrow hawks started to recover.

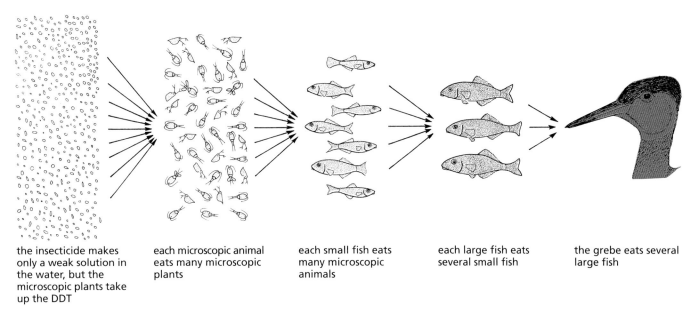

the insecticide makes only a weak solution in the water, but the microscopic plants take up the DDT

each microscopic animal eats many microscopic plants

each small fish eats many microscopic animals

each large fish eats several small fish

the grebe eats several large fish

Figure 21.17 Pesticides may become more concentrated as they move along a food chain. The intensity of colour represents the concentration of DDT.

These new insecticides had been thoroughly tested in the laboratory to show that they were harmless to humans and other animals when used in low concentrations. It had not been foreseen that the insecticides would become more and more concentrated as they passed along the food chain.

Insecticides like this are called **persistent** because they last a long time without breaking down. This makes them good insecticides but they also persist for a long time in the soil, in rivers, lakes and the bodies of animals, including humans. This is a serious disadvantage.

Herbicides

Herbicides are used by farmers to control plants (usually referred to as weeds) that compete with crop plants for nutrients, water and light (see Figure 21.7). If the weeds are not removed, crop productivity is reduced. However, if the herbicides do not break down straight away, they can leach from farmland into water systems such as rivers and lakes, where they can kill aquatic plants, removing the producers from food chains. Herbivores lose their food source and die or migrate. Carnivorous animals are then affected as well.

Leakage or dumping of persistent herbicides into the sea can have a similar effect on marine food chains.

Herbicides tend to be non-specific: they kill any broadleaved plants they come into contact with or are absorbed by. If herbicides are sprayed indiscriminately, they may blow onto surrounding land and kill plants other than the weeds in the crop being treated. This can put rare species of wild flowers at risk.

Nuclear fall-out

This can be the result of a leak from a nuclear power station, or from a nuclear explosion. Radioactive particles are carried by the wind or water and gradually settle in the environment. If the radiation has a long **half-life**, it remains in the environment and is absorbed by living organisms. The radioactive material bioaccumulates in food chains and can cause cancer in top carnivores.

Probably the worst nuclear accident in history happened at Chernobyl in Russia in April 1986. One of the reactor vessels exploded and the resulting fire produced a cloud of radioactive fallout, which was carried by prevailing winds over other parts of the Soviet Union and Europe. The predicted death toll, from direct exposure to the radiation and indirectly from the fallout, is estimated to be at least 4000 people (and possibly much higher), with many others suffering from birth defects or cancers associated with exposure to radiation. The fall-out contaminated the soil it fell on and was absorbed by plants, which were grazed by animals. Farmers in the Lake District in England were still banned from selling sheep for meat until June 2012, 26 years after the contamination of land there first happened.

Another major nuclear disaster happened at the Fukushima nuclear power plant in Japan in March 2011 (Figure 21.18). The plant was hit by a powerful tsunami, caused by an earthquake. A plume of radioactive material was carried from the site by the wind and came down onto the land, forming a scar like a teardrop over 30 kilometres wide. The sea around the power plant is heavily contaminated by radiation. This is absorbed into fish bones, making the animals unfit for consumption.

Figure 21.19 River pollution. The river is badly polluted by the effluent from a paper mill.

In 1971, 45 people in Minamata Bay in Japan died and 120 were seriously ill as a result of mercury poisoning. It was found that a factory had been discharging a compound of mercury into the bay as part of its waste. Although the mercury concentration in the sea was very low, its concentration was increased as it passed through the food chain (see Figure 21.17). By the time it reached the people of Minamata Bay in the fish and other sea food that formed a large part of their diet, it was concentrated enough to cause brain damage, deformity and death.

High levels of mercury have also been detected in the Baltic Sea and in the Great Lakes of North America.

Oil pollution of the sea has become a familiar event. In 1989, a tanker called the *Exxon Valdez* ran on to Bligh Reef in Prince William Sound, Alaska, and 11 million gallons of crude oil spilled into the sea. Around 400 000 sea birds were killed by the oil (Figure 21.20) and the populations of killer whales, sea otters and harbour seals among others, were badly affected. The hot water high-pressure hosing techniques and chemicals used to clean up the shoreline killed many more birds and sea creatures living on the coast. Since 1989, there have continued to be major spillages of crude oil from tankers and off-shore oil wells.

Figure 21.18 Fukushima nuclear power plant, destroyed by a powerful tsunami and fire

Chemical waste

Many industrial processes produce poisonous waste products. Electroplating, for example, produces waste containing copper and cyanide. If these chemicals are released into rivers they poison the animals and plants and could poison humans who drink the water. It is estimated that the River Trent receives 850 tonnes of zinc, 4000 tonnes of nickel and 300 tonnes of copper each year from industrial processes.

Any factory getting rid of its effluent into water systems risks damaging the environment (Figure 21.19). Some detergents contain a lot of phosphate. This is not removed by sewage treatment and is discharged into rivers. The large amount of phosphate encourages growth of microscopic plants (algae).

Discarded rubbish

The development of towns and cities, and the crowding of growing populations into them, leads to problems of waste disposal. The domestic waste from a town of several thousand people can cause disease and pollution in the absence of effective means of disposal. Much ends up in landfill sites, taking up valuable space, polluting the ground and attracting vermin and insects, which can spread disease. Most consumable items come in packaging, which, if not recycled, ends up in landfill sites or is burned, causing air pollution. Discarded rubbish that ends up in the sea can cause severe problems for marine animals.

Figure 21.20 Oil pollution. Oiled sea birds like this long-tailed duck cannot fly to reach their feeding grounds. They also poison themselves by trying to clean the oil from their feathers.

Sewage

Diseases like typhoid and cholera are caused by certain bacteria when they get into the human intestine. The faeces passed by people suffering from these diseases will contain the harmful bacteria. If the bacteria get into drinking water they may spread the disease to hundreds of other people. For this reason, among others, untreated sewage must not be emptied into rivers. It is treated at the sewage works so that all the solids are removed. The human waste is broken down by bacteria and made harmless (free from harmful bacteria and poisonous chemicals), but the breakdown products include phosphates and nitrates. When the water from the sewage treatment is discharged into rivers it contains large quantities of phosphate and nitrate, which allow the microscopic plant life to grow very rapidly (Figure 21.21).

Figure 21.21 Growth of algae in a lake. Abundant nitrate and phosphate from treated sewage and from farmland make this growth possible.

Fertilisers

When nitrates and phosphates from farmland and sewage escape into water they cause excessive growth of microscopic green plants. This may result in a serious oxygen shortage in the water, resulting in the death of aquatic animals – a process called **eutrophication**.

Eutrophication

Nitrates and phosphates are present from a number of sources, including untreated sewage, detergents from manufacturing and washing processes, arable farming and factory farming.

If these nitrates or phosphates enter a water system, they become available for algae (aquatic plants) to absorb. The plants need these nutrients to grow. More nutrients result in faster growth (Figure 21.21). As the plants die, some through lack of light because of overcrowding, aerobic bacteria decompose them and respire, taking oxygen out of the water. As oxygen levels drop, animals such as fish cannot breathe, so they die and the whole ecosystem is destroyed (Figure 21.22).

21 HUMAN INFLUENCES ON ECOSYSTEMS

Figure 21.22 Fish killed by pollution. The water may look clear but is so short of oxygen that the fish have died from suffocation.

Figure 21.23 shows this sequence of events as a flow chart.

```
nitrates or phosphates from raw sewage, fertilisers or
other sources enter a water system (river or lake)
                        ↓
algae absorb the nutrients and grow rapidly
        (called an algal bloom)
                        ↓
algae form a blanket on the surface of the water,
blocking light from the reaching algae below
                        ↓
        algae die without light
                        ↓
bacteria decompose the dead algae, using up
the oxygen in the water for respiration
                        ↓
animals in the water die through lack of oxygen
```

Figure 21.23 The sequence of events leading to eutrophication

The greenhouse effect and global warming

Levels of carbon dioxide in the atmosphere are influenced by natural processes and by human activities. Processes that change the equilibrium (balance) include:

- cutting down forests (deforestation) – less photosynthesis
- combustion of fossil fuels (coal, oil and gas)
- increasing numbers of animals (including humans) – they all respire.

An increase in levels of carbon dioxide in the atmosphere is thought to contribute to global warming. Carbon dioxide forms a layer in the atmosphere, which traps heat radiation from the Sun.

Methane also acts as a greenhouse gas. Its levels in the atmosphere have more than doubled over the past 200 years and its effects on global warming are much greater than carbon dioxide. It is produced by the decay of organic matter in anaerobic conditions, such as in wet rice fields and in the stomachs of animals, e.g. cattle and termites. It is also released from the ground during the extraction of oil and coal.

The build-up of greenhouse gases causes a gradual increase in the atmospheric temperature, known as the **enhanced greenhouse effect**. This can:

- melt polar ice caps, causing flooding of low-lying land
- change weather conditions in some countries, increasing flooding or reducing rainfall – changing arable (farm) land to desert; extreme weather conditions become more common
- cause the extinction of some species that cannot survive in raised temperatures.

Eutrophication

In Chapter 6 it was explained that plants need a supply of nitrates for making their proteins. They also need a source of phosphates for many chemical reactions in their cells. The rate at which plants grow is often limited by how much nitrate and phosphate they can obtain. In recent years, the amount of nitrate and phosphate in our rivers and lakes has been greatly increased. This leads to an accelerated process of eutrophication.

Eutrophication is the enrichment of natural waters with nutrients that allow the water to support an increasing amount of plant life. This process takes place naturally in many inland waters but usually very slowly. The excessive enrichment that results from human activities leads to an overgrowth of microscopic algae (Figure 21.21). These aquatic algae are at the bottom of the food chain. The extra nitrates and phosphates from the processes listed on page 329 enable them to increase so rapidly that they cannot be kept in check by the microscopic animals which normally eat them. So they die and fall to the bottom of the river or lake. Here, their bodies are broken down by bacteria. The bacteria need oxygen to carry out this breakdown and the oxygen is taken from the water (Figure 21.24). So much oxygen is taken that the water becomes deoxygenated and can no longer support animal life. Fish and other organisms die from suffocation (Figure 21.22).

The following processes are the main causes of eutrophication.

Discharge of treated sewage

In a sewage treatment plant, human waste is broken down by bacteria and made harmless, but the breakdown products include phosphates and nitrates. When the water from the sewage treatment is discharged into rivers it contains large quantities of phosphates and nitrates, which allow the microscopic plant life to grow very rapidly (Figure 21.21).

Use of detergents

Some detergents contain a lot of phosphate. This is not removed by sewage treatment and is discharged into rivers. The large amount of phosphates encourages growth of microscopic plants (algae).

Arable farming

Since the Second World War, more and more grassland has been ploughed up in order to grow arable crops such as wheat and barley. When soil is exposed in this way, the bacteria, aided by the extra oxygen and water, produce soluble nitrates, which are washed into streams and rivers where they promote the growth of algae. If the nitrates reach underground water stores they may increase the nitrate in drinking water to levels considered 'unsafe' for babies.

Some people think that it is excessive use of artificial fertilisers that causes this pollution but there is not much evidence for this.

'Factory farming'

Chickens and calves are often reared in large sheds instead of in open fields. Their urine and faeces are washed out of the sheds with water forming 'slurry'. If this slurry gets into streams and rivers it supplies an excess of nitrates and phosphates for the microscopic algae.

The degree of pollution of river water is often measured by its **biochemical oxygen demand** (**BOD**). This is the amount of oxygen used up by a sample of water in a fixed period of time. The higher the BOD, the more polluted the water is likely to be.

It is possible to reduce eutrophication by using:

- detergents with less phosphates
- agricultural fertilisers that do not dissolve so easily
- animal wastes on the land instead of letting them reach rivers.

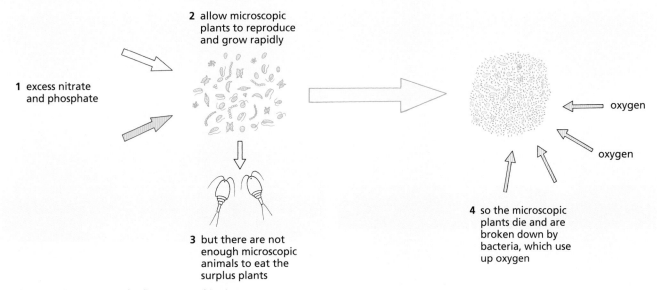

Figure 21.24 Processes leading to eutrophication

21 HUMAN INFLUENCES ON ECOSYSTEMS

Plastics and the environment

Plastics that are non-biodegradable are not broken down by decomposers when dumped in landfill sites or left as litter. This means that they remain in the environment, taking up valuable space or causing visual pollution. Discarded plastic bottles can trap small animals; nylon fishing lines and nets can trap birds and mammals such as seals and dolphins.
As the plastics in water gradually deteriorate, they fragment into tiny pieces, which are eaten by fish and birds, making them ill. When plastic is burned, it can release toxic gases.

Plastic bags are a big problem, taking up a lot of space in landfill sites. In 2002, the Republic of Ireland introduced a plastic bag fee, called a PlasTax, to try to control the problem. It had a dramatic effect, cutting the use of single-use bags from 1.2 billion to 230 million a year and reducing the litter problem that plastic bags create. Revenue raised from the fee is used to support environmental projects.

Air pollution

Some factories (Figure 21.25) and most motor vehicles release poisonous substances into the air. Factories produce smoke and sulfur dioxide; cars produce lead compounds, carbon monoxide and the oxides of nitrogen, which lead to smog (Figure 21.26) and acid rain (Figure 21.27).

Figure 21.26 Photochemical 'smog' over a city

Figure 21.25 Air pollution by industry. Tall chimneys keep pollution away from the immediate surroundings but the atmosphere is still polluted.

Figure 21.27 Effects of acid rain on conifers in the Black Forest, Germany

Sulfur dioxide and oxides of nitrogen

Coal and oil contain sulfur. When these fuels are burned, they release sulfur dioxide (SO_2) into the air (Figure 21.28). Although the tall chimneys of factories (Figure 21.25) send smoke and sulfur dioxide high into the air, the sulfur dioxide dissolves in rainwater and forms an acid. When this acid falls on buildings, it slowly dissolves the limestone and mortar. When it falls on plants, it reduces their growth and damages their leaves.

This form of pollution has been going on for many years and is getting worse. In North America, Scandinavia and Scotland, forests are being destroyed (Figure 21.27) and fish are dying in lakes, at least partly as a result of acid rain.

Oxides of nitrogen from power stations and vehicle exhausts also contribute to atmospheric pollution and acid rain. The nitrogen oxides dissolve in rain drops and form nitric acid.

Oxides of nitrogen also take part in reactions with other atmospheric pollutants and produce ozone. It may be the ozone and the nitrogen oxides that are largely responsible for the damage observed in forests.

One effect of acid rain is that it dissolves out the aluminium salts in the soil. These salts eventually reach toxic levels in streams and lakes.

There is still some argument about the source of the acid gases that produce acid rain. For example, a large proportion of the sulfur dioxide in the atmosphere comes from the natural activities of certain marine algae. These microscopic 'plants' produce the gas dimethylsulfide which is oxidised to sulfur dioxide in the air.

Nevertheless, there is considerable circumstantial evidence that industrial activities in Britain, America and Central and Eastern Europe add large amounts of extra sulfur dioxide and nitrogen oxides to the atmosphere.

Control of air pollution

The Clean Air Acts of 1956 and 1968

These acts designated certain city areas as 'smokeless zones' in Britain. The use of coal for domestic heating was prohibited and factories were not allowed to emit black smoke. This was effective in abolishing dense fogs in cities but did not stop the discharge of sulfur dioxide and nitrogen oxides in the country as a whole.

Reduction of acid gases

The concern over the damaging effects of acid rain has led many countries to press for regulations to reduce emissions of sulfur dioxide and nitrogen oxides.

Reduction of sulfur dioxide can be achieved either by fitting desulfurisation plants to power stations or by changing the fuel or the way it is burnt. In 1986, Britain decided to fit desulfurisation plants to three of its major power stations, but also agreed to a United Nations protocol to reduce sulfur dioxide emissions to 50% of 1980 levels by the year 2000,

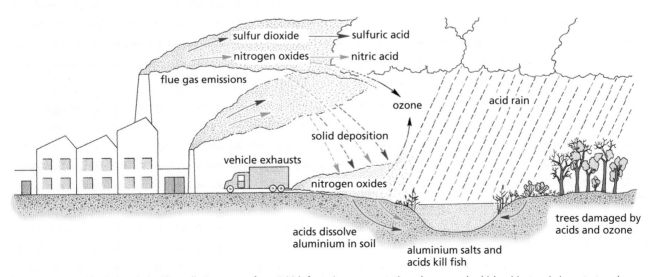

Figure 21.28 Acid rain in Britain. The pollution comes from British factories, power stations, homes and vehicles. Most emissions start as dry gases and are converted slowly to dilute sulfuric and nitric acids.

and to 20% by 2010. This was to be achieved largely by changing from coal-fired to gas-fired power stations.

Reduction of vehicle emissions

Oxides of nitrogen come, almost equally, from industry and from motor vehicles (Figure 21.28). Flue gases from industry can be treated to remove most of the nitrogen oxides. Vehicles can have **catalytic converters** fitted to their exhaust systems. These converters remove most of the nitrogen oxides, carbon monoxide and unburned hydrocarbons. They add £200–600 to the cost of a car and will work only if lead-free petrol is used, because lead blocks the action of the catalyst.

Another solution is to redesign car engines to burn petrol at lower temperatures ('**lean burn**' engines). These emit less nitrogen oxide but just as much carbon monoxide and hydrocarbons as normal engines.

In the long term, it may be possible to use fuels such as alcohol or hydrogen, which do not produce so many pollutants.

The European Union has set limits on exhaust emissions. From 1989, new cars over 2 litres had to have catalytic converters and from 1993 smaller cars had to fit them as well.

Regulations introduced in 1995 should cut emissions of particulates by 75% and nitrogen oxides by 50%. These reductions will have less effect if the volume of traffic continues to increase. Significant reduction of pollutants is more likely if the number of vehicles is stabilised and road freight is reduced.

Protecting the ozone layer

The appearance of 'ozone holes' in the Antarctic and Arctic, and the thinning of the ozone layer elsewhere, spurred countries to get together and agree to reduce the production and use of CFCs (chlorofluorocarbons) and other ozone-damaging chemicals.

1987 saw the first Montreal protocol, which set targets for the reduction and phasing out of these chemicals. In 1990, nearly 100 countries, including Britain, agreed to the next stage of the Montreal protocol, which committed them to reduce production of CFCs by 85% in 1994 and phase them out completely by 2000. Overall, the Montreal protocol has proved to be very successful: by 2012, the world had phased-out 98% of the ozone-depleting substances such as CFCs. However, the chemicals that were used to replace CFCs (HCFCs) are not as harmless as they were first thought to be, as they contribute to global warming.

The 'greenhouse effect' and global warming

The Earth's surface receives and absorbs radiant heat from the Sun. It re-radiates some of this heat back into space. The Sun's radiation is mainly in the form of short-wavelength energy and penetrates our atmosphere easily. The energy radiated back from the Earth is in the form of long wavelengths (infrared or IR), much of which is absorbed by the atmosphere. The atmosphere acts like the glass in a greenhouse. It lets in light and heat from the Sun but reduces the amount of heat that escapes (Figure 21.29).

If it were not for this 'greenhouse effect' of the atmosphere, the Earth's surface would probably be at $-18\,°C$. The 'greenhouse effect', therefore, is entirely natural and desirable.

Not all the atmospheric gases are equally effective at absorbing IR radiation. Oxygen and nitrogen, for example, absorb little or none. The gases that absorb most IR radiation, in order of maximum absorption, are water vapour, carbon dioxide (CO_2), methane and atmospheric pollutants such as oxides of nitrogen and CFCs. Apart from water vapour, these gases are in very low concentrations in the atmosphere, but some of them are strong absorbers of IR radiation. It is assumed that if the concentration of any of these gases were to increase, the greenhouse effect would be enhanced and the Earth would get warmer.

In recent years, attention has focused principally on CO_2. If you look back at the carbon cycle in Chapter 19, you will see that the natural processes of photosynthesis, respiration and decay would be expected to keep the CO_2 concentration at a steady level. However, since the Industrial Revolution, we have been burning 'fossil fuels' derived from coal and petroleum and releasing extra CO_2 into the atmosphere. As a result, the concentration of CO_2 has increased from 0.029 to 0.039% since 1860. It is likely to go on increasing as we burn more and more

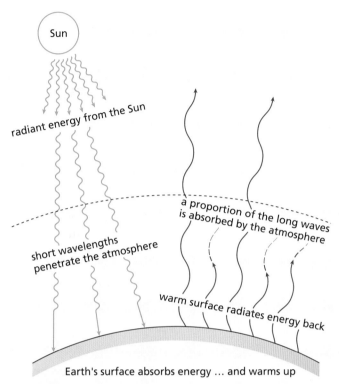

Figure 21.29 The 'greenhouse effect'

fossil fuel. According to NOAA data, CO_2 levels rose 2.67 parts per million in 2012, to 395 ppm. This was the second largest increase since 1959, when scientists first began measuring atmospheric CO_2 levels.

Although it is not possible to prove beyond all reasonable doubt that production of CO_2 and other 'greenhouse gases' is causing a rise in the Earth's temperature, i.e. global warming, the majority of scientists and climatologists agree that it is happening now and will get worse unless we take drastic action to reduce the output of these gases.

Predictions of the effects of global warming depend on computer models. But these depend on very complex and uncertain interactions of variables.

Changes in climate might increase cloud cover and this might reduce the heat reaching the Earth from the Sun. Oceanic plankton absorb a great deal of CO_2. Will the rate of absorption increase or will a warmer ocean absorb less of the gas? An increase in CO_2 should, theoretically, result in increased rates of photosynthesis, bringing the system back into balance.

None of these possibilities is known for certain. The worst scenario is that the climate and rainfall distribution will change, and disrupt the present pattern of world agriculture; the oceans will expand and the polar icecaps will melt, causing a rise in sea level; extremes of weather may produce droughts and food shortages.

An average of temperature records from around the world suggests that, since 1880, there has been a rise of 0.7–0.9 °C, most of it very recently (Figure 21.30), but this is too short a period from which to draw firm conclusions about long-term trends. If the warming trend continues, however, it could produce a rise in sea level of between 0.2 and 1.5 metres in the next 50–100 years.

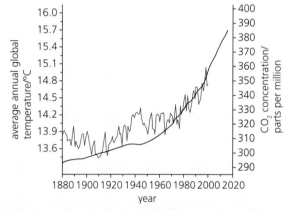

Figure 21.30 Annual average global temperatures and carbon dioxide levels since 1880

The first Kyoto Conference (Japan) in 1997 set targets for the industrialised countries to reduce CO_2 emissions by an average of 5.2% by 2010. Europe, as a whole, agreed to cuts of 8%, though this average allowed some countries to increase their emissions. The countries committed to the Kyoto convention, excluding the USA, eventually modified the targets, but agreed to make cuts of 4.2% on average for the period 2008–2012.

Britain planned to reduce emissions by 20% of 1990 levels by 2010 but really needed an overall cut of 60% to halt the progress of global warming. The big industrialised countries who contribute 80% of the greenhouse gases, particularly the USA, are opposed to measures that might interfere with their industries, claiming that global warming is not a proven fact.

The precautionary principle suggests that, even if global warming is not taking place, our supplies of fossil fuels will eventually run out and we need to develop alternative sources of energy now.

The generation of energy using fossil fuels is the biggest source of CO_2 released by humans into the atmosphere. The alternatives are nuclear power or methods such as wind farms and solar energy. The experiences of Chernobyl and Fukushima have made people around the world very wary of the nuclear option. Not all countries have climates and weather suited to alternative energy and their environmental impact (visual and sometimes through the noise they can create) creates opponents to these methods. The next section discusses this topic in more detail.

Pollution by contraceptive hormones

When women use the contraceptive pill, the hormones in it (oestrogen or progesterone – Chapter 16) are excreted in urine and become present in sewage. The process of sewage treatment does not extract the hormones, so they end up in water systems such as rivers, lakes and the sea. Their presence in this water affects aquatic organisms as they enter food chains. For example, male frogs and fish can become 'feminised' (they can start producing eggs in their testes instead of sperm). This causes an imbalance between numbers of male and female animals (more females than males).

Drinking water, extracted from rivers where water from treated sewage has been recycled, can also contain the hormones. This has been shown to reduce the sperm count in men, causing a reduction in fertility.

It should be noted that the contraceptive pill is not the only source of female hormones in water systems: natural hormones are also present in urine from cattle, for example, and these can enter the water with run-off from farms.

● Conservation

Key definition
A **sustainable resource** is one that is produced as rapidly as it is removed from the environment so that it does not run out.

Non-renewable resources such as fossil fuels need to be conserved because the stocks of them on the planet are finite: coal, oil, natural gas and minerals (including metallic ores) cannot be replaced once their sources have been totally depleted. Estimates of how long these stocks will last are unreliable but in some cases, e.g. lead and tin, they are less than 100 years.

By the time that fossil fuels run out, we will have to have alternative sources of energy. Even the uranium used in nuclear reactors is a finite resource and will, one day, run out.

The alternative sources of energy available to us are hydroelectric, nuclear, wind and wave power, wood and other plant products. The first two are well established; the others are either in the experimental stages, making only a small contribution, or are more expensive (at present) than fossil fuels (Figure 21.31). Plant products are **renewable resources** and include alcohol distilled from fermented sugar (from sugar-cane), which can replace or supplement petrol (Figure 21.32), sunflower oil, which can replace diesel fuel, and wood from fast-growing trees. In

Figure 21.31 Wind generators in the USA. On otherwise unproductive land or offshore, these generators make an increasing contribution to the electricity supply.

addition, plant and animal waste material can be decomposed anaerobically in fermenters to produce **biogas**, which consists largely of methane.

Chemicals for industry or drugs, currently derived from petroleum, will have to be made from plant products.

In theory, fuels produced from plant sources should have a minimal effect on the carbon dioxide concentration in the atmosphere and, therefore, on global warming. The carbon dioxide released when they are burned derives from the carbon dioxide they absorbed during their photosynthesis. They are '**carbon neutral**'. However, the harvesting of the crop and the processes of extraction and distillation all produce carbon dioxide. The net effect on atmospheric carbon dioxide is questionable.

Also, the clearing of forests to make space for fuel crops removes a valuable carbon sink and the burning that accompanies it produces a great deal of carbon dioxide. In addition, the use of land for growing crops for **biofuels** reduces the land available for growing food and increases the price of food.

Currently, the benefit of deriving fuel from plant material is open to question.

When **non-renewable resources** run out they will have to be replaced by recycling or by using man-made materials derived from plant products. Already some bacteria have been genetically engineered to produce substances that can be converted to plastics.

Some resources, such as forests and fish stocks can be maintained with careful management. This may involve replanting land with new seedlings as mature trees are felled and controlling the activities of fishermen operating where fish stocks are being depleted.

Figure 21.32 An alcohol-powered car in Brazil

Recycling

As minerals and other resources become scarcer, they also become more expensive. It then pays to use them more than once. The recycling of materials may also reduce the amount of energy used in manufacturing. In turn this helps to conserve fuels and reduce pollution.

For example, producing aluminium alloys from scrap uses only 5% of the energy that would be needed to make them from aluminium ores. In 2000, Europe recycled 64.3% of the aluminium in waste. Germany and Finland do really well, partly because they have a deposit scheme on cans: they recycle between 95 and 96% of their aluminium waste.

About 60% of the lead used in Britain is recycled. This seems quite good until you realise that it also means that 40% of this poisonous substance enters the environment.

Manufacturing glass bottles uses about three times more energy than if they were collected, sorted, cleaned and reused. Recycling the glass from bottles does not save energy but does reduce the demand for sand used in glass manufacture. In 2007, 57% of glass containers were recycled in Britain.

Polythene waste is now also recycled (Figure 21.33). The plastic is used to make items such as car seat covers, sports shoes, hi-fi headphones and even bridges (Figure 21.34).

Waste paper can be pulped and used again, mainly for making paper and cardboard. Newspapers are de-inked and used again for newsprint. One tonne of waste paper is equivalent to perhaps 17 trees. (Paper is made from wood-pulp.) So collecting waste paper may help to cut a country's import bill for timber and spare a few more hectares of natural habitat from the spread of commercial forestry.

Sewage treatment

Micro-organisms, mainly bacteria and protoctista, play an essential part in the treatment of sewage to make it harmless.

Sewage contains bacteria from the human intestine that can be harmful (Chapter 10). These bacteria must be destroyed in order to prevent the spread of intestinal diseases. Sewage also contains substances from household wastes (such as soap and detergent) and chemicals from factories. These too must be removed before the sewage effluent is released into the rivers. Rainwater from the streets is also combined with the sewage.

21 HUMAN INFLUENCES ON ECOSYSTEMS

Figure 21.33 Recycling polythene. Polythene waste is recycled for industrial use.

Figure 21.34 The world's longest bridge made from recycled plastic. It was constructed in Peeblesshire in Scotland.

Inland towns have to make their sewage harmless in a sewage treatment plant before discharging the effluent into rivers. A sewage works removes solid and liquid waste from the sewage, so that the water leaving the works is safe to drink.

In a large town, the main method of sewage treatment is by the activated sludge process (Figures 21.35 and 21.36).

The activated sludge process

1 **Screening.** The sewage entering the sewage works is first 'screened'. That is, it is made to flow through a metal grid, which removes the solids like rags, plastics, wood and so forth. The 'screenings' are raked off and disposed of – by incineration, for example.
2 **Grit.** The sewage next flows slowly through long channels. As it flows, grit and sand in it settle down to the bottom and are removed from time to time. The grit is washed and used for landfill.
3 **First settling tanks.** The liquid continues slowly through another series of tanks. Here about 40% of the organic matter settles out as crude sludge.

The rest of the organic matter is in the form of tiny suspended particles, which pass, with the liquid, to the aeration tanks.

The semi-liquid sludge from the bottom of the tank is pumped to the sludge digestion plant.
4 **Aeration tanks.** Oxygen is added to the sewage liquid, either by stirring it or by bubbling compressed air through it. Aerobic bacteria and protoctista grow and reproduce rapidly in these conditions.

These micro-organisms clump the organic particles together. Enzymes from the bacteria digest the solids to soluble products, which are absorbed by the bacteria and used for energy and growth.

Dissolved substances in the sewage are used in the same way. Different bacteria turn urea into ammonia, ammonia into nitrates and nitrates into nitrogen gas. The bacteria derive energy from these chemical changes. The protoctista (Figure 21.37) eat the bacteria.

In this way, the suspended solids and dissolved substances in sewage are converted to nitrogen, carbon dioxide (from respiration) and the cytoplasm of the bacteria and protoctista, leaving fairly pure water.
5 **Second settling tanks.** The micro-organisms settle out, forming a fine sludge, which is returned to the aeration tanks to maintain the population of micro-organisms. This is the 'activated sludge' from which the process gets its name. The sewage

Conservation

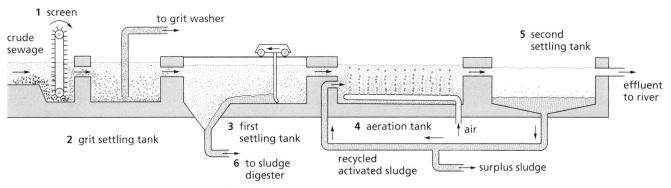

Figure 21.35 Sewage treatment – activated sludge process

stays in the aeration tanks for only 6–8 hours but the recycling of activated sludge allows the micro-organisms to act on it for 20–30 days.
6 When all the sludge has settled, the water is pure enough to discharge into a river and the sludge passes to a digester, which is used to produce methane (biogas).

Figure 21.36 Sewage treatment – activated sludge method. In the foreground are the rectangular aeration tanks.

Biogas production is not confined to sludge. Many organic wastes, e.g. those from sugar factories, can be fermented anaerobically to produce biogas. In developing countries, biogas generators use animal dung to produce methane for whole villages. On a small scale, biogas is a useful form of sustainable alternative energy.

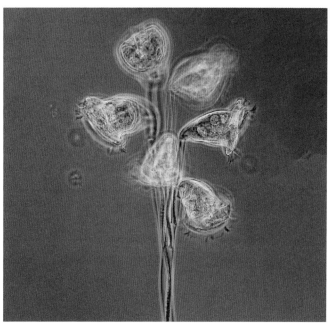

Figure 21.37 Protoctista in activated sludge (×150). These single-celled organisms ingest bacteria in the liquid sewage.

Endangering species and causing their extinction

Anything that reduces the population of a species endangers it (puts it at risk of extinction). Factors that endanger species include habitat destruction, the introduction of other species, hunting, international trade or pollution. Climate change can also put species at risk of extinction.

Species become extinct in the course of evolution. After all, the fossil remains of plants and animals represent organisms that became extinct hundreds of thousands of years ago. There have been periods of mass extinction, such as that which wiped out the dinosaurs during the Cretaceous era, 65 million years ago.

21 HUMAN INFLUENCES ON ECOSYSTEMS

The 'background' extinction rate for, say, birds might be one species in 100–1000 years. Today, as a result of human activity, the rate of extinction has gone up by at least ten times and possibly as much as 1000 times. Some estimates suggest that the world is losing one species every day and within 20 years at least 25% of all forms of wildlife could become extinct. Reliable evidence for these figures is hard to obtain, however.

A classic example is the colonisation of the Pacific islands by the Polynesians. They hunted and ate the larger bird species, and introduced rats, which ate the eggs and young of ground-nesting species. Their goats and cattle destroyed plant species through grazing and trampling. Of about 1000 plant species, 85% has been lost since they were first discovered.

This may be an extreme example but the same sorts of changes are happening all over the world. For example, the World Wide Fund for Nature (WWF) estimated that only about 3200 tigers remained in the wild in 2011. This is less than 5% of their number in 1900 (Figure 21.38). They are hunted for their skins and their bones and some body parts are used in traditional Chinese medicines.

Figure 21.38 In 110 years the tiger population has fallen from 120 000 to 3200.

Some species of animal are not introduced deliberately into different ecosystems, but find their way in due to man's activities and then upset food chains. One example happened in the Great Lakes in Canada and the USA. The lakes were artificially joined together with shipping canals to provide transport links, but sea lampreys found their way into the lakes through the new waterways. The lampreys had no natural predators in the lakes and fed on trout by sticking to them with their circular mouths and boring into their flesh (Figure 21.39). The fisheries in the lakes harvested about 7 million kilograms of trout annually before the lampreys entered the water systems. Afterwards, the harvest dropped to about 136 000 kilograms, so the fisheries collapsed. The lampreys are now controlled to enable the trout population to recover.

Figure 21.39 Sea lamprey feeding on a trout

Climate change is also responsible for a reduction in the number of species. Some people argue that this is a natural, uncontrollable process, but the consensus by scientists is that processes like global warming are made worse by human activity.

Global warming is causing oceans to warm up. Even prolonged temperature increases of just one or two degrees can have a devastating effect. In 1994, coral colonies (see Figure 1.8) in the Indian Ocean were observed to expel food-producing algae they are closely associated with. As the coral rely on the algae, if they lose them they die. The coral reefs became bleached. When the area was surveyed again in 2005, four fish species appeared to be extinct and six other species had declined to the point of being endangered. Increases in CO_2 in the sea also affect coral reefs. The CO_2 dissolves in the water, making it more acidic. The acid dissolves the calcium carbonate deposited in the coral, making it collapse.

Species such as the Atlantic cod are becoming endangered and at possible risk of extinction, partly because of overfishing (see Chapter 19) but also because of climate change. Cod survive in cold water.

As seawater warms up, the cod migrate north. However, the populations of microscopic plankton that cod rely on further down the food chain are also sensitive to temperature change – cod may not have the food supplies they need to survive.

Scientists developed a computer model to study the effect of climate change on fish stocks over the next 50 years. It predicted a large-scale redistribution of species and the extinction of some species, with the disruption of ecosystems and reduction in biodiversity.

Conservation of species

Species can be conserved by passing laws that make killing or collecting them an offence, by international agreements on global bans or trading restrictions, and by conserving habitats (Figure 21.40).

Habitats can be conserved in a number of ways:

- using laws to protect the habitat
- using wardens to protect the habitat
- reducing or controlling public access to the habitat
- controlling factors such as water drainage and grazing, that may otherwise help to destroy the habitat.

In Britain, it is an offence to capture or kill almost all species of wild birds or to take eggs from their nests; wild flowers in their natural habitats may not be uprooted; newts, otters and bats are just three of the protected species of mammal.

Many organisations monitor species numbers, so that conservation measures can be taken if they decline significantly.

CITES (Convention on International Trade in Endangered Species) gives protection to about 1500 animals and thousands of plants by persuading governments to restrict or ban trade in endangered species or their products, e.g. snake skins or rhino horns. In 2013, nearly 180 countries were party to the Convention.

The WWF operates on a global scale and is represented in 25 countries. The WWF raises money for conservation projects in all parts of the world, but with particular emphasis on endangered species and habitats.

The **IWC** (International Whaling Commission) was set up to try and avoid the extinction of whales as a result of uncontrolled whaling, and has 88 members.

Figure 21.40 Trying to stop the trade in endangered species. A customs official checks an illegal cargo impounded at a customs post.

The IWC allocates quotas of whales that the member countries may catch but, having no powers to enforce its decisions, cannot prevent countries from exceeding their quotas.

In 1982, the IWC declared a moratorium (i.e. a complete ban) on all whaling, which was reaffirmed in 2000 and is still in place in 2014, despite opposition from Japan and Norway. Japan continues to catch whales 'for scientific purposes'.

Captive breeding and reintroductions

Provided a species has not become totally extinct, it may be possible to boost its numbers by **breeding in captivity** and releasing the animals back into the environment. In Britain, modest success has been achieved with otters (Figure 21.41). It is important (a) that the animals do not become dependent on humans for food and (b) that there are suitable habitats left for them to recolonise.

Sea eagles, red kites (Figure 21.42) and ospreys have been introduced from areas where they are plentiful to areas where they had died out.

Figure 21.41 The otter has been bred successfully in captivity and released.

21 HUMAN INFLUENCES ON ECOSYSTEMS

Figure 21.42 Red kites from Spain and Sweden have been reintroduced to Britain.

Seed banks

These are a way of protecting plant species from extinction. They include seed from food crops and rare species. They act as gene banks (see the next section). The Millennium Seed Bank Partnership was set up by Kew Botanical Gardens in London. It is a global project involving 80 partner countries. The target of the partnership is to have in storage 25% of the world's plant species with bankable seeds by 2020. That involves about 75 000 plant species.

Conservation of habitats

If animals and plants are to be conserved it is vital that their habitats are conserved also.

Habitats are many and varied: from vast areas of tropical forest to the village pond, and including such diverse habitats as wetlands, peat bogs, coral reefs, mangrove swamps, lakes and rivers, to list but a few.

International initiatives

In the last 30 years it has been recognised that conservation of major habitats needed international agreements on strategies. In 1992, the Convention on Biological Diversity was opened for signature at the 'Earth Summit' Conference in Rio, and 168 countries signed it. The Convention aims to preserve biological diversity ('biodiversity').

Biodiversity encompasses the whole range of species in the world. The Convention will try to share the costs and benefits between developed and developing countries, promote '**sustainable development**' and support local initiatives.

'Sustainable development' implies that industry and agriculture should use natural resources sparingly and avoid damaging natural habitats and the organisms in them.

> **Key definition**
> **Sustainable development** is development providing for the needs of an increasing human population without harming the environment.

The Earth Summit meeting addressed problems of population, global warming, pollution, etc. as well as biodiversity.

There are several voluntary organisations that work for worldwide conservation, e.g. WWF, Friends of the Earth and Greenpeace.

Sustaining forest and fish stocks

There are three main ways of sustaining the numbers of key species. These are:

1 Education

Local communities need to be educated about the need for conservation. Once they understand its importance, the environment they live in is more likely to be cared for and the species in it protected.

In tree-felling operations in tropical rainforests, it has been found that the process of cutting down the trees actually damages twice as many next to them and dragging the trees out of the forest also creates more damage. Education of the men carrying out the operations in alternative ways of tree felling, reduction of wastage and in the selection of species of trees to be felled makes the process more sustainable and helps to conserve rarer species.

In the **tomato fish project** in Germany (see later in this section), the Research Institute involved has an active education programme to inform the public about its work in sustainable development. It has even published a book for children (*Nina and the tomato fish*) to educate them about the topic.

2 Legal quotas

In Europe the Common Fisheries Policy is used to set quotas for fishing, to manage fish stocks and help protect species that were becoming endangered through overfishing (see Chapter 19). Quotas were set for each species of fish taken commercially and also for the size of fish. This was to allow fish to reach breeding age and maintain or increase their populations.

The Rainforest Alliance has introduced a scheme called *Smartlogging*. This is a certification service, which demonstrates that a logging company is working legally and is a sustainable way to protect the environment. The timber can be tracked from where it is felled to its final export destination and its use in timber products. The customer can then be reassured that the timber in the product is from a reputable source and has not been removed illegally.

In some areas of China where bamboo is growing, there are legal quotas to prevent too much felling. Some animals such as giant panda rely on the bamboo for their food.

In Britain it is illegal to cut down trees without permission. The Forestry Commission issues licenses for tree felling. Included in the license are conditions that the felled area must be replanted and the trees maintained for a minimum of ten years.

3 Restocking

Where populations of a fish species are in decline, their numbers may be conserved by a restocking programme. This involves breeding fish in captivity, then releasing them into the wild. However, the reasons for the decline in numbers need to be identified first. For example, if pollution was the cause of the decline, the restocked fish will die as well: the issue of pollution needs to be addressed first. Great care is needed in managing fish farms because they can produce pollution if the waste water from the farms, containing uneaten food and fish excreta, is discharged into the environment.

Organisations such as the Woodland Trust help to conserve areas of woodland and provide funding for restocking where species of trees are in decline. This is important as some animal species rely on certain trees for food and shelter. Large areas of land planted with single species (an example of a monoculture) create little biodiversity. In Britain, the Forestry Commission has been steadily increasing the range of tree species it plants, growing them in mixed woodland, which provides habitats for a wider range of animals.

Sustainable development

This is a complex process, requiring the management of conflicting demands. As the world's population grows, so does the demand for the extraction of resources from the environment. However, this needs to be carried out in a controlled way to prevent environmental damage and strategies need to be put in place to ensure habitats and species diversity are not threatened.

Planning the removal of resources needs to be done at local, national and international levels. This is to make sure that everyone involved with the process is aware of the potential consequences of the process on the environment, and that appropriate strategies are put in place, and adhered to, to minimise any risk.

Tomato fish project

The ASTAF-PRO project – Aquaponic System for (nearly) Emission-Free Tomato and Fish Production – in Germany is run by the Leibniz Institute of Freshwater Ecology and Inland Fisheries. The scientists have developed a way of simultaneously producing fish and tomatoes in a closed greenhouse environment. Both organisms thrive at a temperature of 27 °C. The system is almost emission-free (so atmospheric CO_2 levels are not affected), recycles all the water in the process and does not put any waste into the environment (Figure 21.43). All the energy needed to heat the greenhouses is generated by solar panels. These factors make it a sustainable and climate-friendly method of food production. The scientists recognised that fish and plants have very similar environmental needs for their growth. Nile Tilapia (*Oreochromis niloticus*) is chosen as the fish species, because they survive well in artificial conditions, growing and maturing quickly. Since they are omnivorous as adults, no fish meal diet is needed, and they can be fed with pellets of processed food extracted from plants. Water from the fish tanks is cleaned and the nutrients remaining in it are used as a fertiliser for tomato plants, grown in the same greenhouse (Figure 21.44).

The plants are grown on mineral wool, through which the nutrient-rich water flows. This avoids soil, which can contain pathogens. This method of growing plants, called **hydroponics**, also means that no peat is needed for soil. The removal of peat for use in horticulture is threatening heathland and the organisms living on it.

As the tomato plants transpire, the water vapour is condensed and recycled into the fish tanks. The tomatoes are harvested and sold under the name 'fish tomatoes'. The scientists call the project 'The Tomatofish'. The next goal is to implement the system into global food production systems.

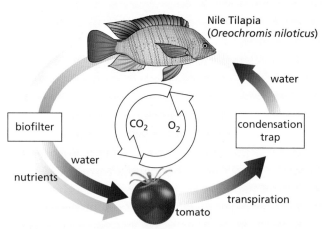

Figure 21.43 The tomato-fish project

Figure 21.44 Tomatoes and fish being grown in the same environment

Conservation programmes

If the population of a species drops, the range of variation within the species drops, making it less able to adapt to environmental change. The species could, therefore, be threatened with extinction. When animal populations fall, there is less chance of individuals finding each other to mate.

In 'Selection', Chapter 18, it was explained that crossing a wild grass with a strain of wheat produced an improved variety. This is only one example of many successful attempts to improve yield, drought resistance and disease resistance in food plants. Some 25 000 plant species are threatened with extinction at the moment. This could result in a devastating loss of hereditary material and a reduction of about 10% in the genes available for crop improvement. 'Gene banks' have been set up to preserve a wide range of plants, but these banks are vulnerable to accidents, disease and human error. The only secure way of preserving the full range of genes is to keep the plants growing in their natural environments.

Conservation programmes are set up for a number of reasons:

Reducing extinction

Conservation programmes strive to prevent extinction. Once a species becomes extinct its genes are lost forever, so we are also likely to deprive the world of genetic resources. Apart from the fact that we have no right to wipe out species forever, the chances are that we will deprive ourselves not only of the beauty and diversity of species but also of potential sources of valuable products such as drugs. Many of our present-day drugs are derived from plants (e.g. quinine and aspirin) and there may be many more sources as yet undiscovered.

Protecting vulnerable environments

Conservation programmes are often set up to protect threatened habitats so that rare species living there are not endangered. Some species of plant require very special conditions to grow successfully, for instance wet, acidic conditions associated with heathland (see Figure 21.46). Some animal species have very limited diets or other needs: the large heath butterfly only feeds on one type of plant called cottongrass. If that plant was allowed to become extinct, perhaps through drainage of the peat bog land on which the cottongrass lives, the butterflies would die out as well.

There are a number of organisations involved with habitat conservation in Britain. English Nature, the Countryside Council for Wales and Scottish Natural Heritage were formed from the Nature Conservancy Council (NCC). They are regulatory bodies committed to establish, manage and maintain nature reserves, protect threatened habitats and conduct research into matters relevant to conservation.

The NCC established 195 nature reserves (Figure 21.45) but, in addition, had responsibility for notifying planning authorities of **Areas of Special Scientific Interest** (**ASSIs**), also known as Sites of Special Scientific Interest (SSSIs). These are privately owned lands that include important habitats or rare species (Figure 21.46). English Nature and other conservation bodies establish management agreements with the owners so that the sites are not damaged by felling trees, ploughing land or draining fens (Figure 21.47).

Figure 21.46 Area of Special Scientific Interest. This heathland in Surrey is protected by a management agreement with the landowner.

Figure 21.45 An English Nature National Nature Reserve at Bridgewater Bay in Somerset. The mudflats and saltmarsh attract large numbers of wintering wildfowl.

Figure 21.47 The Royal Society for the Protection of Birds (RSPB) maintains this wet grassland by Loch Leven, Scotland, for nesting redshanks, snipe, lapwings and ducks.

There are now about 5000 ASSIs, and the Countryside and Rights of Way Act of 2000 has strengthened the rules governing the maintenance of ASSIs.

There are several other, non-governmental organisations that have set up reserves and which help to conserve wildlife and habitats. There are 47 Wildlife Trusts in the UK, managing thousands of sites. The Royal Society for the Protection of Birds (RSPB) has 200 sites, the Woodland Trust looks after over 1100 woods and there are about 160 other reserves managed by other organisations.

The National Parks Commission has set up 15 National Parks covering more than 9% of England and Wales, e.g. Dartmoor, Snowdonia and the Lake District. Although the land is privately owned, the Park Authorities are responsible for protecting the landscape and wildlife, and for planning public recreation such as walking, climbing or gliding.

The European Commission's **Habitats Directive** of 1994 requires member states to designate **Special Areas of Conservation** (**SACs**) to protect some of the most seriously threatened habitats and species throughout Europe. The UK has submitted a list of 340 sites, though many of these are already protected areas, such as ASSIs.

Desirable though ASSIs, National Parks and SACs are, they represent only relatively small, isolated areas of land. Birds can move freely from one area to another, but plants and small animals are confined to an isolated habitat so are subject to risks that they cannot escape. If more farmland were managed in a way 'friendly' to wildlife, these risks could be reduced.

The Farming and Wildlife Advisory Group can advise farmers how to manage their land in ways that encourage wildlife. This includes, for example, leaving strips of uncultivated land around the margins of fields or planting new hedgerows. Even strips of wild grasses and flowers between fields significantly increase the population of beneficial insects.

Certain areas of farmland have been designated as Environmental Sensitive Areas (ESAs), and farmers are paid a subsidy for managing their land in ways that conserve the environment.

Maintaining ecosystem functions

There is a danger of destabilising food chains if a single species in that food chain is removed. For example, in lakes containing pike as the top predators, overfishing can result in smaller species of carnivorous fish, such as minnows, increasing in numbers. They eat zooplankton. If the minnows eat the majority of the zooplankton population, it leaves no herbivores to control algal growth, which can cause an algal bloom when there are sufficient nutrients to support this growth. To prevent such an event happening, the ecosystem needs to be maintained, by controlling the numbers of top predators removed, or by regular restocking.

Ecosystems can also become unbalanced if the nutrients they rely on are affected in some way. Guano is the accumulated droppings of seabirds and bats. It is extremely rich in nitrogen compounds and phosphates, so it makes a valuable fertiliser. In the early 1900s Peru and South Africa both developed guano industries based on sustained-yield production from marine birds. However, overfishing around their coastlines reduced fish stocks, removing the food the seabirds relied on. As the seabird populations diminished, they deposited less guano and the guano industries failed.

The term ecosystem services can be defined as the benefits people obtain from ecosystems, whether they are natural or managed. Humans are affecting ecosystems on a large scale because of the growth in the population (Chapter 19) and changing patterns of consumption. Scientists estimate that around 40% of the Earth's land surface area is taken over by some form of farmed land. Crops are grown for food (directly, or indirectly through their use in feeding animals), extraction of drugs (both legal and illegal) and the manufacture of fuel (see details about biofuels below). Crop growth has major impacts in ecosystems, causing the extinction of many species and reducing the gene pool.

In theory, biofuels produced from plant sources should have a minimal effect on the carbon dioxide concentration in the atmosphere and, therefore, on global warming. The carbon dioxide released when they are burned derives from the carbon dioxide they absorbed during their photosynthesis. They are 'carbon neutral'. However, the harvesting of the crop and the processes of extraction and distillation all produce carbon dioxide. The net effect on atmospheric carbon dioxide is questionable. More details of biofuels are given in Chapter 20.

Also, the clearing of forests to make space for fuel crops removes a valuable carbon sink and the burning that accompanies it produces a great deal of carbon dioxide. In addition, the use of land for growing crops for biofuels reduces the land available for growing food and increases the price of food. Currently, the benefit of deriving fuel from plant material is open to question.

With all these demands on resources from ecosystems, it is a very complicated process to manage them effectively and this makes conservation programmes invaluable to protect species and their habitats.

Questions

Core

1. The graph in Figure 21.8 shows the change in the numbers of mites and springtails in the soil after treating it with an insecticide. Mites eat springtails. Suggest an explanation for the changes in numbers over the 16-month period.
2. What are the possible dangers of dumping and burying poisonous chemicals on the land?
3. Before most water leaves the waterworks, it is exposed for some time to the poisonous gas, chlorine. What do you think is the point of this?
4. If the concentration of mercury in Minamata Bay was very low, why did it cause such serious illness in humans?
5. Explain why some renewable energy sources depend on photosynthesis.
6. In what ways does the recycling of materials help to save energy and conserve the environment?
7. Explain why some of the alternative and renewable energy sources are less likely to cause pollution than coal and oil.
8. What kinds of human activity can lead to the extinction of a species?
9. How do the roles of CITES and WWF differ? In what respects might their activities overlap?
10. How might the loss of a species affect:
    a our health (indirectly)
    b the prospect of developing new varieties of crop plants resistant to drought?
11. What part do micro-organisms (bacteria and protoctista) play in sewage treatment?
12. What do you understand by:
    a biodiversity
    b sustainable development?
13. What is the difference between an ASSI and a nature reserve?

Extended

14. a What pressures lead to destruction of tropical forest?
    b Give three important reasons for trying to preserve tropical forests.
15. In what ways might trees protect the soil on a hillside from being washed away by the rain?
16. If a farmer ploughs a steeply sloping field, in what direction should the furrows run to help cut down soil erosion?
17. What is the possible connection between:
    a cutting down trees on hillsides and flooding in the valleys, and
    b clear-felling (logging) in tropical forests and local climate change?
18. To what extent do tall chimneys on factories reduce atmospheric pollution?
19. What are thought to be the main causes of 'acid rain'?
20. Why are carbon dioxide and methane called 'greenhouse gases'?

Checklist

Food supply

- Modern technology has resulted in increased food production.
- Agricultural machinery can be used on larger areas of land to improve efficiency.
- Chemical fertilisers improve yields.
- Insecticides improve quality and yield.
- Herbicides reduce competition with weeds.
- Selective breeding improves production by crop plants and livestock.
- Monocultures can have negative impacts on the environment.
- Intensive farming has resulted in habitat deterioration and reduction of wildlife.

- Problems with world food supplies contribute to difficulties providing enough food for an increasing human global population.
- Food production in developed countries has increased faster than the population growth.
- Food production in developing countries has not kept pace with population growth.
- Problems that contribute to famine include unequal distribution of food, drought, flooding and an increasing population.

Habitat destruction

- There are a number of reasons for habitat destruction, including:
  – increased area needed for food-crop growth, livestock production and housing
  – the extraction of natural resources
  – marine pollution.
- Through altering food webs and food chains, humans can negatively impact on habitats.
- Deforestation is an example of habitat destruction: it can lead to extinction, soil erosion, flooding and carbon dioxide build-up in the atmosphere.
- The conversion of tropical forest to agricultural land usually results in failure because forest soils are poor in nutrients.

- Deforestation has many undesirable effects on the environment.

Pollution

- We pollute our lakes, rivers and the sea with industrial waste, sewage, crude oil, rubbish, factory wastes and nuclear fall-out.
- Use of fertilisers can result in water pollution.
- Pesticides kill insects, weeds and fungi that could destroy our crops.
- Pesticides help to increase agricultural production but they kill other organisms as well as pests.

- A pesticide or pollutant that starts off at a low, safe level can become dangerously concentrated as it passes along a food chain.
- Eutrophication of lakes and rivers results from the excessive growth of algae followed by an oxygen shortage when the algae die and decay.
- We pollute the air with smoke, sulfur dioxide and nitrogen oxides from factories, and carbon monoxide and nitrogen oxides from motor vehicles.
- The acid rain resulting from air pollution leads to poisoning of lakes and possibly destruction of trees.
- The extra carbon dioxide from fossil fuels might lead to global warming.

- The process of eutrophication of water involves:
  - increased availability of nitrate and other ions
  - increased growth of producers
  - increased decomposition after death of the producers
  - increased aerobic respiration by bacteria, resulting in a reduction in dissolved oxygen
  - the death of organisms requiring dissolved oxygen in water.
- Non-biodegradable plastics can have detrimental effects on aquatic and terrestrial ecosystems.
- Sulfur dioxide, produced by burning fossil fuels, causes acid rain. This kills plants, as well as animals in water systems.
- Measures that might be taken to reduce sulfur dioxide pollution and reduce the impact of acid rain include a reduction in use of fossil fuels.
- Methane and carbon dioxide are building up in the atmosphere, resulting in the enhanced greenhouse effect and climate change.
- Female contraceptive hormones are entering water courses and can cause reduced sperm count in men and feminisation of aquatic organisms.

Conservation

- A sustainable resource is one that can be removed from the environment without it running out.
- Raw materials, such as metal ores, will one day run out.
- We need to conserve non-renewable resources such as fossil fuels.
- When supplies of fossil fuels run out or become too expensive, we will need to develop alternative sources of energy.
- Recycling metals, paper, glass and plastic helps to conserve these materials and save energy.
- Some resources such as forests and fish stocks can be maintained.
- Sewage can be treated to make the water that it contains safe to return to the environment or for human use.
- Some organisms are becoming endangered or extinct due to factors such as climate change, habitat destruction, hunting, pollution and introduced species.
- Endangered species can be conserved by strategies that include monitoring and protecting species and habitats, education, captive breeding programmes and seed banks.

- Sustainable development is development providing for the needs of an increasing human population without harming the environment.
- Forest and fish stocks can be sustained using strategies such as education and legal quotas.
- Sustainable development requires the management of conflicting demands, as well as planning and co-operation at local, national and international levels.
- Although extinction is a natural phenomenon, human activities are causing a great increase in the rates of extinction.
- Conservation of species requires international agreements and regulations.
- These regulations may prohibit killing or collecting species and prevent trade in them or their products.
- Loss of a plant species deprives us of (a) a possible source of genes and (b) a possible source of chemicals for drugs.
- Conserving a species by captive breeding is of little use unless its habitat is also conserved.
- The Earth Summit Conference tried to achieve international agreement on measures to conserve wildlife and habitats, and reduce pollution.
- National Parks, nature reserves, ASSIs and SACs all try to preserve habitats but they cover only a small proportion of the country and exist as isolated communities.
- Incentives exist for farming in a way that is friendly to wildlife.

Examination questions

Do not write on these pages. Where necessary copy drawings, tables or sentences.

Characteristics and classification of living organisms

1 Four of the classes of vertebrates and five possible descriptions of these classes are shown below. Draw a straight line to match each class of vertebrate to its description. [4]

class of vertebrate	description
bird	body with naked skin, two pairs of limbs
fish	body with hair, two pairs of limbs
mammal	body with feathers, one pair of wings
reptile	body with scales, with fins
	body with scaly skin, two pairs of limbs or no limbs

[Total: 4]

(Cambridge IGCSE Biology 0610 Paper 2 Q1 November 2006)

2 a Three characteristics of living organisms and four possible descriptions are shown below. Draw a straight line to match each characteristic to its description. [3]

characteristic	description
respiration	pumping air in and out of the lungs
nutrition	producing new individuals of the same species
reproduction	obtaining organic chemicals for the repair of tissues
	the release of energy from sugars

b State **two** other characteristics of living organisms. [2]

[Total: 5]

(Cambridge IGCSE Biology 0610 Paper 2 Q1 June 2006)

3 Vertebrate animals are grouped into a number of classes.
Complete the sentences by naming each of the vertebrate classes that are described.

a A vertebrate with scaly skin and no legs could be either a _____ or a _____. [2]
b A vertebrate with lungs and hair is a _____ but if it has feathers instead of hair it is a _____. [2]

[Total: 4]

(Cambridge IGCSE Biology 0610 Paper 21 Q1 November 2012)

4 The diagram below shows five mammals.

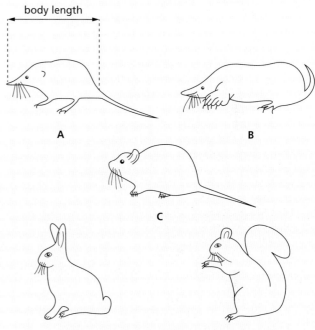

a Use the key to identify each of these mammals. Write the letter for each mammal in the table. [4]

1 { tail more than half that of body length .... go to 2
    tail less than half that of body length .... go to 4

2 { ears at top of head, with thick tail ...... *Sciurus caroliniensis*
    ears at side of head, with thin tail .... go to 3

3 { nose pointed, nose length longer than its depth ...... *Sorex araneus*
    nose blunt, nose length shorter than its depth .... *Clethrionomys glareolus*

4 { front legs as wide or wider than long .... *Talpa europaea*
    front legs longer than wide .... *Oryctolagus cuniculus*

EXAMINATION QUESTIONS

Name of animal	Letter
Clethrionomys glareolus	
Oryctolagus cuniculus	
Sciurus caroliniensis	
Sorex araneus	
Talpa europaea	

b The diagram below shows a young deer feeding from its mother.

State **two** features, visible in the diagram, that distinguish mammals from other vertebrates. [2]

[Total: 6]

(Cambridge IGCSE Biology 0610 Paper 3 Q1 November 2006)

5 The table below shows some of the external features of the five classes of vertebrates. Complete the table by placing a tick (✓) to indicate if each class has the feature. [5]

class of vertebrate	external ear flap	feathers or fur	scaly skin	two pairs of limbs
amphibians				
birds				
fish				
mammals				
reptiles				

[Total: 5]

(Cambridge IGCSE Biology 0610 Paper 21 Q2 June 2010)

6 Vertebrates can be classified by their external features. Complete the paragraph by using the name of a vertebrate class in each space.

Some vertebrates have scales all over their skin. If they also have nostrils that allow air into their lungs and two pairs of legs they are _____.

Some vertebrates have wings. If their body is also covered in feathers they are _____, but if their body has fur they are _____.
Vertebrates that do not have feathers, fur or scales on the outside of their body are _____. [4]

[Total: 4]

(Cambridge IGCSE Biology 0610 Paper 2 Q1 November 2009)

7 Arachnids, crustaceans, insects and myriapods are all classified as arthropods.
Scorpions, such as *Heterometrus swammerdami* shown in the diagram below, are arachnids.

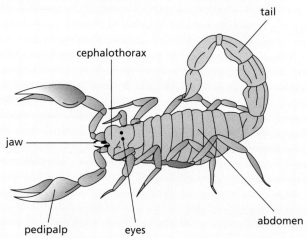

a State **three** features, shown by *H. swammerdami* and **visible** in the diagram above that arachnids share with other arthropods. [3]

b The diagram below shows seven species of arachnid.

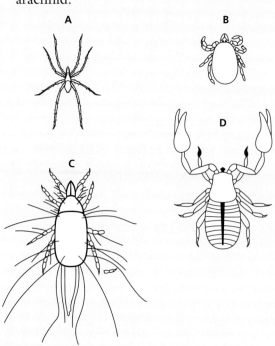

Characteristics and classification of living organisms

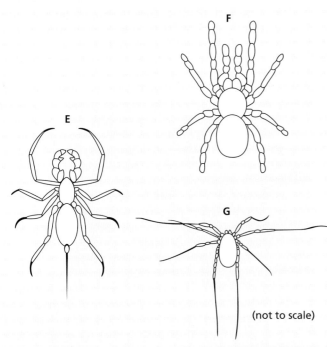

(not to scale)

Use the key to identify each species. Write the letter of each species (A to G) in the correct box beside the key. One has been done for you. [4]

Key

1	a)	Abdomen with a tail	*Abaliella dicranotarsalis*	E
	b)	Abdomen without a tail	go to 2	
2	a)	Legs much longer than abdomen and cephalothorax	go to 3	
	b)	Legs not much longer than abdomen and cephalothorax	go to 4	
3	a)	Hairs on legs	*Tegenaria domestica*	
	b)	No hairs on legs	*Odielus spinosus*	
4	a)	Cephalothorax or abdomen segmented	*Chelifer tuberculatus*	
	b)	Cephalothorax or abdomen not segmented	go to 5	
5	a)	Abdomen and cephalothorax about the same size	*Poecilotheria regalis*	
	b)	Abdomen larger than cephalothorax	go to 6	
6	a)	Body covered in long hairs	*Tyroglyphus longior*	
	b)	Body not covered in hairs	*Ixodes hexagonus*	

[Total: 7]
(Cambridge IGCSE Biology 0610 Paper 31 Q1 November 2012)

8 Non-living things, such as a car, often show characteristics similar to those of living organisms.
  a State which characteristic of a living organism matches each of the descriptions linked to a car.
    (i) burning fuel in the engine to release energy [1]
    (ii) headlights that switch on automatically in the dark [1]
    (iii) filling the car's tank with fuel [1]
    (iv) release of waste gases [1]
  b Identify **one** characteristic of living things that is **not** carried out by a car. [1]

[Total: 5]
(Cambridge IGCSE Biology 0610 Paper 21 Q1 June 2012)

9 The diagram below shows a bacterium, a virus and a fungus.

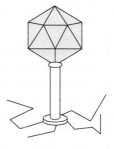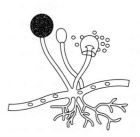

(not to scale)

a Complete the table to compare the three organisms shown in the diagram above by using a tick (✓) to indicate if the organism shows the feature, or a cross (✗) if it does not. The first row has been completed for you. [3]

feature	bacterium	virus	fungus
produces spores	✗	✗	✓
hyphae			
capsule			
nucleus			

b Explain how the fungus shown in the diagram above is adapted to obtain its food. [3]
c Explain how the fungus spreads to new sources of food. [2]

[Total: 8]
(Cambridge IGCSE Biology 0610 Paper 31 Q1 November 2009)

EXAMINATION QUESTIONS

● Organisation and maintenance of the organism

1 Five types of animal and plant cells and five possible functions of such cells are shown below. Draw one straight line from each type of cell to a function of that cell. [5]

type of cell	function of cell
red blood cell	absorption of mineral ions
root hair cell	transport of oxygen
white blood cell	movement of mucus
xylem	protection against pathogens
ciliated cell	structural support

[Total: 5]
(Cambridge IGCSE Biology 0610 Paper 2 Q5 June 2009)

2 The diagram shows a cell from the palisade layer of a leaf.

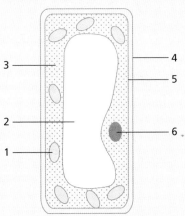

a In the table below tick (✓) the numbers that label the **three** features of the palisade cell which are also found in animal cells. [3]

label number	present in both animal and plant cells
1	
2	
3	
4	
5	
6	

b State and describe the function of **two** features of the palisade cell that are only found in plant cells. [4]

c The photograph below shows some red blood cells, which are animal cells.

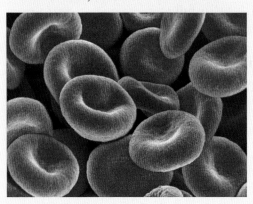

(i) Which feature normally present in an animal cell is absent from a red blood cell? [1]

(ii) State the function of a red blood cell **and** describe **one** way in which the red blood cell is adapted to carry out its function. [2]

[Total: 10]
(Cambridge IGCSE Biology 0610 Paper 21 Q8 November 2012)

3 The diagram below shows two cells.

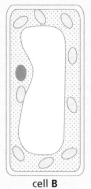

cell **A**   cell **B**

a (i) State where, in a human, a cell of type **A** would normally be found. [1]

(ii) State where, in a plant, a cell of type **B** would be found. [1]

b Use only words from the list to complete the statements about cell **B**. [5]

air   cellulose   chloroplasts   membrane
mitochondria   nucleus   starch   vacuole
wall   cell sap

350

Cell **B** has a thick layer called the cell _____. This is made of _____. The cytoplasm of cell **B** contains many _____ that are used in the process of photosynthesis. The large permanent _____ is full of _____ and this helps to maintain the shape of the cell.

c The diagram below shows structures that produce urine and excrete it from the body of a mammal.

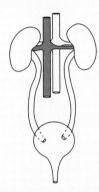

(i) On the diagram, label and name **one** organ. [1]
(ii) Use examples from the diagram to explain the difference between the terms *organ* and *organ system*. [3]

[Total: 11]
(Cambridge IGCSE Biology 0610 Paper 21 Q1 June 2010)

4 a The diagram shows a partly completed diagram of a palisade cell.

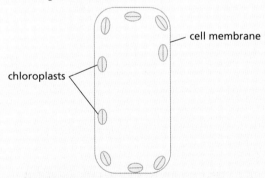

Complete the diagram to show the other major components of this cell.
Label all the components that you have added to the diagram. [4]

b State precisely where palisade cells are found in a plant. [2]

[Total: 6]
(Cambridge IGCSE Biology 0610 Paper 2 Q2 November 2009)

You may find it helpful to study Chapter 9 before attempting this question.

5 The photomicrograph below is of a human blood smear.

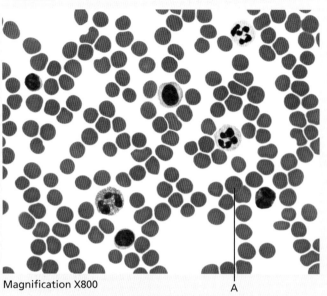

Magnification X800

a (i) On the photomicrograph, draw label lines and name **three** different types of blood cell. [3]
(ii) Name **two** parts of the blood that can pass through the capillary walls. [2]

b (i) Measure the diameter of the blood cell labelled **A**. [1]
(ii) The photomicrograph has been enlarged by × 800, calculate the actual size of cell **A**.
Show your working. [2]
(iii) State the function of cell **A**. [1]

[Total: 9]
(Cambridge IGCSE Biology 0610 Paper 6 Q3 June 2009)

● Movement in and out of cells

1 Thin slices of dandelion stem were cut and placed into different salt solutions and left for 30 minutes.
Figure 1 shows how these slices were cut. Figure 2 shows the appearance of these pieces of dandelion stem after 30 minutes in the different salt solutions.

EXAMINATION QUESTIONS

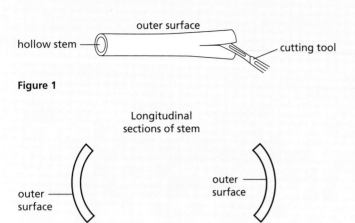

Figure 1

Figure 2

a (i) Describe the appearance of the pieces of dandelion stem in Figure 2. [2]
(ii) Explain what causes the two pieces of dandelion stem to change in the way you have described in a(i). [4]
b Suggest how you could plan an investigation to find the concentration of salt solution which would produce no change from that shown in the original dandelion stem before being cut in Figure 1. [4]

[Total: 10]

(Cambridge IGCSE Biology 0610 Paper 06 Q1 November 2009)

2 a Define *diffusion*. [2]
b The diagram below shows an apparatus that was used to investigate the effect of concentration of a chemical on the rate of diffusion.

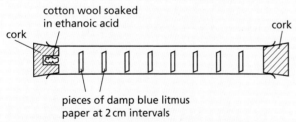

As ethanoic acid diffused along the tube, the pieces of blue litmus paper turned red.
Two different samples of ethanoic acid, **A** and **B**, were used in this apparatus. The two samples had different concentrations. The results are shown in the graph.

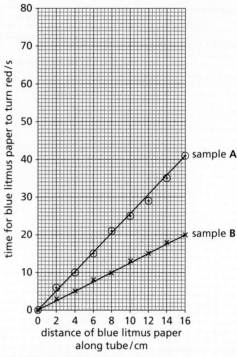

The table shows the results for a third sample, **C**, of ethanoic acid.

distance of blue litmus paper along tube/cm	time for blue litmus paper to turn red/s
2	9
4	18
6	28
8	35
10	45
12	55
14	63
16	72

(i) Complete the graph above by plotting the results shown in the table above. [3]
(ii) State which sample of ethanoic acid, **A**, **B** or **C**, took the longest time to travel 8 cm along the tube. [1]
(iii) State and explain which sample of ethanoic acid was the most concentrated. [2]
c Substances can enter and leave cells by either diffusion or by osmosis.
State **two** ways in which osmosis differs from diffusion. [2]

[Total: 10]

(Cambridge IGCSE Biology 0610 Paper 21 Q3 June 2012)

3 a (i) Define *osmosis*. [3]
  (ii) Osmosis is considered by many scientists to be a form of diffusion. Suggest **two** ways in which diffusion is different from osmosis. [2]
b (i) Explain how root hair cells use osmosis to take up water. [2]
  (ii) The land on which a cereal crop is growing is flooded by sea water. Suggest the effect sea water could have on the cereal plants. [4]

[Total: 11]
(Cambridge IGCSE Biology 0610 Paper 2 Q9 November 2009)

4 The diagram shows an alveolus in which gaseous exchange takes place.

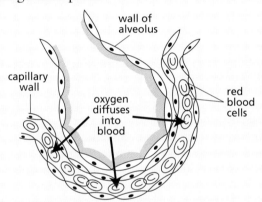

a (i) Define the term *diffusion*. [2]
  (ii) State what causes oxygen to diffuse into the blood from the alveoli. [1]
  (iii) List **three** features of gaseous exchange surfaces in animals, such as humans. [3]
b (i) At high altitudes there is less oxygen in the air than at sea level. Suggest how this might affect the uptake of oxygen in the alveoli. [2]
  (ii) In the past some athletes have cheated by injecting themselves with extra red blood cells before a major competition. Predict how this increase in red blood cells might affect their performance. [2]

[Total: 10]
(Cambridge IGCSE Biology 0610 Paper 21 Q9 November 2006)

Biological molecules

1 The sweet potato, *Ipomoea batatus*, is a different species to the Irish potato, *Solanum tuberosum*.

sweet potato

Irish potato

a (i) Describe **one** similarity, visible in the photo, between the two species of potato. [1]
  (ii) Complete the table to show two differences, visible in the photo, between the two species of potato. [2]

	sweet potato	Irish potato
difference 1		
difference 2		

b Potato crops are grown for their carbohydrate content.
Describe how you could safely test the two species of potato to compare their carbohydrate content.
test for starch
test for reducing sugar [8]

[Total: 11]
(Cambridge IGCSE Biology 0610 Paper 61 Q2 June 2010)

Enzymes

1 Enzymes are used commercially to extract fruit juices. The use of enzymes increases the volume of juice produced.
An investigation was carried out to determine the volume of apple juice produced at different temperatures.
Mixtures of apple pulp and enzyme were left for 15 minutes at different temperatures.
After 15 minutes, the mixtures were filtered and the juice collected.
The diagram shows the volume of juice collected from each mixture.

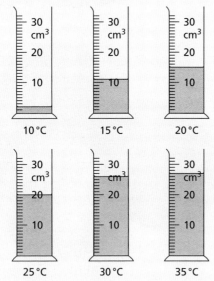

a (i) Record the volume of juice in each measuring cylinder in the table. [3]

temperature/°C	volume of juice collected/cm³
10	
15	
20	
25	
30	
35	

(ii) Present the data in a suitable graphical form. [5]
(iii) Describe the results. [2]

b Describe an investigation to show the effect of pH on the activity of the enzyme that is used to extract apple juice. [6]

[Total: 16]
(Cambridge IGCSE Biology 0610 Paper 61 Q1 November 2010)

2 Catalase is an enzyme that breaks down hydrogen peroxide into water and oxygen.

$$2H_2O_2 \rightarrow 2H_2O + O_2$$

By using small pieces of filter paper soaked in a solution of catalase, it is possible to measure the enzyme activity.
The pieces are placed in a solution of diluted hydrogen peroxide in a test-tube.
The filter paper rises to the surface as oxygen bubbles are produced.
The time taken for these pieces of filter paper to rise to the surface indicates the activity of catalase.

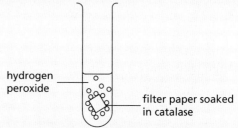

An experiment was carried out to find the effect of pH on the activity of catalase.
Five test-tubes were set up as shown in the diagram, each with a different pH.
The same volume and concentration of hydrogen peroxide was used in each test-tube.
The table shows the results obtained for the experiment as described.

pH	time taken for filter paper to rise/s
3.0	62
4.0	54
5.0	35
6.0	25
7.0	20
8.0	50

a (i) Plot a line graph to show the time taken for the filter paper to rise against pH. [4]
(ii) Describe the relationship between pH and the time taken for the filter paper to rise. [2]

b Suggest **four** ways in which this experiment could be improved. [4]

c Suggest how this experiment could be changed to investigate the effect of temperature on the activity of catalase. [6]

[Total: 16]
(Cambridge IGCSE Biology 0610 Paper 06 Q3 November 2009)

3 a All organisms depend on enzymes. Define the term *enzyme* and describe the function of enzymes in living organisms. [3]

b Samples of an amylase enzyme were incubated with starch at different temperatures. The rate of starch digestion in each sample was recorded and points plotted on the graph shown below.

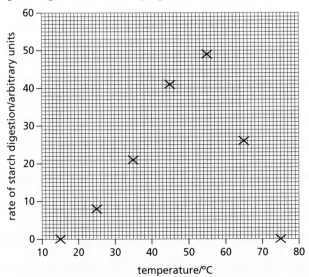

(i) Complete this line graph to show the effect of temperature on rate of digestion of starch by the amylase enzyme by adding the most appropriate line to the points. [1]
(ii) Using your graph estimate the optimum temperature for this enzyme. [1]
(iii) Suggest the rate of starch digestion at 37 °C. [1]
(iv) Describe the effect of temperature on the rate of starch digestion. [2]
(v) The enzymes originally incubated at 15 °C and 75 °C did not digest any starch. These samples were later incubated at the optimum temperature.
Predict what results could be expected in each sample and suggest reasons for your predictions. [3]

[Total: 11]
(Cambridge IGCSE Biology 0610 Paper 21 Q8 June 2012)

4 Catalase is an enzyme found in plant and animal cells. It has the function of breaking down hydrogen peroxide, a toxic waste product of metabolic processes.

a (i) State the term used to describe the removal of waste products of metabolism. [1]
(ii) Define the term *enzyme*. [2]

An investigation was carried out to study the effect of pH on catalase, using pieces of potato as a source of the enzyme.
Oxygen is formed when catalase breaks down hydrogen peroxide, as shown in the equation.

hydrogen peroxide $\xrightarrow{\text{catalase}}$ water + oxygen

The rate of reaction can be found by measuring how long it takes for 10 cm³ oxygen to be collected.

b (i) State the independent (input) variable in this investigation. [1]
(ii) Suggest two factors that would need to be kept constant in this investigation. [2]

The table shows the results of the investigation, but it is incomplete.

pH	time to collect 10 cm³ oxygen/min	rate of oxygen production/cm³ min⁻¹
4	20.0	0.50
5	12.5	0.80
6	10.0	1.00
7	13.6	0.74
8	17.4	

c Calculate the rate of oxygen production at pH 8. Show your working. [2]
d Complete the graph by plotting the rate of oxygen production against pH. [4]

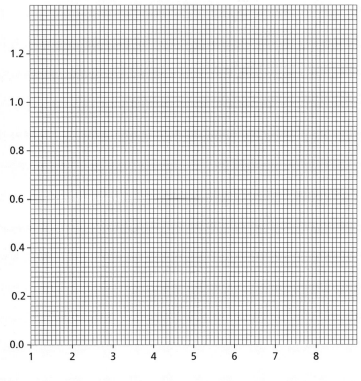

EXAMINATION QUESTIONS

e (i) Using data from the graph, describe the changes in the reaction rate between pH4 and pH8. [2]
   (ii) Explain the change in the reaction rate between pH6 and pH8. [3]

[Total: 17]
(Cambridge IGCSE Biology 0610 Paper 31 Q3 June 2008)

5 a The graph shows the activity of an enzyme produced by bacteria that live in very hot water.

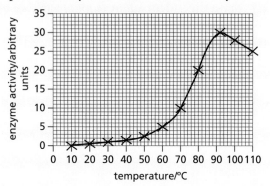

Using the information in the graph, describe the effect of increasing temperature on the activity of the enzyme. [3]

Enzymes extracted from bacteria are used in biological washing powders.

b Describe how bacteria are used to produce enzymes for biological washing powder. [4]
c Food and blood stains on clothes may contain proteins and fats.
   Explain how enzymes in biological washing powders act to remove food and blood stains from clothes. [4]
d When blood clots, an enzyme is activated to change a protein from one form into another. Describe the process of blood clotting. [3]

[Total: 14]
(Cambridge IGCSE Biology 0610 Paper 31 Q3 June 2009)

Plant nutrition

1 The diagram shows four test-tubes that were set up and left for 6 hours at a constant warm temperature.

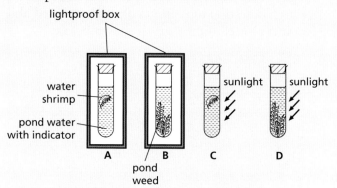

Hydrogencarbonate indicator (bicarbonate indicator) changes colour depending on the pH of gases dissolved in it, as shown below.

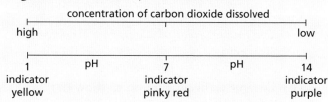

After 6 hours the colour of the indicator in all four tubes had changed.

a (i) Complete the table to predict the colour of the indicator after 6 hours. [4]

tube	colour of indicator at start	colour of indicator after 6 hours
A	pinky red	
B	pinky red	
C	pinky red	
D	pinky red	

(ii) Suggest the reason for the change in colour of the indicator in each of tubes A and D. [4]

b The diagram shows a fifth tube, E, set up at the same time and in the same conditions as tubes C and D.

E

Suggest and explain the possible colour of the indicator in tube E after 6 hours. [3]

[Total: 11]
(Cambridge IGCSE Biology 0610 Paper 2 Q6 June 2009)

Plant nutrition

2 The diagram shows a section through a leaf.

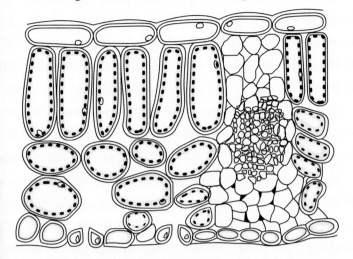

a On the diagram, label a stoma, the cuticle and a vascular bundle. Use label lines and the words 'stoma', 'article' and 'vascular bundle' on the diagram. [3]

b **(i)** The upper layers of a leaf are transparent. Suggest an advantage to a plant of this feature. [1]

(ii) The cuticle is made of a waxy material. Suggest an advantage to a plant of this feature. [1]

(iii) State **two** functions of vascular bundles in leaves. [2]

c Most photosynthesis in plants happens in leaves.

(i) Name the two raw materials needed for photosynthesis. [2]

(ii) Photosynthesis produces glucose. Describe how plants make use of this glucose. [3]

[Total: 12]

(Cambridge IGCSE Biology 0610 Paper 21 Q4 November 2010)

3 A student set up the apparatus shown in the diagram to investigate the effect of light intensity on the rate of photosynthesis of a pond plant.

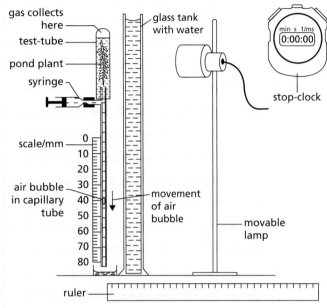

The student maintained the temperature at 20 °C and measured the distance travelled by the air bubble in the capillary tube for a period of 5 minutes on three occasions for each light intensity.

The student's results are shown in the table.

distance of lamp from pond plant/mm	distance travelled by air bubble/mm	rate of photosynthesis/ mm per minute
20	30	6.0
30	26	5.2
40	14	2.8
50	7	
60	3	0.6

a **(i)** Explain why the student included the glass tank and the syringe in the apparatus. [2]

(ii) Explain why the air bubble moves down the capillary tube. [3]

b **(i)** Calculate the rate of photosynthesis when the lamp was 50 mm from the pond plant. [1]

(ii) Plot the student's results from the table on the axes below. Draw an appropriate line on the graph to show the relationship between distance of the lamp from the pond plant and the rate of photosynthesis. [2]

EXAMINATION QUESTIONS

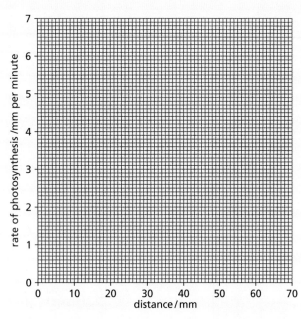

c (i) Using the graph to help you, predict the results that the student would get if the lamp was positioned 15 mm and 70 mm from the pond plant. [2]
(ii) Explain why the rate of photosynthesis **decreases** as the distance of the lamp from the pond plant **increases**. [3]

[Total: 13]
(Cambridge IGCSE Biology 0610 Paper 31 Q3 November 2009)

● Human nutrition

1 The diagram shows the human digestive system and associated organs.

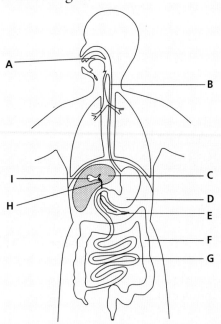

a Use letters from the diagram to identify the structures described. Each letter may be used once, more than once, or not at all.
  (i) One structure where digestion of protein occurs.
  (ii) One structure where bile is stored.
  (iii) One structure where peristalsis happens.
  (iv) One structure where starch digestion occurs.
  (v) One structure where amino acids are absorbed into the blood. [5]
b State **two** functions of each of the structures labelled C and E on the diagram.
  (i) structure C [2]
  (ii) structure E [2]

[Total: 9]
(Cambridge IGCSE Biology 0610 Paper 21 Q9 November 2011)

2 a (i) State what is meant by the term *balanced diet*. [3]
  (ii) Balanced diets should include fat, fibre, mineral salts and vitamins. Name **two** other types of nutrients that should be present in a balanced diet. [1]
b Suggest and explain the effects on a person of a diet with:
  (i) too little fibre, [2]
  (ii) too much animal fat. [2]
c Calcium, a mineral salt, is needed in the diet. Explain the role of calcium in the body and the effect of calcium deficiency. [3]

[Total: 11]
(Cambridge IGCSE Biology 0610 Paper 21 Q2 June 2011)

3 The diagram shows three different types of teeth from a human.

A　　　　B　　　　C

a (i) Name the types of teeth labelled A and B. [2]
  (ii) State where in the jaw tooth type C is found. [1]
b Explain how regular brushing helps to prevent tooth decay. [3]
c Explain the roles of chewing and of enzymes in the process of digestion. [4]

[Total: 10]
(Cambridge IGCSE Biology 0610 Paper 21 Q7 June 2010)

4 a Micronutrients are food materials that are only needed in very small quantities in the human diet. Draw one straight line from each micronutrient to its deficiency symptom. [4]

micronutrient	deficiency symptom
calcium	anaemia
vitamin C	rickets
vitamin D	scurvy
iron	

b Explain how iron, in the diet of humans, is used in the body. [3]

[Total: 7]
(Cambridge IGCSE Biology 0610 Paper 2 Q3 November 2009)

5 a Enzyme activity is vital in human digestion. Complete the table by choosing appropriate words from the list. [6]

amino acids   amylase   cellulose
fatty acids   hydrochloric acid   lipase
protein   starch   water

substrate	enzyme	product
fat		glycerol +
	protease	
		maltose

b Maltose is changed into glucose.
  (i) Which part of the blood carries glucose? [1]
  (ii) Which process, happening in all living cells, needs a constant supply of glucose? [1]
  (iii) Excess glucose is stored. Which carbohydrate is glucose changed into for storage? [1]
  (iv) Which organ is the main store of this carbohydrate? [1]
  (v) Name a hormone that causes glucose to be released from storage. [1]

[Total: 11]
(Cambridge IGCSE Biology 0610 Paper 2 Q4 November 2009)

Transport in plants

1 a Phloem and xylem are two types of tissue in plants. The diagram shows a section through a plant stem, **A**, and a plant leaf, **B**.

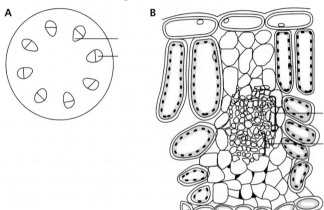

  (i) Label the phloem (P) and the xylem (X) on both **A** and **B** on the diagram. Write the letters P and X on both **A** and **B**. [2]
  (ii) Describe **two** functions of the xylem. [2]

b Translocation takes place in the phloem tissue.
  (i) State which materials are translocated in the phloem. [2]
  (ii) The diagram shows a plant in the sunlight. The three lines are arrows, with no arrow heads, showing the translocation of materials within parts of the plant.

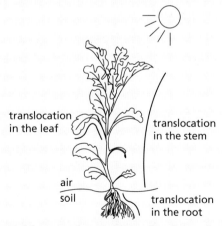

Add arrow heads to **each** of the **three** lines to show the direction of translocation in the organs shown. [3]

[Total: 9]
(Cambridge IGCSE Biology 0610 Paper 21 Q9 June 2012)

EXAMINATION QUESTIONS

2 An investigation of the uptake and loss of water by a plant was carried out over 24 hours. The results are shown in the table.

time of day/hours	water uptake/g per hour	water loss/g per hour
0400	7	2
0700	11	8
1000	18	24
1300	24	30
1600	24	24
1900	20	13
2200	11	5

a (i) The data for water uptake have been plotted on the grid below. Plot the data for water loss on the same grid. Label both curves. [4]

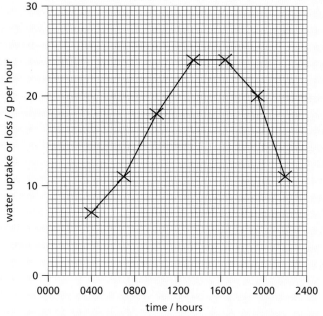

(ii) State the **two** times at which the uptake and loss of water were the same. [1]

b Explain how a **decrease** in temperature and humidity would affect the water loss by this plant.
(i) Temperature [2]
(ii) Humidity [2]

[Total: 9]
(Cambridge IGCSE Biology 0610 Paper 21 Q6 November 2011)

3 a Explain what is meant by the term *transpiration*. [3]
b Describe the effect that **two** named environmental factors can have on the rate of transpiration. [4]

[Total: 7]
(Cambridge IGCSE Biology 0610 Paper 21 Q9 June 2011)

4 The photograph is of a root of radish covered in many root hairs.

a Using the term *water potential*, explain how water is absorbed into root hairs from the soil. [3]

A potometer is a piece of apparatus that is used to measure water uptake by plants. Most of the water taken up by plants replaces water lost in transpiration. A student used a potometer to investigate the effect of wind speed on the rate of water uptake by a leafy shoot. As the shoot absorbs water the air bubble moves upwards. The student's apparatus is shown in the diagram.

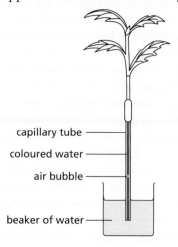

The student used a fan with five different settings and measured the wind speed. The results are shown in the table.

wind speed/ metres per second	distance travelled by the air bubble/mm	time/minutes	rate of water uptake/mm per minute
0	4	10	0.4
2	12	5	2.4
4	20	5	4.0
6	35	5	7.0
8	40	2	

b Calculate the rate of water uptake at the highest wind speed and write your answer in the table. [1]

c Describe the effect of increasing wind speed on the rate of water uptake. You may use figures from the table to support your answer. [2]

d State two environmental factors, **other than wind speed**, that the student should keep constant during the investigation. [2]

e Some of the water absorbed by the plants is not lost in transpiration. State **two** other ways in which water is used. [2]

f Water moves through the xylem to the tops of very tall trees, such as the giant redwoods of North America. The movement of water in the xylem is caused by transpiration. Explain how transpiration is responsible for the movement of water in the xylem. [4]

g Plants that live in hot, dry environments show adaptations for survival. State three **structural** adaptations of these plants. [3]

[Total: 17]

(Cambridge IGCSE Biology 0610 Paper 31 Q4 June 2009)

● Transport in animals

1 The diagram shows the route taken by blood around the body.

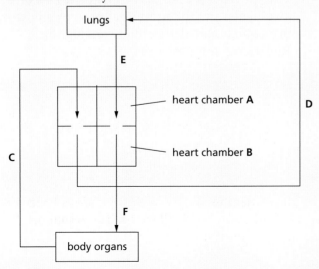

a (i) Name the heart chambers **A** and **B**. [2]
  (ii) Use information shown in the diagram to identify the type of blood vessel **C** as either an artery or a vein. Give a reason for your choice. [2]
b (i) State and explain **two** differences between the contents of the blood flowing in vessels **C** and **E**. [2]
  (ii) Suggest and explain which of the four blood vessels contains blood at the highest pressure. [2]

[Total: 8]

(Cambridge IGCSE Biology 0610 Paper 21 Q8 June 2010)

2 As the heart pumps blood around the human body, a pulse may be felt at certain sites, such as the one shown in the diagram.

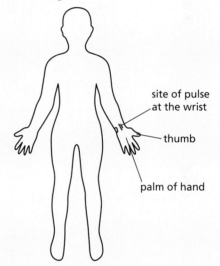

a (i) Label on the diagram, one other site where a pulse may be felt. [1]
  (ii) Suggest why it is possible to feel the pulse at these sites. [2]
b A student counted the number of pulses felt in 15 seconds at the site shown on their wrist. The student did this three times. The results are recorded in the table.

	pulses per 15 seconds	pulses per minute
1st count	18	
2nd count	19	
3rd count	17	
mean		

(i) Complete the right-hand column in the table to show the number of pulses per minute for each count and the mean pulses per minute. [2]

EXAMINATION QUESTIONS

  (ii) Explain why it is advisable to repeat readings at least three times. [1]
  (iii) State **two** factors that may affect heart rate. For each factor explain its effect on heart rate. [4]
 c Body mass and heart rates for a number of different mammals are shown in the table.

Mammal	body mass/kg	heart rate/beats per minute
rabbit	1.0	200
cat	1.5	150
dog	5.0	90
human	60.0	
horse	1200.0	44
elephant	5000.0	30

Copy the mean pulses per minute from the first table into the second table.
  (i) Plot the data in a bar chart to show heart rate for all six mammals. [5]

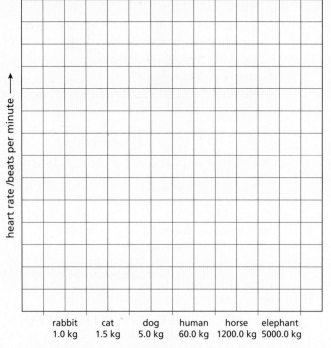

  (ii) Describe the general trend shown by this data plotted on the bar chart. [1]
 d An elephant can live for 70 years, a cat for 15 years and a rabbit for 9 years. Suggest how heart rate and body mass might affect life expectancy of mammals. [1]

[Total: 17]
(Cambridge IGCSE Biology 0610 Paper 61 Q2 June 2009)

3 The diagram shows an external view of the heart.

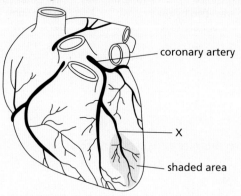

 a A blood clot is stuck at X. Explain what will happen to the heart muscle cells in the shaded area. [3]
 b List **three** actions people can take to reduce the risk of having a blood clot in the coronary arteries. [3]

[Total: 6]
(Cambridge IGCSE Biology 0610 Paper 21 Q3 November 2012)

4 a The human circulatory system contains valves.
  (i) State the function of these valves. [1]
  (ii) Complete the table by placing a tick (✓) against **two** structures in the human circulatory system that have valves. [1]

structure in circulatory system	have valves
arteries	
capillaries	
heart	
veins	

 b Describe how you would measure the heart rates of some students before they start running. [2]
 c The bar chart (opposite) shows the results of an investigation of the heart rates of some students before and immediately after running. Each student ran the same distance.

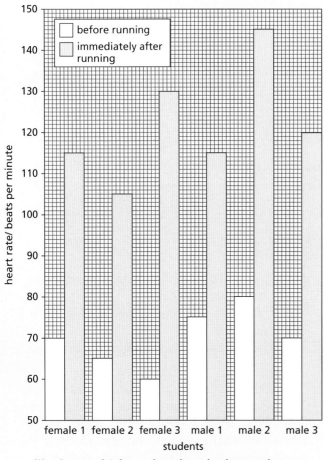

(i) State which student has the lowest heart rate immediately after running. [1]
(ii) State which student has the largest change in heart rate from before to immediately after running. [1]
(iii) Describe any trends that you can see in the results. [2]
d Explain why heart rate changes when you run. [4]

[Total: 12]
(Cambridge IGCSE Biology 0610 Paper 21 Q2 November 2011)

5 The diagram shows a section through the heart.

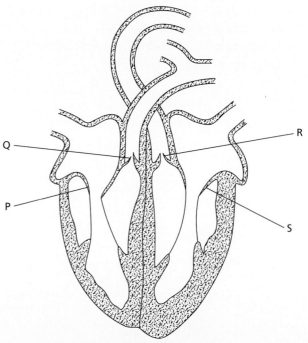

a (i) Name the **two** blood vessels, shown on the diagram, that carry oxygenated blood. [1]
(ii) State the letter that identifies the tricuspid valve. [1]
(iii) State the letter that identifies a semilunar valve. [1]
b Describe how the heart forces blood into the aorta. [3]
c (i) Name the blood vessel that delivers blood to the muscles of the walls of the atria and ventricles. [1]
(ii) Name the **two** blood vessels that deliver blood to the liver. [2]

[Total: 9]
(Cambridge IGCSE Biology 0610 Paper 21 Q8 June 2011)

● Diseases and immunity

1 a Many communities treat their sewage and release non-polluting water into a local river. What is meant by the term *sewage*? [2]
b Sometimes the sewage treatment works cannot deal with all of the sewage and untreated material is released into the river. Suggest the likely effects of releasing untreated sewage into a river. [4]

[Total: 6]
(Cambridge IGCSE Biology 0610 Paper 2 Q2 November 2006)

EXAMINATION QUESTIONS

2 The lymphatic system consists of:
- thin-walled lymph vessels that drain tissue fluid from many organs of the body
- lymph nodes that contain the cells of the immune system.

The fluid in the lymph vessels is moved in a way similar to the movement of blood in veins. The diagram shows part of the lymphatic system.

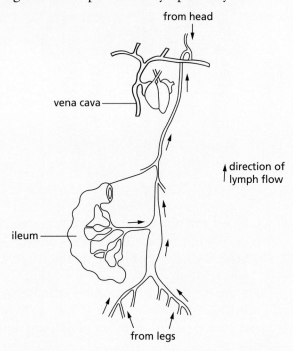

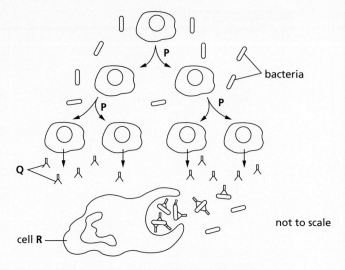

a Suggest how lymph is moved in the lymph vessels. [2]

b After a meal rich in fatty foods, the lymph leaving the ileum is full of fat droplets. Explain why there are fat droplets in the lymph leaving the ileum. [2]

Lymph flows through lymph nodes. The diagram (above right) shows the action of white blood cells in a lymph node when bacteria are present.

c (i) Name the type of nuclear division shown at **P** in the diagram. [1]
(ii) Name the molecules labelled **Q** in the diagram. [1]
(iii) Describe how bacteria are destroyed by cell **R**. [3]

Antibiotics are used to treat bacterial infections. An investigation was carried out into the effect of prescribing antibiotics on antibiotic resistance in 20 countries. The graph shows the results of this investigation. Each point represents the result for a country.

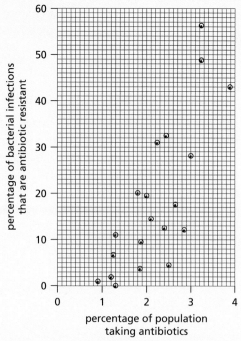

d Describe the results shown in the graph. Credit will be given for using figures from the graph to support your answer. [3]

e Many different antibiotics are used. Suggest why some antibiotics are used less frequently than others. [3]

[Total: 15]
(Cambridge IGCSE Biology 0610 Paper 31 Q4 November 2010)

3 a Describe the function of the immune system, including antibody production and phagocytosis. [9]
  b Outline the problems of organ transplantation and how they can be overcome. [6]

[Total: 15]
(Cambridge IGCSE Biology 0610 Paper 3 Q6 November 2003)

Gas exchange in humans

1 Gaseous exchange takes place while air flows in and out of the lungs.
  a State **three** ways in which inspired air is different from expired air. [3]
  b List **three** features of gaseous exchange surfaces that help to make them more efficient. [3]

[Total: 6]
(Cambridge IGCSE Biology 0610 Paper 21 Q8 November 2009)

2 The ribcage and diaphragm are involved in the breathing mechanism to ventilate the lungs. The flow chart shows the changes that take place when breathing in.

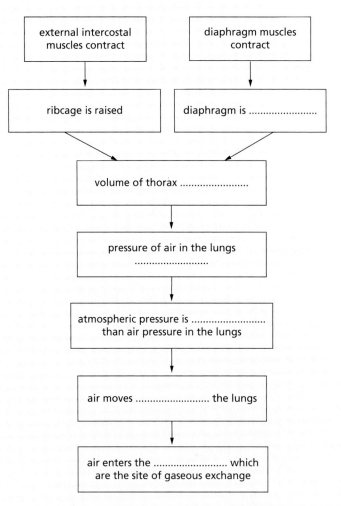

a Complete the flow chart by writing appropriate words in the spaces provided. [6]
b The photograph shows part of the epithelium that lines the trachea.

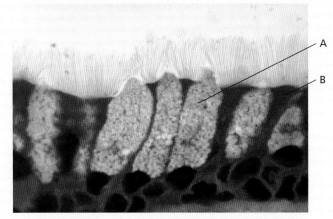

Explain how the cells labelled A and B in the photograph protect the gas exchange system. [4]

[Total: 10]
(Cambridge IGCSE Biology 0610 Paper 31 Q3 November 2012)

EXAMINATION QUESTIONS

3 a Define the term *aerobic respiration*. [2]

During exercise the movement of the ribcage enables air to enter the lungs.

b Describe how the ribcage is moved during inspiration (breathing in) and explain how this causes air to enter the lungs. [4]

c Explain how the ribcage returns to its resting position during expiration (breathing out). [2]

Some students carried out an investigation on a 16-year old athlete. The table shows the results of their investigation on the athlete's breathing at rest and immediately after 20 minutes of running.

Ventilation rate is the volume of air taken into the lungs per minute.

	at rest	immediately after 20 minutes of running
rate of breathing/breaths per minute	12	20
average volume of air taken in with each breath/dm³	0.5	3.5
ventilation rate/dm³ per minute	6.0	

d (i) Calculate the ventilation rate of the athlete immediately after 20 minutes of running. [1]

(ii) Explain why the athlete has a high ventilation rate **after the exercise has finished**. [5]

[Total: 14]
(Cambridge IGCSE Biology 0610 Paper 31 Q3 November 2010)

● Respiration

1 a (i) State the word equation for aerobic respiration. [2]

(ii) Complete the table to show three differences between aerobic respiration and anaerobic respiration in humans. [3]

	aerobic respiration in humans	anaerobic respiration in humans
1		
2		
3		

b Yeast is used in making some types of bread and in brewing.

(i) Explain the role of yeast in bread-making. [3]

(ii) Explain the role of yeast in brewing. [2]

[Total: 10]
(Cambridge IGCSE Biology 0610 Paper 21 Q5 November 2010)

2 a State, using chemical symbols, the equation for aerobic respiration. [3]

A student compared the respiration of germinating mung bean seeds with pea seeds using the apparatus shown in the diagram.

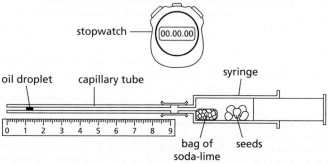

The soda-lime absorbs any carbon dioxide released by the germinating seeds. The student recorded the position of the oil droplet every minute over a period of 6 minutes.

b State **three** variables that should be kept constant in this investigation. [3]

c The table shows the student's results.

time/ minute	germinating mung bean seeds		germinating pea seeds	
	position of droplet/mm	distance moved/mm per minute	position of droplet/mm	distance moved/mm per minute
0	0	0	0	0
1	12	12	10	10
2	23	11	19	9
3	36	13	28	9
4	45	9	33	5
5	48	3	36	3
6	48	0	36	0

(i) State which way the droplet moves **and** explain your answer. [3]

(ii) State what happens to the movement of the droplet after 3 minutes **and** suggest an explanation. [2]

[Total: 11]
(Cambridge IGCSE Biology 0610 Paper 31 Q3 November 2011)

Excretion in humans

1 a The kidney is an excretory organ. Name **two** other excretory organs in humans and in each case state a substance that the organ excretes. [4]

b The table shows the amounts of some substances in the blood in the renal artery and in the renal vein of a healthy person.

substance	amount in blood in renal artery (arbitrary units)	amount in blood in renal vein (arbitrary units)
oxygen	100.0	35.0
glucose	10.0	9.7
sodium salts	32.0	29.0
urea	3.0	0.5
water	180.0	178.0

Suggest what happens in the kidney to bring about the differences in the composition of the blood shown in the table. [4]

[Total: 8]

(Cambridge IGCSE Biology 0610 Paper 21 Q9 November 2010)

2 a Why do most waste products of metabolism have to be removed from the body? [1]

b The diagram shows the human excretory system.

Name the parts that fit each of the following descriptions.
(i) The tube that carries urine from the kidneys. [1]
(ii) The organ that stores urine. [1]
(iii) The blood vessel that carries blood away from the kidneys. [1]

c Outline how the kidneys remove only waste materials from the blood. [3]

d Excess amino acids cannot be stored in the body and have to be broken down.

(i) Where are excess amino acids broken down? [1]
(ii) Which waste chemical is formed from the breakdown of excess amino acids? [1]

[Total: 9]

(Cambridge IGCSE Biology 0610 Paper 2 Q2 June 2009)

3 a Define the term *excretion*. [3]

b The figure below shows a section through a kidney.

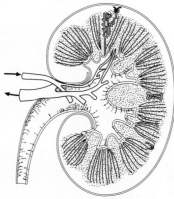

(i) Using label lines and the letters given, label the following on a copy of the figure:
F   where filtration occurs
R   the renal artery
U   where urine passes to the bladder [3]

(ii) Describe the process of filtration in the kidney. [3]

(iii) Name the processes resulting in the reabsorption of
1 glucose
2 water. [3]

[Total: 12]

(Cambridge IGCSE Biology 0610 Paper 3 Q3 November 2007)

Co-ordination and response

1 a Define the term *homeostasis*. [2]

b It has been suggested by some scientists that the iris reflex is an example of homeostasis. Describe this reflex and explain why it might be considered to be a homeostatic mechanism. [3]

[Total: 5]

(Cambridge IGCSE Biology 0610 Paper 21 Q10 June 2008)

EXAMINATION QUESTIONS

2 a Complete the following paragraph using appropriate words.
Sense organs are composed of groups of _____ cells that respond to specific _____. The sense organs that respond to chemicals are the _____ and the _____. [4]

b The eye is a sense organ that focuses light rays by changing the shapes of its lens. It does this by contracting its ciliary muscles.
(i) What links the ciliary muscles to the lens? [1]
(ii) Describe the change in shape of the lens when a person looks from a near object to a distant object. [1]

c The graph shows changes in the contraction of the ciliary muscles as a person watches a humming bird move from flower to flower while feeding on nectar.

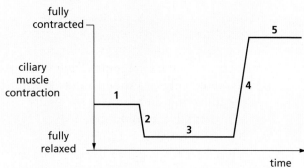

In which period of time, 1, 2, 3, 4 or 5, was the bird
(i) feeding from a flower very near to the person [1]
(ii) flying away from the person [1]
(iii) flying towards the person. [1]

[Total: 9]
(Cambridge IGCSE Biology 0610 Paper 21 Q7 June 2009)

3 a Name **two** sense organs and an environmental stimulus that each detects. [2]
b (i) Tropisms occur in plants. State the meaning of the term *tropism*. [2]
(ii) Complete the table about tropisms in plants. [4]

stimulus	name of tropism	effect on plant shoot
gravity		
light		

[Total: 8]
(Cambridge IGCSE Biology 0610 Paper 21 Q9 June 2010)

4 a The diagram shows the structures involved in a reflex arc.

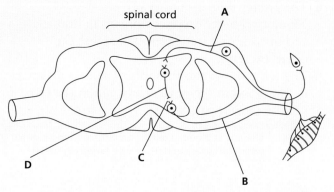

(i) On the diagram label structures **A**, **B**, **C** and **D**. [4]
(ii) Name the two types of tissue in the body that can act as effectors. [2]

b (i) Describe the characteristics of a reflex action resulting from the activity of structures **A**, **B**, **C** and **D**. [2]
(ii) State **one** example of a reflex action. [1]

[Total: 9]
(Cambridge IGCSE Biology 0610 Paper 21 Q4 June 2011)

5 a Plants, like animals, respond to stimuli. Tropisms are an example of a plant response.
(i) Define the term *geotropism*. [2]
(ii) Suggest the advantages of geotropic responses for a seed germinating in the soil. [3]

b State three external conditions necessary for the germination of a seed in the soil. [3]

[Total: 8]
(Cambridge IGCSE Biology 0610 Paper 21 Q3 November 2011)

● Drugs

1 The first diagram shows an organism **W** and the second diagram shows how the reproduction of this organism is affected by an antibiotic.

organism **W**

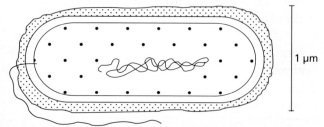

Reproduction

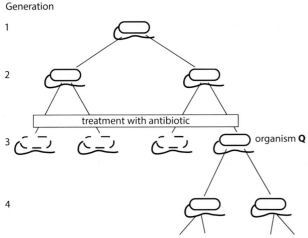

a (i) What type of organism is **W** most likely to be? [1]
   (ii) State **three** reasons for your answer. [3]
b Name the type of reproduction shown by organism **W**. [1]

Q is the only organism surviving the antibiotic treatment.

c Suggest an explanation for the survival of **Q** and its offspring. [2]
d Explain why patients who are treated with antibiotics are always advised to take a complete course of treatment, rather than stop the treatment as soon as they feel better. [3]

[Total: 10]
(Cambridge IGCSE Biology 0610 Paper 3 Q9 June 1998)

● Reproduction

1 Choose words from the list to complete each of the spaces in the paragraph.
   Each word may be used once only and some words are not used at all.
   bright   dry   dull   heavy   large
   light   sepals   small   stamens
   sticky   style
   Flowers of plants that rely on the wind to bring about pollination tend to have _____ petals that have a _____ colour. Their pollen is normally _____ and _____. In these flowers, the _____ and the _____ both tend to be long. [6]

[Total: 6]
(Cambridge IGCSE Biology 0610 Paper 21 Q2 June 2008)

2 The diagram shows the male reproductive system.

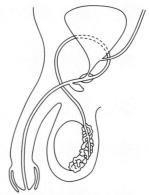

a Using a label line and the letters given, label the diagram.
   (i) **G** where gametes are formed [1]
   (ii) **S** the sperm duct [1]
   (iii) **T** where testosterone is formed [1]
   (iv) **U** the urethra [1]
b Describe **two** secondary characteristics regulated by testosterone. [2]
c Choose words from the list to complete each of the spaces in the paragraph. Each word may be used once only and some words may not be used at all.
   four   diploid   double   half
   haploid   meiosis   mitosis   two
   Gametes are formed by the division of a nucleus, a process called _____. This process produces a total of _____ cells from the original cell. Each of these cells has a nucleus described as being _____ and each nucleus contains _____ the number of chromosomes present in the original nucleus. [4]

[Total: 10]
(Cambridge IGCSE Biology 0610 Paper 21 Q8 June 2009)

3 The diagram shows a section through parts of the male reproductive and urinary systems.

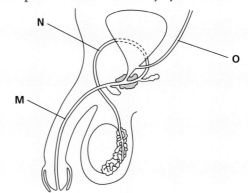

EXAMINATION QUESTIONS

a (i) Name the tubes labelled **M, N** and **O**. [3]
  (ii) Explain the roles of the testes, the prostate gland and the scrotum. [4]
b Humans use a variety of methods of birth control.
  (i) On the diagram, put an **X** where a vasectomy could be carried out. [1]
  (ii) Explain **one** method of birth control, used by males, that can also protect against infection by a sexually transmitted disease. [2]
  (iii) Name **one** sexually transmitted disease. [1]

[Total: 11]
(Cambridge IGCSE Biology 0610 Paper 21 Q3 June 2011)

4 Reproduction in humans is an example of sexual reproduction. Outline what occurs during:
  a sexual intercourse [2]
  b fertilisation [3]
  c implantation. [2]

[Total: 7]
(Cambridge IGCSE Biology 0610 Paper 21 Q8 Nov 2011)

5 The diagram shows an experiment to investigate the conditions needed for germination. Tubes A, B, C and D are at room temperature and tube E is in a freezer.

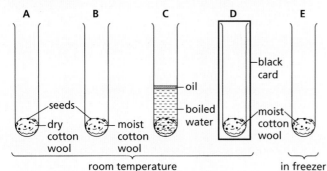

a State **three** of the environmental conditions this experiment is investigating. [3]
b Predict in which **two** tubes the seeds will germinate. [2]
c Nuclear and cell division happen during germination.
  (i) Name the type of nuclear division that takes place during the growth of a seedling. [1]
  (ii) State how the number of chromosomes in each of the new cells compares with the number of chromosomes in the original cells. [1]

d The graph shows the changes in the dry mass of a broad bean seed in the first 5 days after planting.

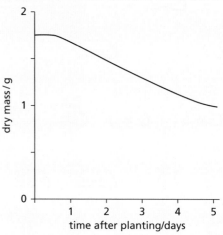

Describe and suggest an explanation for the changes that happen to the dry mass of the seed in the first 5 days after planting. [3]

[Total: 10]
(Cambridge IGCSE Biology 0610 Paper 21 Q5 June 2010)

6 a Using straight lines, match the names of flower parts with their functions. One has been completed for you. [4]

anther	allows the passage of the pollen tube to the ovary
petal	attracts insects for pollination
sepal	produces pollen grains
style	protects the flower when in bud
stigma	the surface on which the pollen lands during pollination

b Describe how the stigmas of wind-pollinated flowers differ from the stigmas of insect-pollinated flowers. Relate these differences to the use of wind as the pollinating agent. [3]
c Discuss the implication to a species of self-pollination. [3]

[Total: 10]
(Cambridge IGCSE Biology 0610 Paper 31 Q1 June 2008)

Reproduction

7 The diagram shows the structure of the placenta and parts of the fetal and maternal circulatory systems.

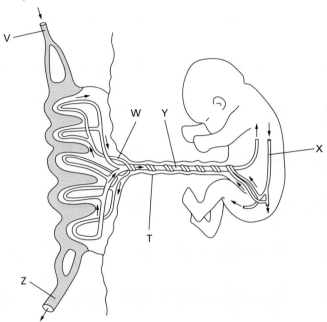

a (i) Complete the table by listing the blood vessels that carry oxygenated blood. Use the letters in the diagram to identify the blood vessels. [2]

circulatory system	blood vessels that carry oxygenated blood
maternal	
fetal	

(ii) Name structure **T** and describe what happens to it after birth. [2]

(iii) The placenta is adapted for the exchange of substances between the maternal blood and the fetal blood. Describe the exchanges that occur across the placenta to keep the fetus alive and well. [4]

b The placenta secretes the hormones oestrogen and progesterone. Describe the roles of these hormones during pregnancy. [3]

[Total: 11]
(Cambridge IGCSE Biology 0610 Paper 31 Q5 June 2012)

8 The diagram shows a human egg cell and a human sperm cell.

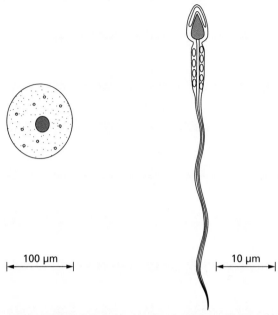

human egg cell   human sperm cell

a (i) What is the name given to the release of eggs from the ovary? [1]
(ii) Sperm cells and egg cells are haploid. State the meaning of the term *haploid*. [1]

b Complete the table to compare egg cells with sperm cells. [4]

feature	egg cells	sperm cells
site of production		
relative size		
numbers produced		
mobility		

c Three hormones that control the menstrual cycle are:
- follicle stimulating hormone (FSH)
- luteinising hormone (LH)
- oestrogen.

(i) Name the site of production and release of oestrogen. [1]
(ii) Describe the role of oestrogen in controlling the menstrual cycle. [2]

d Artificial insemination is sometimes used as a treatment for female infertility. Outline how artificial insemination is carried out in humans. [2]

[Total: 11]
(Cambridge IGCSE Biology 0610 Paper 31 Q3 June 2010)

EXAMINATION QUESTIONS

● Inheritance

1 Flowers from three red-flowered plants, **A**, **B** and **C**, of the same species were self-pollinated.
  a Explain what is meant by the term *pollination*. [2]
  b Seeds were collected from plants **A**, **B** and **C**. The seeds were germinated separately and were allowed to grow and produce flowers. The colour of these flowers is shown in the table.

seeds from plant	colour of flowers grown from the seeds
A	all red
B	some red and some white
C	some red and some white

  (i) State the recessive allele for flower colour. [1]
  (ii) State which plant, **A**, **B**, or **C**, produced seeds that were homozygous for flower colour. [1]
  (iii) Suggest how you could make certain that self-pollination took place in the flowers of plants **A**, **B** and **C**. [2]
  c Complete the genetic diagram to explain how two red-flowered plants identical to plant **B** could produce both red-flowered and white-flowered plants. Use the symbols **R** to represent the dominant allele and **r** to represent the recessive allele. [4]

	parent 1		parent 2
parental phenotypes	red-flowered	×	red-flowered
parental genotypes		×	

gametes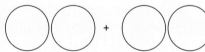

offspring genotypes

offspring phenotypes

[Total: 10]
(Cambridge IGCSE Biology 0610 Paper 21 Q10 November 2011)

2 The diagram shows a family tree for a condition known as nail-patella syndrome (NPS).

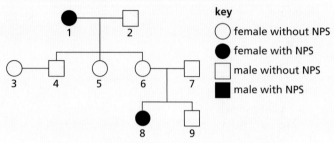

  a (i) State whether NPS is controlled by a dominant or a recessive allele.
    (ii) Explain which evidence from the family tree confirms your answer to **(i)**. [3]
  b Explain what the chances are for a third child of parents 6 and 7 having NPS. You may use a genetic diagram to help your explanation. [3]

[Total: 6]
(Cambridge IGCSE Biology 0610 Paper 21 Q7 June 2008)

3 There is a variation in the shape of human thumbs. The diagram shows the two forms referred to as 'straight' and 'hitch hikers'.

straight   hitch hikers

A survey of thumb shapes was carried out on 197 students. The results are shown in the table.

age/years	number of students with 'straight' thumbs		number of students with 'hitch hiker' thumbs	
	male	female	male	female
12	21	24	4	2
13	18	28	3	5
14	19	15	2	3
15	26	20	3	4
total	84	87	12	14

  a Describe the results shown in the table. [3]
  b Scientists think that thumb shape is controlled by a single gene. What evidence is there from the table to support this idea? [3]

[Total: 6]
(Cambridge IGCSE Biology 0610 Paper 61 Q3 November 2010)

4 Complete the sentences by writing the most appropriate word in each space. Use only words from the list below.

allele   diploid   dominant   gene   genotype
haploid   heterozygous   homozygous   meiosis
mitosis   phenotype   recessive

Wing length in the fruit fly, *Drosophila*, is controlled by a single _____ that has two forms, one for long and one for short wings. The sperm and ova of fruit flies are produced by the process of _____. When fertilisation occurs the gametes fuse to form a _____ zygote. When two long-winged fruit flies were crossed with each other some of the offspring were short-winged. The _____ of the rest of the offspring was long-winged. The short-winged form is _____ to the long-winged form and each of the parents must have been _____. [6]

[Total: 6]
(Cambridge IGCSE Biology 0610 Paper 21 Q6 November 2010)

5 The diagram shows three species of zebra.

Equus burcheli

Equus grevyi

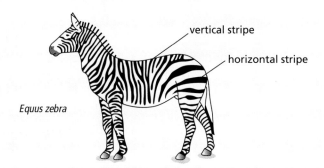
Equus zebra

a Describe one method a scientist could use to show that zebras shown in the diagram are different species. [1]

b Studies have shown that the hotter the environment, the more stripes zebras have.
  (i) State the type of variation which would result in different numbers of stripes. [1]
  (ii) Study the diagram. Suggest which species of zebra lives in the hottest environment. [1]

c Occasionally, zebras are born that are almost completely black. The change in appearance is the result of mutation.
  (i) State the term that is used to describe the appearance of an organism. [1]
  (ii) Define the term *mutation*. [2]

d Tsetse flies attack animals with short fur, sucking their blood and spreading diseases. The diagram shows a tsetse fly. This fly is an insect, belonging to the arthropod group.

  (i) State **one** feature, visible in the diagram, which is common to all arthropods. [1]
  (ii) State **two** features, visible in the diagram, which distinguish insects from other arthropod groups. [2]

e Scientists have discovered that zebras with more horizontal stripes attract fewer tsetse flies.
  (i) Suggest why the stripes on the head and neck of the zebra would be an advantage when it feeds on the grass on the ground. [2]
  (ii) Describe how a species of zebra could gradually develop more horizontal stripes. [3]

[Total: 14]
(Cambridge IGCSE Biology 0610 Paper 31 Q4 June 2008)

EXAMINATION QUESTIONS

6 The flowers of pea plants, *Pisum sativum*, are produced for sexual reproduction. The flowers are naturally self-pollinating, but they can be cross-pollinated by insects.

a Explain the difference between self-pollination and cross-pollination. [2]

b Explain the disadvantages for plants, such as *P. sativum*, of reproducing sexually. [4]

Pea seeds develop inside pea pods after fertilisation. They contain starch. A gene controls the production of an enzyme involved in the synthesis of starch grains. The allele, R, codes for an enzyme that produces normal starch grains. This results in seeds that are round. The allele, r, does not code for the enzyme. The starch grains are not formed normally. This results in seeds that are wrinkled. The diagram shows round and wrinkled pea seeds.

round pea seed   wrinkled pea seed

Pure bred plants are homozygous for the gene concerned. A plant breeder had some pure bred pea plants that had grown from round seeds and some pure bred plants that had grown from wrinkled seeds.

c State the genotypes of the pure bred plants that had grown from round and from wrinkled seeds. [1]

These pure bred plants were cross-pollinated (cross 1) and the seeds collected. All the seeds were round. These round seeds were germinated, grown into adult plants (offspring 1) and self-pollinated (cross 2). The pods on the offspring 1 plants contained both round and wrinkled seeds. Further crosses (3 and 4) were carried out as shown in the table.

cross		phenotype of seeds in the seed pods		ratio of round to wrinkled seeds
		round seeds	wrinkled seeds	
1	pure bred for round seeds x pure bred for wrinkled seeds	✓	✗	1 : 0
2	offspring 1 self-pollinated	✓	✓	
3	offspring 1 x pure bred for round seeds			
4	offspring 1 x pure bred for wrinkled seeds			

d Complete the table by indicating
- the type of seeds present in the pods with a tick [✓] or a cross [✗]
- the ratio of round to wrinkled seeds. [3]

e Seed shape in peas is an example of discontinuous variation. Suggest **one** reason why seed shape is an example of discontinuous variation. [1]

Plants have methods to disperse their seeds over a wide area.

f Explain the **advantages** of having seeds that are dispersed over a wide area. [3]

[Total: 14]
(Cambridge IGCSE Biology 0610 Paper 31 Q6 November 2012)

● Variation and selection

1 One variety of the moth, *Biston betularia*, has pale, speckled wings. A second variety of the same species has black wings. There are no intermediate forms. Equal numbers of both varieties were released into a wood made up of trees with pale bark. Examples of these are shown in the diagram.

374

After 2 weeks as many of the moths were caught as possible. The results are shown in the table.

wing colour of moth	number released	number caught
pale, speckled	100	82
black	100	36

a (i) Suggest and explain **one** reason, related to the colour of the bark, for the difference in numbers of the varieties of moth caught. [1]

(ii) Suggest and explain how the results may have been different if the moths had been released in a wood where the trees were blackened with carbon dust from air pollution. [2]

The table below shows the appearance and genetic make-up of the different varieties of this species.

wing colour	genetic make-up
pale, speckled	GG; Gg
black	gg

b (i) State the appropriate terms for the table headings. [2]

(ii) State and explain which wing colour is dominant. [2]

c State the type of genetic variation shown by these moths. Explain how this variation is inherited. [3]

d Heterozygous moths were interbred. Use a genetic diagram to predict the proportion of black-winged moths present in the next generation. [5]

e (i) Name the process that can give rise to different alleles for wing colour in a population of moths. [1]

(ii) Suggest **one** factor which might increase the rate of this process. [1]

[Total: 17]
(Cambridge IGCSE Biology 0610 Paper 31 Q5 June 2007)

Organisms and their environment

1 a The chart shows the flow of some of the energy through a food chain in an ocean.

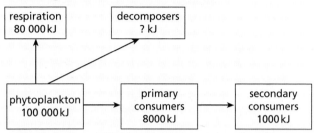

About 1% of the light energy reaching the ocean is converted to chemical energy by the phytoplankton. The phytoplankton produce sugars, fats and proteins.

(i) Name the process that changes light energy to chemical energy. [1]

(ii) Name the chemical in the phytoplankton that absorbs light energy. [1]

(iii) Calculate, using information from the flow chart, how much energy passes from the phytoplankton to the decomposers. [1]

(iv) Name **two** groups of decomposers. [2]

(v) Calculate, using information from the flow chart, the percentage of energy passed from the phytoplankton to the primary consumers. [2]

(vi) About 88% of the energy in the primary consumers does not become part of the secondary consumers. Explain how this energy is lost from the food chain. [3]

b The organisms in this food chain form a community in the ocean. This community is formed of many populations. Explain what is meant by the term *population*. [2]

[Total: 12]
(Cambridge IGCSE Biology 0610 Paper 21 Q6 June 2011)

EXAMINATION QUESTIONS

2 The diagram shows part of a food web for the South Atlantic Ocean.

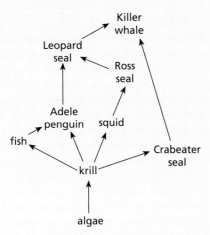

  a (i) Name the top carnivore in this food web. [1]
  (ii) Name a member of this food web that is both a secondary and a tertiary consumer. [1]
  b Use the information from the food web to complete the food chain of five organisms.
  algae → _____ → _____ → _____ → _____ [2]
  c In the future the extraction of mineral resources in the Antarctic might occur on a large scale. This could destroy the breeding grounds of the Ross seal.
  (i) State and explain what effects this might have on the population of Leopard seal. [2]
  (ii) State and explain what effects this might have on the population of fish. [4]

[Total: 10]
(Cambridge IGCSE Biology 0610 Paper 21 Q9 June 2008)

3 The diagram shows a food web.

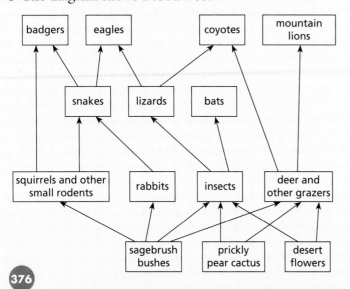

  a Explain the difference between a food web and a food chain. [2]
  b From the food web name:
  (i) a carnivore
  (ii) a producer
  (iii) a consumer from the 2nd trophic level. [3]
  c In some regions, mountain lions have been hunted and face extinction. Suggest how the coyotes might be affected if the mountain lion became extinct. [3]

[Total: 8]
(Cambridge IGCSE Biology 0610 Paper 21 Q9 November 2012)

4 The diagram shows a carbon cycle.

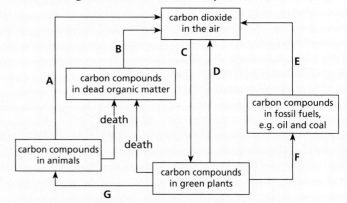

  a (i) Name the process represented by arrow **A**. [1]
  (ii) Name the process represented by arrow **E**. [1]
  b (i) Name **one** group of organisms responsible for process **B**. [1]
  (ii) List **two** environmental conditions needed for process **B** to occur. [2]
  c (i) Which arrow represents photosynthesis? [1]
  (ii) Complete the word equation for photosynthesis.
  _____ + _____ ⟶ oxygen + _____ [2]
  (iii) This process needs a supply of energy. Name the form of energy needed. [1]
  d In an ecosystem the flow of carbon can be drawn as a cycle but the flow of energy cannot be drawn as a cycle. Explain this difference. [3]

[Total: 12]
(Cambridge IGCSE Biology 0610 Paper 21 Q5 November 2012)

5 The diagram shows the water cycle.

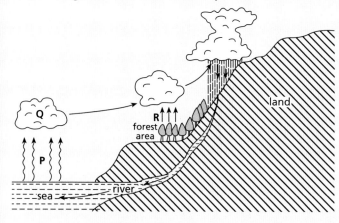

a (i) The arrows labelled **P** represent evaporation. Which type of energy is needed for this process? [1]
  (ii) State what causes the formation of clouds at **Q**. [1]
b (i) What process is represented by the arrows labelled **R**? [1]
  (ii) Name **three** factors that could alter the rate at which process **R** happens. [3]
c A logging company wants to cut down the forest area.
  (i) Suggest what effects this deforestation might have on the climate further inland. Explain your answer. [2]
  (ii) State **two** other effects deforestation could have on the environment. [2]

[Total: 10]
(Cambridge IGCSE Biology 0610 Paper 2 Q4 June 2009)

6 a The diagram shows the carbon cycle.

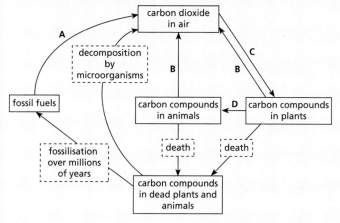

  (i) Name the processes that cause the changes shown by the arrows. [4]
  (ii) Name **one** type of organism that brings about decomposition. [1]
b Over the last few decades, the carbon dioxide concentration in the atmosphere has been rising. Suggest how this has happened. [3]

[Total: 8]
(Cambridge IGCSE Biology 0610 Paper 21 Q7 November 2008)

7 Rabbits are primary consumers. The graph shows changes in the population of rabbits after a small number were released on an island where none had previously lived.

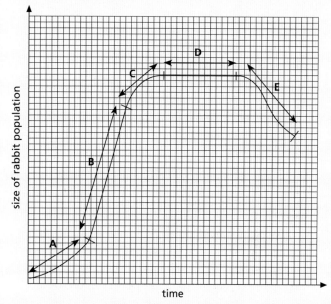

a Which stage, **A**, **B**, **C**, **D** or **E**, shows when the birth rate was
  (i) equal to the death rate [1]
  (ii) slightly greater than the death rate? [1]
b (i) Suggest **two** factors that allowed the change in the rabbit population during stage **B**. [2]
  (ii) Suggest **two** reasons for the change in the rabbit population during stage **E**. [2]

[Total: 6]
(Cambridge IGCSE Biology 0610 Paper 2 Q5 November 2009)

EXAMINATION QUESTIONS

8 The graph shows a population growth graph for a herbivorous insect that has just entered a new habitat.

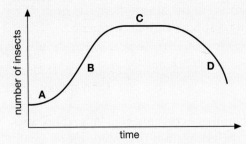

a (i) Which of the four phases, labelled **A**, **B**, **C** and **D**, represents the stationary phase and which the lag phase? [2]

(ii) During which phases will some of this insect population die? [2]

b (i) State **two** factors that could affect the rate of population growth during phase **C**. [2]

(ii) Suggest how these two factors might change. Explain how each change would affect the rate of population growth. [4]

[Total: 10]

(Cambridge IGCSE Biology 0610 Paper 21 Q2 November 2010)

9 An agricultural student investigated nutrient cycles on a farm where cattle are kept for milk. The farmer grows grass and clover as food for the cattle. Clover is a plant that has bacteria in nodules in its roots.

The diagram shows the flow of nitrogen on the farm as discovered by the student. The figures represent the flow of nitrogen in kg per hectare per year.

(A hectare is $10\,000\,m^2$.)

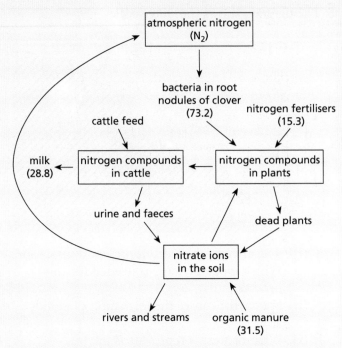

a (i) Name the process in which bacteria convert atmospheric nitrogen into a form that is available to clover plants. [1]

(ii) Name **two** processes that convert nitrogen compounds in dead plants into nitrate ions that can be absorbed by grass. [2]

b The total quantity of nitrogen added to the farmer's fields is 120 kg per hectare per year. Calculate the percentage of this nitrogen that is present in the milk. Show your working. [2]

c State **two** ways in which the nitrogen compounds in the cattle's diet are used by the animals **other than to produce milk**. [2]

d The student found that a large quantity of the nitrogen compounds made available to the farmer's fields was not present in the milk or in the cattle. Use the information in the diagram to suggest what is likely to happen to the nitrogen compounds that are eaten by the cattle, but are not present in compounds in the milk or in their bodies. [5]

e The carbon dioxide concentration in the atmosphere has increased significantly over the past 150 years. Explain why this has happened. [2]

[Total: 14]

(Cambridge IGCSE Biology 0610 Paper 31 Q6 June 2009)

Biotechnology and genetic engineering

1 Penicillin is an antibiotic produced by the fungus *Penicillium chrysogenum*. The diagram shows the process used to produce penicillin.

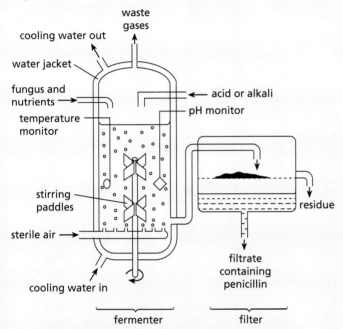

a Enzymes in the fungus are used to make penicillin. Explain why there is a water jacket around the fermenter and why acids and alkalis are added to the fermenter. [6]
The graph shows the mass of fungus and the yield of penicillin during the fermentation process.

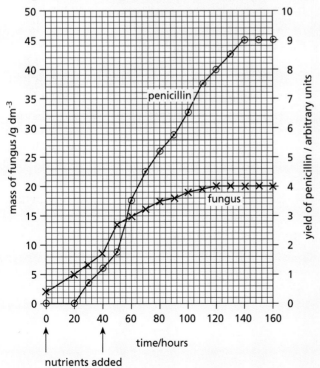

b (i) State the time interval over which the fungus grew at the maximum rate. [1]
 (ii) As the fungus grows in the fermenter, the nuclei in the fungal hyphae divide. State the type of nuclear division that occurs during the growth of the fungus in the fermenter. [1]
 (iii) Explain why the growth of the fungus slows down and stops. [3]
c Penicillin is not needed for the growth of *P. chrysogenum*.
 (i) State the evidence from the graph that shows that penicillin is not needed for this growth. [2]
 (ii) The people in charge of penicillin production emptied the fermenter at 160 hours. Use the information in the graph to suggest why they did **not** allow the fermentation to continue for longer. [1]
d Downstream processing refers to all the processes that occur to the contents of the fermenter after it is emptied. This involves making penicillin into a form that can be used as medicine. Explain why downstream processing is necessary. [3]
e Explain why antibiotics, such as penicillin, kill bacteria but not viruses. [2]

[Total: 19]
(Cambridge IGCSE Biology 0610 Paper 31 Q4 November 2011)

2 The chart shows the change in percentage of disease-causing bacteria that were resistant to the antibiotic penicillin from 1991 to 1995.

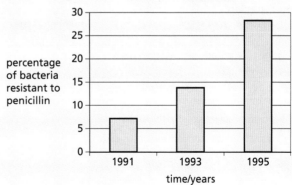

a (i) Describe the change in percentage of bacteria resistant to penicillin between 1991 and 1995. [2]

EXAMINATION QUESTIONS

(ii) Explain how a population of antibiotic-resistant bacteria can develop. [4]

b Although bacteria can cause disease, many species are useful in processes such as food production and maintaining soil fertility.
  (i) Name **one** type of food produced using bacteria. [1]
  (ii) Outline the role of bacteria in maintaining soil fertility. [3]

c Bacteria are also used in genetic engineering. The diagram outlines the process of inserting human insulin genes into bacteria using genetic engineering.

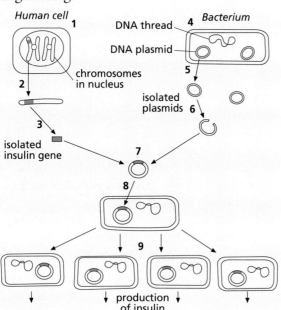

Complete the table below by identifying **one** of the stages shown in the diagram that matches **each** description. [5]

description of stage	number of the stage
the plasmids are removed from the bacterial cell	
a chromosome is removed from a healthy human cell	
plasmids are returned to the bacterial cell	
restriction endonuclease enzyme is used	
bacterial cells are allowed to reproduce in a fermenter	

[Total: 15]
(Cambridge IGCSE Biology 0610 Paper 31 Q4 November 2006)

Human influences on ecosystems

1 Deforestation occurs in many parts of the world.
  a State **two** reasons why deforestation is carried out. [2]
  b (i) Explain the effects deforestation can have on the **carbon cycle**. [4]
    (ii) Describe **two** effects deforestation can have on the soil. [2]
    (iii) Forests are important and complex ecosystems. State **two** likely effects of deforestation on the forest ecosystem. [2]

[Total: 10]
(Cambridge IGCSE Biology 0610 Paper 2 Q2 June 2006)

2 The diagram shows an Arctic food web.

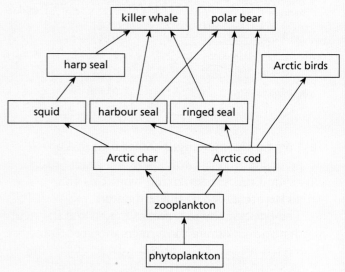

a (i) The phytoplankton are the producers in this food web. Name the process by which phytoplankton build up stores of chemical energy. [1]
  (ii) Name a secondary consumer in the food web above. [1]
  (iii) Complete the food chain using organisms shown in the food web.
    phytoplankton → _____ → _____ → _____ → killer whale [1]

b The polar bear has been listed as an endangered species. Explain what the term *endangered species* means. [2]

c Suggest how the loss of the polar bear from the Arctic ecosystem could affect the population of killer whales. [3]

[Total: 8]
(Cambridge IGCSE Biology 0610 Paper 21 Q5 November 2011)

3 Modern technology can be used to increase the yield of crops.
  a The use of chemicals, such as fertilisers, herbicides and pesticides, is one of the developments used.
    (i) Name **two** mineral ions commonly included in fertilisers. [1]
    (ii) Explain the dangers to the local environment of the overuse of fertilisers on farmland. [4]
    (iii) Suggest how the use of herbicides can be of benefit to crop plants. [3]
    (iv) Suggest **two** dangers of using pesticides on farmland. [2]
  b Artificial selection and genetic engineering can also be used to increase crop yields. Explain the difference between these two techniques. [2]

[Total: 12]
(Cambridge IGCSE Biology 0610 Paper 21 Q9 June 2009)

4 After an accident at a nuclear power plant in 1986, particles containing radioactive strontium were carried like dust in the atmosphere. These landed on grassland in many European countries. When sheep fed on the grass they absorbed the strontium and used it in a similar way to calcium.
  a Explain where in the sheep you might expect the radioactive strontium to become concentrated. [2]
  b Suggest the possible effects of the radiation, given off by strontium, on cells in the body of the sheep. [3]

[Total: 5]
(Cambridge IGCSE Biology 0610 Paper 21 Q3 November 2008)

5 The bar graph shows crop productivity for a range of plants but it is incomplete.

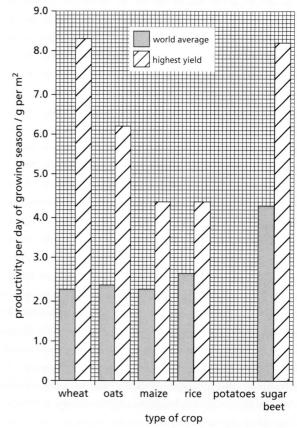

a Complete the graph using the following data. [2]

crop	productivity per day of growing season/g per m²	
	world average	highest yield
potatoes	2.6	5.6

b State which crop has
  (i) the highest average productivity
  (ii) the greatest difference between the average yield and the highest yield. [2]
c Outline how modern technology could be used to increase the productivity of a crop from the average yield to a high yield. [3]
d When the yield is measured, dry mass is always used rather than fresh mass. Suggest why dry mass is a more reliable measurement than fresh mass. [1]
e Maize is often used to feed cows, which are grown to provide meat for humans. Explain why it is more efficient for humans to eat maize rather than meat from cows that have been fed on maize. [3]

EXAMINATION QUESTIONS

f (i) Complete the equation for photosynthesis.

$$6CO_2 + 6H_2O \xrightarrow{\text{light energy}} C_6H_{12}O_6 + \underline{\hspace{2cm}}$$ [1]

(ii) Describe how leaves are adapted to trap light. [2]

(iii) With reference to water potential, explain how water is absorbed by roots. [3]

(iv) Explain how photosynthesising cells obtain carbon dioxide. [2]

[Total: 19]
(Cambridge IGCSE Biology 0610 Paper 31 Q2 November 2008)

6 The Food and Agriculture Organisation (FAO) collects data on food supplies worldwide. The FAO classifies the causes of severe food shortages as either by natural disasters or as the result of human action. Natural disasters are divided into those that occur suddenly and those that take a long time to develop. Human actions are divided into those that are caused by economic factors and those that are caused by wars and other conflicts. The graph shows the changes in the number of severe food shortages between 1981 and 2007.

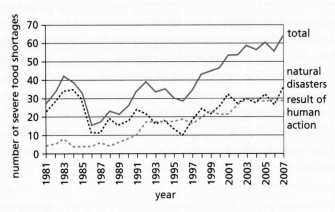

The pie charts show the causes of severe food shortages in the 1980s, 1990s and 2000s.

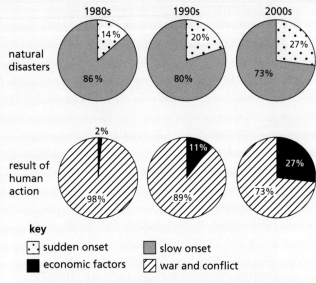

a (i) State **two** types of natural disaster that occur suddenly and may lead to severe food shortages. [2]

(ii) State **one** type of natural disaster that may take several years to develop. [1]

b Use the information in the graph and pie charts to describe the changes in food shortages between 1981 and 2007. [5]

c Explain how the increase in the human population may contribute to severe food shortages. [3]

The quality and quantity of food available worldwide has been improved by artificial selection (selective breeding) and genetic engineering.

d Use a **named** example to outline how artificial selection is used to improve the quantity or quality of the food. [4]

e Define the term *genetic engineering*. [1]

[Total: 16]
(Cambridge IGCSE Biology 0610 Paper 31 Q6 June 2010)

7 The table shows some information about air pollution.

pollutant	source of air pollutant	effect of pollutant on the environment
	combustion of fossil fuels	increased greenhouse effect and global warming
methane		increased greenhouse effect and global warming
sulfur dioxide	combustion of high sulfur fuels	acid rain
nitrogen oxides	fertilisers	acid rain

a Complete the table by writing answers in the spaces. [2]
b Explain how the increased greenhouse effect is thought to lead to global warming. [3]
c The graph shows changes in the emissions of sulfur dioxide in Europe between 1880 and 2004.

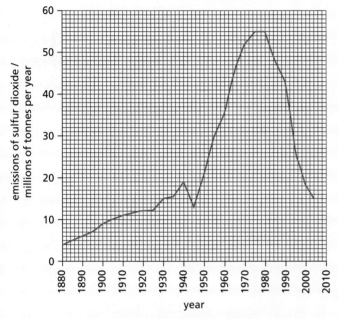

(i) Use the information in the graph to describe the changes in the emissions of sulfur dioxide in Europe between 1880 and 2004. [4]
(ii) Describe the effects of acid rain on the environment. [3]
(iii) Outline the methods that have been used to reduce the emissions of sulfur dioxide. [3]

[Total: 15]
(Cambridge IGCSE Biology 0610 Paper 31 Q5 November 2012)

a (i) State **one** cause of acid rain other than that shown in the diagram. [1]
(ii) Describe **two** effects of acid rain on forest ecosystems. [2]
b Describe **two** different ways to reduce pollution so that there is less acid rain. [2]

The chart shows the pH ranges that some animals that live in lakes can tolerate.

animals		pH							
group	examples	7.0	6.5	6.0	5.5	5.0	4.5	4.0	3.5
fish	trout								
	bass								
	perch								
amphibians	frogs								
	salamanders								
molluscs	clams								
	snails								
crustacean	crayfish								
insects	mayfly larvae								
	blackfly larvae								

c State **one** feature of molluscs that is not a feature of crustaceans. [1]
d Using the information in the chart
(i) name an animal that could be found in a lake with a pH of 4.0 [1]
(ii) name the animals that are most sensitive to a decrease in pH [1]
(iii) suggest why some animals cannot tolerate living in water of pH as low as 4.0. [2]

[Total: 10]
(Cambridge IGCSE Biology 0610 Paper 31 Q4 June 2010)

8 Acid rain is a serious environmental problem in some areas of the world. Lakes in Canada, Norway and Scotland are highly acidic as a result of acid rain. The diagram shows a cause of acid rain.

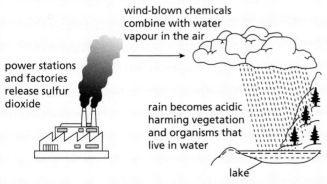

Answers to numerical questions

2 Organisation and maintenance of the organism
5 b (i) 5 +/− 0.5mm
    (ii) 5/800 = 0.00625 or 6.25 × 10⁻³

5 Enzymes
1 a (i)

temperature/°C	volume of juice collected/cm³
10	2
15	11
20	15
25	20
30	26
35	27

3 b (ii) 55 (°C) if point to point curve (+/− half square)
    (iii) 24 or 25 (+/− half square)
4 c 0.57

6 Plant nutrition
3 b (i) 1.4
  c (i) 6.0–7.0
       0–0.6
19 a 1 tonne of wheat per hectare extra
   b 1.8 tonnes of wheat per hectare extra

8 Transport in plants
4 b 20.0

9 Transport in animals
2 b (i) calculation x 4 for rate per minute (72, 76, 68) mean calculated: 72

11 Gas exchange in humans
3 d (i) 70

12 Respiration
14 a (i) 8616.2 kJ
     (ii) 49.248 kJ

19 Organisms and their environment
1 a (iii) 12 000 kJ
    (v) 8000/100 000 × 100 = 8 (%)
9 b 28.8/120 × 100 = 24 (%)

Index

A

abiotic factors 301–2
absorption 95, 97, 103–5
accommodation of the eye 188, 189
acid rain 330, 331
acquired characteristics 270
activated sludge process 336–7
active immunity 149
active sites 43, 61
active transport 48–9, 116
adaptation 274–8, 281
   flowering plants 225–6
   leaves 80–1
adaptive features 274, 277
adenine 54, 56, 252
adipose tissue 91
adolescence 241
adrenal glands 191
adrenaline 130, 174, 180, 191–2
adrenal medulla 191
adrin 318
adventitious roots 16, 114
aerobic respiration 156, 165–9
agricultural machinery 316
agriculture
   energy transfer in 290–1
   intensification of 316–18
   reproduction in 217, 219, 220
   world issues 88–9, 293, 299, 300
AIDS (acquired immune deficiency syndrome) 245–6, 297, 298
air
   breathing and 158, 159, 163
   pollution 330–4
alcohol 208–9, 237–8, 240
alimentary canal 96–8
alleles 259, 260–5, 272–3
alveoli 157
amino acids 53, 73, 81, 92, 105, 175
ammonium nitrate 82
amnion 237
amniotic fluid 237
amoebic dysentery 148
amphibia 8, 13–14, 15
amylase 61, 307
anabolic steroids 211–12
anabolism/anabolic reactions 60, 61, 171
anaemia 93, 94
anaerobic respiration 169–71
anatomy 3–4
angina 88, 128
angioplasty 131
animal cells
   cell division 254
   osmosis in 40–1, 44–5
   structure 24–5, 27, 29
animals
   asexual reproduction 218

classification 6, 7–8, 11–15
transport in 124–39
antenatal care 237
anthers 222, 224, 225, 258, 259
antibiotics 205–7
   bacterial resistance to 205–6, 281, 314
   production 305, 309–10
antibodies 53, 149, 151
antigens 53, 149
anus 97
aorta 126, 133, 134
aqueous humour 186, 187
arachnids 7, 12
archaea 6
Areas of Special Scientific Interest (ASSIs) 343
arteries 124, 132, 134
arterioles 124, 132, 134–5
arthropods 7, 11–12
artificial insemination (AI) 244
artificial propagation 217–18
artificial selection 280–2
asexual reproduction 19, 213–19, 254, 258
assimilation 95, 97, 105, 175
atheroma 127, 128
ATP 168
atria 125, 129
atrioventricular valves 129
autoimmune diseases 152
auxin 199–201

B

back-crosses 264
bacteria
   antibiotic resistance 205–6, 281, 314
   biotechnology and 305, 313–14
   in decomposition 291–2, 293, 294
   mutations in 205–6, 273, 281
   in nitrogen cycle 294, 295
   pathogenic 142
   reproduction 19, 213
   structure 18
bacteriocidal antibiotics 309
bacteriostatic antibiotics 309
balanced diets 86, 87, 91
basal layer 192
basal metabolism 87, 171
'the bends' 37
bicuspid valves 129
bile 102
bilirubin 136, 174
binomial system of naming 2–3
biochemical oxygen demand (BOD) 329
biodiversity 287, 324
biofuels 305–6, 335
biogas 335, 337

biological washing powders 307, 308
biomass 290, 291
biosphere 298
biotechnology 305
biotic factors 301–2
birds 8, 14–15
birth 238–9
birth control 243–4
bisexuality 221
bladder 177
blind spot 186, 187, 189–90
blood 124, 136–7
   circulatory system 31, 32, 125–35, 138–9
   clotting 137–8
   concentration of 41, 175
   gaseous exchange and 156, 157
   in placenta 240
   red blood cells 29, 31, 93, 94, 136
   white blood cells 53, 136, 137, 149
blood groups 264, 272
blood pressure 128, 130, 133–4
blood sugar 194, 196
blood vessels 124, 132–5
blubber 289
B lymphocytes 150
body temperature 13
   control of 45, 135, 193–4, 195, 196–7
botulism 146
brain 182, 194, 195
bread 306
breastfeeding 88, 151, 240–1
breathing 156, 158, 161–3
   exercise and 158, 160–1
breeding in captivity 339
'breeding true' 261
bronchi 157
bronchioles 157
bronchitis 209
buccal cavity 101
bulbs 216
by-pass surgery 131

C

cacti 277
calcium 93
calculus 100
callus 218
camels 274–5
cancer 209, 211, 272–3
capillaries 124, 132–3, 134
capsids 19
capsomeres 19
carbohydrates 51–2, 55
   in diet 91, 92
   in photosynthesis 66
carbon cycle 292–3

Index

carbon dioxide
  in the atmosphere 322, 328, 332–4
  in the carbon cycle 292–3
  in exhaled air 159, 160, 163
  from respiration 36–7, 293
  in photosynthesis 37, 68–9, 71, 72, 74, 75, 292
carbon monoxide 209, 210, 330
'carbon neutral' 335
carcinogens 209
carnivores 285
carpels 221, 222
catabolism/catabolic reactions 60, 61, 171
catalase 62
catalysts 59
catalytic converters 332
cell bodies 181
cell division 25–6, 213, 254–5
  see also meiosis; mitosis
cell membrane 25, 27, 40, 43, 48
cells
  movement into/out of 36–49
  specialisation 29–31, 254
  structure 24–9
  synthesis/conversion in 53, 66, 72–4
cell sap 26
cellular respiration 165
cellulose 51, 52, 59, 91
cell wall 26, 27, 41, 51, 52, 254
cement 99
central nervous system 31, 32, 180, 181–5, 190, 210
centromere 250
cervix 233
Chain, Ernst 207
chemical digestion 95, 97, 100–3
chemical fertilisers 44, 82, 317
chemical waste 326
children, dietary requirements 88
chlorophyll 26, 67, 68, 72
chloroplasts 26, 27, 29, 72, 78, 80, 254
cholera 98
cholesterol 90, 128
choroid 186, 187
chromatids 250, 256
chromosomes 25, 250–1, 256
  function of 257
  number of 220, 253
chronic obstructive pulmonary disorder (COPD) 209
chyme 100
ciliary body 186, 187, 188
ciliary muscle 187, 188
ciliated cells 30, 148, 163
circular muscle 188
circulatory system 31, 32, 125–35, 138–9
cirrhosis 208

CITES 339
cladistics 5
classification systems 2–5, 20
climate
  change 328, 332–4, 338
  deforestation and 323
clinostats 197–8
clones 218
clotting of blood 137–8
co-dominance 264
'cold-blooded' 13, 166, 195
coleoptile 201
collecting ducts 176
colon 97, 103
colostrum 241
colour blindness 265
combustion 293
communities 297, 298
compensation point 74
competition 279, 298, 301
compound eyes 11
concentration
  of the blood 41, 175
  diffusion and 36
  osmosis and 47–8
concentration gradient 37, 38, 39, 48–9
condensation 294
cones 188
conjunctiva 186, 187
conservation 334–44
constipation 90
consumers 285
continuous variation 271, 272
contraception 243–4, 245, 334
contraceptive pills 243–4, 334
contractile vacuoles 44
controlled diffusion 38, 48
controls 60, 67, 69, 168–9
co-ordination 180, 190
copulation 233, 234, 235–6
corms 216
cornea 186, 187
coronary arteries 126
coronary heart disease 88, 127–9, 130–1, 209–10
coronary thrombosis 128
corpus luteum 242
cortex (kidneys) 175
cortex (plants) 113
cotyledons 227, 228, 231
crenated cells 45
Crick, Francis 56, 57
critical pH 99
cross-breeding 220–1
cross-pollination 230–1, 264
crown 99
crustacea 7, 11, 12
cuticle (arthropods) 11
cuticle (leaves) 77, 78, 79

cuttings 217
cytoplasm 25, 27, 41
cytosine 54, 56, 252

D

DDT 324–5
deamination 97, 175, 294
death rate 298–9
decomposers 285, 291–2, 293
decomposition 293, 294
decompression sickness 37
defecation see egestion
deforestation 89, 293, 306, 316, 322–4
dehydration 148
dehydrogenase 61
denaturation 54, 62
dendrites 181
denitrifying bacteria 295
dental decay (caries) 99–100
dentine 99
dermis 192
diabetes 151–2, 196
dialysis 37, 43, 177–8, 179
diaphragm 157, 161
diarrhoea 45, 97–8
dichotomous keys 21–2
dicotyledons 10, 17
dieldrin 318
diet 86–95, 128
  balanced 86, 87, 91
diffusion 36–40, 116
diffusion gradient see concentration gradient
digestion 93, 95, 97, 98–103
digestive enzymes 96
digestive system 33, 96–8, 100–5
diploid nucleus 253
diploid number 220, 258
direct evidence 172
disaccharides 51
discontinuous variation 270–1, 272
disease 142, 296
  coronary heart disease 88, 127–9, 130–1, 209–10
  defences against 148–51, 163
  sexually transmitted infections 143, 245–6
  transmission 143–8
'division of labour' 29–30
DNA
  in classification 4–5
  genetic engineering and 313–14
  structure 54–5, 56–7, 252
  see also chromosomes; genes
dominant alleles 259, 260–1, 272
dopamine 210
dormancy 228
dorsal root 184
double circulation 125
Down's syndrome 272, 273

droplet infection 148
drugs 205
  medicinal 205–7
  misused 44–5, 207–12, 238
dry weight 84
ducts 96
duodenum 97, 101, 103

E
ecosystems 297–8, 344
effectors 181
egestion 95, 97, 103
egg cells
  animal/human 31, 220, 232–3, 234–5, 236, 239
  plant 220
ejaculation 234, 236
electrocardiogram (ECG) 127
embryonic stem cells 257–8
embryos
  human 233, 236–7
  plant 231–2
emphysema 209
emulsification of fats 102
enamel 99
endangered species 337–9
endemic diseases 151
endocrine glands 174, 180, 190
endocrine system 180, 190
energy
  alternative sources 334
  from food 87, 95, 165
  from sunlight 284–5, 289–90
  kinetic 37
  pyramids of 291
  in respiration 165, 166–7, 168, 169
  transfers of 289–91
enterokinase 103
enzymes 25, 53, 59–60
  in digestion 100–1, 103
  pH and 60–1, 62, 63–4, 103
  production 306–7
  rate of reactions 60–4, 194
  in respiration 165–6, 168, 169, 170
  temperature and 60, 62, 63
epidermis (plants) 38, 78, 79, 111
epidermis (skin) 192
epididymis 234
epiglottis 101, 157
epithelial cells 49, 102
epithelium
  digestive system 96, 104, 105
  respiratory system 156, 157
erectile tissue 234
eubacteria 6
eukarya 6
eutrophication 319, 327–9
evaporation 294
evolution 279, 281

excretion 1, 55, 174–9
exercise
  effect on breathing 158, 160–1
  effect on heart/pulse rate 127, 130, 131–2
  heart disease and 129, 130
  respiration and 169, 170
extinction 337–8, 342
extracellular enzymes 61, 306
eyes 186–90

F
F-1 generation 220–1, 261–2, 266
'factory farming' 319, 329
faeces 103
Fallopian tubes 233
family planning 243–4, 299
Farming and Wildlife Advisory Group (FWAG) 320, 344
fats 52–3, 55
  in diet 90, 91, 92, 105
  emulsification 102
  test for 57, 58
fatty acids 52, 90, 100, 104
female reproductive system 233, 234
fermentation 169, 305–6
fermenters 313
ferns 9, 16–17
fertilisation 219, 226, 258
  flowering plants 226, 231
  human reproduction 232–3, 236, 260
fertilisers 44, 76, 82, 317, 327
fertility drugs 244
fertility rate 299
fetus 236
fibre 90, 91, 93
fibrous root systems 114
filaments 222
fish 8, 13, 15, 124
fish stocks 340–2
fission 213
fitness 277, 279
flaccid 44, 119
flagella 18
Fleming, Sir Alexander 206–7, 309
Florey, Howard 207
flowering plants 10, 17
  adaptations 225–6
  reproduction 17, 215–18, 220
  structure 110, 221–4
follicles 235
follicle-stimulating hormone (FSH) 191, 242
food
  classes 90–3
  energy from 87, 95, 165
  genetically modified 89, 310–11, 312, 314

  need for 66, 86
  sources and sinks 112, 121, 122
  supply of 296, 300, 316–20
  world issues 88–9, 319–20
food chains 285, 290, 298
food pyramids 285–6
food tests 57–8
food webs 285, 286–7
foramen ovale 129–30
foreign species 289, 319
forests 89, 293, 306, 316, 322–4, 340–1
fossil fuels 292, 293, 320, 334
fossils 292, 337
fovea 187, 188, 189
Franklin, Rosalind 56, 57
fraternal twins 238
fruits 223, 231–2
fungi 6, 17–18
  asexual reproduction 213–14
  decomposition and 293, 297
  pathogenic 142, 147

G
Galen 138–9
gall bladder 97
gametes 219, 226, 255, 258
  see also egg cells; sperm
ganglion 184
gaseous exchange
  in humans 156–63
  in plants 74–5
gastric juice 101, 103
gene mutation 272–3, 281
genera 2
genes 250, 252, 257, 272
  expression of 253, 257
gene-splicing 313
genetic code 252
genetic engineering 282, 305, 310–14
genetics 250, 254
genetic variations 270, 272
genotypes 259, 261–3
geotropism see gravitropism
germination 168, 227–30
gestation period 238
gingivitis 100
glands 96
global warming 328, 332–4, 338
glomeruli 176
glucagon 196
glucose 51, 91, 100, 105
  in the blood 194, 196
  in plants 72
  test for 57, 58
gluten 306
glycerol 52, 100, 104
glycogen 51, 52, 180, 196
GM crops/food 89, 310–11, 312, 314
goblet cells 163

Index

gonads 255
gravitropism 197–201
greenhouse effect 328, 332–4
grey matter 183
growth 1, 254
   in plants 199–201
growth substances 199–201
guanine 54, 56, 252
guard cells 77, 78, 79–80
gullet 97, 101
gum disease 100
gums 99

H
habitats 298
   conservation of 340, 342–4
   destruction of 320–4
Habitats Directive 343
haemoglobin 31, 93, 94, 136, 252
haemolysis 44
half-life 325
hand lens 33
haploid nucleus 253
haploid number 220, 255, 258
Harvey, William 139
heart 125–7, 129–30, 131
heart attacks 88, 130
hepatitis B vaccine 312
herbicides 311, 318, 325
herbivores 285
heredity 250, 265–7
   see also inheritance
hermaphrodites 221
heroin 185, 207, 210, 240
heterozygosity 259, 261
HIV (human immunodeficiency virus) 143, 240, 245–6
homeostasis 192–7
homiothermy 13, 165, 195
homologous chromosomes 253, 258
homozygosity 259, 261
hormones 180, 190
   growth and 199
   in humans 190–2, 241–2, 244, 245
   performance-enhancing 211–12
   pollution by 334
   sex hormones 191, 241–2, 244, 245
horticulture, propagation in 217, 219
houseflies 147
human population 296–7, 298–300
human reproductive system 233–4
hydrochloric acid 101, 103
hydrophytes 278
hydroponics 82, 342
hypocotyl 227, 228
hypothalamus 194, 195
hypotheses 66, 171–2

I
identical twins 238
ileum 97, 103, 104–5

images 187, 188, 189
immunity 149, 151
implantation 236
impulses 181–2, 185
incomplete dominance 265
indirect evidence 172
infant mortality 297
inflorescences 223–4
ingestion 95, 97, 101
inheritance 250, 259–65, 270
   of sex 250–1
inherited characteristics 270
innate immunity 149
inoculation see vaccination
insecticides 310, 318, 324–5
insect-pollinated flowers 222, 223, 224, 225–6
insects 7, 11–12
insulin 190, 191, 196, 252, 310, 313, 314
intercostal muscles 157, 161
internal respiration 158, 165
internodes 110
intestines 49
intoxication 208
intracellular enzymes 61, 306
invertebrates 7
in vitro fertilisation 244–5
involuntary actions 185
iris 186, 187, 188
iron 93, 94, 136
isotonic drinks 45
IWC (International Whaling Committee) 339

J
Jenner, Edward 152
'junk DNA' 272

K
karyotypes 250, 251
kidneys 37, 49, 174, 175–7, 194
kidney transplants 178–9
kinetic energy 37
'knee-jerk' reflex 182–3
kwashiorkor 94

L
lactase 308–9
lactation see breastfeeding
lacteals 103, 104
lactic acid 170
lactose intolerance 308
lamina 77
large intestine 103
lateral buds 110
leaching 295
'lean burn' engines 332
leaves
   adaptation 80–1
   photosynthesis in 73, 80–1

   structure 73, 77–81, 110
   water loss from 118–19
   see also plants
leguminous plants 294, 295
lens 186, 187
life expectancy 297, 298
light
   effect on eyes 187–8
   germination 228
   photosynthesis and 68, 69–71, 73–4, 75
   plant growth and 200–1
   transpiration and 120–1
light microscope 33–4
lightning 295
lignin 78, 111
limiting factors 75–6
   population growth and 301–2
Linnaeus, Carl 20
lipase 102, 107, 307
lipids 52–3, 91
liver 97, 102, 174, 175, 193, 208
longitudinal sections 24, 26, 111, 112
low density lipoproteins (LDLs) 90
lung cancer 209, 211
lungs 156–8, 159–60, 161–3, 174, 195
lupin flowers 223–4
luteinising hormone (LH) 191, 242
lymph 133, 135
lymphatic system 103, 135
lymph nodes 135
lymphocytes 53, 135, 136, 137, 149, 150, 246
lysozyme 149, 186

M
magnesium 81
magnification 33–4
malaria 143–4, 151, 273, 297
male reproductive system 233–4
malnutrition 88
maltase 103
maltose 100, 103
mammals 8, 15
manometers 166
marasmus 94
marine pollution 321–2
marram grass 278
mastication 98
mating 235–6
mechanical digestion 95, 97, 98–100
medulla 175
meiosis 219, 251, 255, 258–9
melanin 192
memory cells 150
Mendel, Gregor 265–7
menopause 242
menstrual cycle 241–2
menstrual period 242

mesophyll 77, 78, 80
metabolism 170–1
micro-organisms 293
   see also bacteria; fungi
micropyle 231, 232
microvilli 38, 104
midrib 77, 78
minerals
   in diet 92–3
   in plants 37, 73, 81–4, 115–16, 295
mining 320–1
mitochondria 27, 49, 168, 254
mitosis 19, 254–5, 256–7, 258–9
MMR vaccine 150
monocotyledons 10, 17, 231–2
monoculture 317–18
monohybrid inheritance 259–65
monosaccharides 51
morphology 3–4
motor impulses 181
motor neurones 181, 182
mouth 97, 101, 103
movement 1
mRNA 252–3
MRS GREN mnemonic 1
mucus 96, 148, 163
mutagens 271
mutation 205–6, 271, 272–3, 281
myriapods 7, 12

N
narcotics 207–8
natural selection 279–80, 282
negative feedback 134, 195
nephrons 176
nerve cells see neurones
nerve fibres 181
nerves 181, 182
nervous system see central nervous
    system
neurones 30, 181, 182
nicotine 209–10, 240
nitrates 73, 81, 116, 294, 295
nitrification 294
nitrifying bacteria 294
nitrogen 37, 73
nitrogen cycle 294–5
nitrogen fixation 294
nitrogenous waste products 174
nitrogen oxides 331, 332
nodes 110
non-disjunction 273
non-renewable resources 335
NPK fertilisers 82
nuclear fall-out 325–6
nuclei 25–6, 27
nucleotides 54, 252
nutrition 1
   human 86–95
   plant 66–84

O
obesity 90
oesophagus 97, 101
oestrogen 191, 240, 241, 245
oil pollution 320, 321, 322, 326, 327
optic nerve 186
optimum pH 60
oral rehydration therapy 148
organelles 25
organisms 1, 6, 33
organs 31
organ systems 31, 32
osmoregulation 175
osmosis 40–8, 115, 119
osteo-malacia 93, 94
ovaries
   flowering plants 222–3
   human 191, 233, 235, 258
overfishing 288–9
over-harvesting 287–8
oviducts 233, 234
ovulation 234–5
ovules 220, 222, 258
oxidation 165
oxygen
   in breathing 159, 163
   from photosynthesis 37, 69, 74
   in germination 228, 229
   in respiration 36, 166–7
oxygen debt 170
oxyhaemoglobin 136, 158
oxytocin 239
ozone layer 332

P
'pacemaker' 130
palisade mesophyll cells 26, 30, 77, 78, 79, 80
pancreas 96, 97, 102, 191
pancreatic amylase 102, 103
pancreatic juice 102
pandemics 296
partially permeable membranes 40, 43, 48
passive immunity 151
Pasteur, Louis 152–3
pathogens 53, 142
   see also disease
pectinase 307–8
pelvis 175
penicillin 205, 207, 309–10
penis 234, 235
peppered moths 280
pepsin 102, 107
pepsinogen 103
peptidase 103
peptides 100
performance-enhancing
    hormones 211–12
peridontitis 100
peripheral nervous system 181
peristalsis 96–7, 101
pesticides 310, 318–19, 324
petals 221–2
pH
   critical 99
   enzymes and 60–1, 62, 63–4, 103
phagocytes 53, 135, 136, 137
phagocytosis 137, 149
pharynx 101
phenotypes 259, 261–3
phenotypic variations 270
phloem 78, 111, 113, 121–2
phosphorus 73
photomicrographs 24
photorespiration 75
photosynthesis 66–7, 292
   chemical equation 67, 71
   limiting factors 75–6
   process 71–2
   rate of 69–71, 75–6
phototropism 197–201
physical digestion 97, 100
pine trees 277
pith 113
pituitary gland 191, 242
placenta 237, 239, 240
plant cells
   active transport 48–9, 116
   cell division 254–5
   osmosis 41, 43–4
   structure 24–9
plants
   asexual reproduction 215–18
   classification 4, 6, 9–10, 16–18
   gaseous exchange in 74–5
   growth 199–201
   minerals in 37, 73, 81–4, 115–16, 295
   photosynthesis see photosynthesis
   propagation 215–18
   respiration 72, 74–5
   sexual reproduction 221–31
   structure 110–14
   translocation 121–2
   transpiration 116–21, 294
   tropic responses 197–201
   water in 43–4, 55, 114–15, 116–19
plaque 100
plasma 55, 137, 150, 177
plasmids 305, 313
plasmolysis 45, 46–7
plastics 330
plastids 26, 51
platelets 136, 137
pleural fluid 162
pleural membrane 162
plumule 227, 228
poikilothermy 13, 166, 195
polar bears 275
pollen 222, 223

Index

pollen sacs 222
pollen tubes 226, 231
pollination 220, 221, 222, 223, 224–6, 231
pollution 321–2, 324–34
polymers 51
polysaccharides 51
populations 296, 298–9
   population growth 296–7, 299–302
potassium nitrate 82
potometers 116–18
precipitation 294
predators/predation 285, 296, 302
pregnancy 88, 208, 236–8
primary consumers 285, 290
producers 285
products 61
progesterone 240, 242, 245
prokaryotes 6, 18–19, 27
propagation 215–18
prophylactics 144, 151
prostate gland 234
protease 61, 101, 102, 307
proteins 53–4, 55, 175
   in diet 87–8, 91–2
   digestion of 102–3
   manufacture 252–3
   test for 57, 58
protoctista 6, 19
protophyta 19
protozoa 19
ptyalin see salivary amylase
puberty 241
pulmonary artery 126, 133, 134
pulmonary circulation 125
pulmonary vein 125, 133, 134
pulp cavity 99
pulse/pulse rate 126, 127, 131–2
Punnett square 262, 263
pupil 186, 187, 188
pyloric sphincter 101
pyramids of biomass 290, 291
pyramids of energy 291
pyramids of numbers 286, 287

R

radial muscle 188
radicle 197–8, 227
Ray, John 20
receptacles 223
receptors 186
recessive alleles 259, 260–1, 272, 273
recombinant DNA 313
rectum 97, 103
recycling
   in ecosystems 291–2
   waste materials 335, 336
red blood cells 29, 31, 93, 94, 136
reflex actions 182, 184
reflex arcs 182–3, 184

relay neurones 181
renal artery 133, 134, 175, 176
renal capsules 176
renal tubules 175, 176
renal vein 133, 134, 175, 176
renewable resources 335
repair 254
replacement 254
replication 256
reproduction
   asexual 19, 213–19, 254, 258
   in humans 232–41
   sexual 219–41, 254
reptiles 8, 14, 15
respiration 1, 165
   aerobic 156, 165–9
   anaerobic 169–71
   effect of temperature 168, 171
   energy and 165, 166–7, 168, 169
   in plants 72, 74–5
respiratory surfaces 156
respirometers 166, 167
restriction enzymes 313
retina 186, 187, 188
rhizomes 16, 215–16, 217
ribosomes 6, 27, 252
rickets 93, 94–5
rods 188
root cap 113
root hair cells 29, 30, 44
root hairs 113–14, 115
root nodules 294, 295
roots (plant) 16, 110, 113–14
   tropic responses 197–8, 199
roots (teeth) 99
rootstocks 215, 216
rough endoplasmic reticulum (ER) 6, 27
rubella 238, 240

S

saliva 101
salivary amylase 101, 103, 105–6
salivary glands 96, 97, 181
Salmonella food poisoning 144–6
salts see minerals
saturated fatty acids 90
scavengers 285, 293
sclera 186, 187
scrotum 233, 234
scurvy 88
secondary consumers 285, 290, 298
secondary sexual characteristics 241
seed banks 340
seeds 223, 231–2
selection 279–80
selection pressures 280
selective breeding 280–2, 319
selective reabsorption 177
self-pollination 230, 264

semen 234
semi-lunar valves 129
seminal vesicle 234
sense organs 186–90
sensitivity 1
sensory impulses 181
sensory neurones 181, 182
sepals 222
septum 125, 129
sewage disposal 305, 327, 329, 335–7
sex cells see gametes
sex chromosomes 250–1, 265
sex-linked characteristics 265
sexually transmitted infections 143, 245–6
sexual reproduction 219–41, 254
   in humans 232–41
   in plants 221–31
shivering 194
shoots 24, 31, 110
   growth 200–1
   tropic responses 198–9
shunt vessels 135
sickle-cell anaemia 265, 273
sieve tubes/plates 77, 78, 111, 113
sigmoid population growth curves 301
single circulation 124
sink, food 112, 121, 122
size of specimens 33–4
skin 174, 192–3, 195, 196–7
slime capsules 18
small intestine 102, 103
smallpox 151, 152, 299
smoking 128, 209–10, 211, 237–8
soil erosion 322, 323
somatic cells 250, 256
source, of food 112, 121, 122
Special Areas of Conservation (SACs) 343
species 2
sperm cells 31, 220, 232–3, 234, 235, 236, 239
sperm duct 234
sphincter 177
spinal cord 183–4
spinal reflexes 184
spongy mesophyll 78, 79, 80
sporangia 16–17
stains 24
stamens 221, 222
starch 51
   in diet 91
   enzymes and 100–1, 103, 105–7
   in plants 67–8, 72
   test for 57, 58
starvation 88
stem cells 254, 257–8
stems 110–11, 112
stem tubers 216–17
stents 131

Index

stethoscopes 126
stigma 222
stimulus 182, 186
stolons 215, 216
stomach 97, 101–2, 103
stomata 76, 77, 78, 79–80, 120
streptomycin 205, 309
structural proteins 53
style 222
substrates 61
sucrose 90, 91
sugar 90, 91, 103
   see also glucose
sulfanilamides 206
sulfates 73
sulfur 73
sulfur dioxide 330, 331
sunlight 284–5, 289–90
superphosphates 82
surface area
   diffusion and 37–8, 38–9
   gaseous exchange and 156
survival value 280
suspensory ligaments 186, 187
sustainable development 340, 341–2
sustainable resources 334
swallowing 101
sweating 45, 174, 194
sympathetic nervous system 192
synapses 184, 185, 210
synthesis 53, 66, 72–4
systemic circulation 125

T
tap roots 114
target organs 190, 192
tear glands 186, 187
teeth 98–100
temperature
   body 13, 45, 135, 193–4, 195, 196–7
   diffusion and 38, 39
   enzymes and 60, 62, 63
   germination 228, 229–30
   photosynthesis and 71, 75, 76
   respiration and 168, 171
   transpiration and 121
terminal buds 110
tertiary consumers 285
testa 227, 232
test-crosses 264
testes 191, 233–4, 258
testosterone 191, 211, 241
three-domain scheme 6
thrombus 127–8
thymine 54, 56, 252
thyroid gland 190–1
thyroxine 190–1

tinea ('ringworm') 142, 147
tissue culture 217–18
tissue fluid 41, 133, 138, 193, 195
tissue respiration 158, 165
tissues 31, 32
T lymphocytes 150
tomato fish project 340, 341–2
toxins 142, 149, 150
toxoids 150
trace elements 73
trachea 157
translocation 121–2
transmissible diseases 142
transpiration 116–21, 294
transverse sections 24, 26, 112
tricuspid valves 129
trophic levels 290
tropisms 197–201
trypsin 102
trypsinogen 103
turgid 43
turgor pressure 43–4, 45–6, 115, 119
twins 238
Type 1 diabetes 151–2, 196

U
ultrafiltration 177
umbilical cord 237, 239, 240
unsaturated fatty acids 90
uracil 252
urbanisation 320
urea 174, 294
ureter 175
urethra 177
uric acid 174
urine 174, 177
uterus 233, 236–7

V
vaccination 149, 150
vacuole 26, 27, 41, 254
vagina 233, 235–6
valves
   in the heart 124, 126, 127, 129
   in veins 124, 133
variables 169, 230
variation 2, 220, 270–1, 272
vascular bundles 78, 79, 80, 111, 112, 115
vasoconstriction 135, 196–7
vasodilation 196–7, 208
vectors (disease) 143
vegetarian/vegan diets 87–8
vegetative propagation 215–18
vehicle emissions 330, 331, 332
veins
   human 124, 132, 133, 134
   in plants see vascular bundles

vena cava 125, 133, 134
ventilation 156, 158, 161–3
ventral root 184
ventricles 125, 129
venules 124, 132
Venus flytraps 275–6
vertebrates 3–4, 8, 13–15
vessels 111, 113
villi 49, 102, 104, 105
viruses 6, 19, 142, 206
vitamins 53, 92
   vitamin A 88, 89, 311, 314
   vitamin C 53, 57, 88, 93
   vitamin D 93, 94, 104
vitreous humour 186, 187
voluntary actions 185
vulva 233

W
'warm-blooded' 13, 165, 195
waste disposal 147, 326–7
water 53, 55
   contamination 146
   germination and 228, 229
   in human bodies 93, 175
   osmosis 40–8
   plant adaptations to 278
   in plants 43–4, 55, 114–15, 116–19
   treatment 146–7
water cultures 82
water cycle 294
water potential 43–5, 47
Watson, James 56, 57
weaning 241
weedkillers 201
whaling 288–9, 339
white blood cells 53, 136, 137, 149
white matter 183
Whittaker five-kingdom scheme 6
Wilkins, Maurice 56, 57
wilting 41, 44, 119, 120
wind-pollinated flowers 222, 223, 224–6
World Charter for Nature 321
The World Ethic of Sustainability 321

X
xerophytes 277
xylem vessels 30, 77, 78, 111, 113, 114, 115, 121

Y
yeast 170, 171, 306

Z
zona pellucida 235
zygotes 220, 226, 232–3, 236, 254